See inside back cover for summary of IO *and* FILES *libraries.*

Oh My! Modula-2!

Oh My! Modula-2!

by Doug Cooper

W. W. Norton & Company
New York and London

Acknowledgments

Cover: *Three Cold Creams* by Wayne Thiebaud. Private Collection. Photo by Bernard Marque. **Intro**: Photo courtesy of Jane Scherr. **A Few Preliminaries**: Courtesy The Milwaukee Public Library Collection. **Ch. 1**: Nancy Graves, *Diagonal Sequence of Limb Movements in a Newt*, 1971, collection of the artist. **Ch. 2**: Courtesy of Warden Collection. **Ch. 3**: © Paul Miller, all rights reserved. **Ch. 4**: Laura Grisi, *Endless Dialogue—Part 5*, 1977 (Detail). Photo courtesy Leo Castelli Gallery. **Ch. 5**: Paul Gaugin, *Where Do We Come From? What Are We?* Detail, Tompkins Collection, Courtesy Museum of Fine Arts, Boston. **Ch. 6**: Photo courtesy Jane Scherr. **Ch. 7**: Courtesy Bibliotheque Nationale. **Ch. 8**: Photo by Mike Adaskaveg. **Ch. 9**: Courtesy of Library of Congress. **Ch. 10**: Courtesy Michael Bryant. **Ch. 11**: Photo by Christopher Fitzgerald. **Ch. 12**: Christo, *Whitney Museum of Art American Art Packed* (Project for Whitney Museum, New York), 1971. Photo by Geoffrey Clements. Courtesy of Whitney Museum of American Art, New York. **Ch. 13**: Copyright by BIPS. **Ch. 14**: Reuters/ Bettmann Newsphotos. **Ch. 15**: Photo by Charles Phillips. **Ch. 16**: Photo by Douglas Blau, TOPPS Chewing Gum, Inc. **Ch. 17**: Copyright 1988 Sun Microsystems. **Ch. 18**: William Wegman, "Dressed for Ball," 1988. Courtesy Pace/MacGill Gallery. **Ch. 19**: La Huffman, *The Branding Fire—Big Dry, Northern Montana, 1902*. Photo courtesy of the Montana Historical Society, Helena. **Ch. 20**: Courtesy of Museum of Modern Art /Film Stills Archive.

Quotes from the following reprinted by permission of the publisher: *The Zen of Programming*, by Geoffrey James, © 1988, InfoBooks, Santa Monica, CA 90406. *The Tao of Programming*, by Geoffrey James, © 1987, InfoBooks, Santa Monica, CA 90406. *Computer Parables*, by Geoffrey James, © 1989, InfoBooks, Santa Monica, CA 90406. *Debugger's Handbook: Turbo Pascal*, by James McKelvey, Jr. © 1987 by Wadsworth, Inc. *Modula-2: A Complete Guide*, by Kim King, © 1988 D.C. Heath and Company. *Programming in Modula-2*, by Niklaus Wirth, 3rd ed., © 1985 Springer-Verlag Berlin-Heidelberg.

Printed in the United States of America.

Library of Congress Cataloging-in-Publication Data

Cooper, Doug.
Oh my! Modula-2! / Doug Cooper.
p. cm.
Includes index.
ISBN 0-393-96009-9
1. Modula-2 (Computer program language) I. Title.
QA76.73.M63C66 1990

005.13'3—dc20 90-40465

W. W. Norton & Company, Inc., 500 Fifth Avenue, New York, N.Y. 10110
W. W. Norton & Company Ltd., 10 Coptic Street, London WC1A 1PU

4 5 6 7 8 9 0

for my friends

Contents

In Brief

Preface

Welcome to *Oh My! Modula-2!* Let me begin with a confession: when I began to write, I wondered if yet another introductory programming book on yet another general purpose language was needed. Nevertheless, thought I, better *Oh My! Modula-2!* today than *Oh No! Not Ada!* tomorrow. In time, though, I was seduced by Modula-2, and I think you will be too. It is the kind of language that Tony Hoare longingly described years ago—one that would combine its predecessors' merits, rather than their features.

Will Modula-2 make the teacher's life easier? Well, in programming languages, as in hardware, 'progress' means that more can be done for the same basic price. Low-level applications always involve a certain amount of detail, but advanced projects can be attacked more easily. So it is with Modula-2. I do think that teaching Modula-2 is a worthwhile investment that, to the benefit of students and teachers alike, will help keep the non-majors' course within hailing distance of the CS major curriculum.

Oh My! Modula-2! assumes no prior programming experience, and is aimed at one- or two-semester introductory programming courses. The basic chapters, in particular, try to anticipate questions and to be inviting enough to encourage students to do assigned reading before the lecture, instead of before the exam. To make assignments easier, most chapters have four numbered sections:

1. *Strictly Modula-2* deals with Modula-2-specific syntax and details.
2. *Standard Practices* describes basic language-independent programming techniques associated with the statement or type at hand.
3. *Extended Examples* involve longer programs that can often be used as the basis of programming assignments.
4. *Program Engineering Notes* offer programming advice, review, and exercises.

Two additional text features encourage experiment and exploration:

- *Modification Problems*, in the true spirit of Modula-2, follow every example program. They suggest projects that involve modifying the example, or extending it to solve a related problem.
- *Zen Study Guides*, which follow every chapter, try to teach students how to study by making up examination questions.

Nearly one hundred programs and libraries taken from the text can be found on the accompanying disk, along with a handy supply of data, including a 25,000 word dictionary, tens of thousands of prime numbers and digits of π, a long text selection from the *Norton Anthology of American Literature*, a couple of graphic images, and other random bits and pieces. QuickMod, an integrated Modula-2 programming environment produced by Stony Brook Software, is available with the text for a quite modest fee.

Goals

Modula-2 is a nearly ideal vehicle for teaching programming and its issues. 'Computer science = computer programming' doesn't strike me as being terribly unreasonable in the first course, especially when it emphasizes *doing*, and instills a computing maturity that is born of experience. My intention for this text has been to:

1. write a book that is truly introductory because it explains matters clearly, with non-technical examples and extensive review;
2. teach honest-to-goodness programming and reasoning about programs, and avoid buzzword material;
3. use abstraction as a means of consolidating and simplifying what the student has already learned;
4. for the benefit of majors, present the basis of object-oriented design in the first semester; and
5. keep the computer and programming language subordinate to the student's right to explore, test, experiment, and err.

An interesting essay called *The Loss of the Creature** helped set these goals. It compares a lab student who dissects a fish with scalpel, lab manual, and certain knowledge of what she will 'discover,' to a beachcomber who picks up a fish, then uses a pocketknife to grub about inside and see what makes it tick. The author concludes that real scientists are closer kin to the beachcomber than to the student. I agree, and want to avoid making students apprentice priests and priestesses of systems that are beyond their ken and control, and whose controlling incantations they will soon forget.

Presentation

My biggest, but easiest, decision was not to present Modula-2 as it has generally been received: as Pascal-Plus. We are all familiar with textbooks that, using examples dating from language A, explain how the important concepts of language B can be coded in the syntax of language C. As Thoreau tended his beans—'to know beans, [and to make] the earth say beans instead of grass'—so I have tended my programs, and worked to make my examples say Modula-2 instead of Pascal. Consequences of this choice include:

- I supply new standard libraries (IO and FILES) that are simpler and more consistent than the Wirth defaults. They are easily installed as part of any reasonable Modula-2 implementation.

In my opinion, Wirth's libraries strongly reflect the Lilith implementation of Modula-2, were never intended to become generic standards, and have made

* From Walker Percy's *The Message in the Bottle*, Farrar, Straus and Giroux, 1975, pp. 46-63.

it difficult to teach file-handling algorithms. The accompanying disk contains implementations of IO and FILES for all the popular Modula-2 systems.

- Since a Modula-2 program is basically a delivery system for procedure calls, procedures and parameters are introduced in the first chapter.

Using parameters to let general-purpose solutions be applied to highly specific problems is, to me, the key idea of computer-aided problem solving. The spirit of the presentation is *learn by modifying*; detailed comments on scope and precedence are delayed until later.

- 'Goal-oriented' top-down design methods are largely supplanted by object-oriented design, and a layered approach to program coding.

Although I put off introducing library modules until after the structured data types, the design methods they'll rely on are introduced with arrays. I've tried to develop the presence and syntax of libraries in Modula-2 as the consequence of desirable programming practices, rather than their cause.

- I approach loop design using the **LOOP** statement, strictly limited to a single **EXIT** point.

As I employ it, **LOOP** retains the key structured programming feature of single entry point/single exit point, while minimizing fussing over artifacts (like loop priming) that other constructs can impose. I describe **WHILE** and **REPEAT** for those who wish to use them, but feel that the single-exit **LOOP**, with its positively stated exit condition, more accurately reflects an expert programmer's planning process. Ultimately, it encourages better reasoned loops, and clearer, more reliable programs.

Some teachers will consider this an unusual choice, so a few more words are in order. My experience has been that Hoare- and Dijkstra-style program verification methods add little at the introductory level, and even less when they are sweetened up and watered down. Nevertheless, something must be done beyond presenting plausible outlines and comprehensible examples and hoping for the best.

Rather than discussing assertions and variants, I focus on separating a loop's *goal*, or what we want to accomplish, from the *bound* that terminates the loop. By adding a *plan*, or outline of a single step toward the goal and bound, we're able to come up with a formalized loop specification. The instructor may want to proceed directly to Chapter 6 to get a feel for the method; it is revisited in 8-2 in the context of array programming, and is extended to supply a basis for designing recursive programs in Chapter 17.

Other Features

Each feature of a well-designed language implies a certain collection of ideas and methods that deserve to be explored, and which help to build programming maturity. Thus, I have kept the traditional presentation of one statement or type per chapter. Rushing through syntactic structures as unpleasant but unavoidable details is an alternative approach. However, I'm concerned that without an underlying facility at programming, students are easily convinced that they understand more than they really do.

Even in early chapters, I've tried to shift the focus from actions to data. Two unifying concepts that have been particularly helpful are the notion of a data *stream*, and that of program *state*. The standard input is the usual source of a data stream, but we can also generate it internally with statements (like **FOR**) or encapsulated processes (like random number generators). When applied to a stream, even the pedestrian **IF** statement takes on new life, selecting particular items, or rerouting the stream within a program. Where appropriate, I describe actions on the stream as *filters*. A filter turns the stream of raw data into useful information, sometimes by changing it (a *process* filter), and sometimes by learning about it (a *state* filter).

The idea of *state* as 'knowledge about data,' distinct from the data itself, applies naturally to a variety of situations:

- As an approach to problem solving; e.g. counting words by spotting in-a-word/not-in-a-word state transitions.
- In loop programming; e.g. using enumerations to name potential exit states, so that we know precisely why a multi-condition loop was terminated without additional post-loop testing.
- In commenting; developing an after-the-fact style of *confirmation comments* that specify what we believe current program state is (in contrast to traditional, before-the-fact comments about our intentions).

How and where to deal with external modules was the most difficult issue. While writing this text I read dozens of libraries, and came to appreciate that as presented and used, libraries frequently cause as much harm as good. I've tried to make some sense out of chaos by distinguishing between different module applications, and setting specific goals of layering, encapsulation and abstraction.

The text is arranged to make discussion of external modules entirely optional in a one-term course. Chapter 11 contains the basic introduction to external modules, while specific applications are developed in chapters 12 and 16. In an attempt to follow Einstein's dictum 'Make things as simple as possible—but no simpler,' I rely on the following taxonomy of library modules:

- *Interfaces* are 'name slicks.' They encapsulate names, but don't hide types, implementations, or state. They're primarily used to customize calls of existing software—to simplify usage, rather than to provide new functionality. For instance, any would-be 'standard' library that's robust and reliable probably has so many bells and whistles that it needs an interface library (like my own IO) to be usable.
- *Packages* implement algorithms, but aren't meant to hide type information. MathLib0 and string-handling libraries are typical examples.

- *Data modules* encapsulate data abstractions, with further distinctions in regard to internal state, and whether data *types* (colors, fractions) or *structures* (stacks, lists) are being abstracted.

A Taxonomy of Library Modules

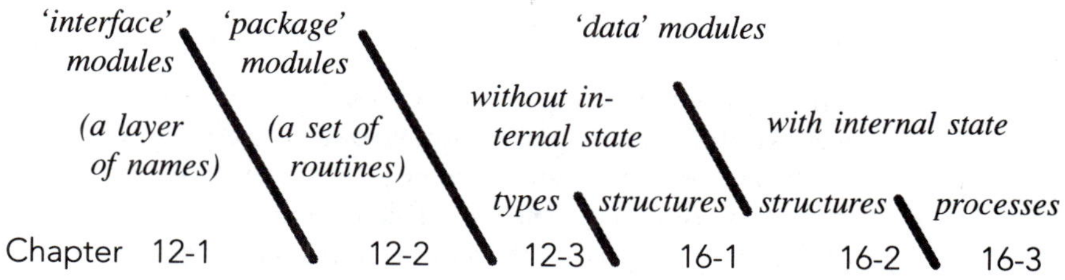

My overall goal is to create a relatively flat module system, so that one module's abstract structure, say, can hold components of another module's abstract type. Distinguishing between abstracted types and abstracted structures (*parameterized type abstractions* in the jargon) is unusual. But what use is teaching 'unified data abstraction theory' if the result is mish-mash? For this reason, I also advocate relying on generic identifiers (In, Out, type name T, etc.) and using fully qualified module names.

Opaque type definitions are not discussed beyond a brief explanation and example. This was a difficult decision; ultimately, I was uncomfortable making a longer, but basically superficial presentation of a topic that is really beyond the scope of this text.

Implementation-Specific Notes

Since this is a teacher's book for teaching, I occasionally rely on a slightly idealized view of Modula-2. First, for consistency and clarity in examples, I line-buffer all input in the I/O modules I've supplied. For all intents and purposes, programs do not recognize input until the user types carriage-return.

Second, in keeping with the fourth edition of Wirth's *Programming in Modula-2* (along with most Modula-2 implementations), CARDINAL is largely banished. I assume that FLOAT and CHR take INTEGER arguments, and that TRUNC, HIGH, ORD, and SIZE all return INTEGER values ('thereby avoiding,' as Wirth puts it, 'nasty type incompatibilities between INTEGER and CARDINAL operands in expressions').

Third, I assume that string variable arrays must have room for one extra value to allow for NUL-termination. Fourth, although I assume that you will have a string-handling library available, the text has just one or two self-explanatory references to string comparison. Finally, although I supply a random number generator on the accompanying disk (it's the one Park and Miller recommend as a minimum standard in CACM, October 1988, p. 1192), it isn't introduced until very late in the text, and you may wish to make use of it earlier.

All examples have been tested on both MS-DOS and UNIX-based compilers. I have tried to favor code that is clear on paper over code that must contain obscurities in order to be universally compatible. It's a good idea to keep an eye open for exceptions (particularly re CARDINAL) that might have to be pointed out to students.

Tools

This book was put together using the following tools. In several cases vendors were kind enough to give me software at reduced or no cost.

The text was written and beaten into shape with various UNIX tools; those created by Bill Joy (vi and csh), Brian Kernighan, Peter Weinberger, and Al Aho (awk), Joseph Ossana (troff), and the authors of grep and sed were invaluable. It was typeset using eroff, from Elan Computer Group, which is the only enhanced version of something (Kernighan's ditroff) I have ever used that really was enhanced, and, moreover, actually runs well under both UNIX and DOS.

PostScript, from Adobe Systems, was incredibly useful, and all hardware and software involved in actual production was built around it. Illustrations were done with Corel Draw, which is so good that my wife Kristin and I built a 386 system just to run it. From the nets, kermit (da Cruz and Catchings) took care of communications, and ispell (Willisson, Buehring) handled spelling correction. LaserTalk PC, from Emerald Software, helped PostScript debugging on the DOS side, while pageview, the NeWS PostScript previewer, was invaluable on the UNIX side. Finally, Kristin wrote the shell scripts that turned program source from the text into pretty-printed, compilable Modula-2 programs.

For hardware, I mainly used Sun Microsystems workstations, ending with the SPARCstation 1 I'm working on now. Yes, it *does* make a big difference to have enough mips available so that fancy pre- and post-processing don't drive you crazy. A Zenith laptop running Mortice Kern Systems UNIX tools atop DOS was very handy for working on my porch, in my kitchen, and at the ski lodge. Output was proofed on an Apple LaserWriter IINT, and finally run on an Linotronic 300 phototypesetter.

Illustrations and DOS program testing were done on the aforementioned 386, which we built in about two hours in the back room of USM Distributors, Fremont, California (they were very sweet and helpful) out of an assortment of parts (Vantage motherboard, Princeton Max-15 monitor, Video-7 VGA card) whose main feature was that no single piece cost more than $350.00. Hints that would have saved us an hour: use Scotch tape to hold washers in place, and wires out of the way; try not to put parts in backward.

Last, but not least, every Modula-2 vendor I dealt with was extremely helpful. They include Oregon Software, Logitech, JPI, Sun, and Stony Brook.

Acknowledgments

Now on to thanks. First, obviously, to Niklaus Wirth, for always remembering that programming languages are tools that should be replaced every so often, and then writing the compilers *and books* that make it happen; to Kim King, whose reference *Modula-2, a Complete Guide*, is an outstanding, exceptionally conscientious piece of work; to Logitech, Inc. for early and continued zeal (never justified by profits, I suspect) in promoting Modula-2; and to Jim Jordan, who reversed the editor's usual role and added the funny parts while cutting out the serious, meaningful (and overblown) sections.

Various friends were patient enough to listen to me explain things until they felt that I understood what I was talking about; they include Hania Gajewska, David Lichtenstein, Bev Woolf, and especially Richard Pattis (despite his bad habit, first noted by Stuart Reges, of having all of our best ideas just before we do). A one-year visit to teach at Smith College, courtesy of Joe O'Rourke, was a refreshing change. Thanks also to a group of people I have never met, but whom I feel I have spoken with many times through their books and articles: Judy Bishop, David Gries, Edsgar Dijkstra, Geoffrey James, Roland Backhouse, Jon Bentley, David Moffat, and many others.

Additional comments on the text came from Richard Rankin, Helene Kershner, Gene Ball, and Randy Bush. Proofreading at W.W. Norton by Deborah Gerish, and technical reading by Steve Cater and Chuck Bilbe, saved me from a variety of mortifying errors. As always, thanks to Ruth Mandel at Norton for help on the photographs.

Old Faithful, hard at work on look and feel.

What Are Computers?

A Software Description

HOW CAN I DESCRIBE A COMPUTER? IT'S ONLY natural to think first of its physical components, or *hardware*. Every computer has, more or less, the same parts:

- A *central processing unit*, or *CPU*, that performs calculations. Its speed is measured in MIPS, or millions of operations per second. The machines you work with are likely to range from .5 to 5 MIPS.
- *Random access memory*, or *RAM*, that holds programs while they're being run. Its size is given in megabytes, or millions of bytes. Your computer is liable to have around a megabyte or so of RAM.

basic hardware

- A *hard disk* that permanently stores information. Disk capacity is also given in megabytes; it's accurate to think of each byte as being equivalent to a single stored character. Your machine might have somewhere between 10 and 100 megabytes of hard disk storage.
- Drives for *removable media*, like floppy disks or tapes, that make it easy to transfer information (including programs) from one computer to another. Your computer can probably work with removable disks that hold from .3 to nearly 1.5 megabytes.
- Some kind of *input* device, like a keyboard or mouse, and an *output* device, like a printer or monitor.

basic software

Knowing all these numbers is, somehow, comforting. But although computer hardware makes a good photograph, it doesn't tell the whole story. A computer's power depends on its *software*, or programs. If I could photograph the software I actually used to write the book, you'd see that UNIX is one computer's *operating system*, PostScript makes the printer go, and MS-DOS ran the nameless 80386 that helped make pictures.

This list of software probably makes less sense to you than a photo of hardware. But it provides a much more accurate picture of the tools I used. For all practical purposes, my software will run on *any* hardware. It sounds strange to think about using a printer's CPU to test example programs, but I could (with effort) manage it. Here's a software 'picture:'

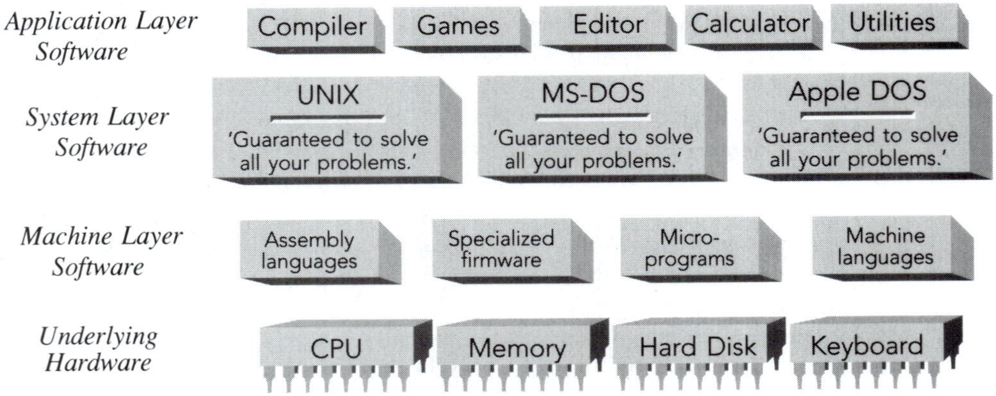

I've obviously taken some artistic license by showing every piece of hardware as a *chip*—a little plastic package that hides an integrated circuit (like the CPU), and has tiny wires sticking out of it. However, that's basically the way that I think about hardware. As long as the wires can be plugged in properly, I can use software to make a system work. I might have to worry about the physical connections *between* parts, but not about the inner workings *of* parts.

Hardware is the bedrock of every computer system. Its strength ultimately determines what can be built above. But hardware, like bedrock, is hidden beneath layer upon layer of software. This layering turns out to be a great scheme because it lets us ignore what we can't see. As long as the top layer meets our needs, we usually don't have to worry about what it conceals.

In this book we're going to ignore hardware, and most of the layers of software as well. But because we're all sometimes like the Princess—no matter how many layers shield us from the pea, it can still make us miserable—let's take a brief look down below before we get into Modula-2.

The Machine Layer

Software in the *machine layer* is closely tied to specific kinds of hardware. The lowest level of software is the *machine language*, or instruction set, the CPU understands. In effect, machine language instructions are built into the computer, just as operations like + and – are built into a hand calculator.

Not every CPU has the same instruction set, or even the same number of instructions. There are two basic strategies used in designing common, single-CPU computers these days. The traditional approach is to provide plenty of instructions (as many as four or five hundred), reasoning that if each instruction does a lot, programs won't require so many of them. A more recent alternative is to provide a small instruction set (only forty or fifty instructions), but to carry each out one very, very quickly.

CISC vs. RISC

CPU's that are built using the first strategy are called *CISC*, for *complex instruction set computer*, machines. The second approach builds a *RISC*, or *reduced instruction set computer.* From our point of view, it doesn't matter which philosophy a CPU hews to. A big, burly CISC command may do six times the work of one fleet RISC instruction, but if it takes six times as long there's no real savings. Indeed, many designers prefer RISC *because* it's simpler—not despite that fact.

In practice, almost nobody programs in machine language, because it's too low-level and inconvenient. A RISC machine, for example, might not even have built-in instructions for multiplication or division. Instead, the lowest practical level of programming is done in *assembly language.* Assembly language programs are written using brief English-like abbreviations, rather than the binary or hexadecimal *op-codes* of machine language. Such programs must be translated (using an *assembler* that was written in machine language) into the lower-level codes of the machine it's going to run on.

assembly language

Compared to machine language, assembly language is great because it lets the programmer ignore many details that are important to the CPU. An assembly language program doesn't have to include the excruciatingly tiny steps the CPU must take in order to process a program. Assembly language abbreviations are sometimes obscure, but at least they're symbols that don't depend on how one wire is attached to another.

In the old days, there was a clear distinction between machine language and assembly language. Indeed, they were referred to as the *first* and *second generation* of programming languages. The distinction has been blurred by CPU's whose machine language, for all practical purposes, *is* an assembly language. These CPU's don't have traditional instructions. Instead, they have *microinstructions* that are permanently *microcoded* into *microprograms* that *emulate*, or act like, built-in operations. Thus, programmers can only run the microinstructions indirectly.

microcode

Assembly languages can also be augmented by *firmware.* Firmware programs come permanently loaded into add-on chips that do special-purpose tasks—managing the screen display, for instance. Although firmware is usually written in microcode and machine language (and is very machine-specific) the commands it makes available are often standardized and relatively machine independent.

firmware

There are, of course, a variety of competing machine languages, microcodes, firmware, and assembly languages. Will they ever be completely standardized? It's neither likely nor desirable. Standards are convenient, but they're rigid. Instead of having fixed standards, programmers have learned to live with layers that can be modified occasionally. It's easier to rewrite just one program—say, to change the way that microcode emulates an assembly language command—than it is to throw away all existing assembly language programs every time a CPU is redesigned.

Ultimately, there are two ways to answer the question *What is machine layer software?* From above, it's usually defined in terms of a set of assembly language commands that all 'higher'-level software uses to operate the computer. But from below, it's generally *implemented*, or made to work, by hidden, extremely CPU-specific machine languages, microcodes, and firmware. Usually, each CPU manufacturer defines one common assembly language that is shared, more or less, by each family of CPU's (of varying speed or capability) that it sells. We'll find that this separation of *what it does* from *how it does it* is one of the recurring themes of working with software.

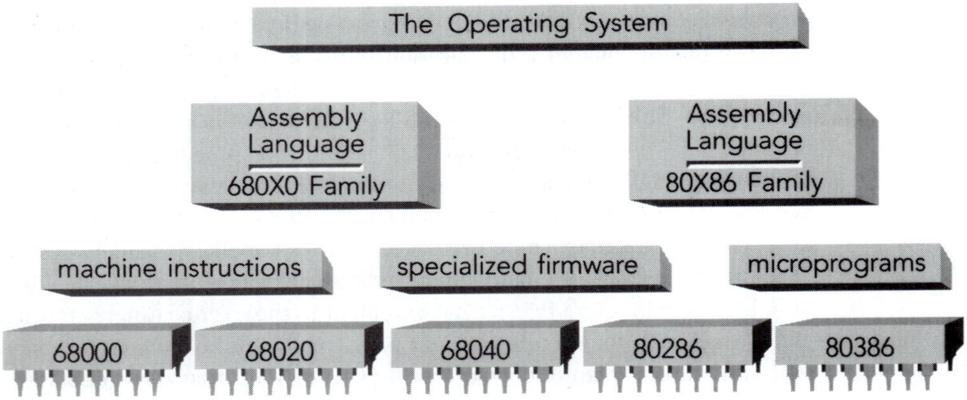

The System Layer

The software you use at the machine layer isn't really a consumer decision. For all intents and purposes, assembly language (often augmented, as noted, by special-purpose firmware) is built into the computer. But although machine-layer software can control everything the computer does, few people want to spend time giving machine-layer instructions. Instead, we rely on a second layer of software, called the *system* layer, to take care of details for us.

The system layer's main feature is a constantly running program called the *operating system*. It's the first program that's installed on a computer, and it usually starts automatically whenever the computer is turned on. The operating system keeps track of the computer's resources—terminals, printers, storage devices, access to memory and the CPU, and so on. As far as we're concerned, the operating system controls the computer. Its jobs can include:

operating system jobs

- 'Listening' to the keyboard or mouse for input, and 'talking' to the screen or to a printer.
- Interpreting the meaning of commands as they're typed in, or letting the user know that it couldn't figure them out.
- Making stored programs available to the user, and letting her install new ones.
- Storing information in *files*, and letting users inspect or change files.
- Controlling access to the computer, especially on multi-user systems, by requiring login names and passwords.

- Automatically splitting the CPU's attention between different jobs, say, so that the computer doesn't grind to a halt every time a file gets printed.
- Communicating with other computers via electronic *network* connections.

Operating systems vary in subtle and not so subtle ways. One system might be limited to a single user, while another allows *timesharing*, so that many users can share a single computer. One operating system could be oriented toward keyboard input, while another relies more heavily on a *graphical interface*, which is a fancy way of saying that you use a mouse more often than you type. Although there are many different operating systems, you'll probably come into contact with one or more of three very common ones: UNIX, MS-DOS, and either Apple DOS or the Macintosh System and Finder.

UNIX

Among computer scientists and Computer Science departments, UNIX is by far the most widely used system. That's partly because UNIX is designed to let many users, or many jobs from one user, run at once. It's also because several hundred programs, for every conceivable purpose, are invariably installed as part of the UNIX system. It's also because the original code of the programs (including the operating system itself) is usually available, and can be studied or customized.

MS-DOS

MS-DOS is probably the most widely used system in the consumer world. It's a very small system, meant for single users on 'personal' computers, that comes with almost no software beyond the bare minimum required to get started. In many ways, MS-DOS is a brain-damaged version of UNIX. Its advantage it that it will run on very small, cheap computers; its disadvantage is that it doesn't run any better on machines that have greater capabilities.

Apple operating systems

The Apple operating systems fall roughly in the middle. A great deal of software is built into the system, and it's very much oriented toward the sort of graphical interface that used to be reserved for high-priced research computers. But as usual, there's a tradeoff involved: we get the benefit of Apple software at the cost of having to buy Apple equipment.

You will find, if you ever have occasion to buy software, that your choice of operating system is the most significant determinant of what you can buy. The software shelf is often marked with the name of an operating system, rather than with a particular manufacturer's CPU or hardware. That's because as programs become increasingly sophisticated, they rely on the standards imposed by an operating system—what *system calls* to underlying hardware does it allow?—rather than dealing with the hardware directly. I could run *any* of the above-mentioned operating systems on some CPU's, but would still have to buy three different versions of any program I might want to use.

The Applications Layer

But I'm drifting into the outermost software layer—the *applications* layer. Application programs are run to do specific jobs: calculate your taxes, play *Leisure Suit Larry*, build a spreadsheet for financial scheming, or even prepare new programs to run as applications. In a way, application programs are like kitchen appliances. Some appliances, like the can opener, do just one job, but do it extremely well. Others, like food processors, attempt to be all things to all cuisines.

tools vs. systems

Application programs, too, can be either special-purpose tools or all-encompassing systems. For example, when I wrote this book, I used separate programs to create the original text, check spelling, fix spelling, check for doubled words, format pages, draw, print each chapter, and so on. However, if I wanted to, I could have used a food processor instead, by buying a full-fledged book composition system that had all the tools built into one humongous program.

a consumer's world view

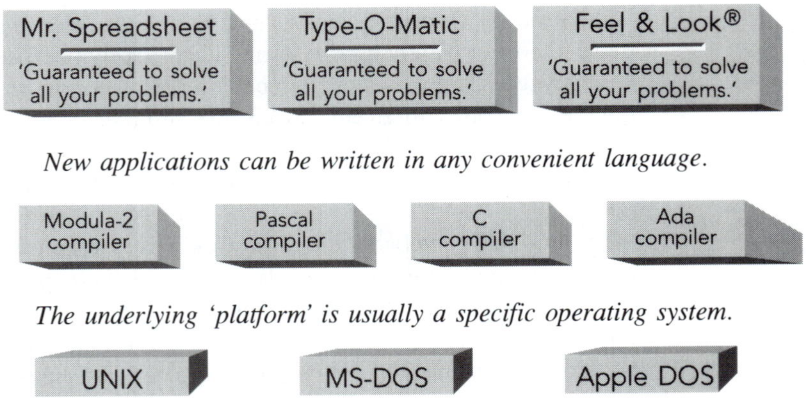

Now, by the time we run them, application programs have already been translated into our computer's machine language. There's really no way to tell what language the application was originally programmed in. However, just to reassure you that you've made the right choice in deciding to study Modula-2, there's no application that *can't* be written in Modula-2 (including the operating system and the Modula-2 compiler itself!).

As programmers, the most important application programs you'll use have to do with preparing Modula-2 programs:

application programs for programming

- A *text editor*, which you'll use to type, save, and modify the English versions of your Modula-2 programs.
- A *compiler*, which will translate your programs into a form that the computer can run, and thus helps extend the application layer.
- Various *utility* programs supplied by your operating system for working with files.

Naturally, I can't say if you'll work with a toolbox of separate programs, or with an all-in-one Modula-2 *programming environment*. Whichever approach you take, you'll learn enough Modula-2 to add new programs to your computer's application layer yourself within a few months.

Nor does it matter what kind of tools you use. When I give you something that I call a Modula-2 program, do I have to know what kind of computer you have? Not at all. All that's necessary is that one particular application program—your Modula-2 compiler—applies the standard rules in translating the English-language Modula-2 program into the form for a lower layer.

Programming Languages

There's a story about an untutored country bumpkin who listened to some big city know-it-alls as they talked about the stars. She was impressed by the number of stars they claimed there were; but she had seen many stars herself on clear nights. She was astonished by their estimates of how far away the stars were, but she could imagine how very accurate navigating tools might lead her to the same figures. What totally amazed her, though, and seemed entirely beyond comprehension, was how the devil they had managed to figure out the stars' *names*!

People often approach programming languages in the same way—as though they were deciphered, rather than invented. In fact, a language is usually conceived when a programmer becomes frustrated with whatever language she's currently working with. Instead of saying everything in this or that machine-layer language, she reasons, why not write one last program (a compiler) that will translate what I have to say (my *source*) into the machine-layer language (my *target*) for me?

Thus are programming languages born. For convenience, languages are often separated into distinct generations, (even though each generation is alive, well, and still reproducing). I've already described first generation machine languages and second generation assembly languages. They are *low-level* or *operation* languages that are closely tied to particular computer instruction sets.

first, second generation languages

From our point of view there is little difference between the first and second generations. Although they're detailed and unwieldy to work with, first and second generation languages are invaluable in two circumstances:

- when a program must use a piece of hardware that can only be controlled by very low-level instructions;
- when a program has to make the best possible use of low-level instructions in order to run very quickly.

Video games are usually programmed in low-level languages for precisely these two reasons—they use special hardware (like trackballs or fancy screens), and they have to blaze along.

The third generation of programming languages, which includes Modula-2, can be described as *high-level* or *implementation* languages. To a limited degree, third generation languages resemble spoken languages, but with severely limited vocabularies. What's important is that they express programming ideas in human-oriented terms, rather than in ways that are convenient for translation into assembly language. Most programs, including operating systems, application programs, and even higher-level languages, are coded in third generation languages. High-level languages are general-purpose in two different ways:

high-level languages

- *They can be used to solve any kind of problem.* It might be more or less convenient to use one language instead of another, just as it could be inconvenient to do chemistry homework with a statistics calculator. Nevertheless, there is a great deal of overlap between high-level language applications.
- *They can run on any kind of computer.* Again, some features might be relatively ineffective on some kinds of hardware. But in general, high-level languages don't require special-purpose machines to run.

Modula-2 is one of the newest members of the family of *imperative* programming languages. 'Imperative' means 'expressing a command,' which pretty accurately describes what the statements of an imperative programming language do. You've probably heard of some other members of the same family—BASIC, FORTRAN, C, Ada, and Pascal are probably the best known. As it happens, Modula-2 was designed by a Swiss computer scientist named Niklaus Wirth, who also designed Pascal.

The languages Niklaus Wirth designs tend toward simplicity. Wirth believes that it's important for a programmer to be able to really understand the entire language, and that this is impossible when a language becomes too big. He also prefers to incorporate the lessons of each language in a new language, rather than in an 'improved,' but more complicated, version of an old one (the original Modula was entirely experimental). He's reluctant to add a new feature without removing an existing one.

why so many?

Why are there so many different third generation languages? Partly because philosophies differ on the best way to make a language general-purpose. In a way, it's *CISC* versus *RISC* all over again. Should a language be extremely simple, even if that makes it hard to perform sophisticated tasks? Choose BASIC. Should a language be extremely powerful, even if that makes it harder to learn? If you think so, choose Ada. Should a language adapt to changing demands, even if it has to grow larger and larger so that old programs will still work too? If you think so, pick the latest version of FORTRAN or Pascal.

fourth generation languages

Finally, fourth generation languages are even more high-level, but they return to being rather specialized. They are *application* languages. Each fourth generation language is meant to help solve just one kind of problem, but to do it very well. For example, PostScript is a page-description language I used to help typeset this book. In effect, PostScript lets me program a book in terms of the curves and shading that every word, letter, and illustration consists of. Within this specialized realm, PostScript is immensely powerful; outside of it, the language is unbelievably tedious to work with.

Database languages, which control storage and retrieval of large amounts of information, are another common fourth generation application. They're meant to let us use information without having to worry about how it's going to be stored or found again. They may rely on special-purpose hardware that lets many people use the system at once, without interfering with each other's actions.

In the final analysis, of course, the choice of programming language to study is up to you. Computer Science departments usually teach a third generation language like Modula-2 in their introductory courses to strike a balance: the language reflects enough of the underlying machine layer to give students a feel for some of the potential and limitations of computing, but it hides the gruesome details of low-level programming.

Modula-2 provides a particularly good vehicle, since its techniques are easily carried over to other programming languages. Once you know it, you'll be able to master BASIC in an afternoon, Pascal in a few days, and C within a week. So, turn the page, and start learning to program!

'New Modula-2 programmers often wonder about libraries. Why are input and output procedures put into a library . . . ?' (p. 11)

A Few Preliminaries

Have you studied programming before? Then try to make your mind totally blank. Haven't ever seen a computer, perhaps? Great—you're already prepared!

The first week of a computer science course, especially the first computer science course, is a madhouse. You've got to learn a lot in order to get started: how to *log on* to your computer, how to *edit* programs, and how to actually *execute* them, not to mention finding out where the computers are, and what *log on*, *edit*, and *execute* mean.

This preliminary chapter will give you a taste of what you'll be studying for the next term. We'll go over a one-line program in detail—in *great* detail—and get an idea of what's going to be involved in studying Modula-2. You'll learn just enough to be able to dig in properly when we get to Chapter 1. Good luck, and have fun!

The First Program

BEHOLD A COMPLETE MODULA-2 PROGRAM. It's about as short and sweet as a program can be, and I'm sure that you can already figure out what it does.

```
MODULE FirstRun;
   (*This is our first example program.*)
IMPORT IO;
BEGIN
   IO.StringOut ("Hello.  I love you.")
END FirstRun.
```

Let's go through FirstRun one line at a time. The heart of every program is a *procedure call* that does something. This procedure call prints things:

```
IO.StringOut ("Hello.  I love you.")
```

If we look at the procedure call carefully, we can break it into two parts. The procedure's name, shaded here:

the procedure call

```
IO.StringOut ("Hello.  I love you.")
```

specifies the exact procedure that we want to call. IO.StringOut is a procedure that's used whenever we want to print a *string*, which is a sequence of characters between quote marks.

The procedure call also includes an *argument*, which goes between parentheses. This particular argument is a string:

its argument

```
IO.StringOut ( "Hello.  I love you." )
```

If we want, we can change the argument in a second call. Here are two different strings (sequences of characters between quotes) as arguments:

```
IO.StringOut ( "Hello.  I love you." );
IO.StringOut ( "Do I know you, or just your type?" )
```

Modula-2 programs have smaller pieces than procedures—we have to build procedures out of something, after all. Programs can also have larger units that are essentially collections of procedures.

However, a procedure call, with its argument, is the basic unit of programming. Procedures are as important to the programmer as sentences with subjects and verbs are to a writer, or functions are to a mathematician. Why?

the Big Idea

> A procedure solves a general problem. Since its argument can change from call to call, one general method can be applied to different specific tasks.

IO.StringOut solves part of the problem of sending program *output* to a program user—it lets us print a string. Soon, we'll see other procedures for output (for printing numbers, say). We'll also learn about some procedures that take care of *input*, and get information *from* a program user.

Importing a Library

Where do procedures come from? By and large, we get them in two different ways. The easy way is to use a procedure that somebody else wrote. Since, as the saying goes, reusing software has the same advantage over writing software that theft has over honest work, I'll use this approach whenever I can! But if we can't find a procedure that does what we want, we can also write new procedures from scratch.

import a library to use its procedures

> Prewritten procedures are collected in *libraries*. A program can use library procedures if it *imports* the library.

Program FirstRun imports, or arranges to use the contents of, a library called IO.

importing a library

```
MODULE FirstRun;
   (*This is our first example program.*)
IMPORT IO;
BEGIN
   IO.StringOut ("Hello.  I love you.")
END FirstRun.
```

The IO library holds a collection of procedures for *I/O*—program input and output. As it happens, I wrote IO myself. Its procedures are a little easier to use than the I/O procedures that might have come with your system. Later on, you'll learn to use those if you want to.

Many programs use procedures from more than one library. Those that do contain more than one **IMPORT**. For instance, a program that needs some mathematical procedures too would probably contain these lines:

importing more than one library

```
IMPORT IO;
IMPORT MathLibØ;
```

One of our main jobs in the next few chapters will be to learn about the procedures in the IO and MathLibØ libraries.

Let's stand back a bit and review. For all practical purposes, a Modula-2 program is just a delivery system for procedure calls. It's convenient to split programs into two parts:

a simple program outline

```
MODULE FirstRun;
   This is the part of the program where procedures are obtained.
BEGIN
   This is the part of the program where procedures are used.
END FirstRun.
```

Reserved Words and Identifiers

Procedures and libraries were hard ideas. Let's relax by going over the *syntax*, or word-by-word construction rules, of our first Modula-2 program. The first line of FirstRun was:

MODULE FirstRun;

the program heading

> A program *heading* serves two purposes: it identifies the program as a *module*, or Modula-2 program, and it names the program.

As you can see, I've printed the words **MODULE** and FirstRun in two different typefaces. I use different typefaces to help you remember to distinguish between *reserved words* and *identifiers*:

- **MODULE** is a *reserved word*—one of the forty words that are an unchangeable part of the Modula-2 language. Reserved words are always capitalized.
- FirstRun is an *identifier*—a word that I chose to use as a name. Identifiers can be capitalized, lower cased, or mixed, as long as each name is always written the same way.

Your terminal and printer will probably just use one typeface, unfortunately.

Appendix A has syntax charts

How can you remember all of the syntax details? Some people like *syntax charts*, like the diagrams that are collected in Appendix A. They can get complicated, so it's usually easier to highlight a particularly clear example or definition. For instance, a program heading is:

MODULE *the module identifier* ;

An identifier, in turn, is:

a letter followed by any sequence of letter or digit characters

Identifiers *can't* contain punctuation, or spell out reserved words. This makes TBone, UB40, and HipHop all legal identifiers, while Run–DMC, Mr.Mr., and 2CoolFoolz aren't.

Program FirstRun's heading is followed by a *comment*:

MODULE FirstRun;
(**This is our first example program.**)

comments

Comments are explanatory notes. A comment can say anything that can be put between *comment brackets* like these: (* *):

(* *The content of the comment* *)

> As far as program operation is concerned, comments are unnecessary—the computer ignores everything that's put between comment brackets. But in practice, comments are used in *every—EVERY*—program you write.

Most programmers follow a philosophy called KISS: 'keep it simple, stupid!' After all, you never know who'll be reading your programs—the grader, your teacher, your boss. Comments help these easily confused people understand your programs, so use them.

The Statement Part

As our program continues, we come to the import of IO. Since almost every program we'll write communicates with the program user, we'll find that just about every program includes this line:

a library import

```
MODULE FirstRun;
   (*This is our first example program.*)
IMPORT IO;          (*Arrange to use procedures from the IO library.*)
```

Like the program heading, an import ends with a semicolon.

The *library identifier* is the name of a file that's part of your Modula-2 system. If, for some reason, you use a library other than IO your teacher will describe its proper name and usage.

We can wrap up our first program with its *statement part*:

the statement part

```
MODULE FirstRun;
   (*This is our first example program.*)
IMPORT IO;
BEGIN
   IO.StringOut ("Hello.  I love you.")
END FirstRun.
```

To help you picture our programs in action, I'll show their output. Arrows like the ones below mean 'The program's output would look like this.' I'll always show the output `this way` because it looks a little like the print you'll see on your screen.

```
Hello.  I love you.
```

A procedure call is just one kind of statement. An outlined statement part shows that a program can contain more than one statement, as long as a semicolon appears between any two statements:

```
BEGIN
    statements, separated by semicolons
END    the module's identifier    .
```

The program ends with its name, repeated from the first line, and a period.

Congratulations on making it this far! Although program FirstRun wasn't too exciting, what we've just managed to do certainly is. A language that perhaps ten minutes ago was barely conceivable (and totally incomprehensible) is starting to mean something. *We're* in charge now! A certain amount of mystery—and maybe a little magic as well—has left our lives forever.

Modification Problems

The best way to become familiar with new programming ideas is to make small changes in programs that already work. These Modification Problems will accompany most of the example programs in the text. All of the examples are available on the disk that comes with this book, so you may be saved the trouble of typing the programs in.

1. Change program FirstRun so that it follows `I love you` with the name of the person sitting next to you.
2. If problem 1 went well, add a second procedure call to FirstRun that issues an invitation to lunch.

Self-Check Questions

Q. Which of these are legal identifiers in Modula-2? Are any legal, but inadvisable?

Start	END	3High
IO	"grade"	Try.Again
365	MODULE	NewName2
String	–Input	Under_Line

A. Start, NewName2, and String are all reasonable names. IO is legal, but it's already being used as a library name. The others either violate syntax rules (like Under_Line) or are reserved words that can't be redefined (like END).

Q. Is there anything wrong with this comment?

```
(*This is a very
 impressive, multi-line comment. *)
```

A. Nothing at all—a comment can extend over several lines. Any program lines caught in the middle will be ignored, though! This would be a mistake, since the heading is part of the comment.

```
(*Here's the beginning of the comment.
MODULE Mistaken;
Here's the end of the comment. *)
```

Background: How Do You Run a Program?

Although the reserved words that make up Modula-2 programs are standardized, the steps you must take to *run* a program will vary from system to system. Because such variations have nothing to do with the Modula-2 language, I intentionally stay away from specific instructions that might be misleading. However, in general terms, there are four parts to the process:

Edit: First, you have to write a program (like FirstRun) and store it in a file on your computer. (The file's name might require a special ending.) This means that you'll probably spend most of this week learning how to use a *text editor*—an application program that lets you create and change files.

Compile: Second, you'll have to compile your program—translate it into a form your computer can understand. The *compiler*—it's just another application program you'll learn to use—is especially helpful, because it will spot and let you know about many kinds of syntax errors. If there is a problem, you'll have to fix it (by returning to the editor) before you go on to step three.

Link: Third, you'll make use of yet another program called a *linker*. The linker 'attaches' your compiled Modula-2 program to any library code it uses. In effect, the **IMPORT** IO line from *your* program will be replaced by the compiled version of the IO library that *I* wrote.

Execute: Fourth and finally, you'll get to execute, or run, your program. Don't worry—it's practically impossible to break anything.

These steps aren't standardized because people who put together computer systems have different ideas about the best way to do things. You might use an integrated, all-in-one system that packages editor, compiler, and the others into a single program, or you may be able to choose individual tools on your own.

Chapter Notes

I CAN'T BE SURE, BUT I'LL BET THAT YOUR GRANDMOTHER wouldn't let you swallow until you'd chewed every mouthful at least twenty times. Personally, I don't want to be responsible for causing you mental indigestion, so I like to go over some important points at the end of every chapter.

debugging

Finding and fixing the bugs or mistakes that creep into programs is called *debugging*. Most debugging should take place while a program is still on paper, well before it's edited, compiled, and run. But since it's always possible to introduce bugs (like misspellings) while typing, even the best-planned programs may require trial runs before they're completely debugged. In fact, allowing time for debugging is an essential part of the programming process.

antibugging

Antibugging means writing in a manner that helps prevent bugs, or makes bugs easy to find if they do occur. It's easy to see that antibugging is a defensive strategy, and we'll often find it employed as part of a general plan of defensive programming.

Bugs aren't inevitable, but they often occur in computer programs because men and women don't think in the literal way that machines operate. Beware—

the computer can't always discover mistakes. People have a remarkable, automatic ability to fix mistakes that is miles beyond anything computers can do.

What is wrong with
with this sentence?

syntax vs. semantics

> Bugs fall into two general categories. *Syntax* bugs are 'grammatical' Modula-2 errors, while *semantic* bugs cause mistakes in the effect of programs.

Syntax errors are found and fixed with the least effort, since the compiler usually detects them automatically as it prepares a program for running. The underlying design of the language helps, because some syntax features are intended to provide markers and restart points the compiler can use when it tries to recover from an error. For example, some words invariably occur in pairs—a **BEGIN** must be matched by an **END** later in the program. A compiler that cannot find the closing **END** prints an error message to point out the problem:

compile-time bugs

```
"END" EXPECTED -- END OF PROGRAM NOT FOUND
```

Punctuation bugs are caught in a similar manner. For instance, the semicolon is Modula-2's statement separator, so the compiler expects to find a semicolon between every two statements. A good compiler may even temporarily insert one so that it can keep working on the program—after all, if the compiler conked out after detecting each minor error, it would take forever to debug a program. In any case, a compiler message keeps the programmer posted.

> Error messages are not standardized, and the ones I use in examples are not based on any particular Modula-2 system. They are meant to be clear and self-explanatory.

An ability to decipher brief, obscure error messages is one of the unadvertised skills that expert programmers have to rely on frequently! Don't be discouraged if the compiler's little missives to you seem curt and unenlightening. You'll come to understand them as time goes by.

The best approach to debugging syntax errors is to read the messages first. If there are more than a few, get a *listing*, or hard-copy printout, of the program and try to find the actual mistakes. Now is a good time to find out how to get a listing that includes the error messages along with your program.

get a listing

If you're just starting to program, I'll give you some surprising advice: Be wary of fixing things that aren't broken. When a short program produces a long series of error messages, look for a simple reason—a misspelled word, or misused feature. Above all, don't despair! As you gain more experience, you'll become adept at separating the real bugs from spurious ones.

Why Use Libraries?

New Modula-2 programmers often wonder about libraries. Why are input and output procedures put into a library, and not built into the language itself? Most programs do have input and output, so why aren't there reserved words for the job? The answer is that libraries are designed to be added to and changed, whereas the Modula-2 language is meant to be permanent and unchanging.

Suppose that our input machines—ordinary keyboards—were replaced by trackballs, paddles, and joysticks. It wouldn't be hard to devise a new library (called GameInput, maybe), and supply it to Modula-2 programmers. However, it would be very difficult indeed to change every Modula-2 compiler so that it could recognize new reserved words. Having to import libraries is a minor hassle now, but they save a lot of trouble in the long run.

Niklaus Wirth (Modula-2's designer) learned a lot about this problem from his earlier language, Pascal. With Pascal, Wirth built input and output procedures into the language. Unfortunately, the machines he used for I/O (punched card readers and line printers) were replaced almost overnight by the sort of terminals we use today. Many compilers did, in fact, have to be rewritten in order to make their I/O work.

Modula-2 Details

Syntax Summary

- comment: an explanatory note that goes between *comment delimiters*:

  (**This is a comment. You'll fail if your programs don't have them.**)

- identifier: a name, or sequence of letters and digits that starts with a letter:

  AnIdentifier ThisIs12

- **MODULE**: the package that Modula-2 programs come in:

  MODULE Example;
  The part where procedures are written or imported.
  BEGIN
  The part where procedures are used.
  END Example.

- library import: makes the contents of a library available:

  IMPORT IO; (**so that we can use* IO *library procedures**)

- procedure calls: invoke or run a procedure. A call typically consists of the procedure's name, followed by an argument given between parentheses:

  IO.StringOut ("This is the argument.")

- string: a series of characters. When it's used as an argument, the string goes between quote marks:

  "A string must be quoted."

Modula-2 Details

Remember . . .

• Libraries have to be imported before their procedures can be used. We'll be working mainly with the IO library.

• Modula-2 uses the semicolon as a separator between procedure calls, other statements, and the different parts of a program.

• The forty or so words that are part of the Modula-2 language proper are called *reserved words*. They can't be reused as identifiers.

• Capitalization matters in Modula-2 programs—**MODULE** isn't the same as **MoDuLe**.

• Syntax errors are mistakes in Modula's 'grammar' rules. They're usually caught by the compiler, and can be corrected. Semantic errors are mistakes in meaning. They can only be found by checking your program's results.

The best way to study for exams is to make up exam questions! The Zen Study Guide that follows each chapter is based on this elementary proposition. I'll always start by pointing out the Big Ideas that sometimes get hidden by programming details, then suggest some ways to guarantee that you won't be surprised at test time.

Zen Study Guide

Big Ideas

The Big Ideas in the introduction were:

1. *A good deal of programming is devoted to avoiding writing program code.* Whenever possible, we'll just call procedures that we can import from a library. If we can't do that, we'll write procedures that are as 'reusable' as possible.

2. *The exact syntax of a programming language is a compromise between precision and simplicity.* When it's possible, good languages (like Modula-2) avoid imposing arbitrary rules. Details that must be required are presented in the most natural, intuitive manner the language designer could muster.

Imagine, if you can, that you're the instructor of your programming class. Your students have obediently read the preliminary chapter, and understand (more or less) program FirstRun. *It's time to write their first exam.*

Zen Study Guide

Questions

1 Make up five identifiers that could be used in the question *Which of these are legal Modula-2 identifiers?*

2 It's picayune, but essential, to remember the exact spelling, including capitalization, of the few Modula-2 reserved words we've seen so far. How can you test your students? By giving them a program (it can be almost identical to FirstRun) that has mistakes. Write the problem.

3 It's always a good idea to exploit potential student panic—especially if you can do it in a way that makes your students feel foolish afterward. Therefore, don't pose the problem 'Copy FirstRun exactly, but have it say *this* instead of *that*.' Instead, say 'Write a program that prints the first line of the Gettysburg Address.' How hard a problem of this sort can you pose, but still answer yourself?

4 What can you do when a Modula-2 rule seems to be too simple to get confused about?

```
IO.StringOut ("Everything between the double quotes is printed.");
```

Improvise! Ask a question like *What's the output of this:*

```
IO.StringOut ("IO.StringOut('IO.StringOut')");
```

Make up a few more questions of this variety.

5 Modula-2 programs are *free-format*, which means that only their contents are meaningful to the computer—layout on the page isn't relevant. Make up a trick *What's wrong with this program* question that uses the fact that Modula-2 programs are free-format to appear incorrect.

6 For filler, ask questions like *Where is this punctuation used: . (; " ?'*

7 Finally, make up three questions that require written answers, a la *What's the difference between a reserved word and an identifier?*

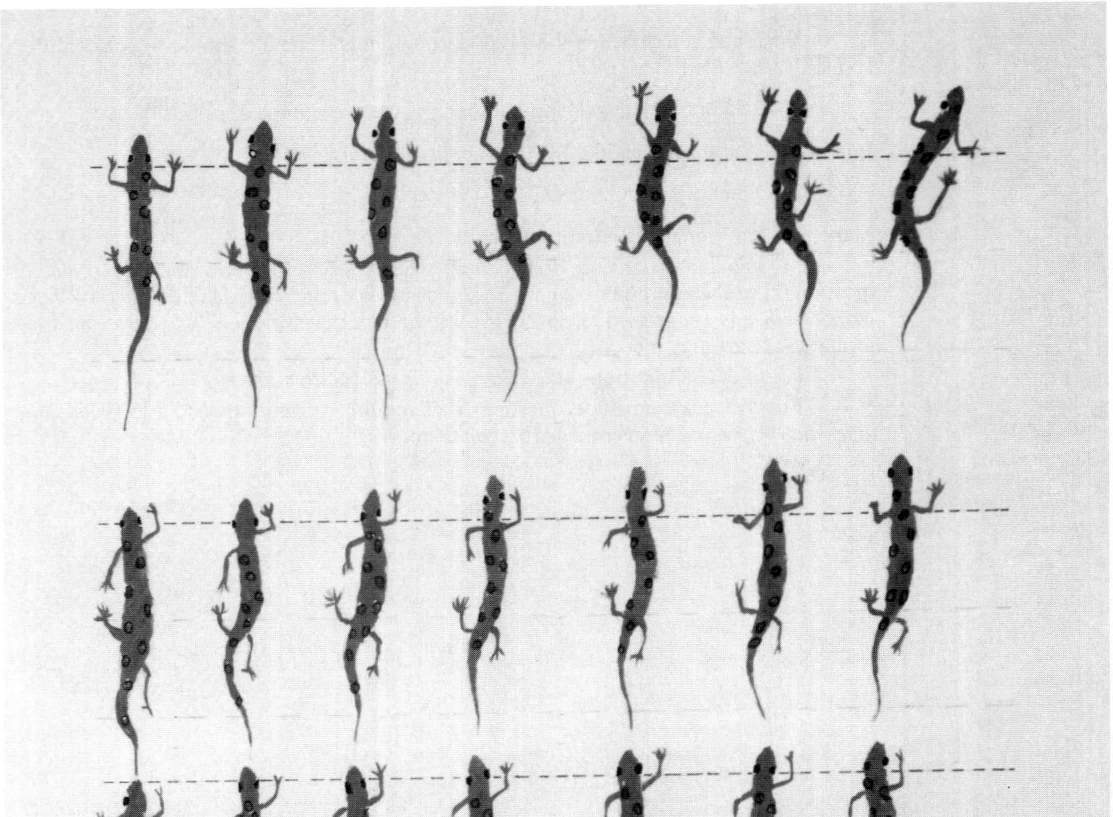

'An algorithm is the sequence of steps required to solve a problem.' (p. 38)

Chapter 1 Input, Output, and Procedures

Let's get started. Of necessity, Chapter 1 contains a lot of information. A few tricks will help keep you from becoming overwhelmed. First, read the chapter twice. Start with a quick skim to see what's really important, and what's just a complicated (but small) detail. Don't try to memorize everything, because there's just too much. If you can handle the Modification Problems, you're doing fine.

Second, try to see how Modula-2 hangs together, rather than taking it as an arbitrary collection of facts. Don't be afraid to question the language designer's wisdom ('Why'd Wirth do it that way?'), or my method of explanation. I got *my* start in writing by complaining about the textbook *I* had to read in my first class, and maybe you will too!

Third, accept the fact that programs are usually experiments. You should expect your programs to work, but don't be put off if they don't. You'll learn just as much from your bugs as you will from getting it right. How do you avoid mistakes? Experience. How do you get experience? Make mistakes.

Finally, the examples in the book are there for you to copy, borrow, alter—in short, to *steal*. Don't be bashful, because I expect you to learn by modifying—not by reinventing the wheel every time you sit down at the terminal. If you do it successfully, you'll finish this chapter able to solve some interesting problems. Almost any program you'll see in the exercises can be built by starting with one of the examples in the text. Good luck, and happy hacking!

Section 1-1
Using Library Output Procedures

HOW ABOUT SOMETHING A LITTLE MORE POETIC than program FirstRun? Program Chatty, below, demonstrates a new procedure called IO.LineOut, and shows how to write a program that's more than one line long.

separating statements

```
MODULE Chatty;
   (*A longer program that needs statement separators.*)
IMPORT IO;
BEGIN
   IO.StringOut ("A fine romance,  ");        (*These lines are printed ...*)
   IO.StringOut ("with no kisses.  ");        (*... on the same output line.*)
   IO.LineOut;                                (*This ends the line.*)
   IO.LineOut;          (*These put two blank lines in the output.*)
   IO.LineOut;
   IO.StringOut ("A fine romance,  ");
   IO.StringOut ("my friend, this is!");   IO.LineOut
   (*I'll occasionally put a second statement on the same line for easy reading.*)
END Chatty.
```

```
A fine romance, with no kisses.


A fine romance, my friend, this is!
```

Program Chatty calls a second procedure, named LineOut, from the IO library. Why can we do that? Well, take a closer look at the procedure we've been using all along—IO.StringOut. It will seem pretty obvious now that the library we imported shows up as part of the procedure name:

```
IMPORT IO;
   ... (*The in-between code isn't important here.*)
   IO.StringOut ("This is the string that StringOut puts out.")
```

the library name joins the procedure name

> To call a procedure that comes from a library, use the library identifier, followed by a period and the procedure name.

The period acts like a possessive apostrophe. IO.StringOut is just a shorthand way of saying IO*'s version of procedure* StringOut. This notation lets some other library have a procedure that's also named StringOut, but which might work differently from IO's.*

* This turns out to be a very convenient scheme. Suppose that a blind programmer has an alternative library for I/O, called BRAILLE, that's designed to work with Braille-oriented input and output devices. She is saved the trouble of having to learn an entirely different set of input and output procedure names. All she has to do is use her library name, BRAILLE, wherever we'd use IO.

procedure LineOut

LineOut prints a carriage return. We need such a procedure because StringOut *can't* print a carriage return. LineOut is a little unusual because it doesn't get an argument. As a result, it does the exact same thing every time; boring, but you can imagine how long lines would get without it.

Because Chatty has more than one statement, it makes heavy use of semicolons to separate one procedure call from the next.

> Notice that Chatty usually has just one procedure call per line. Follow this format, because it will make your programs easier to read and modify.

Modification Problems

1. By adding comment brackets in the right places, change Chatty so that it doesn't print the line `my friend, this is.`
2. Modify Chatty so that it prints a song by somebody you've heard of (instead of nobodies like Dorothy Fields and Jerome Kern).

The Vocabulary of Values

If you've ever studied a foreign language, you know that not only is there a different word for everything, but there are also new sounds to learn for the letters of the alphabet. Remembering new pronunciation is awkward at first, but after a few days it seems perfectly natural.

Programming languages also have a new vocabulary, obviously, and we have to lay a little groundwork before we can go any further. We need to see just how Modula-2 distinguishes between different kinds of *values*—numbers, letters, and so on.

There are five basic kinds of values. The explanations below include examples of the *literal*, or written, values of each sort.

literal values

- A *character* is a single letter, digit, space, or punctuation mark, given between single or double quotes. I generally use double quotes: "X", ";", "8", " ".
- A *string* is a sequence of characters, also given between quotes, that can't contain a carriage return: "This is a string."
- An *integer* is a whole number, either positive, negative, or zero: 23, –15, 28363.
- A *cardinal* value is also a whole number, but it can't be negative: 0, 18.
- A *real* number is what we usually refer to as a decimal: 98.6.

Note that commas never appear in numbers: 2001, not 2,001.

There is an alternative way to write literal strings: put them between single, not double, quote marks. This lets us use any kind of quote mark in a string:

quoting quotes

"You wouldn´t, I couldn´t, and she won´t."
´The tape shows that I said: "But it would be wrong." ´

There's also an alternative way to write literal real numbers: use a *scale factor* as an exponent. The letter E means 'times 10 raised to the power' that follows:

using scale factors

10.0 *is also* 1.0E1, *since it's* 1 *times* 10 *to the first.*
325.0 *is also* 3.25E2, *since it's* 3.25 *times* 10 *squared.*
0.003 *is also* 3.0E–3, *since it's* 3 *times* 10 *to the negative third.*

Now, in Modula-2, every value is said to have a particular *type*. I'd ignore this fact for now, except for one thing: we'll be learning about other output procedures in a moment, and each one will have to have arguments (between parentheses) of the right type. The *type identifiers*, or names of the types, are:

standard type identifiers

- CHAR is the type name for single character values.
- INTEGER is the type name for whole numbers.
- CARDINAL is the type name for non-negative whole numbers.
- REAL is the type name for real or decimal numbers.

Note that strings don't have an official type name. Instead, a string is described (don't be frightened) as an **ARRAY OF** CHAR. I'll be making more formal definitions of these type identifiers later.

Self-Check Questions

Q. What will the output of this procedure call be?

```
IO.StringOut (´ "IO.LineOut; IO.StringOut ("Printout."); ´ );
```

A. Here, the string includes everything between the single quote marks. The output is:

```
"IO.LineOut; IO.StringOut ("Printout.");
```

Q. What are the types of 3 and "3"?

A. The number 3 is a literal INTEGER or CARDINAL value, while the digit character "3" is a CHAR value that might also be used as a string.

Background: Why Have Types?

Why does Modula-2 use types? They protect the programmer from the computer. Since values are stored internally as binary numbers, computers would be perfectly content to add INTEGERS to characters, or to subtract TRUE from 45.378.

Errors of this sort (for example, where a programmer forgot that she was using a variable to store characters, and instead treated the stored value as a number), used to cause serious errors in large programs. Even simpler mistakes—inadvertently letting a positive variable take on a negative value—could be just as devastating. It wasn't that the mistaken results couldn't be spotted, mind you. Rather, the source of an error was almost impossible to find—in its context, the incorrect statement seemed perfectly appropriate.

Modula's type separation helps detect, and prevent, nonsensical activity. Most kinds of usage can be checked for type consistency at compile-time, even before programs are run. To be sure, everybody complains about type checking sometimes—there are occasions where 'errors' detected by type checks are inconsequential. But on the whole, type checking is so useful that almost every high-level language employs it.

Other Output Procedures

In the preliminary introduction, I pointed out that IO.StringOut solves one kind of problem—it prints strings. The IO library contains solutions to a few other problems that you didn't even know you had—printing other types of values.

other output procedures from IO

To print a	*Use IO's procedure*	*With arguments*
return	LineOut	*none*
string	StringOut	(*string*)
CHAR	CharOut	(*character*)
INTEGER	IntOut	(*integer, width*)
CARDINAL	CardOut	(*cardinal, width*)
REAL	FixOut	(*real, width, places*)
REAL	FloatOut	(*real, width, places*)

(*'width' and 'places' values are always whole numbers*)

There are a few points to notice about this list of procedures:

- Each procedure is specialized, because its first argument is always a value of a particular type.
- Some procedures get more than one argument, because there may be a way to vary the exact way output is printed.
- When a procedure gets more than one argument, the arguments are separated by commas.

Each of the four numerical output procedures (there are two ways to print REAL output) gets a *width* argument. This refers to the width of the space the number is printed in, and not to the number itself. The purpose of the *width* argument is to make printing tables easy:

changing *width*

- A large *width*—a wide space—means that there will be blanks before the number when it's printed. Using the same large *width* each time makes numbers evenly spaced, so that they'll line up in columns.

- A small *width*—a narrow space—means that there will be few, or even no, blanks printed before the number. A width of 0 or 1 prints numbers in exactly the space they require (which is how strings are printed).

Thus, digits aren't ever lost, even if the *width* you supply is too small. For example, here's a narrow width:

small widths

```
IO.StringOut ("All  ");   (*note the extra blanks*)
IO.IntOut (444, 1);
IO.StringOut ("  foolish fools played  ");
IO.IntOut (52, 1);
IO.StringOut ("-card pickup.");
```

```
All 444 foolish fools played 52-card pickup.
```

And, for the accountants in the audience, an equally exciting table of numbers.

large, constant widths

```
IO.IntOut (234890, 8);  IO.IntOut (77, 8);  IO.IntOut (891, 8);
IO.LineOut;
IO.IntOut (16, 8);  IO.IntOut (-3394, 8);  IO.IntOut (6, 8);
IO.LineOut;
IO.IntOut (10, 8);  IO.IntOut (7892349, 8);  IO.IntOut (6832, 8);
IO.LineOut;
```

```
  234890      77     891
      16   -3394       6
      10 7892349    6832
```

There are two ways to print REAL numbers:

- *Fixed point* notation is the most common way—the number is written with its decimal point fixed firmly between the 'ones' and 'tenths' columns.
- *Floating point* (also called *scientific*) notation puts the decimal immediately to the right of the first non-zero digit. A scale factor supplies the appropriate magnitude.

In both fixed and floating point output, the *places* argument specifies the number of digits that will be printed on the right of the decimal. Fixed point notation never uses scale factors, while floating point notation always does.

REAL output examples

```
IO.FixOut (3.14159, 10, 2);
IO.FixOut (3.14159, 10, 4);
IO.FixOut (3141.59, 10, 2);  IO.LineOut;
IO.FloatOut (3.14159, 10, 2);
IO.FloatOut (0.0314159, 10, 4);
IO.FloatOut (3141.59, 10, 2);  IO.LineOut;
```

```
 3.14      3.1415     3141.59
3.14E0  3.1415E-2      3.14E3
```

Let me sneak one last fact into this section: an argument can be an *expression*. In English, this means that the argument can include arithmetic—or that we can *do* arithmetic as part of an output procedure call. The symbols to use are:

some arithmetic operators

Symbol	*Operation*	*Example*
+	*addition*	1.0 + 0.03E–17
–	*subtraction*	00.3 – 6.2
*	*multiplication*	32.87 * 6.5E–02
/	REAL *division*	5.0E20 / 2.0

Only the multiplication sign (*) is unusual. An asterisk is used instead of a cross (×) to avoid possible confusion with an identifier named X.

In Chapter 2, we'll take a detailed look at mathematical operations, and see some additional operators for INTEGER division. For now, you won't go wrong as long as all values have the right type for the output procedure you're using:

expressions as arguments

```
IO.IntOut (3 + 5, 1);  (*Prints the sum of three and five.*)
IO.LineOut;
IO.FixOut (10.0 / 3.0, 5, 3);  (*Prints the quotient to three places.*)
```

```
8
3.333
```

Self-Check Questions

Q. What's the output of each of these procedure calls? Don't forget about leading (initial) blanks.

a) IO.FixOut (7.0, 1, 3);
b) IO.FloatOut (3.0E20, 11, 5);
c) IO.FloatOut (16.928, 6, 2);
d) IO.CharOut ("$"); IO.FixOut (1.599E2, 1, 2);
e) IO.FixOut (3.2958777E3, 2, 8)
f) IO.StringOut ("3*2"); IO.CharOut ("="); IO.IntOut (3*2, 1);

A. I've marked each blank space with an underline.

a) `7.000`　　*b*) `_3.00000E20`
c) `1.69E1`　　*d*) `$159.90`
e) `_3295.87`　　*f*) `3*2=6`

Section 1-2 Writing New Output Procedures

I'VE LET YOU TAKE ADVANTAGE OF LIBRARY PROCEDURES long enough. It's time to throw away the training wheels and learn to write some new procedures of your own. We'll do it in two steps: first, we'll write a procedure that prints the same thing every time (it's the intellectual equivalent of IO.LineOut), then we'll see how to arrange for a procedure to get arguments.

Consider program Doggerel, below. It contains a procedure that's named Spout and is called twice:

a simple output procedure

```
MODULE Doggerel;
   (*Shows the basic form of a procedure declaration.*)
IMPORT IO;  (*This IMPORT is used by both procedure and program.*)
PROCEDURE Spout;
   (*Spouts the following bit of doggerel, or bad poetry.*)
   BEGIN
      IO.StringOut ("Celery, raw, develops the jaw;");
      IO.LineOut;
      IO.StringOut ("but celery, stewed, is more quietly chewed.")
   END Spout;
BEGIN     (*the 'main' program, Doggerel.*)
   IO.StringOut ("Here is a famous poem by Ogden Nash.");
   IO.LineOut;   (*Call LineOut (from the IO library).*)
   Spout;   (*Call Spout (from this program—not from a library).*)
   IO.LineOut;
   IO.StringOut ("Did you want to hear that again?");
   IO.LineOut;
   Spout;
   IO.LineOut
END Doggerel.
```

```
Here is a famous poem by Ogden Nash.
Celery, raw, develops the jaw;
but celery, stewed, is more quietly chewed.
Did you want to hear that again?
Celery, raw, develops the jaw;
but celery, stewed, is more quietly chewed.
```

new procedures are declared

Recall that programs are split in two halves. Procedure Spout is *declared* in the first half of program Doggerel, before the statement part. Once it has been declared, it can be called in the statement part. Spout is called twice.

Spout demonstrates the basic ideas of procedure declarations. A procedure is similar to a program, but there are three important differences.

1. A procedure declaration begins with the reserved word **PROCEDURE**. It ends with a semicolon instead of a period.

differences between procedures and programs

2. Procedures don't import libraries. However, as long as the 'main' program imports a library, a procedure can use the library's contents.
3. A procedure can be given arguments (although Spout isn't).

What's the difference between library procedures like StringOut and procedures we write, like Spout?

difference between library and non-library procedures

> A library procedure can be imported and used by *any* Modula-2 programmer, in any program. In contrast, procedures like Spout belong to the programs they're part of. They can only be called there.

Since Spout isn't imported, it can't be preceded by a library name. In contrast, LineOut must be called as IO.LineOut.

Modification Problems

1. Modify program Doggerel so that all the IO.LineOut calls are in procedure Spout.
2. Put the entire statement part of Doggerel into a procedure called Declaim, then have Doggerel call Declaim. Does it matter if the declaration of Declaim precedes or follows that of Spout? Try it both ways.

Starting Arguments

Spout, like IO.LineOut, is cute but dull. It does the same thing every time. If we are going to make a procedure general-purpose, we have to be able to give it one or more arguments. That way, the procedure's action can change slightly every time it's called. Let me pose a more general problem, then:

problem: printing two characters

> Write a procedure called PrintInitials. Make it similar to IO.CharOut, but instead of printing just one character, have it print two different characters—its two arguments.

Solving this problem requires a new tool: a way to *name* the two characters that are going to be arguments. Names will let us talk about the values within PrintInitials, even though we won't know exactly what they are. In effect, each name will *represent* an argument. Program ValueArg, below, shows how it's done in Modula-2.

(*program follows*)

giving arguments to a procedure

```
MODULE ValueArg;
  (*Shows how to arrange for a procedure to have arguments.*)
IMPORT IO;
PROCEDURE PrintInitials (First, Last : CHAR);
  (*Within this procedure, the names First and Last represent
   the argument values. We'll call IO.CharOut to print them.*)
  BEGIN
    IO.StringOut ("My initials are:    ");
    IO.CharOut (First);
    IO.CharOut (Last);
    IO.LineOut
  END PrintInitials;
BEGIN
  PrintInitials ("D", "C");   (*Any two characters can be arguments.*)
  PrintInitials ("Q", "T");   (*Like strings, characters are quoted.*)
  PrintInitials ("&", "#")
END ValueArg.
```

```
My initials are:  DC
My initials are:  QT
My initials are:  &#
```

When PrintInitials is first called, the literal values "D" and "C" are its arguments. Within PrintInitials, the identifiers First and Last *represent* these literal values. They can be used in place of the literal values when we call IO.CharOut.

On each successive call, PrintInitials is given two new arguments. And, within PrintInitials each time, First and Last represent the two new arguments.

How does PrintInitials know that we want First to represent "D", "Q", and "&", while Second represents "C", "T", and "#"? Because there is a *one-to-one correspondence*, by position, between the arguments in the calls, and the identifiers First and Last in the heading, or first line, of PrintInitials.

one-to-one correspondence

```
PROCEDURE PrintInitials (First, Last : CHAR);
                          ↘      ↘
                  PrintInitials ("D", "C");
```

Self-Check Questions

Q. Suppose that these are all supposed to be calls to procedure PrintInitials. Are any legal? What's wrong with the ones that aren't?

a) PrintInitials ("3", "6");
b) PrintInitials ("Z");
c) PrintInitials ("A", "B", "C");
d) PrintInitials (72, –186);
e) PrintInitials (".", ",");

A. Count the arguments, then check their types.

a) Legal, since the arguments are the digit *characters*.
b) Illegal, because there's only one argument.
c) Illegal, because there are too many arguments.
d) Illegal, because the arguments have the wrong types.
e) Legal, because punctuation marks are legitimate characters.

Value Parameters and Their Types

Take a closer look at the PrintInitials heading. What purpose do you think the type identifier CHAR serves? It indicates that First and Last will be representing CHAR values, which is to say they will be characters.

value parameter defined

Formally, First and Second are called *value parameters*. A value parameter can represent a literal value—its argument—of any given type. It can also be used as an argument in place of a literal value.

That explains why First and Last are used without quotes in the calls of IO.CharOut—they are value parameters that represent literal characters:

```
(*These arguments are literal CHAR values.*)
IO.CharOut ( "O" );
IO.CharOut ( "K" );

(*These arguments are value parameters that represent CHAR values.*)
IO.CharOut ( First );
IO.CharOut ( Last );
```

Value parameters can have any type. A procedure that printed two numbers would have a heading like this:

```
PROCEDURE PrintAgeAndIQ (Age, IQ: INTEGER);
  (*When PrintAgeAndIQ is called, its arguments must be INTEGERs.*)
```

declaring value parameters

There can also be any number of value parameters:

```
PROCEDURE PrintReadings (Out1, Out2, Out3: REAL);
  (*This procedure expects three REAL arguments when it's called.*)
```

A procedure might also need value parameters that have different types. Notice how semicolons are used to separate the declarations of three value parameters in the DoItAll heading:

```
PROCEDURE DoItAll (One: INTEGER; Two: REAL; Three: CHAR);
  (*This procedure expects three arguments with different types.*)
```

> Never forget that there is a one-to-one correspondence, *by position*, between the value parameters in a procedure heading, and the arguments in a procedure call. The computer will *not* sort out mixed-up calls.

That means that in the examples below, the left-hand calls are legal (each argument is in the proper position for its value parameter), while the right-hand calls are illegal. The original DoItAll heading is few lines back.

legal and illegal calls

```
(*legal calls*)                    (*illegal calls*)
DoItAll (17, 6.2, "E");            DoItAll ("E", 17, 6.2);
DoItAll (-6, 2.47E9, "+");         DoItAll (-6, "+", 2.47E9);
DoItAll (0, 0.0, "!");             DoItAll (0.0, 0, "!");
```

A Harder Example

Now that we can declare value parameters, let me reprise the informal definitions of the standard type identifiers. The most commonly used types are:

CHAR values include all of the single characters that appear on the keyboard, both lower case and capitalized. They also include *control characters* that backspace, ring bells, work the carriage return, etc. In this text, we use the standardized ASCII (ask´-ee) set of 128 characters.

the standard types

INTEGER values are the counting numbers: –2,–1, 0, 1, 2, and so on. INTEGER values usually fall into one of two ranges, depending on your computer: either –32768 . . . 32767, or –2147483648 . . . 2147483647.

REAL values are decimals and positive or negative powers of 10: 96.08, or 16.7E44, or –2.7E–03. The number of places of decimal accuracy of REALs will vary from computer to computer.

These types have standard names. A type that doesn't have a one-word standard name, but which we will frequently use, is:

ARRAY OF CHAR (the type name consists of all three words) is the type name for parameters that represent string arguments. A string is any sequence of CHAR values between quote marks: "A string".

value parameters represent values

A value parameter that represents a string doesn't have to be put in quote marks. For example, procedure TwoStrings, below, prints its two arguments in reverse order:

strings as value parameters

```
PROCEDURE TwoStrings (First, Last: ARRAY OF CHAR);
  (*Will be called with two string arguments.*)
  BEGIN
    IO.StringOut (Last);
    IO.StringOut (First)
  END TwoStrings;
```

The strings that serve as arguments must be separated by a comma—they are two distinct arguments. A legal call of TwoStrings is:

```
TwoStrings ("one string" , "another");
```

Two types that we will employ in later chapters are:

BOOLEAN (boo´-lee-an) values, sometimes called logical values, have just two values—FALSE and TRUE. We won't use them until Chapter 5.

CARDINAL values are the non-negative counting numbers: 0, 1, 2, etc. The largest CARDINAL will, in general, be twice as large as the largest legal INTEGER, plus 1.

ordinal types

Types CHAR, INTEGER, BOOLEAN, and CARDINAL are sometimes called *ordinal* types. This implies that every value of the type can be consistently ordered and numbered, regardless of computer system. REAL can't be an ordinal type because computers vary in the way they deal with mathematical reals. In effect, one computer may have 'more' reals between 1.0 and 2.0 than another one does, so trying to number them consistently is hopeless.

I'll close our discussion of types by writing one last output procedure. Program DickC, below, shares with you one of my many insights, usually of dubious accuracy, into the history of rock and roll. As you can see, procedure BandStats uses two kinds of value parameters: Name represents a string argument, while Hits and Year represent INTEGER arguments.

multiple value parameters demonstration

```
MODULE DickC;
    (*Demonstrates more procedure declarations.*)
IMPORT IO;
PROCEDURE BandStats (Name: ARRAY OF CHAR;
                        Hits, Year: INTEGER);
    (*Uses several value parameters. When Name is used, below, it
      doesn't need quotes, because it already represents a string.*)
    BEGIN
        IO.StringOut ( Name );  (*Name represents the string argument.*)
        IO.StringOut ("  had  ");
        IO.IntOut ( Hits , 1);  (*Hits represents one INTEGER argument.*)
        IO.StringOut ("  top-10 hits in  ");
        IO.IntOut ( Year , 1);  (*Year represents the other.*)
        IO.LineOut
    END BandStats;
BEGIN
    BandStats ( "Elvis Presley", 4, 1964 );
    BandStats ( "Elvis Costello", 4, 1984 );
    IO.StringOut ("Isn't this a remarkable coincidence?");
    IO.LineOut
END DickC.
```

(output follows)

```
Elvis Presley had 4 top-10 hits in 1964
Elvis Costello had 4 top-10 hits in 1984
Isn't this a remarkable coincidence?
```

Modification Problems

1. Change the heading, and calls, of procedure BandStats so that there are two string arguments (say, Name and RealName) instead of one.
2. Break BandStats into two separate procedures—one that prints only string arguments, and one that prints only INTEGER arguments. It doesn't matter which procedure declaration goes first.

Self-Check Questions

Q. Suppose that procedure BandStats, above, contained these two calls. What would be printed?

```
IO.StringOut ("Name");
IO.StringOut (Name);
```

A. The first call prints the string `Name`, while the second prints its value. Their output (on the first call of BandStats) would be:

```
NameElvis Presley
```

Q. What do you think is wrong with this heading?

```
PROCEDURE Surf (Turf: INTEGER; Smurf: REAL; Turf: CHAR);
```

A. Surf uses the same identifier, Turf, as a value parameter identifier twice. Within Surf, how will the computer know which one we're talking about? It must be illegal to use an identifier more than once in a single heading.

Section 1-3 Using Input Procedures

AS FRAN LEBOWITZ POINTS OUT, THE OPPOSITE of talking isn't listening—it's waiting. So far our programs have managed this very well. They've talked, but they've said the same thing every time, and they certainly haven't listened. If program users are going to enter values for our programs to play with, we need a way to store the values, and a way to change them.

Program WiseGuy, below, shows how it's done. As you read it, pay particular attention to three shaded declarations: a variable, Score, a variable parameter, IQ, and a value parameter, Value. I think that you can figure out what procedure IO.IntIn does.

```
MODULE WiseGuy;
   (*Demonstrates the declaration and use of a variable.*)
IMPORT IO;  (*so that we can use input and output procedures*)
VAR Score: INTEGER;  (*the variable declaration*)
PROCEDURE GetIQ ( VAR IQ: INTEGER );  (*the variable parameter*)
   (*Inside GetIQ, the variable parameter IQ renames the argument. It's a
    variable parameter because when IQ changes, the argument changes.*)
   BEGIN
      IO.StringOut ("Please enter your IQ:   ");  (*Prompt for input.*)
      IO.IntIn (IQ)  (*'Read' the value entered for IQ.*)
   END GetIQ;
PROCEDURE PrintValue ( Value: INTEGER );  (*the value parameter*)
   (*Inside PrintValue, the value parameter Value gets its argument's current
    value. We'll use the value (to print it), but we can't permanently change it.*)
   BEGIN
      IO.StringOut ("You wish!  I bet you can't even count to  ");
      IO.IntOut (Value, 1)
   END PrintValue;
BEGIN
   GetIQ (Score);  (*Score brings a value back from GetIQ.*)
   PrintValue (Score)  (*Score takes a value to PrintValue.*)
END WiseGuy.
```

```
Please enter your IQ:  159
You wish!  I bet you can't even count to 159
```

declaring and using a variable

(I'll always use the **heavy** typeface to show values the user types in.)

A *variable* is a temporary name for a value. The value a variable represents can, and usually will, be changed when a program runs. The idea of a variable isn't very hard. After all, even most hand calculators have variables (with names like *Sto* or *Mem1*) that hold numbers. Modula-2's variables are more interesting, though. We can have as many of them as we want, name them as we wish, and use them to store all kinds of values.

A *variable parameter* is a temporary name for a variable. Just as a value parameter lets a value be given a temporary new name so it can be *used* within a procedure, a variable parameter lets a variable be given a temporary new name so it can be *changed* inside a procedure.

a variable parameter is an alias

Thus, variable parameters are a little more sophisticated than value parameters. A variable parameter names a name—it's an *alias*. Variable parameters serve the same general purpose as value parameters, because they let us write general-purpose procedures. But whereas value parameters let different values be sent to a single procedure to be printed, variable parameters let different variables be sent to a single procedure to be given new values.

> Variables store values, and carry them from procedure to procedure. When a procedure *uses* a variable's value, the variable should be the argument of a value parameter. When a procedure *changes* a variable's value, the variable should be the argument of a variable parameter.

Another way of stating the rule is to say:

- A value parameter's argument can be a value *or* a variable, because a variable represents a value.
- A variable parameter's argument *must* be a variable, because it is going to be changed.

Remember that a Modula-2 program is basically a delivery system for procedure calls. Declarations in the first half of the program 'populate' the environment that the procedure calls will work in:

MODULE Sample;

IMPORT
VAR
PROCEDURE

Imports, declarations, and definitions create the environment . . .

BEGIN
. . . that the statement part operates in.
END Sample.

In WiseGuy, we first created a working environment of procedures and variables:

- We imported a library (IO) of input and output procedures.
- We declared a variable (Score) that could hold an INTEGER.
- We declared two procedures that solved general problems:

how we built WiseGuy

GetIQ solved the general problem *Get a value and store it in the variable that comes as your argument.*

PrintValue solved the general problem *Make a wisecrack and print the value of the variable or value that comes as your argument.*

Then, with variable and procedures in hand, we used the working environment to do something:

GetIQ (Score) had the effect *get the data (argument)*
PrintValue (Score) had the effect *return the results (argument)*

Modification Problems

1. Change program WiseGuy so that it asks for initials (i.e. reads and prints two CHAR values instead of one INTEGER). Have it say 'You wish! I bet you can't even spell . . .'
2. Add a procedure to program WiseGuy that gets the user's age. Then, call PrintValue to make the same wisecrack about her age.

Self-Check Questions

Q. Read each of these procedure calls carefully, then guess which arguments go to value parameters, and which go to variable parameters. Can't be done, you say? Sure it can!

a) Divide (Divisor, Dividend, Quotient);
b) ReadName (First, Last);
c) GetCommand ("-->", Command);
d) PrintClue ("Warm!");
e) ReadSum (17);
f) Process (Data);

A. The rule, of course, is that if the argument is going to be changed, it should be going to a variable parameter. If it's just going to be used—or if it's a literal value—it's going to a value parameter.

a) The names imply arguments to value, value, and variable parameters.
b) Both variable parameters.
c) Value parameter and variable parameter.
d) Must be a value parameter because the argument is a value.
e) The procedure name is misleading—it has to have a value parameter.
f) There's no way to tell if Process uses or changes Data.

Variable Parameters and Input Procedures

Let's take a close look at GetIQ, from program WiseGuy:

declaring a variable parameter

```
PROCEDURE GetIQ ( VAR IQ: INTEGER );
  (*Demonstrates a variable parameter.*)
  BEGIN
    IO.StringOut ("Please enter your IQ:   ");  (*This is a prompt.*)
    IO.IntIn (IQ)
  END GetIQ;
```

GetIQ is different from the other procedures we've written:

- Its heading contains a variable parameter, rather than a value parameter.

- It uses an *input* procedure to get a value from the program user, rather than an output procedure to print a value for the user.

GetIQ takes two actions. First, it *prompts* the program user to enter a value. Then, it reads the value the user entered. Because it reads a number, IO.IntIn actually changes the value of IQ.

prompt for user input

> If a program expects to read a value, it should prompt the user. Otherwise, she has no idea that she's supposed to type one in—and a computer will silently wait forever.

GetIQ literally does get a value for its variable parameter IQ, since the call IO.IntIn ends up changing the value of IQ. At the very same time, GetIQ provides a value for a variable in the main program. Since modifying a variable parameter modifies its argument, a main program variable—IQ's argument—changes too.

Procedure IO.IntIn is just one of the input procedures supplied by the IO library. The complete set is:

other input procedures from IO

To read a	*Use IO's procedure*	*With an argument of type*
CHAR	CharIn	(CHAR)
INTEGER	IntIn	(INTEGER)
REAL	RealIn	(REAL)
string	StringIn	(**ARRAY OF** CHAR)
CARDINAL	CardIn	(CARDINAL)

(*the argument must always be a parameter or variable*)

We'll take a deeper look at all the procedures from IO in the next chapter.

It may seem obvious to you, but I'll say it anyway: an input procedure's argument may not be a literal value! A literal value like 5, say, is always busy enjoying its 'fiveness.' It can't be anything else. An input procedure can't change it. Instead, an input procedure's argument has to be something that can store a literal value—something like a variable parameter or variable.

Self-Check Questions

Q. Even though we import IO.IntIn and IO.IntOut from the IO library, they're procedures that are written in the same way as our own procedures. Do you think that IO.IntIn and IO.IntOut use value parameters or variable parameters?

A. Just ask yourself the usual question: do they change the argument, or just use it? Since IO.IntIn changes its argument's value, it must use a variable parameter. But because IO.IntOut just uses its two arguments—a value to print, and the number of spaces printing should take—they must go to value parameters.

Variable Declaration Syntax

It's time to nail down the exact syntax of declarations. As WiseGuy shows, variable declarations and parameter declarations are quite similar:

```
VAR Score: INTEGER; (*declare a variable*)
  . . .
PROCEDURE GetIQ ( VAR IQ: INTEGER ); (*declare a variable parameter*)
  . . .
PROCEDURE PrintValue ( Value: INTEGER ); (*declare a value parameter*)
```

what?

What? is our first question. What are we going to create? The centerpiece of each definition is the association of an identifier with a type.

> In Modula-2, every variable and parameter has a type. The variable or parameter can only represent values of that type.

Where? is the next question. Where are variables and parameters declared?

where?

- Variables get their names and types in the first part of a Modula-2 program. Imports *must* come first. Then, although it isn't a fixed rule, variable declarations usually come before procedure declarations. My programs (and our examples) are almost invariably written that way.
- Value parameters and variable parameters get their names and types in the heading of the procedure they're used in. A procedure can have both value and variable parameters, although we haven't seen any example of this yet.

how?

Finally, we have a special *How?* rule to remember. Since variables and variable parameters are intended to *change*, their definitions are preceded by the reserved word **VAR**.

Variables and parameters are dealt with in slightly different ways. A program can have more than one variable declaration part, but ours won't for some time. Eventually, though, we'll want to 'cluster' variable declarations with the procedures that use them. It's easiest to look at examples. In a variable declaration, the reserved word **VAR** only has to appear once, no matter how many different kinds of variables are declared. For example:

sample variable declarations

```
VAR Counter: INTEGER;       (*Any positive or negative whole number.*)
    ShoeSize: REAL;         (*Any decimal value.*)
    FirstInitial: CHAR;     (*Any single character value.*)
    AptNumber: CARDINAL;    (*Any non-negative whole number.*)
    OutToLunch: BOOLEAN;    (*Either FALSE or TRUE.*)
```

We can declare any number of variables, of any type, and in any order, as long as commas separate variables and semicolons separate types. This is also a legal variable declaration:

```
VAR TimingMark: INTEGER;            (*One INTEGER variable.*)
    GPA, BattingAverage: REAL;
    Counter, Degrees: INTEGER;      (*More INTEGER variables.*)
```

Because strings don't have a standard type identifier, they're handled in a special way. We've already seen how to declare a string as a value parameter—its type is **ARRAY OF** CHAR. Variable parameters are declared the same way (except, of course, for the reserved word **VAR** that precedes the whole shebang). A variable declaration, in contrast, requires special handling.

specifying string length

> When you declare a string variable, you must specify the maximum number of characters it can contain by describing it as an **ARRAY** [0..*length*] **OF** CHAR. The string variable can hold from one to *length* characters.*

For example, suppose that a program reads and prints a string that's no more than twenty characters long. We'd see these declarations:

declaring strings

```
VAR Name: ARRAY [0..20] OF CHAR;
  (*declares a variable that holds up to 20 characters*)
  . . .
PROCEDURE GetName (VAR First: ARRAY OF CHAR);
  (*declares a variable parameter*)
  . . .
PROCEDURE PrintName (Last: ARRAY OF CHAR);
  (*declares a value parameter*)
```

Self-Check Questions

Q. Which of these is a legal variable declaration?

a)
```
VAR First: INTEGER;
    Second: CHAR;
    Third: INTEGER;
```

b)
```
VAR First, Third: INTEGER;
    Second: CHAR;
```

c)
```
VAR First, Third: INTEGER;
VAR Second: CHAR;
```

A. All three declarations are legal. Ease of understanding should be your main guideline in choosing which format to use.

Q. Suppose that I declare this string variable:

```
VAR Group: ARRAY [0..5] OF CHAR;
```

* Yes, even though, say, 0..5 is really six characters, the string only holds a maximum of five. The system (not *every* system, but many) uses the extra space to store an end-of-the-string character.

Which of these names could Group hold? Which ones did I make up?

`UB40`	`XTC`	`B-52's`	`GoGos`
`Devo`	`InXS`	`1+1`	`Prince`
`===`	`MrMr`	`Sting`	`Guy`

A. With the exception of **`B-52's`** and **`Prince`**, which are six letters long, all the names qualify. Note that the stored string is any sequence of characters—the string can start with digits, or include punctuation, as 1+1 does.

Parameter Declaration Syntax

The syntax of a procedure heading is relatively complicated, since it allows for headings that are much more elaborate than any we'll actually need. Although the procedures that we've used have had either value parameters *or* variable parameters, any procedure can have one or more of both kinds. How can you tell them apart?

telling value and variable parameters apart

Each variable parameter declaration is preceded by the reserved word **VAR**.

There can be more than one variable parameter of a given type—that doesn't require a new **VAR**. But the semicolon at the end of each value or variable parameter declaration marks the end of that kind of declaration. The next declaration needs a new **VAR** if it's going to be a variable parameter.

Let's look at some examples. First, note that a new **VAR** is needed whenever there's a transition from one type to another:

```
PROCEDURE ReadAgeAndIQ (VAR Age, IQ: INTEGER);
    (*Two INTEGER variable parameters.*)
PROCEDURE GetDifferentStuff (VAR First: INTEGER;
                             VAR Second: CHAR;
                             VAR Third: REAL);
    (*There must be a new VAR for each new type.*)
```

Then, be sure to understand how to intermingle value and variable parameter declarations. For example, it's clear in the calls below that arguments are going to be used in different ways:

some sample calls . . .

```
GetValues (Name, RawScore, Percentile);
        (*Changes the values of all three arguments.*)
UseValues (Name, RawScore, Percentile);
        (*Uses Name and RawScore, but change Percentile.*)
PrintResults (Name, RawScore, Percentile);
        (*Uses all three arguments without changing any of them.*)
```

The procedure headings that back up the calls above are a little more complicated than others we've seen, but are still readily decipherable.

. . . and their procedure headings

```
(*Heading of GetValues, with variable parameters.*)
PROCEDURE GetValues (VAR String: ARRAY OF CHAR;
                     VAR Raw, Percent: INTEGER);

(*Heading of UseValues, with mixed parameters.*)
PROCEDURE UseValues (String: ARRAY OF CHAR;
                     Used: INTEGER;
                     VAR Changed: INTEGER);

(*Heading of PrintResults, with value parameters.*)
PROCEDURE PrintResults (Name: ARRAY OF CHAR;
                        Raw, Percent: INTEGER);
```

Background: Three Questions About Parameters

Even though I encourage you to learn to program as I did—by copying and modifying—I usually like to explain the facts behind our programs, too. However, there are occasions that call for simplification, and the subject of parameters is one of them. But since I'm sure some of you are asking these questions, let me provide brief answers:

1. *Can a variable parameter be used without being changed?* Yes, a procedure might simply print a variable parameter's value. *However*, choosing to declare a variable parameter implies that it *will* change, and we'll always follow consistent practice.

2. *Can a value parameter be changed inside a procedure, in addition to being used?* Surprisingly, yes. *However*, the value parameter's argument, if it's a main program variable, *won't* change. Any change to the value parameter is just temporary—it just lasts for the duration of the procedure call.

3. *Do variables and parameters have to have different names?* No. The compiler recognizes that a name declared in a procedure heading refers to a parameter, and not to a variable. *However*, people are confused more easily than compilers, so I make a habit of using different names for now.

As you can see, there are three big '*However*...'s in the answers! Just because Modula-2 allows something doesn't mean that we'll want to do it; there's a reason that we do things the way we do. There will be fuller explanations in Chapters 2 and 3.

Structure, Salvation, and a Final Example

We now have the essential tools for writing a real program—one that can show a little variation between runs.

problem: personalized multiplication

> Write a program that asks for, multiplies, and prints the product of two numbers. Make it a little more personal by finding out (and referring to) the program user's name.

In a way, this is the fundamental computer programming problem—figure something out, but do it with a little more panache or pizzaz than a hand calculator could handle. However, we want to end up with more than just the answer.

Instead, we want a *program*—something that can be run, admired, and perhaps improved and extended later.

I malign BASIC

Programs don't just 'happen' to turn out this way, of course. You may be surprised to hear this news (especially if you've taken a course in BASIC) but most programming is *not* a matter of sitting back and giving commands to an obedient computer. Instead, it's planning and organizing:

- *Planning* the program in terms of the goals it's supposed to meet.
- *Coding* the program in goal-oriented procedures.
- *'Parameterizing'* the procedures so that they can be as general-purpose as possible.
- *Commenting* the program in terms of the goals its procedures achieve.

goal-oriented design

> This approach to programming is called *goal-oriented design.* The key word is *design*; programs are designed to have desirable characteristics—to be easy to write, understand, and change—as well as to find answers.

Like most common-sense approaches, goal-oriented design has been invented many times under many names. Perhaps you've heard of some of them:

the ideas behind goal-oriented design

- Breaking a program down into procedures that carry out particular tasks is sometimes called *procedural abstraction.*
- Stating a 'big' goal first, then restating it in terms of smaller subgoals is called *top-down design.*
- Restating the outline of a single procedure in more and more precise, code-like terms is called *stepwise refinement.*
- Using procedures (and, we shall see, statements) that have just one starting 'entry' point, and just one ending 'exit' point, is called *structured programming.*

> The names are not important, but the underlying ideas are: design procedures before you write statements, and think in English before you code in Modula-2. The sooner you start to code, the longer it will take to finish.

If we had only simple goals, then a simple methodology—sit at the terminal and code, and a simple language—like good old BASIC—would suffice. But we are ambitious! We want to be able to solve problems that are too big to carry around in our own modest heads.

Our deliverance, as Niklaus Wirth points out, is found in organization:

'details are the jungle'

'After all, complicated tasks usually do inherently require complex algorithms, and this implies a myriad of details. And the details are the jungle in which the devil hides. The only salvation lies in structure.'*

An *algorithm* is the sequence of steps required to solve a problem.

As we've already seen, our main tool for realizing algorithms is the procedure. We write procedures to solve specific program goals. When possible, we arrange for the procedures to accept arguments—this makes the procedures more general-purpose. But we always try to solve programming problems by designing solutions to specific goals. We implement our algorithms as non-specifically as we can.

Let's get back to our programming problem—writing a friendly multiplication program. What subgoals can I set?

subgoals in English

Get the user's name and her two numbers.
Mention her name, and print the product.

I can restate these subgoals in a more procedure-specific way by writing them in *pseudocode*—a pidgin Modula-2 that has the 'feel' of code, but plays fast and loose with exact syntax rules:

subgoals in pseudocode

get the information (name, first number, second number)
print the results (name, first number, second number)

Indeed, I can even do a little bit better—I can make mock procedure calls, and comment their results:

subgoals in Modula-2

```
GetTheInfo (Name, First, Second);
  (*Gets values for Name, First, and Second.*)
PrintTheNews (Name, First, Second);
  (*Prints First times Second, along with a comment about Name.*)
```

It's clear from the comments what kind of parameters are required. The first procedure *changes* its arguments, so they must go to variable parameters. The second procedure, in contrast, only *uses* its arguments, so they can go to value parameters.

A completed version of program Multiply is shown below. I call it *a* completed version, rather than *the* completed version, because it's not the only program that can solve our problem. However, it is one of the few programs that really fulfills our extra desire—to write a program that is as easy to read, write, and change as it is to run.

* Wirth, Niklaus, *Programming in Modula-2*, Fourth Edition, Springer-Verlag, 1988, pg. 87.

a program with variables

```
MODULE Multiply;
   (*Gets a user's name (up to 20 characters long), and two numbers to multiply.*)
IMPORT IO;
VAR First, Second: INTEGER;
      Name: ARRAY [0..20] OF CHAR;
PROCEDURE GetTheInfo (VAR Name: ARRAY OF CHAR;
                      VAR Number1, Number2: INTEGER);
   (*Gets the program user's name and favorite numbers.*)
   BEGIN
      IO.StringOut ("Hi!  What's your first name?    ");
      IO.StringIn (Name);
      IO.StringOut (Name);
      IO.StringOut (",  give me two whole numbers to multiply.    ");
      IO.IntIn (Number1);
      IO.IntIn (Number2)
   END GetTheInfo;
PROCEDURE PrintTheNews (String: ARRAY OF CHAR;
                        Number1, Number2: INTEGER);
   (*Prints the product of its arguments.*)
   BEGIN
      IO.StringOut ("Oh, no!  ");
      IO.StringOut (String);
      IO.StringOut (".  Their product is  ");
      IO.IntOut (Number1 * Number2 , 1);
      IO.StringOut (".  How auspicious!")
   END PrintTheNews;
BEGIN
   GetTheInfo (Name, First, Second);  (*These arguments are changed.*)
   (*We have our data.*)
   PrintTheNews (Name, First, Second)  (*These arguments are used.*)
   (*We've printed the output.*)
END Multiply.
```

```
Hi!  What's your first name?  Angelica
Angelica, give me two whole numbers to multiply.  37  18
Oh, no! Angelica.  Their product is 666.  How auspicious!
```

Modification Problems

1. Change Multiply so that instead of multiplying two INTEGER values, it prints the quotient of two REAL numbers.
2. Without changing the effect of Multiply, change the order of arguments to its procedures—have Name be the second argument to each call, rather than the first.

Section 1-4
Program Engineering Notes

LARGE-SCALE SOFTWARE SYSTEMS, LIKE LARGE-SCALE transportation systems, require engineers: people who can put together enough experience, calculation, and common sense to build systems and to keep them running.

Little programs take experience, calculation, and common sense, too. I might even say that we have to become *program engineers*. We have to learn skills and ideas that are language-independent, but are still necessary for writing clear, working programs.

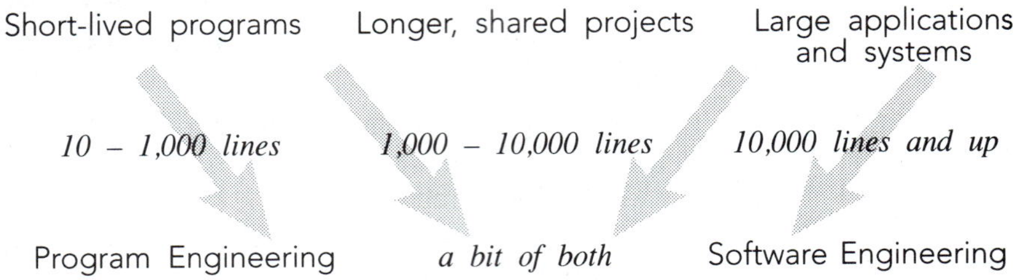

Now, you don't have to be an expert to be a good engineer. Experts know facts. On one hand they know the syntax of a programming language, on the other, they understand the details of a particular algorithm or application. However, expertise at the extremes doesn't always lead to good programming ability. A programmer has to be able to put different talents together, and write programs that:

- are easy to check and test,
- can be finished or improved by somebody else if need be,
- are completed in a reasonable amount of time,
- and, last but not least, work.

The basic guideline is simple:

the golden rule

> Every program should be as easy for a human being to read and understand as it is for a computer to execute.

write comments

The first piece of advice I'll pass along under the heading of program engineering is *write comments*. Commenting is the first and best line of defense against semantic errors. How many comments? Enough so that the person sitting next to you can figure out your program.

but don't always believe them

Piece of advice number two may seem strange, but it's equally important—*Don't believe your comments*. Usually, new programmers write comments that say what they *intend* to do: *get the data, print the results*, and so on. Later, when they debug their programs, novices tend to take their comments at face value. They assume that the code that follows actually carries out the intention of the comment.

Experts, in contrast, write comments that document their accomplished *goals*: *we have data, we've printed results*, etc. When experts debug programs, they are able to look at their claims a little more suspiciously: Do we *really* have the data? Did that procedure *actually* print results? Remember—you trust your mother, but you cut the cards. I'll talk about commenting styles often.

write short test programs

Tip number three is *try it!* The best way to clarify confusing points about syntax, semantics, or the details of your own Modula-2 system (if thinking about it gets you nowhere) is to experiment. The computer won't blow up or break down if you make a mistake, so don't worry about crashing the whole system. Soon, if you don't understand a bug or feature, you'll be able to write a half-dozen line program that tests only the point in question. Be sure to echo, or reprint, any input you supply to make sure that input is really what you think.

The best piece of advice of all is to ask somebody for help, and to be willing to help somebody yourself. Practice in spotting and explaining bugs benefits everyone.

Antibugging and Debugging

Let's begin with a few details. Spelling errors come first and foremost. Anybody who says that math is essential for success in computing doesn't know what she's talking about—typing is much more important!

Reserved words must always be entirely capitalized. Moreover, identifiers must always be spelled in exactly the same way, including any use of capitalization. Otherwise, the compiler can't recognize that an identifier you use is the same one that you declared. Can you spot the bugs in this little segment:

a few spelling errors

```
Module Tester;
IMPORT io;
VAR Trial:  INTEGER;
PROCEDURE TestOutput (Print:  INTEGER);
      . . .     program continues
BEGIN
   TestOutput (Trail);
      . . .
END TESTER.
```

The bugs I inserted were:

a) **MODULE** must be entirely capitalized;

b) I imported io, instead of IO;

c) I declared Trial, but then used Trail in the TestOutput call;

d) I began program Tester but ended program TESTER.

count the arguments

Counting is as important as spelling. When you write a procedure call, be careful to have the right number of arguments! The numerical output procedures all take more than one argument, and the arguments must all be there.

It's easy to leave an argument out when the number that's being printed is being supplied as an expression, for instance:

```
(*Incorrect — missing a width argument.*)
IO.IntOut (3 + 4);
(*Correct — the width argument is included.*)
IO.IntOut (3 + 4 , 1);
```

Parameter Syntax Bugs

The syntax of value and variable parameters may be confusing, but they have a simple goal—to distinguish between 'in' arguments and 'out' arguments. Wirth could have used an easier syntax and given us **INVALUE** and **OUTVARIABLE** parameters, but he didn't, so that's that. Learn parameters now while programs are short and simple, and they'll help you later when our code gets more complicated.

At first, you may find it's easiest to learn to use parameters the same way that you learn to conjugate verbs in a foreign language—by rote, rather than by logic. The two most common syntax errors are:

1. having a mismatch between the type of a parameter, and the type of its argument, and
2. forgetting that the argument of a variable parameter must be a variable.

To learn to avoid these problems, I suggest that you go back into the chapter, and deliberately introduce errors of this sort into sample programs. Ask a classmate if she can spot the bugs, then ask her to do the same for you.

work from a listing

Something that I do when I'm debugging programs is to take a listing (a printout of the program) and a pencil, then draw lines that connect the arguments in a procedure *call* to the parameters in the procedure *declaration*. Remember that arguments and parameters are associated by their positions. The compiler can't reorder a call so that it makes sense.

Background: Are Bugs Inevitable?

A group of programmers were presenting a report to the Emperor. "What was the greatest achievement of the year?" the Emperor asked.

The programmers spoke among themselves and then replied, "We fixed 50% more bugs this year than we fixed last year."

The Emperor looked on them in confusion. It was clear that he did not know what a 'bug' was. After conferring in low undertones with his chief minister, he turned to the programmers, his face red with anger. "You are guilty of poor quality control. Next year there will be no 'bugs'!" he demanded.

And sure enough, when the programmers presented their report to the Emperor the next year, there was no mention of bugs.

From *The Zen of Programming*, by Geoffrey James, © 1988, InfoBooks, Santa Monica, Ca. 90406

Semantic Bugs

It's time for an important point about semantic errors, which are mistakes in the meaning, rather than the syntax, of code.

syntax precedes semantics

> The fact that a program compiles, and runs, is only evidence of syntactic correctness. It says nothing about program semantics—whether or not the program does what we expect it to.

This statement seems obvious, but it does happen that, in the process of making syntactic fixes, a program's semantics can be discombobulated. It's always good to bear the ultimate question—*Is the output right?*—in mind.

By far the most common sort of program bugs come from problems with parameters. The root cause of most problems are semantic errors, or problems with meaning, that are syntactically correct. This pitfall is so universal that program engineering guideline Number One might well be:

> Are you sure that your value parameters are value parameters and that your variable parameters are variable parameters?

For example, consider a procedure that prompts the program user to enter arguments. In outline, the procedure is:

```
PROCEDURE PromptAndRead (···);
  (*How would you complete the heading?*)
  BEGIN
    IO.StringOut ("Please enter  ");
    IO.IntOut (Count, 1);
    IO.String ("  numbers:    ");
    IO.IntIn (First);
    IO.IntIn (Second)
  END PromptAndRead;
```

Note how each of these headings screws up:

two incorrect examples

```
PROCEDURE PromptAndRead (Count, First, Second: INTEGER);
  (*First and Second are mistakenly declared as value parameters.
   Changing them won't change the arguments they rename.*)

PROCEDURE PromptAndRead (VAR Count, First, Second: INTEGER);
  (*Count is wrongly declared as a variable parameter. We couldn't
   call the procedure with a literal Count value.*)
```

The correct version of the heading is:

the correct version

```
PROCEDURE PromptAndRead (Count: INTEGER;
                               VAR First, Second: INTEGER);
  (*Count will be used, and First and Second will change.*)
```

Let me ask a question about program *style*: Can a declaration be wrong even when it's right? Sure. What's wrong with this heading?

```
PROCEDURE PromptAndRead (VAR First: INTEGER;
                             Count: INTEGER;
                         VAR Second: INTEGER);
```

style

What's the matter? The procedure is likely to be called incorrectly.

> Parameters should be ordered so that a reasonably obvious form of the procedure call is correct.

There's a story about the programmer who wrote programs with calls like:

```
AddGin (ToVermouth, Making, Martini);
```

but I'm not going to tell it here.

Modula-2 Details

Syntax Summary

- types: are associated with every value, variable, and parameter in Modula-2. The types we'll use most frequently are:

INTEGER	*whole positive and negative numbers*
REAL	*positive and negative decimals*
CHAR	*single character values*
CARDINAL	*non-negative whole numbers*
ARRAY [0 .. *Length*] **OF** CHAR	*a string* variable *of up to Length characters*
ARRAY OF CHAR	*a string* parameter *of any length*

- variable declaration: names variables, and specifies their types:

```
VAR SomeInteger: INTEGER;
    ACharacter: CHAR;
    Real1, Real2, Real3: REAL;
    String: ARRAY [0..10] OF CHAR;  (*Holds 10 characters.*)
```

- procedure declaration: similar to a module, but there can be *parameters*, between parentheses, in the heading:

```
PROCEDURE Sample (parameter declarations);
    Definitions and declarations,
    but no imports
BEGIN
    Procedure calls and other statements
END Sample;  (*Finish with a semicolon, not a period.*)
```

• value parameter declarations: have syntax similar to variable declarations, but go in the procedure heading:

```
PROCEDURE Sample (Number: INTEGER;
                  Letter1, Letter2: CHAR;
                  Word: ARRAY OF CHAR);
```

Each value parameter takes the value of its argument at the time the procedure is called. Value parameters are employed when a value must be *used* (e.g. printed) by a procedure.

• variable parameter declarations: include the word **VAR** for each new type of variable parameter:

```
PROCEDURE Sample (VAR Value: INTEGER;
                  VAR Ch1, Ch2: CHAR;
                  VAR Name: ARRAY OF CHAR);
```

A variable parameter renames a variable supplied in a procedure call. Value parameters are used when the variable is going to be changed (e.g. by having a value typed in for it) in a procedure. Changing the variable parameter *inside* the procedure permanently changes the argument *outside* the procedure.

• string array parameters: whether value or variable, are declared as **ARRAY OF** CHAR:

```
PROCEDURE GetName (Handle: ARRAY OF CHAR);
```

• input and output procedures from the IO library are:

To print a	*Use* IO*'s procedure*	*With arguments*
return	LineOut	*none*
string	StringOut	(*string*)
CHAR	CharOut	(*character*)
INTEGER	IntOut	(*integer, width*)
CARDINAL	CardOut	(*cardinal, width*)
REAL	FixOut	(*real, width, places*)
REAL	FloatOut	(*real, width, places*)

(*All arguments can be values or variables.*)
(*The 'places' and 'width' values are always whole numbers.*)

To read a	*Use* IO*'s procedure*	*With an argument of type*
CHAR	CharIn	(CHAR)
INTEGER	IntIn	(INTEGER)
REAL	RealIn	(REAL)
string	StringIn	(**ARRAY OF** CHAR)
CARDINAL	CardIn	(CARDINAL)

(*The argument must always be a parameter or variable.*)

Modula-2 Details

Remember . . .

- Planning and organizing is the key to successful programming. Set specific goals for each procedure you write. You should be able to lay out your program in *pseudocode* procedure calls, complete with arguments, before you write it in Modula-2.
- Libraries have to be imported before their procedures can be used. Each procedure must be preceded by the library's name: IO.IntIn.
- Every library imported in the main program is automatically available in new procedures.
- A *literal* value is a written (as opposed to stored) value, like "T" (a literal CHAR), or 73 (a literal INTEGER).
- Every value, every variable, and every parameter has a specific type. Types can't be mixed—one result is that every type has its own input and output procedures.
- There is always a one-to-one correspondence, by position, between the parameters declared in a procedure heading, and the arguments supplied in a procedure call. The argument of a variable parameter *must* be a variable or parameter—it can't be a literal value.
- When an argument is going to be changed by a procedure, pass it to a variable parameter. When it's just going to be used, pass it to a value parameter.
- REAL values can be input and output in either *fixed point* or *floating point* notation. Procedures that print numbers are always given a *printing width* value that can provide extra blanks on the left for lining up columns of figures. A minimal width value (like 1) means 'print the number in the exact space it requires.'
- Modula-2 uses the semicolon as a separator between procedure calls, other statements, and the different parts of a program.
- Interactive programs communicate with the program user. If the program expects input, it must *prompt* for it—the computer won't ask questions by itself.
- Modula-2 programs are *free format* as far as the compiler is concerned, but they should be laid out with indentation and spacing that makes them easy for people to read.

Zen Study Guide

Big Ideas

The Big Ideas in this chapter were:

1. *Procedures and parameters are programming's main contribution to problem solving.* The idea of packaging a specialized solution step with 'hooks'—parameters—that make it general-purpose is enormously powerful. Learning to set up solutions as parameterized single steps is the most valuable thing you'll learn in a programming course.

2. *Every complicated program has a simple, straightforward program buried inside of it.* Most programs take just three steps: get data, do something with it, then print the results. Programs differ in the details of

what happens inside procedures, rather than in their basic approach to problem solving.

3. *Program by modifying, not by working from scratch.* If possible, change an existing program by changing one of its procedures; if you can, change an existing procedure by modifying one of its parameters.

Zen Study Guide
Questions

It's time to return to the Zen Study Method, in which we tackle the problem of learning to answer questions by learning to ask them.

1-1 Let's get the examination off to an easy start by making up two *Write a statement that produces this output. . .* questions. Naturally, the output you want printed should contain both single and double quote marks.

1-2 Recognizing the types of literal values (CHAR, INTEGER, and so on) is a grind, but it's a standard test question. How would you ask the question *What values can be assigned to which variables?* By starting with two columns: a list of literal values, and a list of variables of different types. Try it.

1-3 *Correct this statement* is always fun. Make up five statements that are wrong because they:

a) mix up the number of parameters required for CharOut, IntOut, and FixOut;

b) misorder the parameters of a call of FloatOut;

c) call the wrong IO procedure for an argument of a particular type;

d) use the wrong symbols in an assignment;

e) appear in the wrong place in a program.

1-4 A literal string is any sequence of characters between double quotes. Convince yourself that 3.1415 can be a string. Then, write the question *Which of these* IO.StringOut *calls isn't legal?* (If you're like me, you'll annoy your students by making them all perfectly legitimate, of course.)

1-5 How can you tell if people are getting the basic idea of grouping statements into procedures? One way is to require them to write a program that has lots of repetitious output. If they've caught on, they'll write the repeated parts as procedures, since it's faster to *call* a procedure several times than it is to write out what it does again and again. Come up with a *Write a program that prints this . . .* problem. Hint: listen to a Top-40 radio station if you need some repetitious output.

1-6 In lieu of essay questions, a few basic definitions are always nice. Look for half a dozen words or phrases that you'd just as soon not see on a test, and ask *What do these mean?*

1-7 Parameters, like angry civits, are best dealt with one at a time. Start with value parameters by considering procedure PrintInitials, from program ValueArg. Although that procedure printed the values of its two CHAR arguments, you might just as easily ask for a procedure that prints three INTEGERs, or five REALs, or four words, or a few of each. Make up three questions of this sort.

1-8 Procedure GetIQ, from program WiseGuy, reads a value for one INTEGER variable. As above, we might just as easily read five strings, or a REAL and two INTEGERs. Make up two questions that begin: *Write a procedure that reads values for . . .*

1-9 *Modify this program* is a great exam question, because you can test hard ideas without requiring the student to spend a lot of time writing. For instance, program DickC printed values that were 'hardwired,' or written into the program. If you were so inclined, you might ask that it be modified to get names and dates from the program user, instead. Make up two *Modify this program* questions for DickC, WiseGuy, and Multiply.

1-10 Students can be led to believe that the computer understands the names they use, and will reorder the parameter lists of procedure calls to match the parameter lists of procedure declarations. Write a few incorrect headings that thus mislead, and pose the problem *Fix these headings*.

1-11 When is a procedure correct, but wrong? When its syntax is correct, but its semantics are not. Here is the prototype of a question that is syntactically correct, and compiles just fine. However, it is semantically incorrect, because it doesn't do what the user really wants:

```
(*This procedure compiles perfectly. What's wrong with it?*)
PROCEDURE ReadAge (Age: INTEGER);
  (*Get the user's age for use in the main program.
  BEGIN
    IO.StringOut ("How old are you?  ");
    IO.IntIn (Age)
  END ReadAge;
```

Procedure ReadAge mistakenly uses a value parameter instead of a variable parameter. Write three similar questions that confuse the usage of value and variable parameters.

1-12 Finally, this chapter contained six complete programs: Chatty, Doggerel, ValueArg, DickC, WiseGuy, and Multiply. Each one answers a *Write a program that. . .* question. For each, come up with a new disguise for the same program.

Exercises

Warmups and Test Programs

For me, the key to learning a new language or system is writing many short programs. Right now even a five line test program will seem like a lot of work, but in time you'll be able to knock them out in minutes. Most importantly, you'll feel much better once you realize that you can figure things out by experimenting.

1-13 Can a comment contain a comment? Write a test program to find out.

1-14 Can a program use tabs for indenting? What is the tab character?

1-15 Suppose that you mung (ask your teacher) one of your program files. Is there any way to get it back to the way it was before?

1-16 How meaningful are the decimal digits produced by IO.FixOut and IO.FloatOut? What happens when you print decimals that go on and on (e.g. 1.0/3.0)?

1-17 Recall that as a variable, the declaration of a string looks like this:

```
VAR TestData: ARRAY [0..10] OF CHAR;
```

TestData can hold a word of up to ten characters. As a value parameter, the declaration of a string of *any* length is:

```
PROCEDURE Heading (AnotherTest: ARRAY OF CHAR);
```

Write test programs to answer these questions about string variables and value parameters. Before you begin, though, answer this: Which questions can be answered simply by trying to compile the test program?

a) What happens if the argument of a 10-character string is 11 characters long? How about 12 characters long?

b) Is there a maximum length for a string variable?

c) Suppose that the input of a 10-character string is only 5 characters long. When you print the string (with IO.StringOut) will five extra blanks follow it?

d) Can you store a number in a string variable? Can it be printed with a numerical output procedure?

e) Can a string variable be just one character long? If not, can its input be just one character long? Can the actual parameter of a string-type value parameter be just one character long?

1-18 Write the statements needed to produce the output shown below. For *d* and *e*, use an INTEGER and REAL variable.

a) `They can't take that away from me.`

b) `I said "read my lips," creep.`

c) `"Don't you want another chance?"`

d) `? and the Mysterians sang '96 Tears.'`

e) `The price is $1.98.`

1-19 I'm sure that you recall the scene, in *Rebel Without a Cause*, in which Marlon Brando pithily answers the question *What are you against?* by asking *What do you got?* Let's break some rules just to see what happens.

a) What happens when you type in a character for an INTEGER variable?

b) What happens if you use a value (rather than a variable) as the argument of a variable parameter?

c) What happens if you call StringOut, rather than IO.StringOut?

d) What happens when a procedure call has an extra (or missing) argument?

e) What happens when you declare a procedure that has parameters, but doesn't contain any statements?

1-20 Write a program that prompts its user to enter INTEGER values for the month, day, and year. Then, print the date in *month/day/year* format. Naturally, the printing should be taken care of by a procedure that takes the actual values as parameters. *Option*: let the separator character (the /) be supplied by the user, and passed to the procedure too.

1-21 Procedures can contain procedures. Just for practice, write a program in which the main program gets two INTEGER values from the program use, then uses them as arguments to a procedure that contains, and calls, two internal procedures, and gives one of its arguments to each.

Exercises Programming Problems

1-22 Before the word 'bit' was taken over by computer scientists, a bit used to represent a Spanish *real*—better known as a piece-of-eight, or one-eighth of an old peso. In American slang, a bit is twelve and a half cents, two bits is a quarter, and so on.

In keeping with the tradition established by the early Spanish plunderers, the price of stocks is also recorded to the nearest eighth of a dollar. Write a program that prompts its user for the price of a stock, in dollars and eighths, as well as the number of shares she owns. Print the value of her holding in dollars and cents, using fixed-point notation. *Options*: prompt for today's change as well, and print before and after values.

1-23 The canonical, or standard, first application of procedures is to print the word `MISSISSIPPI` in block letters. It rapidly becomes obvious that writing just four procedures—PrintM, PrintI, PrintS, and PrintP—is much easier than writing the dozens of IO.StringOuts that would otherwise be required.

Make your first contribution to computer science by coming up with a new word for this example. Then, write a program that prints the word, using one procedure for each letter. If you have no shame, you can fall back on old reliable `MISSISSIPPI`.

1-24 The ordinary *statute mile* is based on the Roman mile, which is one thousand Roman paces long. Since it is often difficult these days to find a Roman who is willing to take a thousand paces, two alternative distance measures have been developed: the *kilometer*, which is one ten-thousandth of the distance between either Pole and the Equator, and the *knot* or *nautical mile*, which equals one minute of arc (there are 90 degrees, with 60 minutes of arc per degree, between the Pole and the Equator).

Write a program that lets you know how many kilometers are in a nautical mile, and vice versa.

1-25 California summer campers often take long bus rides. To occupy the campers' short attention spans, camp counselors encourage the singing of endless songs. Although the timeless classic '99 Bottles of Beer on the Wall' has been deemed to be politically incorrect (for its glorification of macho beer-swilling rituals), its replacement is not difficult to learn:

> *99 bottles of Perrier on the wall,*
> *99 bottles of Perrier.*
> *If you should drop a bottle of Perrier,*
> *You might get hit in the derriere.*
> *(and have only 98 bottles of Perrier)*

Write a program that contains a procedure that prints one verse of the song. It should require one value parameter—the number of bottles that currently remain on the wall. *Option*: let a second parameter provide a substitute name for *Perrier*.

1-26 *Ad libitum* is Latin for 'at one's pleasure.' Early Roman travelers would while away the hours by playing a game called *Mad Libitumi*. Working from a prepared text, the leader would ask players to supply missing nouns, adjectives, verbs, and adverbs as necessary. Then, she would read back the text with the new words inserted.

Here is a sample bit of text, translated from Latin, with missing words indicated in italic:

> 'Gosh, Professor *name*, I don't know what happened to my *noun*. My pet *animal name* must have *verb*ed it.'

Write a program that includes at least half a dozen string variables. Get appropriate words from the user in the main program, then use them as actual parameters to an output procedure that prints the text sample with words inserted.

1-27 Write a program that gets a single digit INTEGER from its user. Then, pass the digit to a procedure that multiplies the digit by 9*12345679 (note that 8 is missing). What's the output?

1-28 Write a program that contains separate procedures to:

a) get two REAL numbers from a program user;

b) print their sum and product using floating point notation;

c) print the sum and product in fixed point notation.

Option: Ask the program user how many decimal places she wants the fixed point procedure to print. Then, use the variable that holds her response as a third parameter to the procedure.

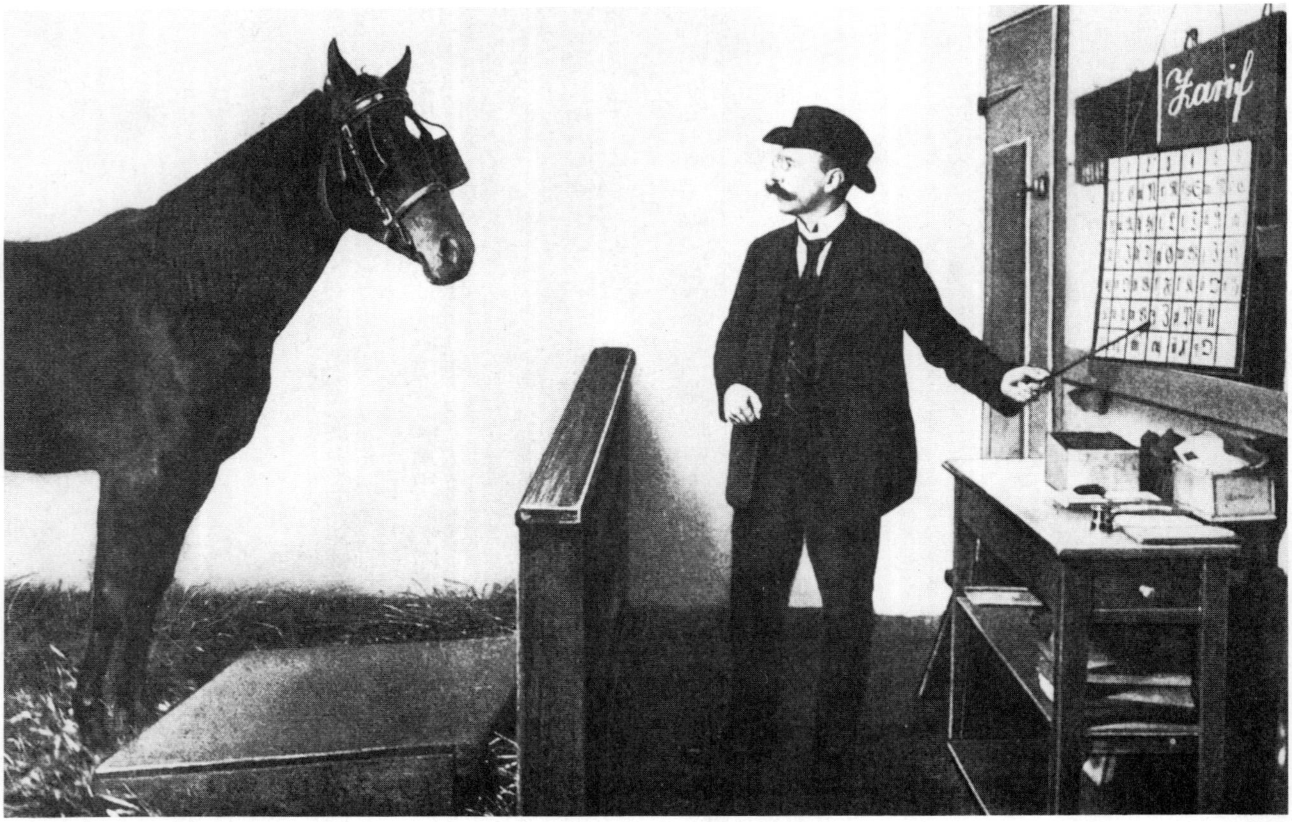

' . . . you'll use expressions in programs pretty much the same way you've always used them on paper.' (p. 58)

Chapter 2 Changing Values

Chapter 1 was tough, so let's take a breather. Chapter 2 deals with a single basic idea—values. You'll see how to save existing values, and how to calculate new ones. You'll only learn one new statement, assignment, but it will open a world of programming.

Quite a bit of our discussion deals with learning how to read *documentation*, or the written explanation of programming language details. As you've certainly gathered by now, I like to present a variety of tools. However, I don't insist that you memorize all the details of each one. Instead, see what's available, then return to peruse the fine points when you need them.

If you can, bolster the contents of this chapter by finding out what your own system has to offer. Of necessity, I minimize our dealings with libraries. If you hope to make a million bucks writing Nintendo games, you'd better learn to poke around.

Finally, a word to the wise: Chapter 2 is a golden opportunity to practice some of the harder ideas from Chapter 1. Take it—don't leave without nailing down the details of procedures and parameters.

Section 2-1 Saving and Changing Values

LET'S START WITH A PROGRAM THAT USES EVERYTHING we know about procedures and parameters, and adds a new statement: assignment.

program Adder

```
MODULE Adder;
   (*Prompts for and adds two numbers.*)
IMPORT IO;
VAR First, Second, Sum: INTEGER;
PROCEDURE GetTheNumbers (VAR One, Two: INTEGER);
   (*Prompts for and reads two INTEGER values.*)
   BEGIN
      IO.StringOut ("Please enter two numbers to add:   ");
      IO.IntIn (One);
      IO.IntIn (Two)
   END GetTheNumbers;
PROCEDURE AddThem (FirstValue, SecondValue: INTEGER;
                            VAR Result: INTEGER);
   (*Adds its first two parameters.*)
   BEGIN
      Result := FirstValue + SecondValue  (*Stores the sum.*)
   END AddThem;
PROCEDURE PrintTheResults (Answer: INTEGER);
   (*Prints its parameter without extra spacing.*)
   BEGIN
      IO.StringOut ("Their sum is:   ");
      IO.IntOut (Answer, 1);
      IO.LineOut
   END PrintTheResults;
BEGIN
   GetTheNumbers (First, Second);
   AddThem (First, Second, Sum);
   PrintTheResults (Sum)
END Adder.
```

```
Please enter two numbers to add:  14  18
Their sum is:  32
```

Before we go any further, let's take a minute to review parameters by looking at each procedure's heading and call. GetTheNumbers has two variable parameters. It changes the values of its arguments by reading input:

```
PROCEDURE GetTheNumbers (VAR One, Two: INTEGER);

     GetTheNumbers (First, Second);
```

AddThem has two value parameters that represent numbers it's going to use (by adding them) and one variable parameter that's going to change (in an assignment statement):

```
PROCEDURE AddThem (FirstValue, SecondValue: INTEGER;
                                  VAR Result: INTEGER);

        AddThem (First, Second, Sum);
```

Finally, PrintTheResults has just one value parameter. It uses its argument by printing its value:

```
PROCEDURE PrintTheResults (Answer: INTEGER);

        PrintTheResults (Sum)
```

Be sure that you can follow the connections between arguments and parameters. Not every program follows this three-procedure outline:

a very common pseudocode

get the data (arguments);
process it (arguments);
print the results (arguments);

but most of the programs we'll write do.

Modification Problems

1. Change Adder so that it multiplies three numbers instead of adding two of them.
2. The distance around a circle is about 3.1416 times its diameter. Modify Adder so that it asks for the distance between two especially unpleasant cities, and lets you know how big a fence is needed to enclose them both.

Self-Check Questions

Q. Why would these calls of GetTheNumbers be illegal?

```
    (*illegal calls*)
a)  GetTheNumbers (First);
b)  GetTheNumbers (Grade, Rank, Service);
c)  GetTheNumbers (7, 59);
d)  GetTheNumbers (RealVariable, IntegerVariable);
```

A. You might try making up a similar exercise for the other calls.

a) There aren't enough arguments.
b) There are too many arguments.
c) A variable parameter's arguments have to be variables.
d) The argument's types have to match the variable parameters.

Assignment Statements

Assignment changes the value of a variable. It can give the variable a starting value, or alter its current value. In Modula-2, assignment takes the form:

variable identifier := *the value represented by an expression*

Program Adder's assignment statement, from procedure AddThem, was:

```
Result := FirstValue + SecondValue  (*Stores the sum.*)
```

making assignments

From the computer's point of view, assignment is a two-step process. First, it performs a calculation by *evaluating* the *expression* on the right-hand side of the statement: what does it equal? Then, the *result*, or calculated value, is given to the variable on the left-hand side.

Most people read the assignment operator as 'gets.' For example:

```
LuckyNumber := 7 + 5;
```

would be read: 'the variable LuckyNumber gets assigned the result of adding 7 and 5,' or, more succinctly, 'LuckyNumber gets 7 plus 5.'

> A variable must be *initialized*, or given a starting value, before it can appear in an expression. The variable is *undefined* until it is initialized.

```
Age := 17;           (*Once Age is initialized ...*)
LuckyNumber := Age - 2;        (*... it can appear in an expression.*)
```

initializing variables

Reading new values entered by the program user is an alternative way of initializing variables. But one method or the other must be used, since an uninitialized variable is like an unmade bed—you can never be sure of what the computer put there when it started to run your program.

numerical assignments

Assignments must involve values and variables of appropriate types. If Age is an INTEGER variable, and Fraction is of type REAL, then these are valid assignments:

```
(*legal assignments*)
Age := 16;      Fraction := 16.5;
```

But these assignments contain type clashes between Age, 16.5, and Fraction:

```
(*illegal assignments*)
Fraction := Age;     Age := 16.5;     Age := Fraction;
```

character assignments

Assignments to CHAR variables are special. Since a variable's name might be just one letter long, we have a distinctive way to indicate a one-letter CHAR value—it's enclosed within quote marks. For example:

```
(*assigning CHAR values*)
Initial := "E";     SeventhDigitCharacter := "6";
```

Quote marks prevent confusion between the literal value "E" and a variable named E, or between the literal value "6" and the INTEGER value 6. Note that a CHAR variable can only represent a single character's value—assigning "XTC" to a CHAR would be an error.

Like characters, literal string values are quoted. As long as it isn't too long, any sequence of characters can be assigned to a string variable. If a variable called Name has been declared as an **ARRAY** [0..5] **OF** CHAR, all assignments of five characters or fewer are legal:

string assignments

```
Name := " Elvis ";        (*five letters*)
Name := "    ";           (*two space characters*)
Name := " Go Go ";        (*two words, for a five-character total*)
```

why the extra character?

Why does the string variable need room for an extra character? Because the Modula-2 compiler will automatically store an 'end-of-string' character in every string variable. This character is usually the *null* character. It doesn't print; instead, it makes it easy for string-handling procedures to find the end of stored strings that might include blanks. Not every compiler depends on things working this way, but since it simplifies life, I'll leave the extra space.

Self-Check Questions

Q. Correct these assignment statements:

a) 2nd := 1st + Error; *b*) 3+9 := Sum;
c) FirstLetter := A; *d*) Initial := "WS";
e) TaxRate := 5%; *f*) Start := Init := 0;

A. The rewritten assignments are:

a) Second := First + Error; *b*) Sum := 3+9;
c) FirstLetter := "A"; *d*) A CHAR variable holds just one letter
e) TaxRate := 0.05; *f*) Start := 0; Init := 0;

Expressions

An expression *represents*, or shows, a value. The very simplest expressions are literal values like 17 or "t". Once a variable has been initialized, or given a starting value, it's also an expression, because now it also represents a value.

Expressions can be complicated, too. For instance, think about 1 + 1, which you probably thought was pretty simple. In fact, it has:

dissecting 1 + 1

- an *operator*, the plus sign,
- two *operands*, the 1's,
- a *result* or *value*, the answer, 2, and, finally,
- a result *type*, which is the answer's Modula-2 type.

Fortunately, even though I have to introduce these terms in order to write precisely, you'll use expressions in programs pretty much the same way you've always used them on paper. The main thing to remember is that no matter how simple or complicated an expression is, it always has a Modula-2 type, just like a variable. This makes sense, because the computer is usually going to evaluate the expression so we can assign it to a variable.

Most of the expressions we'll write for a few chapters involve arithmetic. There are two basic kinds of arithmetic operators—those that take REAL operands, and those that take INTEGERs. The REAL operators are:

REAL operators

REAL *Arithmetic Operators*

+	*addition*	1.0 + 0.03E–17
–	*subtraction*	0.03 – 6.2
*	*multiplication*	32.87 * 6.5E–02
/	*division*	5.0E20 / 2.0

The INTEGER operators are the same except that the slash has been replaced by two special operators for division—**DIV**, and **MOD**. They do the 'whole and remainder' division that you learned in first grade.

INTEGER operators

INTEGER *Arithmetic Operators*

+	*addition*	3 + 2
–	*subtraction*	–7 – 1
*	*multiplication*	52 * 17
DIV	*'whole number' division*	10 **DIV** 3 (*is* 3)
MOD	*'remainder' division*	10 **MOD** 3 (*is* 1)

DIV and MOD

The first of the new division operators, **DIV**, gives us the quotient of a division without any remainder, as though the quotient had been rounded toward zero. **MOD** (whose operands must be positive) does just the opposite. It ignores the 'whole' part of the quotient, and provides only the remainder.

9 **DIV** 5 *is* 1	24 **DIV** 9 *is* 2	–9 **DIV** 5 *is* –1
9 **MOD** 5 *is* 4	24 **MOD** 9 *is* 6	9 **MOD** 24 *is* 9

DIV and **MOD** are reserved words that can't be redefined. They represent operations, just as + and – do.

types of operands

> The operands in an arithmetic expression are both required to have the same type. Naturally, this type must be right for any operators that are being employed.

Modula-2 type rules mean that some kinds of expressions aren't legal, even though they make sense to us. For example:

a few illegal expressions

Expression	*Why it's illegal*
1.0 + 1	*The operands can't have different types.*
5 / 2	*Operands of / can't be* INTEGER.
4.0 **DIV** 2.0	*Operands of* **DIV** *can't be* REAL.
1.0E+02 **MOD** 50	*Operands of* **MOD** *can't be* REAL.

Obviously there must be *some* way to add 1.0 to 1. There is: by using standard *conversion functions* that are part of Modula-2.

FLOAT and TRUNC

FLOAT(1) *equals the* REAL *value* 1.0
TRUNC(1.0) *equals the* INTEGER *value* 1

Any expression that represents a whole number could be an argument in place of 1, while any REAL-valued expression could replace 1.0. FLOAT gets its name because it temporarily represents a whole number as a floating point number. TRUNC is so named because it *truncates*, or temporarily chops off, the fractional portion of a REAL. I'll discuss both of them in section 2–2, along with Modula-2's other standard functions.

Self-Check Questions

Q. Which of these pairs of assignments could be replaced by a single assignment statement? What is it?

a) IQ := SAT – 600;
IQ := 125 + Age;
b) Quiz := 4 – Test;
Quiz := Quiz * 3;
c) Weight := Length + 62;
Measure := Length * Measure;
d) Margin := Margin + 6;
Margin := 32 – Margin;

A. One statement can replace two either when the second statement makes the first one irrelevant (as in *a*), or when the second statement uses the variable assigned in the first statement (as in *b* and *d*).

a) IQ := 125 + Age;
b) Quiz := 12 – (3 * Test);
c) Can't be simplified.
d) Margin := 26 – Margin;

Operator Precedence

In Modula-2, as in ordinary arithmetic, we can combine small expressions into a chain of operations. As long as the rule against intermingling types is obeyed, expressions can be as complicated as you like.

evaluating expressions

However, when expressions have several operators, the problem of *operator precedence* comes up. What part of the expression will take precedence, and be evaluated first? Does 2 * 3 + 4 equal 10 (do the multiplication first) or 14 (do the addition first)?

Programming languages solve this problem by decree: there is a hierarchy, or ordering, of different operations. Expressions that contain more than one operator from a given level of the hierarchy are usually evaluated from left to right. Parentheses can change the order of evaluation, or make it clearer.

In Modula-2, arithmetic operations have two precedence levels—addition and subtraction have less precedence than the other operations. The effect of the operations doesn't change, just the order in which operations are carried out:

operator precedence

* / **DIV** **MOD**	*these operations are completed . . .*
+ –	*. . . before these operations.*

As a result:

2 * 3 + 4	*is* 10, *not* 14
9 – 4 **MOD** 2	*is* 9, *not* 1
2.5 + 5.0 / 2.0	*is* 5.0, *not* 3.75

When an expression contains more than a single operator from any one level, we do the multiplications and divisions from left to right first, then perform the lower level additions or subtractions.

5.5–3.375/1.125	*is* 5.5–3.0	*is* 2.5
5*3+14 **MOD** 4	*is* 15+2	*is* 17
4.5/1.125–3.325*6.5	*is* 4.0–21.6125	*is* –17.6125
7–6*2–33 **DIV** 4–3	*is* 7–12–8–3	*is* –16

parentheses

Parentheses change this order of evaluation, because the part of the expression between parentheses gets evaluated before the rest of the expression does. When we *nest* parentheses—use multiple levels of parentheses—calculations are done from the inside out.

(5+3)*(8–2)	*is equivalent to*	8*6
(6 **DIV** 3)*(2–4)	*is the same as*	2.0*(–2)
2.5*(1.25+0.25)	*is like*	2.5*1.5
(8 **MOD** (2 * (5 – 4)))	*is like*	8 **MOD** 2

You can appreciate that there's usually more than one way to write an expression. You'll often feel that you're faced with a choice between relying on the operator precedence rules and using parentheses.

> Don't choose. Instead, use the combination that makes a clear, readable expression. This usually means that you should add a few parentheses even if they're not required. Write expressions that can be understood by human beings, and the computers will take care of themselves.

An expression can be confusing even though it does just what you want it to. Compare:

```
Income := Receipts + Tips * Share; (*A Share of what?*)
```

write expressions clearly

to the expression below, which adds redundant parentheses to make our intention clear:

```
Income := Receipts + (Tips * Share); (*A Share of Tips.*)
```

Self-Check Questions

Q. Evaluate these expressions:

a) 25 **MOD** 5	*b*) 25 **DIV** 5	*c*) 5 **MOD** 5
d) 5 **DIV** 5	*e*) 27 **DIV** 18	*f*) 33 **MOD** 19
g) 12.0E02 / 2.0	*h*) 1.25 * 1.0E02	*i*) 12.0E–02 / 2.0
j) 1.25 * (–1.0E02)	*k*) 5 – 9 * 3 + 2	*l*) 7 * 8 – 9 + 12 **DIV** 5

A. For more practice, try joining examples with parentheses.

a) 0	*b*) 5	*c*) 0
d) 1	*e*) 1	*f*) 14
g) 600.0	*h*) 125.0	*i*) 0.06
j) 0.0125	*k*) –20	*l*) 49

Constant Definitions

A constant represents a value permanently. For example, consider program Trader, below. It is concerned with the difference between International pounds, which weigh 0.45359237 kilograms, and American pounds, which are slightly heavier at 0.4535924277 kilograms. It solves this problem:

problem: pounds to carats

> A diamond dealer has gone to Amsterdam to buy a ton—2,000 pounds—of stones. She is thrilled to discover that although she buys diamonds in American pounds, her company only expects to receive International pounds, which are a shade lighter. As a little bonus, she can keep any difference! Write a program that lets her know, in carats, just how large her bonus will be. There are 5,000 carats per kilogram.

It's good to be cautious when doing REAL arithmetic, incidentally. Computers are good with numbers, but the number of digits they use for REAL calculations is limited. Program Trader avoids the problem by restricting its result to the first non-zero digit. To be honest, I arrived at the proper 'decimal points' arguments for IO.FixOut by trial and error.

> IO.FixOut and IO.FloatOut can print figures far beyond the computer's accuracy. An impressive sequence of digits can be misleading and wrong.

assignment statement example program

```
MODULE Trader;
   (*Finds the difference (in carats) between weights in International pounds
    and American pounds.*)
IMPORT IO;
CONST INTERNATIONAL = 0.45359237;
      USA = 0.4535924277;          (*All constants show weight in kilos.*)
      ONEcarat = 1.0/5000.0;
VAR Pounds, KgToBuy, KgToDeliver, KgToKeep, CaratsToKeep: REAL;
BEGIN
   IO.StringOut ("How many pounds are you buying? (to nearest 0.1)    ");
   IO.RealIn (Pounds);
   KgToBuy := Pounds * USA;
   KgToDeliver := Pounds * INTERNATIONAL;
   KgToKeep := KgToBuy – KgToDeliver;
   CaratsToKeep := KgToKeep / ONEcarat;
   IO.StringOut ("Difference in pounds:    ");
   IO.FixOut (KgToKeep / USA, 12, 4);
   IO.LineOut;
   IO.StringOut ("Difference in carats:    ");
   IO.FixOut (CaratsToKeep, 12, 1)
END Trader.
```

```
How many pounds are you buying? (to nearest 0.1)  2000.0
Difference in pounds:        0.0003
Difference in carats:           0.6
```

To her dismay, her bonus on the one-ton sale would barely suit her as an engagement ring.

> A constant is an identifier that represents a value. The value can't be changed during program execution—that would be more appropriate for a variable.

Constant definitions belong in the first part of a program, along with the **IMPORT**s and declarations. For example:

defining constants

```
CONST PLANCK = 6.63E–34;
      LASTletter = "Z";
      THISyear = 1990;
      NEXTyear = THISyear + 1;
      TRUElove = "Nicole";
      FIVEblanks = "     ";
```

Although a constant's value can be calculated as part of its definition (as the definition of NEXTyear shows), variables can't be used to supply the value—only literal values or previously defined constants may appear.

> Notice that I capitalize the first word of constant identifiers as a matter of programming style—it makes them instantly distinguishable from variables.

Constants can be defined to represent numbers, characters, or strings. String constants are printed with procedure IO.StringOut, but don't have to be put in quote marks—the identifier represents the actual string.

```
IO.StringOut ( "I long for you,  " );    (*The string goes in quotes ...*)
IO.StringOut ( TRUElove );               (*... but the constant doesn't need them.*)
```

```
I long for you, Nicole
```

Modification Problems

1. As you can see, the calls of IO.FixOut in Trader use actual numbers to specify the number of digits to print, and the size of the printing field. Define appropriate constants to take the place of 10 and 12.

2. Break Trader into three separate procedures—one that gets the user's input, one that does the calculations, and one that prints the results. You can, and should, put the constant definitions into the calculation procedure. Be sure that all three procedures use parameters properly.

Self-Check Questions

Q. Each of these constants can be described as a member of at least one type. Name them.

a) WIDTH = 5	*b*) CORRECTION = –3
c) NUMBER = "15"	*d*) FACTOR = 85.0E–02
e) PRODUCT = –4 * 6	*f*) TENspaces = " "
g) CORRECTION = SIZE	*h*) KILLcharacter = "X"

A. Note that 'string' is not a predefined Modula-2 type.

a — INTEGER *or* CARDINAL
a, b, e, g — INTEGER
d — REAL
c, f — *string*
h — CHAR

Section 2-2
Documenting Modula-2: Standard Functions

THE WRITTEN EXPLANATION THAT ACCOMPANIES software is called *documentation*; it documents, or explains, how the software works. Learning how to find and read documentation is part of learning how to program. Programming isn't medical school—you don't have to memorize the Latin names of the fifty muscles of the lower limb just because they're there.

Instead, programmers rely on documentation over memorization because the facts have a tendency to change. Despite efforts at standardization, details may vary over time, and from system to system. Frequently, a little doubt about the facts is a good thing.

One set of documentation every programmer must deal with involves *standard functions*. Functions calculate values; programming languages, like hand calculators, almost invariably have some set of functions built into the language. I have to present Modula-2's standard functions here because they help write longer expressions. But don't memorize all the details now—instead, find out what's available, and return later if you have to.

Some programming languages are like all-purpose calculators, and include many standard functions. In Modula-2, only a relatively small group of functions are built into the language proper. The rest are contained in specialized libraries that can be imported, as we've imported IO. We'll deal with the standard functions here, then look at some library documentation in section 2–3.

Some Terminology

If you pick up any pocket calculator, you'll almost invariably find function keys (for squares, square roots, reciprocals, and the like) accompanying the arithmetic operators. Function keys save us from the death of a thousand cuts—a single keystroke takes the place of the operations of a commonly used routine.

It's reasonable to assume that a written computer language will be at least as powerful as the 'touch' language of a hand calculator. In fact, Modula-2 has three sources of functions:

where functions are found

- A few functions are built into the language. For example, the CAP function finds the upper-case equivalent of a lower-case letter.
- Other functions are supplied in libraries that can be imported. For example, a library called MathLibØ provides a sqrt function that calculates square roots.
- Finally, we can write new functions of our own, just as we wrote new procedures. We'll see how in Chapter 3.

Here's a typical function call. CAP's argument, given between parentheses, can be any single CHAR value. The entire call—function *and* argument—represents a CHAR value that's assigned to Upper:

```
Upper := CAP(Letter);  (*Assigns the upper-case version of Letter.*)
```

If Letter is "g", then CAP(Letter) represents "G". The function call's value is calculated according to the rule: 'If the argument is a lower-case letter, represent the upper-case version of the same letter. If the argument isn't a lower-case letter, represent the argument itself.'

I'll be able to document the built-in functions more precisely if you know a few definitions first.

calling and evaluating functions

- A function *call* invokes a function, just as a procedure call invokes a procedure. CAP(Letter) is a call.
- Most function calls will require an argument, between parentheses. The argument can be any expression of the proper type, although *which* type will vary from function to function. Letter is CAP's argument.
- The function call is *evaluated*—the call is an expression that represents a value.
- Once it has been evaluated, the function *returns* a *result*, or value. This means that every function has a type. CAP has type CHAR, since it returns a CHAR value.

> A function call, complete with its argument, represents a value that can be assigned to a variable, or used as an argument itself.

That means I could use CAP like this, too:

```
IO.CharOut ( CAP(Letter) );  (*Print the upper-case version of Letter.*)
```

There are three points about calling functions that people sometimes find confusing. First, the type of a function's argument is quite separate from the type of its result. It happens that they're the same for CAP, but that won't always be the case.

Second, the function call doesn't change its argument. The argument's value is used—not modified. In fact, we could have used a literal value instead of a variable in the examples above:

function arguments can be literals

```
Letter := CAP( "h" );
IO.CharOut ( CAP( "a" ) );
```

Finally, even though a function call looks like a procedure call, it isn't one. A procedure call takes the place of a series of statements. A function call, in contrast, takes the place of an expression—it represents a value.

Self-Check Questions

Q. Suppose that a program contains these statements:

a) Initial := CAP("G");
b) Letter := CAP("b");
c) Middle := CAP(Letter);
d) IO.CharIn (CAP(Last));

What are the final values of Initial, Letter, Middle, and Last?

A. A slightly tricky question. Initial is "G", while Letter and Middle have both been assigned "B". The last call, *d*, is illegal. The argument of IO.CharIn has to be a variable. CAP(Last), in contrast, is an expression that represents a value. It can't be used as a variable *or* as a statement.

Standard Functions

Standard functions in Modula-2 usually serve two purposes: either they do tasks that are out of the programmer's reach, or (by taking advantage of hardware instructions), they provide exceptionally efficient versions of routines that could be written independently.*

Modula-2's built-in functions, and the kinds of arguments they expect, are:

miscellaneous	ABS (*number*)	(**Absolute value of an* INTEGER *or* REAL.*)
	CAP (*char*)	(**Capitalized lower-case letter.**)
	ODD (*number*)	(**Is a whole number odd?**)
value conversion	FLOAT (*integer*)	(**Represent an* INTEGER as a REAL*)
	TRUNC (*real*)	(**Represent a* REAL *as an* INTEGER.*)
	MAX (*type ID*)	(**The largest value in a particular type.**)
	MIN (*type ID*)	(**The least value in a given type.**)
ordering	ORD (*value*)	(**Ordinal position of a given value.**)
	CHR (*position*)	(**Character in a particular ordinal position.**)
	VAL (*type, position*)	(**Value in a particular ordinal position.**)
special purpose	HIGH (*array*)	(**Used with arrays (see Chapter 8).**)
	SIZE (*type ID*)	(**Used with pointers (see Chapter 13).**)

Let's look at the individual functions in more detail.

miscellaneous functions

ABS(x) Returns the absolute value |x| of x, which may be of either type INTEGER or REAL. The function's result type is the same as its argument's type.

ABS(–10) *is* 10 ABS(–3.5) *is* 3.5E00 ABS(5) *is* 5

CAP(*char*) Returns the upper-case equivalent of its argument, if there is one.

CAP("d") *is* "D" CAP("A") *is* "A" CAP("!") *is* "!"

ODD(*number*) Returns a BOOLEAN value—TRUE if its argument is an odd whole number, and FALSE otherwise. (We'll use ODD in Chapter 5.)

value conversions

FLOAT(*number*) Represents its INTEGER argument as a REAL value.

FLOAT(4) *is* 4.0 FLOAT(220) *is* 2.20E2

* The rare application of type identifiers as functions is discussed in Chapter 20.

TRUNC(x) Represents the 'whole' part of its REAL argument as an INTEGER. Any portion of the parameter that is a fraction less than 1 is truncated—cut off. In effect, the parameter is rounded to the nearest INTEGER toward zero.

TRUNC(4.8) *is* 4 TRUNC(0.22573E+02) *is* 22

ordering functions

MAX(T), MIN(T) These functions take an ordinal type name T as an argument, and represent the greatest and least value of that type.

ORD(x) Represents the ordinal position, or 'place number,' of a value within its entire type. In particular, if x is a CHAR value, ORD(x) can be expected to supply the position of x in the computer's character set.

VAL(T, x) Represents the value of type T located in ordinal position x. The first value is the 'zeroth,' rather than the 'first,' though.

CHR(x) Represents the CHAR value located in ordinal position x. CHR is just a special-purpose case of VAL, since CHR(x) = VAL(CHAR, x).

The ordering functions provide several capabilities that will turn out to be convenient in later chapters. MAX and MIN are particularly useful for avoiding arithmetic errors; for instance, MAX(INTEGER) and MIN(INTEGER) provide the boundaries for correct INTEGER calculations.

We might also take note of the fact that ORD and VAL are *inverse functions*; what one does, the other undoes. Thus, VAL(T, ORD(x)) equals x itself.

> In some Modula-2 systems, certain functions may require CARDINAL arguments (FLOAT, CHR) or return CARDINAL result values (TRUNC, ORD, HIGH, SIZE).

Learning to live with this sort of inconsistancy is a fact of life of programming—every language has a few details that aren't fully worked out until the language is in wide use.

Self-Check Questions

Q. Write statements that would let you know:

a) the smallest INTEGER value your system can deal with.
b) the whole part of a REAL variable called GPA.
c) the place number of the letter "A" amongst the CHAR values.
d) the REAL equivalent of the number 7.
e) the absolute value of Number.
f) the value of Decimal rounded off to the nearest INTEGER.

A. Note the difference between the types of a function's argument and its result:

a) IO.IntOut (MIN(INTEGER), 1);
b) IO.IntOut ((TRUNC(GPA), 1);
c) IO.IntOut (ORD("A"));
d) IO.FixOut (FLOAT(7), 3, 3);
e) We can't write this unless we know the type of Number.
f) IO.IntOut (TRUNC (Decimal + 0.5), 1);

A Note About Characters

A lot of the programs we'll write involve characters, so I'd like to add a few additional comments about type CHAR and the things we can do with it. I've mentioned that Modula-2 uses the ASCII character set. This is a standard set of 128 characters that's used on many computers, and by many languages.

The ASCII character set begins with 32 non-printing *control characters* that control terminals, printers, and communications in various ways. The very last character in the set is also a control character. The ninety-five printing characters in between are shown below. The very first character is a space:

the printable ASCII characters

```
 !"#$%&´()*+,-./
0123456789:;<=>?@
ABCDEFGHIJKLMNOPQRSTUVWXYZ[\]^_`
abcdefghijklmnopqrstuvwxyz{}~
```

What about the non-printing characters? Can we get a program to output them, so that *we* can control a terminal or printer? The answer is yes, but we need a special way to refer to them.

CHAR constants

Any character can be represented by its *octal* (base 8) position in the ASCII set, followed by a capital C. It's used like a literal or constant value of type CHAR.

Luckily, you don't have to count in base 8—Wirth's use of octal is an anachronism that we have to deal with. The back cover has a table of all the ASCII positions. Some of the more interesting ASCII control characters are:

some interesting CHAR values

0C	the NUL character—an invisible character that does nothing
7C	the BEL character (for programs with bells and whistles)
8C	the BS, or backspace, character
15C	the CR, or carriage return, character

Now that you know something about ASCII, you can better understand the ordering functions described earlier. You know what ordinal numbers are—they indicate position. In most computer applications, the ordinal numbers begin with the 'zeroth' place. Thus, ASCII 0 is the 'zeroth' ASCII character—

0C. As it happens the letter "A" is the sixty-fifth ASCII character, "B" is the sixty-sixth, and so on. Thus:

ORD("A") *is* 65
CHR(65) *is* "A"
VAL(CHAR, 65) *is* "A"

Self-Check Questions

Q. If ordinal counting starts with the zeroth place, and the ASCII character set has 128 characters, what is the ordinal position of the last ASCII character?

A. It's the one hundred twenty-seventh. Confusing, but true.

Q. How far away from the letter "A" is the letter "T"? Write an expression that lets you know.

A. We subtract one letter's ordinal position from the other's:

```
Distance := ORD("T") – ORD("A");
```

Note that the actual values the ORD calls represent are irrelevant.

Q. Quick—if "A" is the first letter of the alphabet, what position is "S" in? No fair counting.

A. We can find the position using ORD, too. We can't just subtract, though, because that would be off by one:

ORD("A") – ORD("A") *is* 0

However, I said that "A" is the first letter of the alphabet, not the zeroth. We need to correct the position by adding 1:

```
Position := ( ORD("S") – ORD("A") ) + 1;
```

Again, the actual values returned by ORD are irrelevant.

Background: Is Documentation Inescapable?

Two programmers were arguing about user interfaces. 'Significant inroads are being made in 'ease-of-use',' said the first programmer. 'Soon people will no longer need to read tedious manuals before they can use a computer. Programs will be self-evident.'

The second programmer thought about this for a moment and then said, 'Last weekend I decided to chop some wood for a fire, but my old axe was dull and worn. So I went to the hardware store and purchased a new one.'

'That's all very interesting,' said the first programmer, 'but what does it have to do with user interfaces?'

'The new axe came with an eight-page instruction booklet,' he replied.

From *The Zen of Programming*, by Geoffrey James, © 1988, InfoBooks, Santa Monica, Ca. 90406

Section 2-3 More Documentation: Reading Libraries

MOST OF MODULA-2'S PROCEDURES AND FUNCTIONS are found in libraries that can be imported. We have already imported the IO library several times.

In Modula-2, libraries are written in a way that's meant to make them *self-documenting*. The explanation of what the library does is part of the library itself. Modula-2 manages this trick by breaking the library's definition into two separate parts: a *definition* file that explains what its procedures and functions do, and an *implementation* file that holds the declarations and does all the work.

We're not going to worry about how to write these separate files for a long time. However, the definition file—strictly speaking, it's called a *definition module*—is easy enough to read and understand. Definition modules contain procedure and function headings, along with comments that explain their operation. On occasion, a definition module will also include a constant definition or two.

> The definition module's contents determine what importing a library lets us do. We can only use the names that are defined in the definition module.

The best way to learn about a library is to read its definition module. Definition modules are ordinary files; you can read them the same way that you can read your own programs. Most Modula-2 systems have a central filesystem that holds definition modules, and you may be able to inspect files there. If not, you'll also find that every Modula-2 text and reference book (including this one) devotes an appendix or two to reprinting definition modules' contents.

A Library of Functions: MathLibØ

Let's begin with a typical definition module—MathLibØ. Since a library of mathematical functions may make you nervous, I'll introduce MathLibØ's contents before we look at the definition module proper. MathLibØ makes these functions available:

MathLibØ's functions

Function	*Argument*	*Result Type*	*Purpose*
sqrt(x)	(*RealValue*)	REAL	square root
ln(x)	(*RealValue*)	REAL	natural log
exp(x)	(*RealValue*)	REAL	exponential (e^x)
sin(x)	(*RealValue*)	REAL	sine
cos(x)	(*RealValue*)	REAL	cosine
arctan(x)	(*RealValue*)	REAL	arctangent
entier(x)	(*RealValue*)	INTEGER	like TRUNC

Any program that contained the import:

```
IMPORT MathLibØ;
```

could refer to MathLibØ.sqrt, MathLibØ.exp, and the others. Don't worry about the details of what each function does—we'll get to examples of them later.

Now, though, let's look at MathLibØ's definition module, below. As you read it, notice that each heading ends with a colon and a type name. In effect, Modula-2 functions are procedures that represent values—this is how the value's type is specified. The function headings provide the specification of how the function is used, while comments (and meaningful identifiers) explain what the functions do.

the MathLibØ definition module

```
DEFINITION MODULE MathLibØ;
  (*A collection of mathematical functions.*)
  PROCEDURE sqrt (Argument: REAL) : REAL;
    (*Returns the square root of Argument.*)
  PROCEDURE ln (Argument: REAL) : REAL;
    (*Returns the natural logarithm (to the base e) of Argument.*)
  PROCEDURE exp (Power: REAL) : REAL;
    (*Represents e raised to the real exponent Power.*)
  PROCEDURE sin (Theta: REAL) : REAL;
    (*Returns the sine of the radian-valued angle Theta.*)
  PROCEDURE cos (Theta: REAL) : REAL;
    (*Returns the cosine of the radian-valued angle Theta.*)
  PROCEDURE arctan (Argument: REAL) : REAL;
    (*Returns the inverse tangent, in radians, of Argument.*)
  PROCEDURE entier (Argument: REAL) : INTEGER;
    (*Returns the greatest INTEGER less than or equal to Argument.*)
END MathLibØ.
```

Example: Power Consumption

Let's see MathLibØ in action. The thoroughly frivolous problem we'll solve is:

problem: power consumption

> If you were a lightbulb, how bright would you be? Although you probably think this is an existential question, there's actually a formula you can use to get a pretty good estimate of how much power your body generates:
>
> $$\textit{power production, in watts} = (9 \times \textit{weight})^{0.751}$$
>
> where *weight* is supplied in pounds. This formula is remarkably accurate for warm-blooded creatures, regardless of size. Write a program that lets you know about how much power it takes you to keep going.

In outline, program Power doesn't hold any surprises. We have to get a fact (the user's weight), perform a calculation, and print the results:

pseudocode

get the input;
perform the calculation;
print the results;

If you're at all uncertain about parameters, it's useful to outline the program with would-be arguments in place, like this:

mock procedure calls

get the input (weight); (*use the weight*)
perform the calculation (weight, result); (*use weight, change result*)
print the results (result); (*use the result*)

How can we raise a number to the 0.751 power? With exp and ln, like this:

```
RaisedNumber := MathLib0.exp ( 0.751 * MathLib0.ln(Number) );
```

Don't worry if you've never heard of exp and ln, by the way. I've deliberately chosen a couple of functions that are probably unfamiliar so that you'll have an example of them in action. I'll talk more about MathLib0 function details in a moment.

a MathLib0 import example

```
MODULE Power;
  (*Demonstrates an IMPORT of the MathLib0 library.*)
IMPORT IO;          (*Each library requires its own IMPORT.*)
IMPORT MathLib0;
VAR Weight, Watts: INTEGER;
PROCEDURE GetFacts (VAR Weight: INTEGER);
  (*Finds the program user's weight, in pounds.*)
  BEGIN
    IO.StringOut ("How much do you weigh, in pounds?    ");
    IO.IntIn (Weight)
  END GetFacts;
PROCEDURE Calculate (Weight: INTEGER; VAR Watts: INTEGER);
  (*Applies the formula for estimating power consumption.*)
  CONST MULTIPLIER = 9;
           EXPONENT = 0.751;
  VAR Temp: REAL;
  BEGIN
    Temp := FLOAT (MULTIPLIER * Weight);
    Temp := MathLib0.exp ( EXPONENT * MathLib0.ln(Temp) );
    Watts := TRUNC(Temp)
  END Calculate;
PROCEDURE PrintResults (Watts: INTEGER);
  (*Prints the user's approximate power consumption.*)
  BEGIN
    IO.StringOut ("If you were a lightbulb, you'd consume about  ");
    IO.IntOut (Watts, 1);
    IO.StringOut ("  watts.")
  END PrintResults;
```

(program continues)

```
BEGIN
    GetFacts (Weight);
    Calculate (Weight, Watts);
    PrintResults (Watts)
END Power.
```

```
How much do you weigh, in pounds?  172
If you were a lightbulb, you'd consume about 248 watts.
```

local variables

> Notice the use of a temporary *local* variable in procedure Calculate. It's there to make the calculation clearer—it helped me write the procedure correctly.

Yes, variables can be declared in procedures too. However, they only last for the duration of a procedure call. A local variable goes away when the procedure call is ended.

I'll close with a few comments about MathLib0's functions, just in case any of them are unfamiliar. MathLib0.sqrt, at least, should ring a bell. Can you think of how to use it to find out if a number is a perfect square? Well, if it isn't, its square root can't be rounded off and still multiply correctly. In other words, it's a perfect square if:

perfect squares

```
Number equals TRUNC( MathLib0.sqrt(Number) )
                * TRUNC( MathLib0.sqrt(Number) )
```

The trigonometric functions sin, cos, and arctan may be a bit unusual to you because they take their arguments in radians, rather than degrees. However, there's a simple conversion factor: $180° = \pi$ radians (3.14159 radians), $360° = 2\pi$ radians, $90° = \pi/2$ radians, etc.

trig review time

You may wonder why functions like tangent, arcsine, and the others aren't part of the MathLib0 library. The answer is economy: any trigonometry or calculus text will have a table that yields all the trigonometric relationships from just sin, cos, and arctan. For instance:

```
TangentOfX := MathLib0.sin(x) / MathLib0.cos(x);
```

Many chapters from now, when we start to write our own library modules, a set of degree-valued trig functions is one of the first we'll create.

Functions ln and exp are used more often than they're understood. ln(x), the natural log function, is the natural logarithm (base *e*) of x. exp(x), the exponential function, represents *e* (the base of the natural log system, about 2.718) raised to the x power (i.e. e^x). I won't bore you with a review of logarithms. I will point out, though, that ln and exp help overcome a Modula-2 shortcoming that often annoys programmers—the lack of a specific *exponentiation* operator (there's no predefined function for raising a number (call it *a*) to some power (call it *n*) as in a^n). As we saw in program Power, this formula does the job:

exponentiation

a^n = MathLibØ.exp (n *MathLibØ.ln (a))

Modification Problems

1. Modify program Power so that it tells you the difference in energy production between a half-ton moose and a half-ton of mice. Assume that each mouse weighs about ⅓ ounce.

Self-Check Questions

Q. Write each of these as an assignment statement.

a) Assign Answer negative 7 times the variable Result.
b) Let Score get the absolute value of Time minus Place.
c) Have Operation get the character that represents REAL division.
d) Give Root the square root of InValue, plus the REAL equivalent of 6.
e) Give RightAngle half the sine of PI squared.
f) Assign the fourth root of 4.70237895 to Fourth.
g) Let FifteenthPower represent nine to the fifteenth power.
h) Make LogCosine equal the logarithm of the cosine of 41.9.
i) Have InverseSquare hold the reciprocal of 16 squared.
j) Let Round be the INTEGER nearest to RealValue.
k) Have Later represent the fourth character after Initial.

A. Yes, it is hard at first to remember if a function is standard, or must be imported from MathLibØ. You'll get used to it sooner or later.

a) Answer := –7 * Result;
b) Score := ABS(Time – Place);
c) Operation := "/";
d) Root := MathLibØ.sqrt(InValue) + 6.0;
e) RightAngle := MathLibØ.sin(PI*PI) / 2.0;
f) Fourth := MathLibØ.sqrt (MathLibØ.sqrt (4.70237895));
g) FifteenthPower := MathLibØ.exp (15.0 * MathLibØ.ln(9.0));
h) LogCosine := MathLibØ.ln (MathLibØ.cos(41.9));
i) InverseSquare := 1.0 / FLOAT (16*16);
j) Round := TRUNC (RealValue + 0.5);
k) Later := CHR (ORD(INITIAL) + 4);

Documenting IO's Input Procedures

Finally, let's take a closer look at the input procedures that come from IO. Input is one area of documentation that can be particularly difficult the first time you learn a programming language, so I won't just throw you IO's complete definition module. However, if we were to look at the headings of the input procedures, here's a somewhat simplified version of what we'd see:

```
DEFINITION MODULE IO;
   (*The main library of input and output routines.*)
   . . .   (*IO includes other definitions and declarations in addition to
             the ones we're interested in now.*)
   PROCEDURE CharIn (VAR Ch: CHAR);
      (*Reads a single input character and assigns it to Ch.*)
   PROCEDURE StringIn (VAR String: ARRAY OF CHAR);
      (*Skips blanks, then reads and saves characters until a space or control
        character is read. Ends the stored string with the 0C character.*)
   PROCEDURE IntIn (VAR Value: INTEGER);
   PROCEDURE RealIn (VAR Value: REAL);
   PROCEDURE CardIn (VAR Value: CARDINAL);
      (*All three procedures skip blanks, then read a string, as above. If the
        string forms a number (e.g. they're digits), returns the number as Value.
        If the input doesn't obey any number syntax, Value stays undefined.*)
END IO.
```

I'll rephrase the comments in prose:

- IO.CharIn reads any single CHAR value—a letter, a digit, a space, etc.

input procedure details

- IO.StringIn skips blanks (including tabs), then reads characters until a blank or control character (e.g. a carriage return) is encountered.
- IO.IntIn, IO.CardIn, and IO.RealIn each skip blank spaces, then read the longest value (of the appropriate type) they can. A blank or control character must end the value. If a number ends with a letter instead, the number is ignored—you'll have to prompt for more input.

When it reads a string, IO.StringIn follows two simple rules: skip any initial blank spaces, then read non-blanks until another blank space *or* a control character is reached. Thus (I've underlined each string in these examples):

string input examples

```
Here are 4 strings.
"This is a 5-string sequence."
But;;here,,is..only%$one--string.
```

As you can see, quote marks are just ordinary characters when they are supplied as input. They lose the special meaning they have within programs (i.e. marking the beginning and end of a string that may include blanks).

When numbers are read, leading blanks are ignored except as value separators. Thus, the statements:

```
IO.StringOut ("Please type in two numbers.");
IO.IntIn (First);
IO.IntIn (Second);
```

numerical input

treat the inputs below identically, even though the first separates the numbers by a single space, and the second by many spaces:

equivalent INTEGER inputs

a) **53 174**
b) **53   174**

Letters, punctuation, and other printing characters generally cause problems when they appear immediately before or after numbers (i.e. without an intervening space). The basic rule of thumb is that if an improper character pops up before, during, or after a number, the entire sequence will be discarded and ignored.

> Don't forget that the IO library we use is convenient, but it's not the only possible input/output library. None of the details we take for granted (for instance, that characters appear on the screen as you type, or that typing carriage return 'sends' a line of input to a running program) are carved in stone. Every Modula-2 system has lower-level libraries that provide greater control over I/O.

Self-Check Questions

Q. What value is read, and what value is about to be read, after each of these sequences? Each statement's input is on the left. You can assume that each input procedure's argument has the proper type (otherwise the program wouldn't compile). Any space is shown as an underline:

a)	**__53_A**	IO.IntIn (Number);
b)	**__53AB**	IO.IntIn (Number);
c)	**__53_42**	IO.IntIn (Number); IO.CharIn (Letter);
d)	**oops_27**	IO.IntIn (Number); IO.CharIn (Letter);
e)	**___53_**	IO.CharIn (Letter); IO.IntIn (Number);

A. Notice that sometimes a digit is a letter, and sometimes a number.

	Number	Letter	*Still remaining*
a)	53	*none*	**_A**
b)	*none*	*none*	*nothing*
c)	53	_	**42**
d)	*none*	_	**27**
e)	5	3	**_**

Section 2-4 Program Engineering Notes

A CONSIDERABLE AMOUNT OF PROGRAM ENGINEERING is devoted to turning programs into what is commonly referred to as *plain vanilla.* In a way, we're constantly resisting two impulses—first, to make programs too clever, and second, to believe that they're already clear. James McKelvey puts his finger on the reason:

why resist cleverness?

> 'An even larger problem with unclear work is that eventually someone is going to have to read what you have written (if your work is of any consequence at all). The reader, whether teacher, colleague, or superior, must understand it. A confused reader is not likely to say, "Gosh, this is so complicated you must be a genius to understand it. How would you like a raise?" More likely, you will get a response like, "This program is so contorted and incomprehensible we'll never be able to maintain it. How good are you at filing and making coffee?" Anybody who understands enough to be reading your code in the first place is intelligent enough to appreciate a clearly written program.'*

I followed this advice in program Power a few pages back. The variable Temp declared in Calculate isn't necessary to make a syntactically correct calculation. This convoluted statement also works:

```
Watts := TRUNC(MathLib0.exp(EXPONENT *
                   MathLib0.ln(FLOAT(MULTIPLIER * Weight))));
```

However, Temp makes the assignment comprehensible, and gives it a fighting chance at being correct:

a comprehensible version

```
Temp := FLOAT (MULTIPLIER * Weight);
Temp := MathLib0.exp ( EXPONENT * MathLib0.ln(Temp) );
Watts := TRUNC(Temp)
```

Mnemonics

Now, some of the methods we use to make programs clear are obvious, for instance, breaking them down into special-purpose procedures that have specific goals.

But there are also less obvious techniques that are sometimes taken for granted. The first is *clear naming.* Names have a tremendous potential for projecting misinformation, as a selection from Abbott and Costello's famous *Who's On First?* routine on the next page shows. A program could have this kind of variable declaration:

```
VAR A, B, C, D, E, F: INTEGER;
```

use meaningful names

But what's is all about? Contrast the single letters with:

* James McKelvey Jr., *Debugger's Handbook: TURBO Pascal*, Wadsworth Publishing Co. 1987. Wrong language, but a very funny book with very good advice.

(Lou Costello is considering becoming a ballplayer. Bud Abbott wants to make sure he knows what he's getting into).

Abbott: Strange as it may seem, they give ball players nowadays very peculiar names.
Costello: Funny names?
Abbott: Nicknames, nicknames. Now, on the St. Louis team we have Who's on first, What's on second, I Don't Know is on third—
Costello: That's what I want to find out. I want you to tell me the names of the fellows on the St. Louis team.
Abbott: I'm telling you. Who's on first, What's on second, I Don't Know is on third—
Costello: You know the fellows' names?
Abbott: Yes.
Costello: Well, then who's playing first?
Abbott: Yes.
Costello: I mean the fellow's name on first base.
Abbott: Who.
Costello: The fellow playin' first base.
Abbott: Who.
Costello: The guy on first base.
Abbott: Who is on first.
Costello: Well, what are you askin' me for?
Abbott: I'm not asking you—I'm telling you. Who is on first.
Costello: I'm asking you—who's on first?
Abbott: That's the man's name.
Costello: That's who's name?
Abbott: Yes.

~ ~ ~ ~ ~ ~

Costello: When you pay off the first baseman every month, who gets the money?
Abbott: Every dollar of it. And why not, the man's entitled to it.
Costello: Who is?
Abbott: Yes.
Costello: So who gets it?
Abbott: Why shouldn't he? Sometimes his wife comes down and collects it.
Costello: Who's wife?
Abbott: Yes. After all, the man earns it.
Costello: Who does?
Abbott: Absolutely.
Costello: Well, all I'm trying to find out is what's the guy's name on first base?
Abbott: Oh, no, no. What is on second base.
Costello: I'm not asking you who's on second.
Abbott: Who's on first!

~ ~ ~ ~ ~ ~

Costello: St. Louis has a good outfield?
Abbott: Oh, absolutely.
Costello: The left fielder's name?
Abbott: Why.
Costello: I don't know, I just thought I'd ask.
Abbott: Well, I just thought I'd tell you.
Costello: Then tell me who's playing left field?
Abbott: Who's playing first.
Costello: Stay out of the infield! The left fielder's name?
Abbott: Why?
Costello: Because.
Abbott: Oh, he's center field.
Costello: Wait a minute. You got a pitcher on the team?
Abbott: Wouldn't this be a fine team *without* a pitcher?
Costello: Tell me the pitcher's name.
Abbott: Tomorrow.

~ ~ ~ ~ ~ ~

Costello: Now, when the guy at bat bunts the ball—me being a good catcher—I want to throw the guy out at first base, so I pick up the ball and throw it to who?
Abbott: Now, that's the first thing you've said right.
Costello: I DON'T EVEN KNOW WHAT I'M TALKING ABOUT!
Abbott: Don't get excited! Take it easy.
Costello: I throw the ball to first base, whoever it is grabs the ball, so the guy runs to second. Who picks up the ball and throws it to what. What throws it to I don't know. I don't know throws it back to tomorrow—a triple play.
Abbott: Yeah, it could be.
Costello: Another guy gets up and it's a long ball to center.
Abbott: Because.
Costello: Why? I don't know. And I don't care.
Abbott: What was that?
Costello: I said I DON'T CARE!
Abbott: Oh, that's our shortstop!

```
VAR Acres, Bales, Cords, Decks, Ears, Flocks: INTEGER;
```

The second list is *mnemonic* (nih-mahn´-ick). Each identifier is a memory aid that make the variable's usage obvious.

It's possible to argue that brief variable names are easy to remember in a short program. The problem is that as time goes on, you won't have just one short program—you'll have dozens of them. As programming assignments become more involved, you'll frequently try to incorporate previously written code in your new work. Indeed, Modula-2 is designed to encourage you to do so. Then you'll see just how hard it is to remember what X, Y, and Z mean.

> As a rule of thumb, use names that are clear enough for your roommate, who may not be studying computer science, to understand!

Constants and Program Style

why use constants?

Constant definitions are a second tool of the plain programming style. Why define constants? No program needs defined constants in order to work, but programs often require constants in order to be good programs. Suppose that a 2000-line program calculates property taxes like clockwork for a few years, and then—horrors! tax reform. Must we search the entire program to update every instance of the old tax rate? Not if we had made the definition:

```
CONST TAXrate = 0.003;
```

for style

Changing the value of the constant TAXrate updates the whole program. Could TAXrate have been declared as a variable? Yes, but that would open the possibility of accidentally changing its value within the program. It's also misleading to call TAXrate a variable instead of a constant, because declaring something as a variable implies that its value will change frequently, or be obtained from the program's user.

as documentation

A second style motivation for constants is less obvious. Writing a program is a little like writing an instruction booklet. Just including all the facts isn't enough—they have to be presented in a manner that even a casual reader can follow. Now, comments provide a running documentation that explains what's happening in a program. Defined constants go further to help make a program *self-documenting*. A statement that subtracts 5 from x doesn't say much. Better variable names, and a comment, make this assignment easier to understand:

```
IO.IntOut (Speed - 5, 1);
    (*Find true speed by subtracting the fixed speedometer error.*)
```

for bug prevention

But mnemonic identifiers and a defined constant manage no-hands commenting—they document *without* additional comments, and are best of all.

```
IO.IntOut (IndicatedSpeed - SPEEDOMETERerror, MINwidth);
```

Type Issues

strong typing

One of the more difficult areas of Modula-2 programming for beginners is types. Why is there so much fuss about keeping types separate? In fifty years the topic may be recalled as a tempest in a teapot, but for the past decade, an enormous amount of verbiage has gone to discussing the relative benefits of strong typing—associating every value and variable with a named type, and barring mixtures of types in expressions or assignments.

As I pointed out in Chapter 1, Modula-2 places a great premium on strong typing—the strict segregation of types. If separating clearly disparate values like characters and numbers helps prevent gross program errors, the argument goes, then partitioning values according to finer distinctions may help avoid subtler mistakes. Nobody (as yet) has gone so far as to say that upper- and lower-case letters should belong to different types, but most language designers agree that forcefully distinguishing between real and integer values is essential. Modula-2, in differentiating between non-negative, and possibly negative integers (i.e. between CARDINAL and INTEGER) goes one step further.

To be honest, nobody claims that obeying picayune Modula-2 type restrictions is always convenient. The type rules represent one of the classic tradeoffs of computer science—a little syntactic inconvenience now instead of a lot of semantic inconvenience later.

It's important to remember that operators are restricted to operands of a particular type. This expression is obviously wrong:

```
"T" + "S"   (*an illegal expression*)
```

because addition doesn't have any meaning for CHAR values. However, you may be surprised when you find that this expression isn't any good either—you can't mix types:

```
1.0 + 1   (*another illegal addition*)
```

operand types

> The operands of an arithmetic operator have to have the same type.

You can't add a REAL and an INTEGER, no matter how reasonable it seems. The values must be added as INTEGER values:

```
TRUNC(1.0) + 1
```

or as two REAL values:

```
1.0 + FLOAT(1)
```

In general, steer clear of problems by remembering that:

1. **DIV** and **MOD** must have INTEGER operands, while / must have REALs.

keep in mind

2. The type of the value that a function call represents is often different from the type of the function's argument.
3. A value must be of the proper type to be assigned to a variable.

The rules that govern the usage of types in expressions and assignments are slightly different, but it's not crucial for us to draw fine distinctions just yet. We'll be discussing types in more detail in section 7-5. Don't get frustrated—in time, you'll no more be tempted to add a REAL to an INTEGER than you would be to wear a necktie to class.

Modula-2 Details

Syntax Summary

• assignment statement: uses the assignment operator to give a variable the value that an expression represents. The variable and value must be assignment-compatible.

Variable := *Value*;

• INTEGER and REAL operators: take numerical *operands* and form *expressions*:

+	*addition*
−	*subtraction*
*	*multiplication*
/	REAL *decimal division*
DIV	INTEGER *whole number division*
MOD	INTEGER *remainder division*

• CHAR constants: the character's octal position in the ASCII set (there's a table in the back) followed by a capital C:

```
IO.CharOut (7C);   (*Makes the terminal beep.*)
```

• operator precedence: fixes the order of evaluation in arithmetic expressions. When two operators have the same precedence, the expression is usually evaluated from left to right:

* / **DIV** **MOD** *all come before* + −

Parentheses can be used to change the relative order of evaluation:

(A + B) * C *is different from* A + B * C

• function: a subprogram that has parameters, just like a procedure. However, a *function call* is an expression—it represents a value.

• standard functions: are built into Modula-2. They include:

ABS (*number*)	(**Absolute value of an* INTEGER *or* REAL.*)
CAP (*char*)	(**Capitalized lower-case letter.**)
ODD (*number*)	(**Is a whole number odd?**)
FLOAT (*integer*)	(**Represent an* INTEGER as a REAL.*)
TRUNC (*real*)	(**Represent a* REAL *as an* INTEGER.*)
MAX (*type ID*)	(**The largest value in a particular type.**)
MIN (*type ID*)	(**The least value in a given type.**)
ORD (*value*)	(**Ordinal position of a given value.**)
CHR (*position*)	(**Character in a particular ordinal position.**)
VAL (*type, position*)	(**Value in a particular ordinal position.**)
HIGH (*array*)	(**Used with arrays (see Chapter 8).**)
SIZE (*type ID*)	(**Used with pointers (see Chapter 13).**)

Your own system's documentation will have the final word on fine details of these functions' operations.

- MathLibØ: a library of mathematical functions. It must be imported, just like IO. Its contents include:

Function	*Argument*	*Result Type*	*Purpose*
sqrt(x)	(*RealValue*)	REAL	square root
ln(x)	(*RealValue*)	REAL	natural log
exp(x)	(*RealValue*)	REAL	exponential (e^x)
sin(x)	(*RealValue*)	REAL	sine
cos(x)	(*RealValue*)	REAL	cosine
arctan(x)	(*RealValue*)	REAL	arctangent
entier(x)	(*RealValue*)	INTEGER	like TRUNC

Modula-2 Details

Remember . . .

- Expressions represent values. The simplest expressions are numbers and characters, but variables, user-defined constants, and function calls are also expressions. More complex expressions can be stated by using mathematical operators.
- A variable can only be assigned values of the same type. Before a variable has been assigned a value, it is *undefined*, and shouldn't be used in an expression.
- Arithmetic operators have types as well. In particular, / can only be used with REAL operands, while **DIV** and **MOD** must have INTEGER (or CARDINAL) operands.
- Function calls are expressions; they represent values. The *result*, or value, of a function call could, for instance, be assigned to a variable, or passed to a value parameter. A function call could *not* take the place of a statement, or be passed to a variable parameter.
- Functions are obtained in three ways: a few are built into the language, others are found in libraries (like MathLibØ) that can be imported; finally, new functions can be written in programs (although we won't until Chapter 3).

Zen Study Guide
Big Ideas

The Big Ideas in this chapter are:

1. *Programming languages, unlike foreign languages, come with documentation.* Tourists might want to avoid carrying their phrasebook in one hand, but programmers should learn to take advantage of written and on-line resources.

2. *As you come to rely on the Law of Least Astonishment, you'll learn to obey it as well.* Procedures should do what they seem to do—you should always be able to assume that the least astonishing alternative is correct without *having* to look it up.

Zen Study Guide
Questions

2-1 Operators must have operands of the proper type. Write a half-dozen expressions to accompany the question *Are these legal expressions in Modula-2?*

2-2 When a variable appears on the right-hand side of an assignment statement, it must be defined, and it must have the same type as the variable on the left-hand side. Write a half-dozen assignments to accompany the question *Are these legal assignments in Modula-2?*

2-3 The third set of self-check questions asks you to rewrite pairs of assignment statements as single assignments. Sometimes assignments are independent, sometimes they're not; sometimes merged assignments can be simplified, sometimes they can't. Write three pairs to accompany the question *Does the order of these assignments matter?*

2-4 Some functions are built into Modula-2, while others are imported from libraries. Write a half-dozen expressions that involve function calls to accompany the question *Which of these functions are:*

a) built into Modula-2;

b) imported from the MathLib0 library;

c) figments of my imagination?

2-5 Program Adder used addition as an excuse to demonstrate the basic sequence of procedure calls—get values, do something to them, then print the results. Make up three alternative excuses that might appear on an exam.

2-6 A procedure that adds two numbers and returns their sum (similar to procedure AddThem from program Adder) has three parameters. Since each can be either a value parameter or a variable parameter, there is a potential for 2*2*2, or eight, variations on a correct heading. Make up three incorrect examples for the trick question: *Which of these three examples is correct?*

2-7 *Evaluate these expressions.* Need I say more?

2-8 Like procedures, functions take parameters, which means that, like procedures, functions can be given parameters of the wrong type. Write *Which function call is correct?* pairs (one is, one isn't) for four standard functions, and four functions from MathLib0.

2-9 It's always fun to inflict mathematics. Find the thickest math book you can, then write the question *Which Modula-2 expression matches which snippet of math?*

2-10 It's tough to test a sense of style, but it can be done. Three points of attack are:

a) *Clarify this assignment.* Write a few assignments that could be clarified by use of constants, extra parentheses, or temporary variables and extra assignments.

b) *Simplify this sequence.* Write a series of assignments that contains a needless statement—one that is undone or ignored by later statements.

c) *Rename these variables.* Supply an assignment (like the complicated assignment to Watts in program Power) that uses one-letter identifiers, along with a clear comment about what the statement does, and ask that it be rewritten.

Exercises

Warmups and Test Programs

2-11 Characters in the ASCII character set are numbered, starting with 0. If you know a character's number, you can use the CHR function (as the argument of a call of IO.CharOut) to print it. What happens when you print character −1?

2-12 The MAX and MIN functions let you find out the largest and smallest values of any type. Find the bound values of INTEGER and CARDINAL.

2-13 What happens when you try to print the number that follows the largest INTEGER? The number that precedes the smallest one?

2-14 What happens if you try to truncate, and try to represent as a whole number, a REAL that is very, very large?

2-15 Can a constant definition require a calculation? Can it involve a call of one of the standard functions? What about a call of an imported function?

2-16 Can you step outside the boundaries of INTEGER during the course of an arithmetic operation? In other words, can you print the final value of an expression that adds 5, then subtracts 6, from the maximum INTEGER?

2-17 The ordinal position of a positive INTEGER is the integer itself. What's the ordinal position of a negative number?

2-18 Write a procedure that returns its three INTEGER variable parameters as positive numbers.

2-19 Write a procedure that reads two letters, and sends them to the main program as capital letters.

2-20 Write the procedure that's called like this

```
PrintCharacterBefore (Character);
```

2-21 Write the procedure that's called like this:

```
PrintNthLetterAfter (Letter, N);
```

2-22 Now, improve the procedure above so that the letter after "Z" is "A". Hint: **MOD** will come in handy.

2-23 What is the Law of Least Astonishment?

Longer exercises and programming problems follow Chapter 3.

'Divide each of the difficulties under study into as many parts as is possible, or are required to solve them best.' (p. 100) Some advice from René Descartes.

Chapter 3 Program Divisions, Program Design

So far we've taken a cut-and-paste approach to Modula-2 programming. It hasn't been a bad way to get up and running. Now, though, it's time to tie up the loose ends of working with procedures.

We'll start by learning something new—how to declare a subprogram that represents a value, rather than a series of actions. *Functions*, like procedures, help make life simpler by separating *what* we want from *how* we go about getting it. And, when we can reduce the number of facts we have to remember, our programs are usually the better for it.

Next, we'll go through a number of details that I've managed to gloss over so far. *Scope* and *precedence* are ideas that may seem hard at first. Eventually, though, you'll see that they're just meant to help establish a level playing field for creating and using identifiers.

Finally, we'll come to the important ideas of the chapter in section 3-3. I'll begin with a look at some of the problems Wirth tried to deal with when he designed Modula-2. Then, we'll look at program design methods that will be increasingly useful as we try to write longer programs, and solve harder problems.

Section 3-1
Declaring New Functions

MODULA-2 HAS TWO KINDS OF SUBPROGRAMS—PROCEDURES (which we've already used and written), and functions (which we've used, but haven't written yet). The new functions we declare will probably be a little more interesting than standard functions and library functions. We can name them as we wish, tell them what to compute, and give them as many arguments as necessary.

functions represent values

> However, all functions share one important feature: a function call represents a *value*. It must be used as part of an expression, and not as a statement.

For example, consider a call of a function that implements the one formula everybody knows—that distance equals rate times time:

a function call

```
AmountTraveled := Distance (Speed, ElapsedTime);
```

Before we can use AmountTraveled, it has to be declared. As always, declarations belong in the first half of a program:

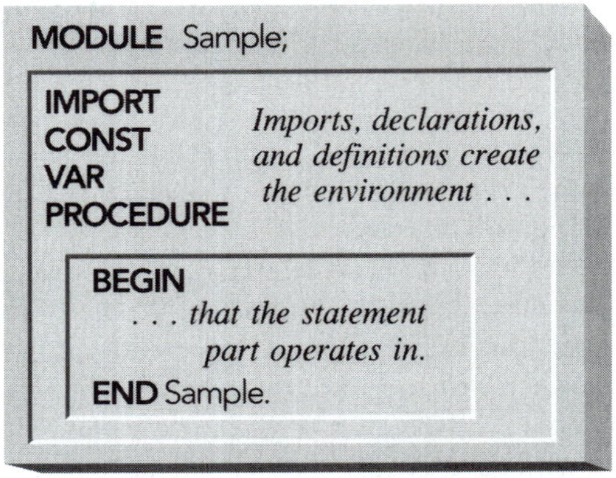

You can read the declaration of Distance, below. Look for the two hallmarks of every function declaration—the specification of the function's type (REAL here), and the **RETURN** of its value.

its declaration

```
PROCEDURE Distance (Rate, Time: REAL) : REAL ;
   (*Calculates Distance given Rate and Time.*)
   BEGIN
      RETURN Rate * Time
   END Distance;
```

Now, any reader with her wits about her could and should immediately object to my calling Distance a function. After all, its heading plainly describes Distance as a **PROCEDURE**.

when are procedures functions?

To put matters bluntly **PROCEDURE** is a homonym, or word with two different meanings, that causes confusion for everybody. As a rule, when a **PROCEDURE** is used to represent a value, I will always refer to it as a function, and not as a procedure.

When he defined Modula-2, Wirth decided to make Modula-2 slightly more compact by booking just one reserved word into a dual role. In any case, according to Wirth's formal terminology Modula has *proper procedures* and *function procedures*. In this text, I'll ignore these stilted, little-used phrases and rely on the vernacular *procedure* and *function*.

With this in mind, I present examples of function headings. Note that the parentheses of a parameter list are always required, even if no parameters are declared. This serves as a reminder that we're writing and using a function, rather than a procedure.*

PROCEDURE *identifier* ( ) : *the function's type* ;
PROCEDURE *identifier* (*parameters*) : *the function's type* ;

Self-Check Questions

Q. Would it be possible for Distance to have INTEGER-valued Rate and Time arguments, but still return a REAL value? If so, how is it done?

A. Sure. What's important is that the **RETURN** statement be given a REAL value, like this:

```
RETURN FLOAT (IntegerRate * IntegerTime)
```

Function Declaration Rules

I've managed to point out repeatedly that a function (unlike a procedure) represents a value. This has two consequences:

First, since the function represents a value, we have to specify its type. Second, the function has to contain a **RETURN** statement that actually gives it the value, and ends the function call.

function type

A function call can always represent any simple type of value, ordinal or REAL. The function's *result type* is specified at the end of the function heading—a colon is followed by the type identifier. The example function Distance was of

* That last statement is merely a plausible lie. The parentheses are required for technical reasons that matter to compiler-writers and advanced programmers.

type REAL. (It isn't uncommon, incidentally, to find Modula-2 implementations that let functions return values of the structured types we'll study in chapters 8 and beyond.)

the **RETURN** statement

Having set the stage by giving the function a type, we need a way to give it a value. The statement part of a function must contain a **RETURN** statement that ends the call and gives it a value. In Distance this **RETURN** was the function's entire statement part. Functions may be very complicated, of course, but a function's last action is to give itself a value.

We'll almost invariably find that functions use one or more value parameters to compute a value—a value that the function then represents. It is extremely unusual to give a function any variable parameters. The general rule is:

- If a subprogram calculates a single value, write it as a function.
- If a subprogram calculates more than one value, write it as a regular procedure that has variable parameters.

Some typical function headings are:

```
PROCEDURE Cube (Number: REAL) : REAL ;
PROCEDURE Decode (Letter: CHAR; CodeKey: INTEGER) : CHAR ;
PROCEDURE Highest (First, Second, Third: INTEGER) : INTEGER ;
PROCEDURE NoParameters ( ) : INTEGER;
```

In each case the heading ends with a colon and the function's type. Note that parentheses are used in function headings *and* calls even if there are no parameters. A call of the last function, above, would be NoParameters ( ).

parameter correspondence

As with procedure calls, there must be a one-to-one correspondence between the arguments of a function call, and the value parameters of a function declaration. Thus, any given function will always be called with the same number of parameters.

```
(*Function heading, which usually has value parameters*)
PROCEDURE Yield ( Investment: INTEGER;
                  Interest: REAL;
                  Days: INTEGER ) : REAL;

Income := Yield ( 1000, 0.097, 365 );
(*Function call, with its arguments*)
```

Variable parameters are seldom declared in function headings because the avowed purpose of a function is to compute and return a *single* value. If a function has variable parameters, it will wind up returning more than just one value. This creates a situation that can confuse an unwary program *re*writer. If a subprogram is supposed to calculate more than one value for the main program, then write it as a procedure.

Self-Check Questions

Q. Which of these problems should be solved with functions? Which with procedures?

a) Exchange the value of two CHAR variables.
b) Find the difference between two numbers.
c) Find the letter that's N places after Initial.
d) Multiply Offset by 17.
e) Multiply 33 by Distance.
f) Print the value of an argument.

A. Since we'll never expect functions to change their arguments, I'd solve problems *b*, *c*, and *e* with functions. *a* and *d* are best for procedures because the arguments are changing. *f*, in contrast, doesn't return a value at all, so writing it as a function wouldn't be appropriate.

A Few Function Examples

I'll demonstrate functions with a few examples that are standard in some other computer languages, and can be found in libraries of most Modula-2 implementations. A reciprocal function is found on most hand calculators:

reciprocals

```
PROCEDURE Reciprocal (Number: INTEGER) : REAL ;
   (*Represents the reciprocal of Number.*)
   BEGIN
      RETURN 1.0 / FLOAT(Number)
   END Reciprocal;
```

importing routines

Harder functions may rely on library functions. Now, when we used library routines from IO in earlier procedure examples, I pointed out that once a library is imported into the program environment, its contents become available to new procedures. Consider function Power, below. It's meant to let us forget, forever, exactly how to do exponentiation. Since it uses functions from MathLibØ, they must be imported into the program Power is part of. In other words, the import statement:

```
IMPORT MathLibØ;    (*in the main program*)
```

must appear before the declaration of Power. Although it is tempting, they cannot be imported *within* Power itself—**IMPORT** statements can't appear in functions or procedures.

exponentiation

```
PROCEDURE Power (Base, Exponent: REAL) : REAL;
   (*Raises Base to the Exponent power.*)
   BEGIN
      RETURN MathLibØ.exp (Exponent * MathLibØ.ln (Base) )
   END Power;
```

A tangent function, also missing from Modula-2, is equally easy to create. We make use of the fact that tangent ϕ = (sine ϕ/cosine ϕ) in writing function RadianTan. As above, though, we must assume that MathLibØ has been imported into the program that contains the new function definition.

tangents

```
(*IMPORT MathLibØ;*)
PROCEDURE RadianTan (AngleInRadians: REAL) : REAL
   (*Represents the tangent of its argument.*)
   BEGIN
      RETURN MathLibØ.sin(AngleInRadians) / MathLibØ.cos(AngleInRadians)
   END RadianTan;
```

Modification Problems

1. How can you test a function? By putting it into a program that doesn't include much besides the function's definition, and a call or two. Such programs are called *test harnesses* or *program shells*. Write a test harness that lets you test functions Distance, Reciprocal, and Power.

Section 3-2
Subprogram Nitty-Gritty

IT'S TIME TO DELVE INTO THE DETAILS THAT GOVERN the construction and use of subprograms. You'll find that many details won't necessarily be relevant for some time, so you may wish to read this section for its basic ideas now, and refer back later if you have to.

As you know, a program has two distinct levels or environments:

separate environments...

- The main program contains *global* declarations of variables that persist for an entire program run.
- Every subprogram can contain *local* declarations of variables that are only used within the subprogram, and don't exist between calls.

But the environments can communicate. A flow of data from procedure to procedure is supported by parameters:

...that can communicate

- Value parameters give names, meaningful within the subprogram, to values that are passed to the subprogram when it is called.
- Variable parameters provide temporary *aliases*, or renamings, of variables that are passed to the subprogram when it's called.

Value parameters let data flow into a subprogram; variable parameters let data flow both into and out of a subprogram.

We can see the two levels, and communication between them, by looking at a classic example program. Passer, below, contains three procedures. The main module variables First and Second help establish the environment that procedures GetTheNumbers, SwitchThem, and PrintTheResults work in.

Pay particular attention to procedure SwitchThem. Just as the global variables First and Second help establish the environment of the module as a whole, the local variable Temporary extends the working environment of SwitchThem. We'll talk more about local and global declarations in a moment.

parameter passing example program

```
MODULE Passer;
   (*Reverses two input INTEGERs.*)

IMPORT IO;

VAR First, Second: INTEGER;

PROCEDURE GetTheNumbers (VAR Primero, Segundo: INTEGER);
   (*Reads values for the variable parameters Primero and Segundo.*)
  BEGIN
     IO.StringOut ("Please enter two integers:   ");
     IO.IntIn (Primero);
     IO.IntIn (Segundo)
  END GetTheNumbers;

PROCEDURE SwitchThem (VAR Premier, Deuxieme: INTEGER);
   (*Swaps the values of two variable parameters.*)
  VAR Temporary: INTEGER;
  BEGIN
     Temporary := Premier;          (*Temporary is undefined until*)
     Premier := Deuxieme;           (*this assignment takes place.*)
     Deuxieme := Temporary
  END SwitchThem;

PROCEDURE PrintTheResults (Primo, Secondo: INTEGER);
   (*Prints its value parameters.*)
  CONST PRINTSPACE = 1;             (*Provides the minimum output field.*)
  BEGIN
     IO.StringOut ("In reversed order, the numbers are  ");
     IO.IntOut (Primo, PRINTSPACE);
     IO.StringOut ("  and  ");
     IO.IntOut (Secondo, PRINTSPACE)
  END PrintTheResults;

BEGIN  (*The main program, Passer.*)
   GetTheNumbers (First, Second);
   SwitchThem (First, Second);
   PrintTheResults (First, Second)
END Passer.
```

```
Please enter two integers:  27  -935
In reversed order, the numbers are -935 and 27
```

As SwitchThem shows, subprograms can include *local* declarations and definitions that customize the subprogram's internal environment. These include the definition of constants, and declaration of variables and additional subprograms. Indeed, the only thing that a subprogram *can't* contain is an **IMPORT** statement—that's permitted only at the module level.

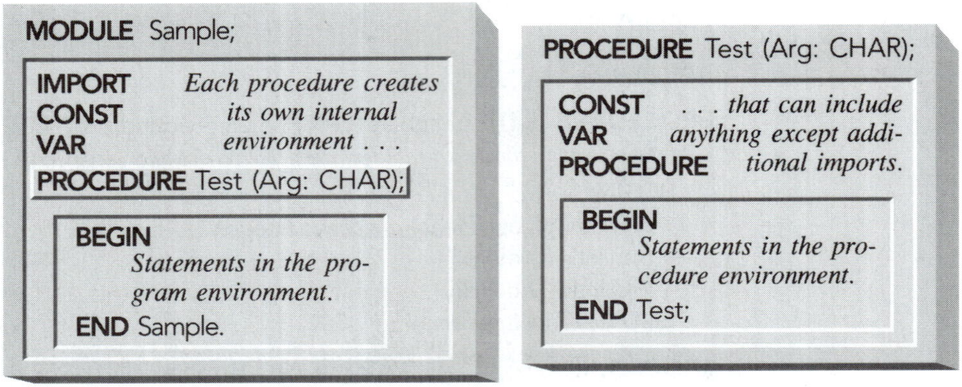

Scope of Identifiers

Once we allow a program to establish identifiers in different locations, two problems arise:

1. Where will identifiers be recognized (the *scope* problem)? Can an identifier defined in one procedure be recognized in another?

2. How will we decide what to do when two identifiers are the same (the *precedence* problem)? Can an identifier from the main program be reused in a procedure without causing confusion?

I sidestepped this problem in program Passer by practically exhausting my knowledge of foreign languages. But most programmers have enough to do without being required to learn a new language whenever they need new identifiers. One efficient way to minimize potential conflicts is to limit the area in which names are known.

> Modula-2's basic limiting rule is that a procedure's local identifiers and parameters are known only within that procedure.

For example, consider program ScopeOut, below. It shows that identifiers created within two procedures are unknown outside of their procedures, and hence can be used again without causing problems.

```
MODULE ScopeOut;
IMPORT Any necessary imports are made here.
VAR A, B, C, D, E, F, G, H: REAL;
PROCEDURE Sooner (A, B: INTEGER);
  VAR C, D: CHAR;          (*Main program identifiers can be reused.*)
  BEGIN
    Do whatever Sooner does, using the local A, B, C, and D.
    Later can be called, but its local variables aren't recognized.
  END Sooner;
PROCEDURE Later (E, F: INTEGER);
  VAR G, H: CHAR;          (*Main program identifiers can be reused.*)
     A, C: REAL;
  BEGIN
    Do whatever Later does, using the local E, F, G, H, A, and C.
    Sooner can be called, but its local variables aren't recognized.
  END Later;
BEGIN (*main program*)
  Here, A, B, C, D, E, F, G, and H are the main-program variables.
  Only the procedure names Sooner and Later are known here—not
   their contents.
     . . .
END ScopeOut.
```

identifiers in procedures are limited

scope

More formally, the word *scope* describes the realm of meaning of an identifier that names a constant, variable, or subprogram. The scope of an identifier is the portion of a program—called a *block*—in which it's recognized as representing a particular value or action.

A module is a block, and procedures form blocks within the module. Along with deciding where identifiers aren't known, the scope rules specify where they *are* recognized, and how conflicts will be resolved.

blocks

You can see the scope rule at work in every **IMPORT**. A module's block is the whole module, including all of its procedures, as well as its statement part. Thus, an imported library is known throughout the entire module that imports it—even within procedures declared within that module.

Procedures create new, enclosed blocks. The procedure's name lies outside the new block, while its parameters, local variables, and so on are inside the new block. Thus, the procedure's name has the same scope as, say, a main program constant. Just as a main program constant is recognized within a procedure, procedures also recognize each others' names—as long as they're declared at the same level (as are Sooner and Later, above). A procedure declared *inside* a procedure, in contrast, is known inside the procedure, but not outside.

Self-Check Questions

Q. The following procedure declaration contains an error that should be easy to spot. What is it?

```
PROCEDURE Wrong (A: INTEGER; VAR B: INTEGER);
  CONST A = 100;
  VAR B: REAL;      etc.
```

A. Example Wrong ignores the fact that parameter declarations share the scope of local constant and variable declarations. Procedure Wrong tries to use two identifiers (A and B) in equally local places. Whether the parameters and local variables are of identical or different types is irrelevant. It's as incorrect a pair of declarations as this would be:

```
VAR A: INTEGER; A: REAL;
```

Precedence of Definitions

Now, a consequence of the scope rule is that local identifiers may be declared within the scope of identical, but non-local, identifiers. When this happens, the most local definition or declaration has precedence over an identical, but relatively global, definition.

For example, you shouldn't have any trouble following program Music, below. Notice that the main program and its procedure Tune have like-named identifiers—the constant SCALE and the variable JohnnyOneNote.

Music begins by importing the contents of IO. As you can see, the three imported procedures we use (IO.StringOut, IO.LineOut, and IO.CharOut) are known throughout Music's block—its statement part, and the statement part of procedure Tune. Since Tune is part of Music's block, library routines that are imported into Music can be used in Tune without further ado.

local precedence

In contrast, the identifiers SCALE and JohnnyOneNote are used twice. They are defined in Music, and then redefined within Tune. But when they appear within procedure Tune, only the local definitions are recognized. Thus, the assignment of "A" to the local JohnnyOneNote has no effect on the value of the global variable with the same name. The local identifier takes precedence over a like-named, but relatively global, identifier.

name precedence example program

```
MODULE Music;
  (*Shows how conflict between identifiers is resolved.*)
IMPORT IO;
CONST SCALE = "Bass Clef ";  (*This defines a text constant.*)
VAR JohnnyOneNote: CHAR;
PROCEDURE Tune;
  (*Note the identically-named local identifiers Tune contains.*)
  CONST SCALE = "Treble Clef ";
  VAR JohnnyOneNote: CHAR;
  BEGIN
    JohnnyOneNote := "A";
    IO.StringOut (SCALE);
    IO.CharOut (JohnnyOneNote)
  END Tune;
BEGIN  (*Music*)
  JohnnyOneNote := "D";
  IO.StringOut (SCALE);  IO.CharOut (JohnnyOneNote);
  IO.LineOut;
  Tune;
  IO.LineOut;
  IO.StringOut (SCALE);  IO.CharOut (JohnnyOneNote)
END Music.
```

```
Bass clef D
Treble clef A
Bass clef D
```

block structure

The illustration below provides a more general example of *block structure*. The scope of a global constant is the entire program—the largest block. Since identifiers are imported from library modules to this main block, they're known everywhere too (although identifiers that aren't explicitly imported *can't* be used).

A local identifier's scope is limited to the block it's declared in—'its' subprogram, and other subprograms declared within that subprogram. As far as the internal subprograms are concerned, an outer local identifier might as well be global. Not all languages have a block structure like Modula-2's; some have only local identifiers, some only globals, and some don't allow subprogram declarations at all.

Identifiers defined in block:	*Are recognized in blocks:*
A (X, B, C)	A, B, C, D, E, F
B (D)	B, D
D (D)	D
C (E, F)	C, E, F
E (E)	E
F (F)	F

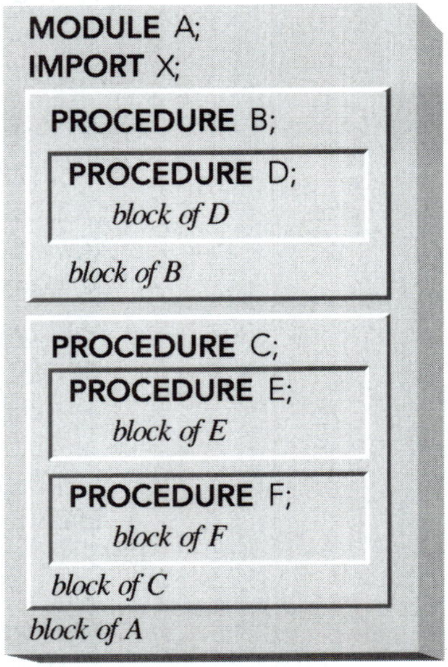

Self-Check Questions

Q. What will the output of this program be?

```
MODULE Confound;
  (*Comments? Nope—that would be telling.*)
IMPORT IO;
VAR A, B, C, D: INTEGER;
PROCEDURE Confuse (C,A: INTEGER; VAR D: INTEGER);
  VAR B: INTEGER;
  BEGIN
    A := 5;  B := 6;  C := 7;  D := 8;
    IO.IntOut (A, 10);  IO.IntOut (B, 10);
    IO.IntOut (C, 10);  IO.IntOut (D, 10)
  END Confuse;
BEGIN
  A := 1;  B := 2;  C := 3;  D := 4;
  IO.IntOut (A, 10);  IO.IntOut (B, 10);
  IO.IntOut (C, 10);  IO.IntOut (D, 10);
  IO.LineOut;
  Confuse (B, A, D);
  IO.LineOut;
  IO.IntOut (A, 10);  IO.IntOut (B, 10);
  IO.IntOut (C, 10);  IO.IntOut (D, 10);
  IO.LineOut
END Confound.
```

A. This intentionally muddled program deliberately tries to confuse value parameters, variable parameters, and local variables. Its output is:

```
1    2    3    4
5    6    7    8
1    2    3    8
```

Side-Effects

Suppose there isn't any ambiguity in definitions. Can an identifier defined in the main program be used in a procedure or function? The answer is yes. Typically, we will only take advantage of this on three occasions:

taking advantage of scope

1. when the main program has imported a library, or
2. when a subprogram uses a main-program constant (this is one reason that constant identifiers are usually capitalized—so that they immediately stand out to the program reader), or
3. when one procedure calls another.

A more interesting question is whether or not we can make an assignment directly to a global variable from within a subprogram.

side-effects

> Technically, nothing in Modula-2 prohibits making a direct assignment to a global variable from within a procedure or function. However, this sort of assignment is known as a *side-effect*. It is strongly discouraged; side-effects lead to some of the most difficult to find bugs in programming.

subprogram scope

As noted above, procedure and function identifiers follow the rules of scope. Once a procedure or function has been declared, its name has meaning in other parts of the program, including the main program's statement part, the statement parts of other subprograms, and the statement part of the procedure or function itself.

Usually we'll invoke procedures or functions from the main program, or from subprograms declared later on. A third case, in which a subprogram calls itself (or two subprograms call each other), won't concern us until we explore recursion in Chapter 17. In general, subprograms are declared in an order that makes sense to the program reader. They can be called in any sequence, for any number of times.

Section 3-3 Program Engineering Notes

PEOPLE USUALLY APPROACH PROGRAMMING languages with the expectation that everything they learn will be a command of one sort or another. It comes as a surprise that a good deal of Modula-2 is devoted to methods of organizing programs, rather than doing things. It may come as an even greater surprise to find out that a good deal of *programming* revolves around planning how programs will be put together.

I've already pointed out that if we restricted ourselves to doing simple tasks, then a simple methodology (sit at the terminal and start coding) would be sufficient. However, our goals are more ambitious. If we're going to solve big problems, then we have be able to organize our thoughts well enough to write big programs.

The search for organization is nothing new. Three hundred and fifty years ago René Descartes, in his *Discourse On Method*, laid down four rules 'for rightly conducting one's reason and for seeking truth in the sciences.' They were:

Descartes´ method for science

1. Never accept as true anything that is not self-evidently so.
2. Divide each of the difficulties under study into as many parts as is possible, or are required to solve them best.
3. Conduct thoughts in an orderly fashion, commencing with the simplest and easiest-to-know items, then rising by degrees to the most composite things.
4. Make enumerations so complete and reviews so general that you can be sure of having omitted nothing.

Descartes was remarkably prescient: well before electricity was discovered, he managed to set out the approach we take to writing programs today! However, his prescriptions have been hard to follow. Programming, unfortunately, is a rather unscientific business, and one programmer's idea of orderly thinking may be her neighbor's definition of *cruft* (an unpleasant substance that is found between the lines of poorly written programs).

Subprograms and Abstraction

Wirth´s goal: let structure help `science´

In any case, Wirth was keenly aware of the difficulty of programming on a strictly scientific basis. When he designed Modula-2, Wirth tried to let mechanics come to the rescue when science fell short. Modula-2 isn't a powerful language in the sense that difficult operations are built into the language. On the contrary, Modula-2 is rather compact and limited.

> Modula-2 is very strong because of its facilities for controlling the internal packaging and structure of programs. We actually can divide our difficulties into parts, and reduce them to Descartes' 'easy-to-know items.'

Subprograms and *external modules*, which we'll refer to as *libraries*, are the main structuring tools. Over the years, subprograms have served three main purposes:

how subprograms structure programs

1. Subprograms replicate code. The detailed code of a subprogram can be written once, then invoked repeatedly just by using the subprogram's name.
2. Division into subprograms provides an organizing principle, along functional lines, for program code. Here, 'functional' means 'by purpose or application.'
3. Subprograms can precisely specify the subactions of a program. The final program can actually be described, before coding, as a well-defined collection of specific procedures and functions.

Libraries, in turn, are to subprograms what subprograms are to programs. Libraries have already made a substantial contribution to the structure of our programs, because:

library structures

1. Libraries let packages of related procedures be removed from a program entirely, without giving up use of the procedure's abilities. An easy import mechanism encourages the reuse of existing code.
2. Libraries hide details. They let us use procedures and functions without having to deal with the details of their algorithms or implementation.
3. Libraries separate usage from implementation. This helps make their contents portable—the details of a particular procedure's coding may vary from system to system, but its usage can stay constant.

Although we haven't written any new library modules (and won't until Chapter 11), we've already taken advantage of the library system in writing programs: each **IMPORT** of the IO library has let us work as though IO's procedures were part of our own programs.

abstraction

> Alone or in libraries, subprograms encourage *abstraction* in programming. We separate application from implementation—*what* gets done is removed from *how* it gets done.

Goal-Oriented Design and Stepwise Refinement

The key, of course, is to think about 'what?' before 'how?' So far, we've relied on an intuitive notion of goal-oriented design, and tried to write programs in terms of procedures and functions that have had limited, specific goals.

But as programming problems become more complex, good intentions won't always be enough. A methodology called *stepwise refinement* becomes useful for deciding just what our limited and specific goals are going to be. The basic idea of stepwise refinement is to repeatedly restate problems (and then parts of problems) in more and more detailed terms. Each restatement is a refinement; each step takes us down to a greater level of detail. For example:

Outline of a Problem

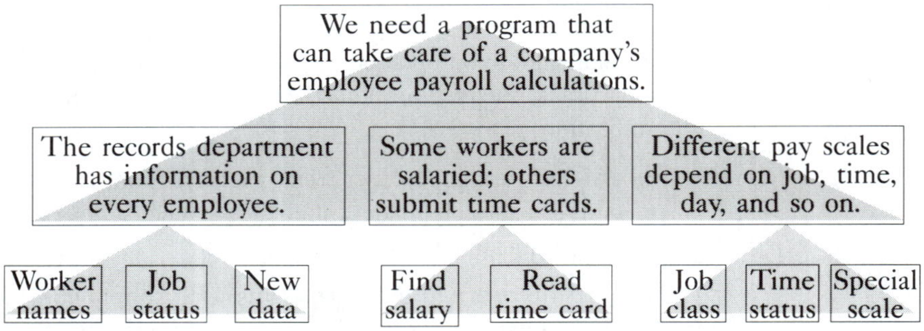

As you can see, we move from the general to the specific. Obviously, we try to stop before we get *too* specific. The stepwise refinement of a problem should get us to the point of procedures and functions, rather than to the level of individual Modula-2 statements.

Now, the interesting thing about using stepwise refinement to restate problems is that it practically gives us an outline for designing programs:

Outline of a Program

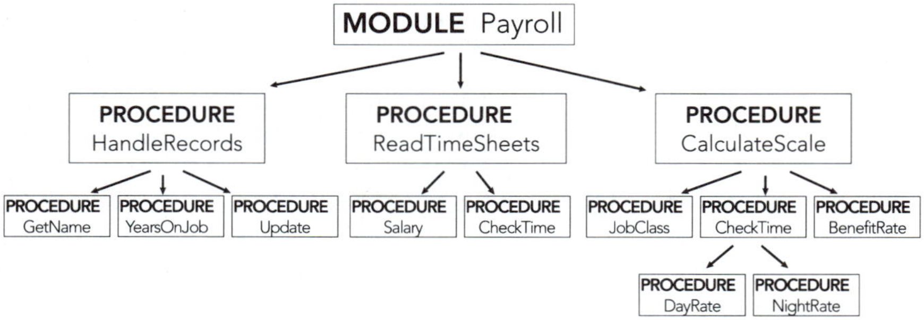

functional decomposition

> This approach can be described as *functional decomposition*, which means that the program is broken down along functional, *how-does-this-bring-us-closer-to-the-solution?* lines. It's best applied in programs that are heavy on algorithms and operations.

There are several reasons why stepwise refinement and goal-oriented design in general are useful. Most obvious is that we try to conquer problems by dividing them. A formidable programming problem may turn out to be a combination of easily solved subproblems—in like a lion, and out like four or five lambs.

But even if a crucial procedure is beyond our abilities at present, we can still work on the program because we've made independent subdivisions in the problem. As a result, the final program is modular. It's composed of separate pieces that can be written and tested by themselves, and then eventually merged together.

When the separate pieces are ultimately joined, procedures and functions help make programs easier to understand. As we've seen, even relatively simple programs are usually broken into sequences of procedure calls like this:

```
BEGIN
    GetInput (arguments ...);
    ProcessTheData (arguments ...);
    PrintResults (arguments ...)
END.
```

transparency

because procedure calls make the program's inner operations *transparent.* We know *what* the program does, but we can ignore *how* it does it. The detailed code of the procedure doesn't obscure our view of the program—it's invisible unless we go looking for it.

thinking in English

Stepwise refinement also lets a programmer plan most of a program without actually writing in Modula-2. It's easier to think in English than in any sort of computerese, and tackling a problem from the top puts off the nitty-gritty of encoding for as long as possible. Outlining a program in terms of its procedures (and procedures and functions *within* procedures) provides a transitional phase between words and code.

finding the seams

In real-life programming projects, the ability to find the seams of a problem or program is the earmark of a good programming manager. If several people are to work on a single program effectively, each must have a clearly outlined task. Each individual has to know exactly how her piece of the program interacts with the whole. Stepwise refinement and goal-oriented design slice up a problem in a natural, intuitive manner. The plan for breaking down a problem goes hand-in-hand with the guide for building a program.

Stepwise Refinement of a Problem

1. State the problem simply, decomposing it into its logical subproblems.
2. If you can immediately figure out how to encode all of the subproblems, you're done. These will be the major procedures of your program. If the subproblems are too complex...
3. Refine the subproblems into smaller, more basic subproblems. Their solutions are written as procedures or functions within procedures.

Goal-Oriented Program Design

1. Write the statement part of the main program first. In a program of any size, this will mainly consist of procedure calls. Each procedure should solve one part of the original problem.

2. The main program's statement part should be simple enough for a non-programmer to read and understand, yet detailed enough to give a programmer an idea of how the program works.

3. If a procedure is particularly complicated, or does more than one job, it should probably be broken down into sub-procedures.

length of subprograms

Remember that a procedure isn't a rug for sweeping code under! One page (or screenful) of code is enough for any human to try to read and understand. If the procedure is longer, *try to break it down.*

Object-Oriented Design

I think that you can get a better idea of what goal-oriented design is if we take a brief look at an alternative approach. After all, keeping our eye firmly fixed on the ultimate goal (or stepping stones to it) isn't the only approach we can take.

We can subdivide a program on a radically different basis by temporarily ignoring the demands laid out by a problem statement. Instead, we can focus on how the program will deal with information, and break it down along *data-oriented* lines. This is the basis of *object-oriented* design—a data *object* is a variable or structure that holds data.

For example, the goal-oriented approach to writing a program that manages a public library might focus on the library's *users*, and on the tasks they set:

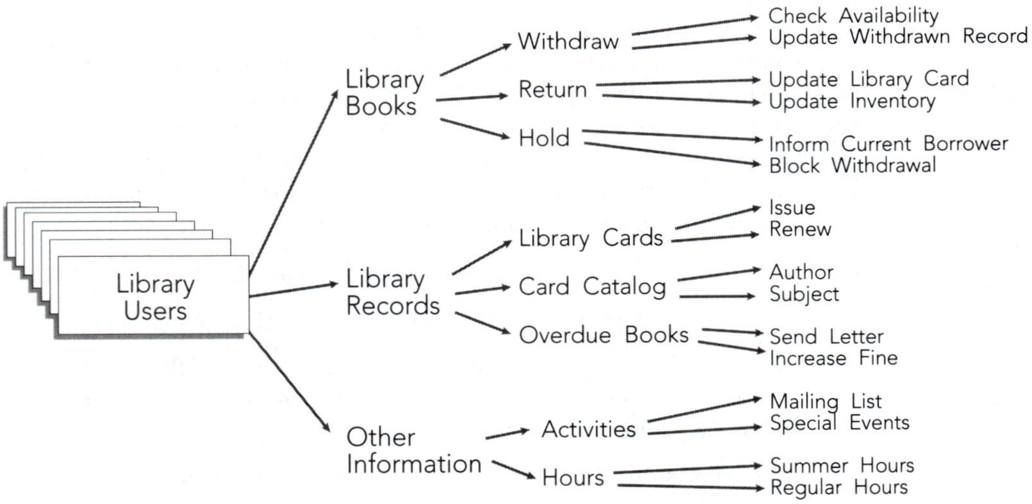

As you can see, the procedures we plan are dominated by an *algorithm*, or sequence of steps toward a solution. You can almost imagine that they were laid out by following in the footsteps of a librarian or library patron.

But what if we were to follow a *book* instead? From a book's point of view, the operation of a library is rather different:

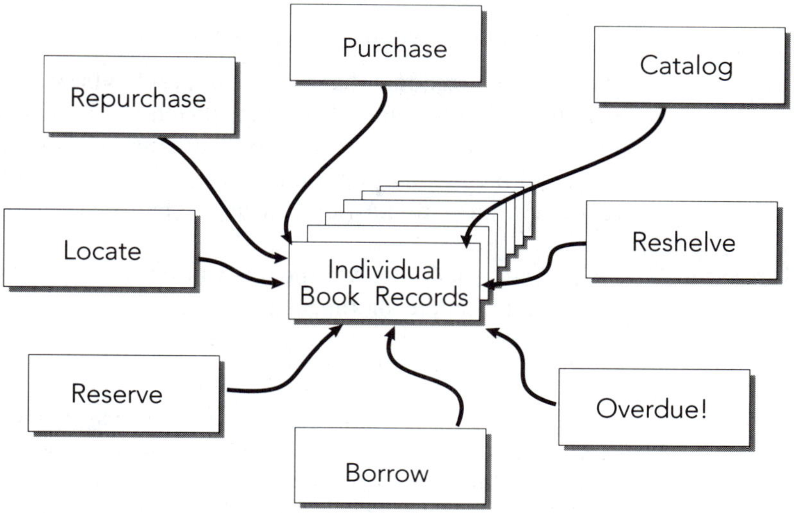

As we learn about more sophisticated means of variable creation and data storage, we'll find that writing data-handling procedures and functions first, even if they don't solve any particular problem now, makes it easier to deal with program goals later. We'll come to rely on the object-oriented approach to program design in later chapters.

Antibugging and Debugging

As always, there are a few common problems that accompany each new Modula-2 feature. Most ordinary function problems stem from confusing them with procedures. Some points to watch out for are:

- Functions have types, procedures don't.

Each function heading must end with a colon and a type identifier:

```
PROCEDURE Function (parameters ) : type identifier ;
```

- Each function must return a value that has the function's type.

If the function's type is CHAR, the function must return a CHAR. The types of any function arguments are irrelevant:

return the proper type

```
PROCEDURE NthLetter (Distance: INTEGER) : CHAR ;
   (*Returns the Nth letter following "a".*)
   BEGIN
      RETURN CHR (Distance + ORD("a"))
   END NthLetter;
```

- Function headings and calls need parentheses, even if the function has no parameters.

Get used to seeing function calls with parentheses in place.

- Functions shouldn't inadvertently call themselves.

A function or procedure that calls itself is said to be *recursive*. This is perfectly legal, but the subprogram has to be written in a manner that lets the series of subprogram calls end eventually. We'll discuss recursion further in Chapter 17.

As I mentioned in passing in one of the Modification Problems, *shell* or *test harness* programs provide a useful test bed for subprograms.

driver programs

> A program shell or test harness is written to supply the simplest environment a procedure or function call can get away with—usually just the declaration, a variable or two, and a test call.

Driver programs give you a chance to test new code in an uncluttered arena. You can assure yourself that a subprogram works before subjecting it to the pressure of having to cooperate with the rest of your main program. At first, it's hard to shake the feeling that writing a driver involves oodles of extra work. But, with practice, you'll get the time needed to type something like this shell down to less than a minute:

```
MODULE Test;
   (*A test bed for procedure DryRun.*)
IMPORT IO;
IMPORT MathLib0;  (*If needed.*)
VAR as necessary
PROCEDURE DryRun with its parameters, etc.
   BEGIN
      whatever DryRun does
   END DryRun;
BEGIN
   Give DryRun a workout
END Test.
```

Driver programs are about as close as we can come to approximating the 'failure' tests of engineering. For example, here's how Rolls Royce airplane engines used to be tested:

> 'The engine would be driven to full power; this would normally result in a component failure. The component would then be strengthened and the test restarted. This time another component would fail. This process was continued until no component failure occurred.'*

Have fun turning up the throttle on your code!

* Ince, Darrel, *Software Development: Fashioning the Baroque*, Oxford University Press, 1988., pg. 44.

Modula-2 Details
Syntax Summary

- function: a subprogram that calculates and represents a value.

```
PROCEDURE Sample ( value parameters ) : INTEGER;
  Definitions and declarations if necessary,
   but no imports.
  BEGIN
    statements that calculate the function's value, followed by:
    RETURN a value
  END Sample;  (*Finish with a semicolon.*)
```

- Parentheses are necessary in a function declaration and call even if the function doesn't have parameters:

```
PROCEDURE Example ( ) : CHAR;
      (*declaration . . .*)
      . . .
  Temp := Example ( );
      (*. . . and call*)
```

- **RETURN**: generally the last statement in a function. It's followed by a value of the function's type.

Modula-2 Details
Remember . . .

- A function call is an expression that represents a value. Although the function's declaration contains statements, the function call itself is *not* a statement.
- Since functions are used to represent values, they're rarely given variable parameters (which would imply changing arguments). Instead, they're generally given value parameters.
- Identifiers created in the main program are said to be global, while subprogram identifiers are described as being local.
- An identifier's scope is the range, within a program, in which it's recognized. Every subprogram creates a new scope, in which its own local identifiers take precedence. In general (procedures can be the exception), declaration should precede use.
- A side-effect is an assignment to an identifier that's defined outside the local scope. Technically, they're legal. In practice, though, they're almost never used. If you must change a relatively global identifier from with a procedure, pass it to a variable parameter.
- Programming in terms of subprograms is one method of abstraction. Subprograms, sometimes collected in libraries, help separate usage from implementation. Designing programs in abstract terms helps keep us from getting bogged down in details.

• Goal-oriented design focuses on *actions*—what steps we must take in the process of carrying out an algorithm. Stepwise refinement helps goal-oriented design by breaking a problem—and, implicitly, a program—into procedure-sized chunks.

Zen Study Guide Big Ideas

The Big Ideas in Chapter 3 are:

1. *Scope rules keep programs flexible by avoiding restrictions on the use, and reuse, of identifiers.* At first the scope rules seem like a new set of details to remember. In the long run, though, you'll see that they let you avoid remembering details.

2. *No matter how easy programming in Modula-2 is, planning in English is easier.* But because statements can be packaged into subprograms that are called by name, we can use the same terms to describe steps in an algorithm, and procedure calls in a program. Abstraction bridges the gap between high-level problem solving steps and low-level computer statements.

Zen Study Guide Questions

3-1 A popular biology experiment in the old days was to teach pigeons or mice to distinguish between circles and ovals, then gradually make the ovals more circular and vice versa. The question, of course, was *How long does it take to drive a pigeon crazy?* In these more enlightened times, we pose the question *Should you write it as a function or procedure?*, and supply a half-dozen problems (like those in the second set of Self-Check Questions). Try it.

3-2 There are so many little things to leave out of a function declaration: a colon, a type, a **RETURN**. It's a wonder that any get written at all. Make up three *Where's the bug?* questions that might test understanding of function declarations.

3-3 How can you test a student's understanding of the scope rules? Well, for examination purposes only, all kinds of crazy assignments are liable to appear. Starting with an example program like Passer, make up the question *Which assignments in this program are a) legal; b) illegal; c) legal but inadvisable?*

3-4 Another way to approach scope is to write a short piece of code that names each and every identifier X, then ask that as few identifiers as possible be renamed. For example:

```
MODULE X;
IMPORT X;
CONST X = . . .;
VAR X: . . .;
PROCEDURE X (X: . . .);    and so on.
```

Try it.

3-5 One of the Self-Check Questions dealt with a program named Confound. Create a CHAR-oriented version that would be suitable for a *What's the output?* test problem.

3-6 The test question I'm most proud of is *Here's the output, what was the input?* After all, it's no harder to grade than *Here's the input, what's the output?*, but it's *so* much harder to do! Seriously, anybody can grind through *what's the output?* eventually, but doing *what's the input?* really requires understanding. Modify a program like Confound so that it has input, then ask *Here's the output, what's the input?*

3-7 *Global/local? Argument/parameter?* Find three more word pairs that would be suitable for *What's the difference?* questions.

3-8 The answer to *Fix the bug* problems are usually discovered by carefully reading comments. Write two procedures or functions that are buggy because they use the wrong kind of parameter. Be sure to include a comment that gives your victim a clue.

3-9 *Write a function that calculates . . .* Need I say more?

3-10 Finally, can you think up a single example that might be suited as a parallel to the public library program described in the Program Engineering Notes? In other words, come up with a problem to accompany the question *Briefly describe a top-down (or object-oriented) approach to writing this program.*

Exercises
Warmups and Test Programs

3-11 Making an assignment directly to a global variable from within a procedure is a very bad programming practice (renaming it with a variable parameter is the preferred way of accomplishing the same goal). What is one of these direct assignments called?

3-12 Suppose we've written procedures to solve these problems. Which ones will require at least one local variable?

a) Swapping the values of two variables.

b) Doubling the value of an INTEGER.

c) Reading four characters, and returning just the fourth one to the main program.

d) Printing a REAL argument along with its square root.

3-13 Procedures can be defined within procedures. Is there any limit to how long this can go on?

3-14 Suppose that a program contains two procedures named A and B, declared in that order. Not every Modula-2 system will actually let A call B, because some compilers strictly require you to *define before you use*. Does your system allow the out-of-order call? If not, ask your teacher how such calls can be made.

3-15 Suppose that procedure Order is meant to put its three CHAR arguments into alphabetical order. Write a test harness for Order. What test data will you use?

3-16 It's time to explore error messages. What happens if you write a function but forget its **RETURN**? What happens if you declare a function that has no parameters, and forget to include parentheses? What if you forget the parentheses in the call of such a function?

3-17 Write a function that represents *a*) pounds as kilograms, *b*) degrees Fahrenheit as centigrade, *c*) dollars as Italian lire, *d*) hours as minutes, *e*) its REAL argument as the next higher INTEGER.

3-18 *Officer: Did you know that you were going 75 miles per hour?*
Offender: Well, I wasn't planning on being out that long.
The last time (perhaps I should say the first and last time) I got a speeding ticket I was informed that the fine was $75 plus $5 for each mile over the limit. Write a procedure or function that takes a limit and an actual speed, and returns a fine.

3-19 Write a function that returns the 'next' letter, with the understanding that the letter after "z" (and "Z") is "a" (or "A").

Exercises
Programming Problems

3-20 Have you ever had time on your hands? Was it messy? Write a program that adds and subtracts time. It should ask the user for two input values—a time (like 3:57) and a duration (such as 1:05), then print the sum (here, 5:02) and difference (2:52). To make sure that you get practice writing functions, you might want to write Hour and Minute functions that return the 'whole' hour portion and 'fractional' minute portion of adding two minute values.

3-21 What stops people from ordering goods by telephone, then giving totally fictitious credit card numbers? Well, the thought of San Quentin must have some deterrent effect, but preventive measures are usually built into the cards themselves.

For example, one simple check would be to add the digits of a number, then tack on a zero or one, as required, to make their sum an even number. Thus, 52771 would be a legal number, since its digits sum to 22. In contrast, 49930 isn't legal, because 25 is odd.

Write a program that computes a *check character* according to the following rules:

a) Start with a ten-digit number, e.g. 1122334455.

b) Add the five pairs it contains, e.g. 11 plus 22 and so on.

c) Take the sum **MOD** 26.

a) Return, as the check character, the letter in that position in the alphabet. You should assume that 0 represents "A".

Don't forget to test your program!

3-22 Product expiration dates are often encoded to discourage consumers from complaining. One common technique is to use letters rather than numbers in showing dates. Suppose that the White Bread Mfg. Co. encodes the months as the letters "A" through "L", each digit of the day's date as the letters "Q" through "Z", and the year as the single letter "Z" through "A", plus 1990, where "Z" equals 1 and "A" is 26.

Write a program that reads an encoded date, then prints the decoded version. Be sure to use separate procedures for input, output, and calculation.

3-23 My example of goal-oriented versus object-oriented programming styles contrasted following library patrons to following library books. Try sketching a few steps of the two styles as they'd apply to each of these problems:

a) Operating a record store.

b) Writing a course grading program.

c) Running a bowling alley.

d) Operating a car's ignition/fuel system.

3-24 Mathematicians are always searching for new ways to be mean. Three of my favorites are:

a) The *arithmetic mean* of two numbers is their sum, divided by two.

b) The *geometric mean* of two numbers is the square root of their product.

c) The *harmonic mean* of two numbers is the arithmetic mean of the reciprocals.

Write the three functions, then write a test harness to try them out. Does it matter if you write the functions inside a procedure?

3-25 Even if you can't remember the names of the planets (I vaguely recall something about mother and jelly sandwiches) there's an easy way to remember how far from the Sun they are. It's *Bode's Law*, ascribed to J.E. Bode (1747-1826). It isn't really a law; rather, it's an empirically derived rule that describes the distances from most of the bodies in the solar system to the Sun pretty accurately. It is:

$$\text{Distance of } n\text{th planet} = (4 + 3\times 2^{n-2}) / 10 \text{ astronomical units}$$

with the exception that Mercury has no $3\times 2^{n-2}$ term. An astronomical unit is the distance between the Earth and the Sun, or about 93,000,000 miles. How well does Bode's Law predict the actual distances shown below? How do you account for a certain missing planet (at n=5)?

Mercury	0.39	Mars	1.52	Uranus	19.18
Venus	0.72	Jupiter	5.20	Neptune	30.07
Earth	1.0	Saturn	9.54	Pluto	39.67

3-26 Three professors of Classics were arguing over their coffee—or, more precisely, about their coffee. Apparently, three temperatures were available: boiling, 50°, and room temperature (which happens to be 20°). The professors all agreed that the room temperature cup wouldn't cool any further, but they disagreed about how long it would take the other cups to cool one degree apiece.

'It makes no difference—a degree's a degree,' claimed the professor of Greek, who had many degrees herself. 'Stuff and nonsense,' retorted her colleague, the Latin specialist. 'The boiling one is twice as hot, so it cools twice as quickly.' 'Hold your tongue,' snapped the third professor. 'What's important is the difference from room temperature—the hot one will shed one degree 8/3 times faster.'

The engineering student who was serving coffee (and failing all three Classics courses) knew that the real answer was proportional to the black body (coffee, in this case) temperature^4, less the room temperature^4, with all temperatures in degrees Kelvin (Celsius plus 273), of each cup. Having grown to appreciate the Socratic method, though, she didn't speak up.

Well, how much faster *does* the hotter cup cool off 1 degree? In other words (ignoring convection and conduction), solve:

$$\frac{\textit{hot coffee}_{kelvin}{}^4 - \textit{room temperature}_{kelvin}{}^4}{\textit{warm coffee}_{kelvin}{}^4 - \textit{room temperature}_{kelvin}{}^4}$$

3-27 A physics professor who hoped to teach her students creative methods of problem solving gave each a barometer, and the task of measuring the height of an office building. Most of the students came up with clever, innovative approaches: dropping the barometer from the top of the building and timing its fall, boiling the barometer at the top of the building and checking the water temperature, selling the barometer to the building superintendent in exchange for a glimpse at the blueprints, and so on.

Naturally, there's a dullard in every crowd. One student actually checked the pressure at ground level (30.28 inches of mercury), and on the roof (30.16 inches), then wrote a program that applied the formula:

$$\textit{height in feet} = 25{,}000 \ln \frac{\textit{ground pressure}}{\textit{roof pressure}}$$

How tall is the building? Write a general-purpose program to find the answer for any two pressures.

3-28 Ten seconds remain on the Super Bowl clock when a horrible accident befalls a broadcaster stationed high overhead in the Goodyear blimp. Just as play begins, the broadcaster drops his hairpiece. Suppose that the game is being played at Mile-High Stadium in Denver, so air resistance is negligible. If the blimp is hovering at 1,350 feet over the line of scrimmage, and the descent time is given by:

$$\textit{time} = \sqrt{\frac{\textit{twice the height}}{32.174 ft/sec^2}}$$

is it possible that the hairpiece will be mistaken for a penalty flag before the game ends?

3-29 Tracks and roadways are *banked*, or tilted, to help vehicles negotiate curves at high speeds. You can see banked curves at highway exits, on bicycle racing tracks, and at Roller Derby rinks. Sometimes (on bicycle tracks, for instance) the banking is so steep that the rider seems to be equally threatened by sliding off the track as by flying off of it.

However, when the roadbed is properly banked a driver is in no danger of skidding—as long as she travels neither much faster nor slower than the posted speed. The proper angle of banking for a given radius and velocity is given by:

$$\tan\theta = \frac{velocity^2}{radius} \times 32 feet/sec^2$$

For instance, if a road is posted for 40 MPH, a 400-foot curve should be banked at a 16-degree angle. Write a program that asks a user for a radius and speed, then prints the proper banking radius.

3-30 An average person's surface area, in square meters, can be roughly calculated with:

$$7.184^{-3} \times \text{Weight}^{0.452} \times \text{Height}^{0.725}$$

where the weight is given in kilograms, and height is supplied in centimeters.

Use this formula to write a program that will help you solve one of the great problems of our day: *How thick is a layer of suntan lotion?* In case you don't have one around to check, a tube of suntan lotion holds about fifty milliliters. You'll have to estimate how long it lasts by yourself.

3-31 The image of a procedure or function as a black box that magically computes an answer isn't far from the truth. The choice of one such box over another may depend on an issue as small as the form of its arguments.

Suppose that we have two representations of triangles:

1. as three pairs of points (X_a, Y_a, X_b, Y_b, X_c, Y_c) that mark the triangle's vertices;

2 as a triple (a, b, c) that represents the length of the triangle's sides.

Write and test two black boxes that calculate area. Formulas to use are:

$$area = \frac{(X_aY_b - Y_aX_b) + (X_bY_c - Y_bX_c) + (X_cY_a - Y_cX_a)}{2}$$

$$area = \sqrt{(a+b+c)(a+b)(a+c)(b+c)} \qquad (Heron's\ formula)$$

3-32 Journey with me now to the wilds of New Mathico, where the sun rises on the x-axis, and most everything can be described by an equation.

A cowgirl is riding the range one day when she finds herself, hot and thirsty, at some point x, y. The position of the nearest river can be described by the point-slope equation $y = mx+b$. How far is she from the river? Write a program that asks for her position and the river's equation, then prints the distance.

3-33 A traveling salesperson travels between five cities. Suppose that the x, y coordinates of each city are supplied as input to a program. What is the total distance the salesperson travels? What is the direct distance between the first and last cities on her route?

'The **FOR** *loop is designed to generate a particular series of data values . . .'*
(p. 117)

Chapter 4 The **FOR** Statement

For better or worse, the first three chapters held pretty concentrated doses of facts, details, and drill. In the next few chapters, we'll kick back and change gears. Each chapter will introduce a single statement and the programming techniques associated with it.

Most of the statements we'll learn about actually appear in many programming languages, perhaps with slight syntactic differences. I've changed the chapter organization slightly to help you get an idea of what is Modula-2 specific, and what is generally accepted programming practice in *any* language.

The first section of each chapter, *Strictly Modula-2*, concentrates on the new statement's syntax. Then, a *Standard Practices* section describes programming and problem-solving techniques that use the new statement, but are not really language-specific. Finally, *Extended Examples* solve more difficult programming problems, and often provide the basis of chapter exercises.

Section 4-1

Strictly Modula-2

THE **FOR** STATEMENT MAKES THINGS HAPPEN—again and again and again. Here's an example:

```
MODULE ForTrial;
  (*Prints a popular California bottled-water drinking song.*)
IMPORT IO;
VAR Count: INTEGER;
PROCEDURE Verse (Current: INTEGER);
  (*Prints the Currentth verse of a song.*)
  BEGIN
    IO.IntOut (Current, 1);
    IO.StringOut ("  salmon fillets on the lawn,  ");
    IO.IntOut (Current, 1);
    IO.StringOut ("  salmon fillets.");    IO.LineOut;
    IO.StringOut ("If one of those salmon should happen to spawn,");
    IO.LineOut;
    IO.IntOut (Current + 1, 1);
    IO.StringOut ("  salmon fillets on the lawn.")
  END Verse;
BEGIN
  FOR Count := 1 TO 100 DO
    Verse (Count);
    IO.LineOut;
    IO.LineOut
  END
END ForTrial.
```

```
1 salmon fillets on the lawn, 1 salmon fillets.
If one of those salmon should happen to spawn,
2 salmon fillets on the lawn.

2 salmon fillets on the lawn, 2 salmon fillets.
If one of those salmon should happen to spawn,
3 salmon fillets on the lawn.
```

. . . *many tedious lines of output follow, until . . .*

```
100 salmon fillets on the lawn, 100 salmon fillets.
If one of those salmon should happen to spawn,
101 salmon fillets on the lawn.
```

Whew! Program ForTrial contains a single, highly productive **FOR** statement. If we were to 'unfold' the statement—replace it with the statements it invokes—we'd see this sequence of calls instead:

`unfolding´ the **FOR**

```
Verse (1);   IO.LineOut;   IO.LineOut;
Verse (2);   IO.LineOut;   IO.LineOut;
      . . .
Verse (100);   IO.LineOut;   IO.LineOut;
```

FOR is a loop

> Statements that make actions repeat are called *loops*. A **FOR** loop can include any number of statements, but the sequence is repeated as a whole.

The general form of the **FOR** loop includes an option (shaded, here) that we haven't seen yet:

> **FOR** *counter variable* := *initial value* **TO** *limit* **BY** *amount* **DO**
> *sequence of statements*
> **END**;

The shaded **BY** *amount* portion is optional. If it's not included, **BY** 1 is the default, or assumed, value.

Generating Data

Modula-2 has other loops that repeat statements. The **FOR** loop has an exceptional property, though—it generates *data*. For instance, the loop in program ForTrial generated the numbers 1 through 100.

Thus, while there's a natural tendency to think of the **FOR** statement as a controller of *actions*, I like to start with the premise that **FOR** is meant to harness *information*. The **FOR** loop is designed to:

FOR as a data generator

a) generate a particular series of data values, or to

b) get a specific number of data values from the program user, or to

c) return to particular data items—one or two variables—repeatedly.

The **FOR** statement's syntax is carefully crafted to support this view of the **FOR**'s purpose. A loop *might* have been designed to take a number, then repeat an action that many times. But the **FOR** loop relies on a different mechanism; one that's a bit more complicated, but is much more useful.

> Instead of giving the **FOR** statement the exact number of times we want an action to occur, we provide the *initial value* and *limit* of a *counter variable*.

the counter variable is extremely handy

The counter variable starts out with the initial value. Each time the loop is reentered, the counter variable is *incremented*, or assigned the next higher value. As long as the counter variable doesn't exceed the limit, the loop will be entered—and its action will be repeated—again and again.

Now come the neat ideas—the ones that make the **FOR** loop so useful for generating data:

- The counter variable can have *any* ordinal (counting) type, including INTEGER and CHAR.
- No matter what type the counter variable is, it can be incremented by any positive or negative INTEGER amount.
- We can inspect (but not *change*) the counter variable's value.

Let's look at a few examples. First, we'll generate the numbers 1 through 5:

generating INTEGERs

```
FOR IntegerCount := 1 TO 5 DO
    IO.StringOut ("    Loop:  ");
    IO.IntOut (IntegerCount, 1)
END;
```

```
  Loop: 1   Loop: 2   Loop: 3   Loop: 4   Loop: 5
```

However, we can 'count' by characters just as easily. Here, I assume that Char-Counter is a CHAR-valued variable, and generate the lower-case letters:

generating CHARs

```
FOR CharCounter := "a" TO "z" DO
    IO.CharOut (CharCounter)
END;
```

```
abcdefghijklmnopqrstuvwxyz
```

Supplying a **BY** *amount* value accelerates the count. The first time through the loop the counter variable is assigned the initial value, as before. On subsequent trips, though, it is increased by the value of *amount*. Eventually, the counter variable reaches the limit, the loop's action doesn't take place again, and the program moves along to the next statement.

increasing the increment

For example, let's generate numbers, by tens, from ten to fifty:

generating every tenth number

```
FOR LoopCount := 10 TO 50 BY 10 DO
    IO.StringOut ("    Loop:  ");
    IO.IntOut (LoopCount, 1)
END;
```

```
Loop: 10  Loop: 20  Loop: 30  Loop: 40  Loop: 50
```

The *amount* number can be negative, too—even if the counter variable has type CHAR! We can generate the capital letters in reverse with:

generating CHARs in reverse

```
FOR CharCounter := "Z" TO "A" BY -1 DO
    IO.CharOut (CharCounter)
END;
```

```
ZYXWVUTSRQPONMLKJIHGFEDCBA
```

If the initial value and limit are equal, the **FOR** loop's action takes place just once.

Naturally, if the initial value is greater than the limit (and vice versa if a negative *amount* is supplied) the loop's action is skipped entirely.

Modification Problems

1. Write a short-trip version of ForTrial. In other words, have it count up by twos or threes instead of one salmon at a time.
2. Make up an alphabet-oriented version of ForTrial, then modify the program appropriately.

Self-Check Questions

Q. The first lines of some **FOR** statements are shown below. What sequence of values does each loop generate? Assume that each of these assignments have been made beforehand: Letter := "J"; Number := 7; Jump := 2; and Bound := –3.

a) **FOR** Counter := 1 **TO** Number **DO** . . .
b) **FOR** Index := 1 **TO** Number **BY** 2 **DO** . . .
c) **FOR** Marker := Bound **TO** Number **DO** . . .
d) **FOR** LetCount := Jump **TO** Bound **DO** . . .
e) **FOR** Value := 20 **TO** 10 **BY** –3 **DO** . . .
f) **FOR** Ch1 := "D" **TO** "A" **BY** –2 **DO** . . .
g) **FOR** Ch2 := "T" **TO** "T" **DO** . . .
h) **FOR** Ch3 := "A" **TO** Letter **DO** . . .

A. In each instance, 'through' means 'through and including.'

a) The numbers 1 through 7.
b) The odd numbers 1, 3, 5, and 7.
c) The numbers –3 through 7.
d) No output—the initial value is greater than the limit.
e) The numbers 20, 17, 14, and 11.
f) The characters "D" and "B".
g) The character "T".
h) The number 7.
i) The letters "A" through "J".

Summary of **FOR** Rules

Before we move on to more examples, let me summarize a few **FOR** statement rules.

REAL counters are forbidden

> 1. The counter variable, initial value, and limit must have the same ordinal (counting) type. REAL is forbidden.

This rule lets the compiler carry out the **FOR** loop's action in an efficient and consistent manner.

the counter is local

> 2. The counter variable should be locally declared.

People are sometimes tempted to declare one or two global 'scratch' variables to use whenever a counter is necessary. Don't—it's confusing, and doesn't save anything.

bounds are evaluated on entry

> 3. The expressions that give the counter variable's initial value and the limit are evaluated when the statement is first entered.

As a result, the number of repetitions can't be changed by the **FOR** loop's action. Assignments to variables that might represent the initial value and limit will not affect the counter variable or its limits.

the increment is a constant whole number

> 4. If a **BY** *amount* is provided, it can't be a variable. It must be an INTEGER or CARDINAL value.

The increment has to be a constant value, or an expression that can be calculated without actually running the program. Although it can't be a CHAR value, a numerical increment can be used with a CHAR counter variable.

no assigning the counter

> 5. It's an error to make an assignment to the counter variable from within the loop's action.

The counter variable, like any other variable, represents a value within the action of the **FOR** statement. But although it may be *used* within the **FOR** loop's action, it may not be *changed* there. Thus, the documentation implicit in the control statement's first line—that it will repeat for a certain number of times—cannot be undermined inside the loop.

the counter is undefined on exit

> 6. The value of the counter variable is undefined on exit from the **FOR** statement.

In effect, the variable is in the pristine condition it held when it was first declared. It must be reinitialized before being used in an expression.

Self-Check Questions

Q. What is the output of this program segment? Are there any illegal or improper statements in it?

```
Lower := 1;
Upper := 3;
FOR Counter := Lower TO Upper DO
    Lower := 0;
    Upper := 1000;
    Counter := 5000;
    IO.StringOut ("Are we having fun yet?")
END;
IO.IntOut (Counter, 5)
```

A. Although the assignments to Lower and Upper within the loop are legal, the initial and limit values of the loop's counter variable were calculated and stored away when the **FOR** loop was first entered. The segment prints `Are we having fun yet?` three times, rather than 1,001 times.

On the other hand, the assignment to Counter inside the loop is illegal. In addition, attempting to print Counter's value after the loop is improper because it's undefined.

Section 4-2 Standard Practices

AFTER READING THE **FOR** STATEMENT RULES ABOVE, I hope that you can appreciate that they represent a certain kind of bargain. **FOR** *could* have been simpler. But when he designed Modula-2, Wirth decided that adding a bit of syntax would greatly increase the expressive power of his new language. He traded a little extra *how do you do it?* for a lot more *what can you do easily?*

We'll see this by exploring the **FOR** statement's typical applications. They include:

typical **FOR** applications

- Generating a particular sequence of values—the alphabet, a series of fractions, the counting numbers, etc.
- Obtaining a sequence of input values from the program user—numbers to add or use in doing calculations.
- Generating patterns that are based on counts. For instance, a checkerboard is an eight by eight pattern of squares, while a right triangle is a sequence of steadily lengthening rows.
- Creating tables of values, usually by combining a sequence generator with a pattern maker, to produce a table of squares and cubes.

We've already seen several instances of the first application—generating sequences. We'll look at a few more examples as soon as I introduce an important technique—nesting statements.

Nested Statements

Putting statements *within* statements is a necessary and common practice. A loop statement may contain other loops, just as it contains assignments or procedure calls.

> Statements inside a loop are said to be *nested*. Nested loops are found wherever one loop includes another.

For example:

a nested loop

```
FOR Outer := 1 TO 3 DO
    FOR Inner := 1 TO 2 DO
        IO.StringOut ("Hip,  ")
    END;  (*inner loop*)
    IO.StringOut ("Hooray!");
    IO.LineOut
END  (*outer loop*)
```

```
Hip, Hip, Hooray!
Hip, Hip, Hooray!
Hip, Hip, Hooray!
```

Here, the inner **FOR** loop has a simple job: it prints `Hip, Hip, `. Together with the other statements of the output loop, a single `Hip, Hip, Hooray!` is produced. Since the outer loop repeats three times—**FOR** Outer := 1 **TO** 3—we wind up with three lines of output.

When loops are nested, it's useful to distinguish between the number of times the inner loop is *reached*, and the number of times it is *entered*. How many times will the inner loop execute, and how many times will the statements *inside* the inner loop execute?

reaching vs. entering inner loops

- An inner loop is *reached* once for each turn of the outer loop.
- The inner loop is *entered*, and its contents executed, once for each turn of the outer loop *times* each turn of the inner loop.

Thus, the inner loop above is reached just three times, but its contents are executed 3*2 or six times.

Nesting is largely responsible for the appearance of programs. To help keep programs readable, inner loops are indented more than outer loops; the effect in a deeply nested program is of the paired wings of a flying bird.

indent and comment inner loops

> The more deeply nested a program is, the more important it is to follow two guidelines: indent inner loops evenly, and comment the **END** of each loop.

Nested loops are more interesting when the inner loop actually *uses* the sequence of values generated by the outer loop. For example, what is the effect of this program segment?

a little problem

```
FOR OuterChar := "a" TO "e" DO
   FOR InnerChar := "a" TO OuterChar DO
      IO.CharOut (InnerChar)
   END (*inner loop*)
END (*outer loop*)
```

I can help you analyze the segment by 'unfolding' part of the inner loop. I'll replace the limit by the actual value that OuterChar represents each time the inner loop is reached:

unfolding the inner loop

```
FOR InnerChar := "a" TO "a" DO . . .
FOR InnerChar := "a" TO "b" DO . . .
FOR InnerChar := "a" TO "c" DO . . .
FOR InnerChar := "a" TO "d" DO . . .
FOR InnerChar := "a" TO "e" DO . . .
```

The segment prints `aababcabcdabcde`.

Self-Check Questions

Q. Suppose the segment above printed OuterChar instead of InnerChar. What would its output be?

A. It would print `abbcccddddeeeee`.

Q. Rewrite this nested statement as a single **FOR** statement:

```
FOR Number := 1 TO 5 DO
   FOR Value := 6 TO 8 DO
      FOR Ch := "d" TO "a" BY -1 DO
         IO.StringOut ("Just testing.")
      END (*Ch loop*)
   END (*Value loop*)
END (*Number loop*)
```

A. To find the answer, calculate the number of times the most deeply nested statement will be executed—five (for the outer loop) times three (for the middle loop) times four (for the inside loop), or sixty. The rewritten statement is:

```
(*The semantic equivalent of the triply nested loop above.*)
FOR Answer := 1 TO 60 DO
   IO.StringOut ("Just testing.")
END
```

Pseudocode

As programming problems become more involved, pseudocode program outlines are increasingly useful. For example, let's consider the *pattern problem*:

problem: print a pattern

> Write a procedure that is given width and depth arguments, then prints a pattern of stars with that many columns and rows. For example, here's a pattern of stars that's six wide and four deep:

```
* * * * * *
* * * * * *
* * * * * *
* * * * * *
```

Try to control the overwhelming temptation to start writing code! Instead, let's just describe what we want in pseudocode—a dialect of English that looks a lot like Modula-2.

> Should you floss all of your teeth? No—just the ones that you want to keep. Should you pseudocode every one of your programs? Nope—only the programs that you hope will run.

I like to begin pseudocoding by getting an idea of where I'm headed, with the pseudocode of the final procedure call:

goal pseudocode

(**What I'd ultimately like to be able to say is . . .**)
print pattern (width, depth);

The pseudocode call serves two purposes—it makes an explicit statement of my goal, and it helps clarify the parameters of the solution. Could any other facet of the solution be 'parameterized?' Yes, we might vary the character (and print something besides a star) and spacing between the characters. Can you think of anything else?

Now let's turn to the problem. What does a single line consist of?

one-line pseudocode

FOR *the correct number of columns*
    print a star;
print a carriage return;

Notice indenting separates the contents of the pseudocoded **FOR** loop (*print a star*) from the statement that follows the loop (*print a carriage return*).

Suppose that I pose another question for pseudocoding: How will you print several lines?

several-lines pseudocode

FOR *the correct number of lines*
    print one line;

Miraculously—or maybe not so miraculously—we already know how to *print one line*:

printing the pattern

```
FOR the correct number of lines
    FOR the correct number of columns
        print a star;
    print a carriage return;
```

Translating the final pseudocode into Modula-2 is routine. You can already see how the nested **FOR** loops are going to be used in procedure PrintPattern, below:

pattern-printing procedure

```
PROCEDURE PrintPattern (Width, Depth: INTEGER);
    (*Prints a Width by Depth pattern of stars.*)
    VAR Row, Column: INTEGER;  (*The loop counter variables.*)
    BEGIN
        FOR Row := 1 TO Depth DO
            FOR Column := 1 TO Width DO
                IO.StringOut ("*   ")
            END; (*Column loop*)
            IO.LineOut
        END  (*Row loop*)
    END PrintPattern;
```

The pattern problem is interesting because it provides the background of more difficult table-printing problems. Tables, after all, are usually just columns and rows of calculated figures—taxes, squares and cubes, conversion between systems of measure, and so on. The inside step may be more complicated than *print a star*, but the pseudocode design approach is exactly the same.

Modification Problems

1. Modify procedure PrintPattern so that it takes the character to be printed, and the space between characters, as additional arguments.
2. Add one additional argument to PrintPattern—the number of spaces to indent before printing each line.

Internally Generated Data

Where does program data come from? From two places—either it's created by the program, or it is supplied by the program user. **FOR** loops turn up in both applications. Let's look at two brief examples that typify dozens of similar tasks. One uses a **FOR** loop to generate data, while the other relies on **FOR** to read a user's input.

Our first example comes from a story about the eighteenth-century mathematician Carl Gauss, whose teacher one day required his class to add a long sequence of consecutive numbers.

'Shortly after his seventh birthday Gauss entered his first school, a squalid relic of the Middle Ages run by a vile brute, one Büttner, whose idea of teaching the hundred or so boys in his charge was to thrash them into such a state of terrified stupidity that they forgot their own names. More of the good old days for which sentimental reactionaries long! It was in this hell-hole that Gauss found his fortune.'*

problem: adding consecutive numbers

To make a long story short, Büttner had barely finished stating the problem when Gauss piped up with the answer. His insight was that the first and last numbers in any long sequence (say, 1 and 100) have the same sum as their neighbors (2 and 99) and so on (3+98, 4+97, ...), with 50 such pairs in all. As a result, the entire sum of 1 through n can be expressed as the sum $(n+1)$, times the $(n/2)$ pairs. This is the kind of elegant solution we all wish *we'd* thought up!

But suppose that young Gauss had been corrupted by access to a computer. His train of thought might have run along very different lines: why not keep a running sum of the numbers generated by a **FOR** loop? In pseudocode:

series-summing pseudocode

```
FOR every number from 1 through 100
    add the number to a running sum;
```

The variable that holds the *running sum* is usually called an *accumulator*, because it accumulates values. There's one important fact to remember about accumulators—they must be initialized. Returning to the pseudocode:

Gauss in pseudocode

```
initialize a Sum variable to 0;
FOR every number from 1 through 100
    add the number to Sum;  (*Clear indenting makes an END unnecessary.*)
print the final value of Sum;
```

The pseudocode outline can't be run, but it plainly points to our final code:

Gauss in Modula-2

```
MODULE Gauss;
   (*Adds the sequence of numbers 1 through 100.*)
IMPORT IO;
VAR Sum, Count: INTEGER;
BEGIN
   Sum := 0;
   FOR Count := 1 TO 100 DO
      Sum := Sum + Count
   END;  (*We need the END in Modula-2.*)
   (*Sum represents the added sequence.*)
   IO.StringOut ("The sum of 1 . . . 100 is:  ");
   IO.IntOut (Sum, 1)
END Gauss.
```

```
The sum of 1 . . . 100 is: 5050
```

---

* E.T. Bell, *Men Of Mathematics*, Simon and Schuster 1937, p. 221.

Modification Problems

1. Rewrite Gauss so that the sum is calculated by a procedure that takes two arguments—the lower and upper bounds of the sequence. Don't forget that the counter variable must be local.
2. Here's a wild claim for you: the sum of the first *n* odd numbers equals *n* squared. Modify Gauss to find out if it's true.

Externally Obtained Data

Gauss generated its own data. A programming problem that's equivalent, but which gets data from the program user, is:

problem: averaging numbers

> Write a program that finds the average of a series of REAL numbers typed in by the program user. The very first number will be an INTEGER that tells how many numbers are available.

The averaging program uses an accumulator, just as Gauss did. However, it also requires a *counter* to keep track of the number of values that have been read in. It doesn't take too much math to recognize that the average will equal the sum divided by the final count.

How can I break up the program's work? A pseudocode outline might be:

averaging pseudocode

give instructions;
find the number of values;
add the input (number of values, the running sum);
print the results (running sum / number of values);

The *add the input* step is written as procedure AddInput, below.

As you read program Averages, note that its comments do not merely parrot the pseudocode outline. Instead, they're aimed at a program reader who may have cause to doubt our programming prowess. Unlike pseudocode, which states our intentions for what we want to do, the program comments are meant to confirm what we've actually done. I'll be talking more about *confirmation commenting* in the Program Engineering notes.

(*program follows*)

number averaging program

```
MODULE Averages;
  (*Averages a sequence of REAL numbers. The first input value
   should be an INTEGER that gives the length of the sequence.*)
IMPORT IO;
VAR Length: INTEGER;  (*the number of numbers to be added*)
    Result: REAL;  (*the final sum*)
PROCEDURE AddInput (Limit: INTEGER; VAR Total: REAL);
  (*Reads and sums Limit values.*)
  VAR Counter: INTEGER;
      Next: REAL;  (*the current input value*)
  BEGIN
    FOR Counter := 1 TO Limit DO
      IO.RealIn (Next);
      Total := Total + Next
    END
    (*Total represents the sum of the input values.*)
  END AddInput;
BEGIN
  IO.StringOut ("Please enter the number of values to average,");
  IO.StringOut ("  then the values.");   IO.LineOut;
  IO.IntIn (Length);
  Result := 0.0;
  (*We know how many numbers there are, and the accumulator is initialized.*)
  AddInput (Length, Result);
  (*Result represents the sum of the input numbers.*)
  IO.StringOut ("The average is:   ");
  IO.FixOut (Result / FLOAT(Length), 10, 3 );
  (*We've printed the average of the input values.*)
END Averages.
```

```
Please enter the number of values to average, then the values.
8  23.3  8234.702  893.3  23.56  43.0  245.1  2.452  430.003
The average is:   1236.927
```

Modification Problems

1. Modify program Averages so that it find the product of a sequence of INTEGERs. Why might you want Result to stay REAL?
2. Change the program so that Length is a local variable of AddInput, rather than a global variable.

Self-Check Questions

Q. Four dipsomaniacs were sharing a fifth of hundred-proof (50% alcohol) gin. Unfortunately, the first one spit up two-thirds of her share. How much alcohol did she manage to imbibe? In other words, how much is half of a third of a quarter of a fifth?

A. Note that Result must be initialized to 1.0, rather than 0.0.

```
Result := 1.0;
FOR Denominator := 2 TO 5 DO
    Result := Result * (1.0 / FLOAT (Denominator))
END;
IO.FixOut (Result, 10, 5);
```

```
    0.00833
```

Aha! Programming Riddles

The final **FOR** programming practice I'll describe is having fun! A *programming riddle* is a problem that never quite grew up. If a riddle is practical, it's by accident. Aside from providing a little mental entertainment, they should be entirely useless.

Riddles that have **FOR** loop solutions typically involve sequences and patterns. The more nesting they require, the better. A riddle that is too simple for you might be:

How can you generate the sequence 1/2, 1/3, 1/4 . . . 1/*n*?

Of course, that one is too easy! Something a bit more worthy of your talents is:

a sequence riddle

Generate the sequence 1/2, 1/3, 1/4, 1/5, 2/3, 2/4, 2/5, 3/4, 3/5, 4/5.

That's a little harder. Suppose that LIMIT equals 5. Then:

generating the fractions

```
FOR Top := 1 TO LIMIT – 1 DO
    FOR Bottom := Top + 1 TO LIMIT DO
        IO.IntOut (Top, 3);
        IO.CharOut ("/");
        IO.IntOut (Bottom, 1)
    END (*Bottom*)
END (*Top*)
```

```
1/2  1/3  1/4  1/5  2/3  2/4  2/5  3/4  3/5  4/5
```

Suppose that we try a pattern riddle. How would you generate this triangle?

a pattern riddle

```
    A
   BB
  CCC
 DDDD
EEEEE
```

Hmm. Why don't we forget about printing different letters for a moment, and see if we can handle a triangle of stars? It looks like the number of figures we print *increases* as the number of blanks we do *decreases*. Overall, we want to:

pattern pseudocode

FOR *five lines*
  print the right number of blanks;
  print the right number of figures;
  print a carriage return

But what's the *right number*? Well, we can relate it to the line number. Together, the two *right numbers* add up to the right number; separately, the number of figures equals the line number while the number of blanks is five minus the line number. Hmmmm! Suppose we say that LIMIT equals five:

printing the pattern with stars

```
FOR Line := 1 TO LIMIT DO
    FOR Blanks := 1 TO LIMIT – Line DO IO.CharOut (" ") END;
    FOR Figures := 1 TO Line DO IO.CharOut ("*") END;
    IO.LineOut
END;
```

I'll leave it up to you to figure out how to turn the stars into the proper letters.*

why bother?

> Programming riddles usually rely on one fundamental technique—basing the initial value, limit, or increment of an *inner* loop on the value of the counter variable in an *outer* loop.

This turns out to be more than an enormously useful trick. Solving programming riddles helps you learn to say 'Aha!'—to spot an underlying pattern that holds the key to a program solution. We'll see more of them in the exercises.

Self-Check Questions

Q. Suppose that you wanted to print every three-letter combination of the letters "a", "b", and "c". How would you use nested **FOR** loops to do it? How many combinations are there?

A. The appropriate code is shown below. We can figure out that there are twenty-seven different combinations by calculating the number of times the inner loop is entered: 3*3*3.

* Better yet, why not toss out my solution entirely, and have a loop that counts by letters, "A" to "E", and takes care of spacing on the basis of the current letter?

```
FOR Ch1 := "a" TO "c" DO
    FOR Ch2 := "a" TO "c" DO
        FOR Ch3 := "a" TO "c" DO
            IO.CharOut (Ch1); IO.CharOut (Ch2); IO.CharOut (Ch3)
        END
    END
END
```

Q. Suppose that you wanted to print the pattern of numbers shown below. How would you do it?

```
 1   2   3   4
 5   6   7   8
 9  10  11  12
13  14  15  16
```

A. It requires nested **FOR** loops, of course. The pattern is created with:

```
FOR Out := 0 TO 3 DO
    FOR Put := 1 TO 4 DO
        IO.IntOut ((Out*4) + Put, 4)
    END;
    IO.LineOut
END
```

Section 4-3 Extended Examples

problem: adding odd numbers

LET'S BEGIN WITH A QUICK EXAMPLE THAT WAS SUGGESTED by the story about Gauss I told earlier. Although the historical records that support this claim have, unfortunately, been lost, you can take it from me that Professor Büttner came back the next day seeking revenge. 'I want,' he said, 'the sum of the first *n* odd numbers—and let's make *n* something nice and round, like 183.' One of Gauss's lesser classmates, desperate to score points, and hoping that a formula was involved, took a wild guess: 'That's easy—it's 183^2.' Büttner was apoplectic; the answer is correct.

Sounds too good to be true, doesn't it? Let's write a program that checks the rule. The heart of our program will be a call that looks something like:

goal pseudocode

add N *odd numbers (the value of* N, *the final sum);*

We have to find out N—how many numbers to add, and also calculate and print N^2. I can outline the complete program like this:

program pseudocode

get N;
print N^2;
initialize a Sum *variable to* 0;
FOR *the first* N *odd numbers* (**what procedure* SumOdds *will do**)
    add the number to Sum;
print the final value of Sum;

You should remember how to generate only odd numbers, but getting *N* of them, rather than half-*N*, might take you a minute. Try to figure it out before you read program OddAdder, below.

odd-number addition program

```
MODULE OddAdder;
   (*Checks the rule that the sum of N odd integers equals N^2.*)
IMPORT IO;
VAR HowMany, Answer: INTEGER;
PROCEDURE SumOdds (Limit: INTEGER; VAR Sum: INTEGER);
   (*Adds the first Limit odd numbers.*)
   CONST INCREMENT = 2;
   VAR Count: INTEGER;
   BEGIN
      Sum := 0;
      FOR Count := 1 TO 2*Limit BY INCREMENT DO
         Sum := Sum + Count
      END (*FOR*)
      (*Sum represents the added sequence.*)
   END SumOdds;
BEGIN
   IO.StringOut ("How many odd numbers shall I add?    ");
   IO.IntIn (HowMany);
   IO.IntOut (HowMany*HowMany, 10);  (*Prints the square.*)
   SumOdds (HowMany, Answer);
   IO.IntOut (Answer, 10)  (*Prints the sum for comparison.*)
END OddAdder.
```

```
How many odd numbers shall I add?  174
     30276     30276
```

Modification Problems

1. Modify OddAdder so that it finds the sum of the first *N* even numbers. Is there a formula that gives you the same result?
2. Change SumOdds so that it sums odd fractions (like 1/3, 1/5, 1/7 and so on), rather than whole numbers.

Problem: Multiplication Tables

Let's turn to a more difficult problem that requires nested **FOR** statements.

problem: multiplication tables

> Write a program that prints a 'times' table. Have it show all multiples, from $1 \times N$ to $N \times N$, of any number.

It's probably been some time since you've had to rely on such a table for help with multiplication. A pseudocode procedure call shows that there's just one parameter for the problem as stated:

pseudocode goal

print a times table (N);

But to be perfectly frank, the pseudocoded call isn't a whole lot of help. Suppose that I restate the pseudocode in a way that's a bit more useful:

first pseudocode

get N, the Limit;
print every multiple from 1 × 1 *to* Limit × Limit;

Print every multiple might be clear English, but it's far too vague to be the basis of Modula-2 code. A refinement of the pseudocode is needed:

a refinement

get the Limit;
FOR *every number from* 1 *to* Limit
    print all multiples of that number;

Now, what should the output look like? Why not just copy a grade-school math book? In a simple picture, the rows and columns each go from 1 to N, while row/column intersections will hold the current row/column product.

```
1  2  3  4  5  6
2  4  6  8 10 12
3  6  9 12 15 18
4  8 12 16 20 24     etc.
```

Where will the rows and columns—the program's data—come from? We can use **FOR** loops to generate them internally. In fact, we can also use the counter variables to label the rows and columns:

labeling the rows and columns

```
  |  1  2  3  4  5  6
---------------------
 1|  1  2  3  4  5  6
 2|  2  4  6  8 10 12
 3|  3  6  9 12 15 18
 4|  4  8 12 16 20 24     etc.
```

Each row and column is numbered. Wouldn't it be nice to be able to change printing fonts on the lineprinter! More seriously, wouldn't it be nice to suppress some of the output, so that duplicates wouldn't print at all?

simplifying the table

```
  |  1  2  3  4  5  6
---------------------
 1|  1
 2|  2  4
 3|  3  6  9
 4|  4  8 12 16           etc.
```

Let's return to our pseudocode. A third refinement adds the last feature I suggested, above. Can you see how it's done? Notice that as pseudocode becomes more complicated, incidentally, I'll occasionally use **END**'s to keep my intentions clear.

yet another refinement

```
get the Limit;
label the columns;
FOR every Row from 1 to Limit
    label the current row;
    FOR every Column from 1 to the current Row value
        print the value Column times Row;
    END the current row          (*Here, the pseudocoded ENDs*)
END the whole series of rows     (*help keep the plan clear.*)
```

The completed program ShowTime is shown below. As before, notice that its comments are different from its pseudocode—pseudocode helped us write the program, but comments help us understand and correct it.

multiplication table program

```
MODULE ShowTime;
    (*Prints a multiplication table of Limit values.*)
IMPORT IO;
VAR Limit, Row, Column: INTEGER;
BEGIN
    IO.StringOut ("Enter the upper limit for this times table.");
    IO.LineOut;
    IO.IntIn (Limit);
    (*Limit has been initialized.*)
    IO.StringOut ("    ");
    FOR Column := 1 TO Limit DO
        IO.IntOut (Column, 3)
    END; (*END*)
    IO.LineOut;
    (*The columns have been labeled.*)
    FOR Row := 1 TO Limit DO
        IO.IntOut (Row, 3); (*Label every row.*)
        FOR Column := 1 TO Row DO   (*Going to Limit would print duplicates.*)
            IO.IntOut (Row*Column, 3)
        END; (*the inner FOR*)
        IO.LineOut
        (*Each row has been printed.*)
    END (*the outer FOR loop*)
END ShowTime.
```

(output follows)

```
Enter the upper limit for this times table.
9
     1  2  3  4  5  6  7  8  9
  1  1
  2  2  4
  3  3  6  9
  4  4  8 12 16
  5  5 10 15 20 25
  6  6 12 18 24 30 36
  7  7 14 21 28 35 42 49
  8  8 16 24 32 40 48 56 64
  9  9 18 27 36 45 54 63 72 81
```

Modification Problems

1. Change ShowTime so that instead of printing numbers on and *below* the center diagonal, it prints numbers on and *above* the diagonal.
2. ShowTime is missing a feature—the little lines that separate the row and column labels from the actual table. Add them to the program. Be sure that the horizontal line under the column labels isn't fixed at any particular Limit.
3. Put the table-printing code from ShowTime into a procedure. Give it extra parameters, though—don't require a 'square' table.

Problem: Income Projection

The nested **FOR** loop often appears when a calculation must itself be repeated. Let's look at a problem that will appeal to all students who are planning careers in business: making income projections.

problem: income projection

> An editorial assistant who is learning just what publishers mean by 'voracious reading' asks for an increase in her $18,000 salary. Her sadistic boss makes the following offer:
>
> *Plan A*: a 10% raise each year, commencing immediately, along with an immediate $1,500 bonus that won't be counted in base pay.
>
> *Plan B*: a raise of 1/12 of 10% each and every month, with no bonus.
>
> Which deal is better? She has ten minutes to decide.

Note that the more frequent raises of Plan *B* will work like compound interest, and lead to somewhat higher yearly raises. All the same, the $1,500 bonus of Plan *A* is nothing to be sneezed at either. We can anticipate that, over time, one deal will replace the other as the best offer.

What will the program's output be? A novice coder might be tempted to calculate and print only the exact date on which the better early deal becomes the long-run loser. But the program engineer's approach is more sophisticated.

people read output

> From the viewpoint of human decision making, limiting output to a single figure can be misleading. Factors that have not been explicitly mentioned in a problem statement may turn out to be influential when a final decision has to be made. Programs should supply *information*, and not just answers.

In the case of this particular problem, the worse long-run choice might be favorable if the employee needs quick money, and the long-run cost isn't too great. Since 'too great' is one of those concepts that's hard to program, projecting the actual figures involved over some period will be the best approach.

Let's turn to pseudocode. We want to produce a table that will let us make an intelligent comparison. Two kinds of calculations will be involved—those of Plan *A*, and those of Plan *B*. Our first 'goal' pseudocode will be a little different:

goal pseudocode

```
initialize variables as necessary;
FOR some reasonable number of years
    label the table;
    do Plan A (this year's A income, running total A income);
    do Plan B (this year's B income, running total B income);
```

Instead of highlighting the goal-oriented steps, I want to draw your attention to the background material. By now, we're getting smart enough to anticipate having to do some program initialization, along with some labeling of output.

The pseudocode of *do Plan A* is pretty straightforward:

Plan *A* refinement

```
CurrentPayA := a 10% increase;
TotalPayA := increased by CurrentPayA;
```

Plan *B* is a little harder to calculate:

Plan *B* refinement

```
FOR each month of the current year
    CurrentPayB := one twelfth of a 10% increase;
TotalPayB := increased by CurrentPayB;
```

The final version of program Project is shown below. Note that we're using the **FOR** loop counter variables in the most limited way possible—they don't generate data, and they're not part of any calculation. Instead, the loops let us revisit and linger over particular variables, adjusting and readjusting their values as we make our income projections.

income projection program

```
MODULE Project;
  (*Develops an income projection for two alternative salary offers.*)
IMPORT IO;
CONST YEARS = 10;          (*We'll run this projection for ten years.*)
VAR Year: INTEGER;
    CurrentA, CurrentB,          (*The current yearly pay rates.*)
    TotalA, TotalB: REAL;          (*The running totals for each plan.*)
PROCEDURE PlanA (VAR Current, Total: REAL);
  (*Calculates the 10% yearly raise of Plan A.*)
  BEGIN
    Current := 1.1 * Current;
    Total := Total + Current
  END PlanA;
PROCEDURE PlanB (VAR Current, Total: REAL);
  (*Calculates the monthly pay raises of Plan B.*)
  VAR Month: INTEGER;
  BEGIN
    FOR Month := 1 TO 12 DO
      Current := Current + ((0.10/12.0) * Current)
    END;
    Total := Total + Current
  END PlanB;
BEGIN
  CurrentA := 18000.00;
  CurrentB := 18000.00;
  TotalA := 1500.00;          (*Plan A starts with the $1,500 bonus.*)
  TotalB := 0.00;
  (*All variables have been initialized.*)
  IO.StringOut ("      A: Yearly    Total    B: Yearly    Total");
  (*The output table has been labeled.*)
  FOR Year := 1 TO YEARS DO
    IO.LineOut;
    IO.StringOut ("Year  ");
    IO.IntOut (Year, 2);
    PlanA (CurrentA, TotalA);
    IO.FixOut (CurrentA, 12, 2);
    IO.FixOut (TotalA, 12, 2);
    (*We've calculated and printed a year according to Plan A.*)
    PlanB (CurrentB, TotalB);
    IO.FixOut (CurrentB, 12, 2);
    IO.FixOut (TotalB, 12, 2)
    (*We've calculated and printed a year according to Plan B.*)
  END
END Project.          ( output follows )
```

	A: Yearly	Total	B: Yearly	Total
Year 1	19800.00	21300.00	19884.84	19884.84
Year 2	21780.00	43080.00	21967.04	41851.88
Year 3	23958.00	67038.00	24267.28	66119.15
Year 4	26353.80	93391.80	26808.38	92927.52
Year 5	28989.18	122380.98	29615.56	122543.09
Year 6	31888.10	154269.08	32716.71	155259.80
Year 7	35076.91	189345.98	36142.57	191402.38
Year 8	38584.60	227930.59	39927.18	231329.56
Year 9	42443.06	270373.66	44108.08	275437.66
Year 10	46687.37	317061.03	48726.76	324164.41

Modification Problems

1. Suppose that our editorial assistant were to make a counteroffer—a bonus of $750, with a daily raise of 1/365 of 10%. Under what circumstances should her editor accept?
2. Procedure PlanB contains a calculation that's usually regarded as a no-no. It's the division within the loop. Since the answer is always the same (0.10/12.0) is a constant, why bother to calculate it again and again? Modify the code to remove the constant from the loop.
3. Plans *A* and *B* are equivalent to compound interest—additional increases come on top of raises already received. However, many companies give yearly bonuses that are not included so that they won't have to raise base pay—they're like simple interest. Add a PlanC procedure to Project that analyses the effect of a yearly bonus of $10,000.

Background: Do Computers Supply Answers?

A programmer went to a business meeting. The first speaker threw a slide upon the overhead projector, displaying many columns of figures. 'This spreadsheet shows,' he said, 'that we will make a great deal of money.'

The programmer asked: 'What assumptions did you use when you prepared the spreadsheet?'

The speaker scratched his head. 'I produced this slide on the computer.'

The programmer gritted his teeth. 'Have you never heard the saying 'Garbage In, Garbage Out'?'

The speaker threw his head back and laughed. 'You misquote it. The correct saying is 'Garbage In, Gospel Out'!'

From *Computer Parables*, by Geoffrey James, © 1989, InfoBooks, Santa Monica, Ca. 90406

Section 4-4

Program Engineering Notes

THE LONG-TERM GOAL OF THIS COURSE may be to teach you faith in computers, but our immediate goal is to learn how to doubt software. One of the hardest aspects of program engineering is to recognize that intentions are not always accomplishments. What a loop is supposed to do, and what it actually does, are often quite different.

Confirmation Commenting

The program engineer's main weapon is doubt: debugging the code, rather than the comments. After all, what a program is supposed to do is not nearly as important as what it actually did. For example, consider this program segment. Try to find the bug before you go past the example:

the loop average bug

```
(*Calculate the average of HowMany input values.*)
IO.IntIn (HowMany);  (*Find out how many numbers there are.*)
FOR Count := 1 TO HowMany DO
    Total := 0;  (*Initialize the running total.*)
    IO.IntIn (Number);  (*Get the next number.*)
    Total := Total + Number  (*Add it to the running sum.*)
END;
IO.StringOut ("The average was:   );
IO.FixOut ( FLOAT(Total) / FLOAT(HowMany), 2, 5)
```

Where's the bug? The Total assignment is in the wrong place. Total should be initialized to zero before the loop begins, rather than once each iteration.

How can we avoid such problems? Well, there are three safety barriers.

avoiding coding bugs

- *Plan ahead.* Use pseudocode to outline code before you write it.
- *Code from comments.* As you write code, make a notation of what you intend to write.
- *Confirm the outcome.* Confirm what you think your code has done.

> A *confirmation comment* follows a code segment. By summarizing what we've done, it sets the stage for the program's next actions.

confirmation comments

Confirmation comments don't have to detail everything we've done—regular commenting still helps to explain a tricky piece of code. However, confirmation comments do describe the current program *state*—the facts that *must* be true if the rest of the program is going to work properly.

The 'loop average' bug is exceptionally educational because it boldly walks past the first two checks. By itself, *planning ahead* isn't enough. This is perfectly reasonable pseudocode:

planning alone doesn't prevent bugs

initialize as necessary;
sum the input values;
calculate and print the average;

Does it prevent the bug? No. Speaking realistically, programmers often gloss over the initialization step when they're planning—they make a mental note to add the appropriate statement later, then never get around to it.

commenting alone doesn't prevent bugs

Nor has *coding from comments* worked as an effective error-prevention tool in the buggy segment above. It has *lots* of comments! In a literal sense, the comment *initialize the running total* isn't incorrect—Total is being initialized. Unfortunately, the comment is only correct when it causes us to read the statement out of context.

Only the third step, *confirm the outcome*, can save us, because it subjects Total to a second look. Let's take another look at the code segment, with 'confirmation commenting' in place of 'intention commenting':

the bug is still here

```
IO.IntIn (HowMany);
   (*We know how many numbers to read.*)
FOR Count := 1 TO HowMany DO
   Total := 0;
   IO.IntIn (Number);
   Total := Total + Number
END;
   (*Total represents the sum of HowMany values.
    HowMany represents the number of values read.*)
IO.StringOut ("The average was:   );
IO.FixOut ( FLOAT(Total) / FLOAT (HowMany), 2, 5)
   (*We've printed the average.*)
```

Can you see how the confirmation comments work? They don't describe everything the program does, but they do explicitly set out the facts that the rest of the program relies on.

Confirmation commenting won't prevent bugs either, of course! Nevertheless, it helps to mirror the process an expert programmer goes through in checking her work—she checks what she's done, rather than what she intended to do.

The First Law of Debugging

It ain't what you don't know that hurts you. It's what you *do* know that just ain't so.

Program Testing and the Scientific Method

Confirmation commenting is a *bench test* method. It's used to help show that the program will work before it's compiled or run. However, most programmers also rely on actual program tests. There is one rule, though, that must always be remembered.

> *The Second Law of Debugging*
> Testing can show the *presence* of program bugs, but it says nothing about their *absence*.

A program test is very much like the test of a scientific theory. Science doesn't advance by looking for evidence that theories are correct—that was a hot approach in the sixteenth century, but it isn't science. Instead, theories are literally put to the test. You devise the best tests you can to prove that the theory is wrong, then hope that the tests fail.

Similarly, programs aren't tested by demonstrating that they work. For example, consider the buggy number-averaging segment we've been working with. You can show that the program 'works' just by averaging a single number—try it.

But a real program test tries to show that a program *won't* work. A true test tries to demonstrate a program's problems. Averaging a single number will certainly be one trial, but it can't be the only trial. A good test, in this case, would also try to average at least two numbers.

Will every test necessarily be passed? No. Suppose that we rely on this code, which is correct for a HowMany value of one or more, and is the best we can do with the Modula-2 we know so far:

yet another segment with a bug

```
Total := 0;
IO.IntIn (HowMany);
FOR Count := 1 TO HowMany DO
    IO.IntIn (Number);
    Total := Total + Number
END;
IO.StringOut ("The average was:   );
IO.FixOut ( FLOAT(Total) / FLOAT (HowMany), 2, 5)
```

What would happen if we tried to find the average of no numbers? The program would crash when it tried to divide zero by zero. We'll see how to avoid this problem in the next chapter.

Bugs

Let's turn to a review of potential **FOR** statement bugs that have a syntactic flavor. Most bugs result from forgetting that the statement's very first line makes a promise that a certain sequence of values will be created, or an action will take place a certain number of times. Anything that undermines this implicit documentation is likely to be disallowed or ignored.

In fact, it might be easiest to view all three or four identifiers and values that can appear in the **FOR** statement as parameters that are *used*, but not *changed*. Thus, in the statement:

```
FOR Count := Low TO High BY INCREMENT DO ...
```

all four identifiers—Count, Low, High, and INCREMENT—can be inspected. This is correct:

the control values can be inspected

```
FOR Count := Low TO High BY INCREMENT DO
    IO.IntOut (Count);
    IO.IntOut (Low);
    IO.IntOut (High);
    IO.IntOut (INCREMENT)
END
```

On the other hand, attempting to make changes is either illegal or fruitless. Both of these statements are illegal because the counter variable can only be changed before or after—and never within—the **FOR** statement:

the control variable bug

```
FOR Count := Low TO High BY INCREMENT DO
    IO.IntIn (Count);         (*Invalid attempts at input and assignment.*)
    Count := Count + 1
END
```

the increment bug

In effect, the counter variable is *read-only*. It can be inspected, but it can't be changed. The INCREMENT can't be changed either. In fact, it can't even be a variable—it can only be an INTEGER (or CARDINAL), or a constant, or an arithmetic expression.

Trying to modify the values of Low and High fall into a different category. If they're variables, Low and High can be changed, but the change won't affect the statement. In other words, this statement:

the bound bug

```
Low := 1;
High := 5;
FOR Count := Low TO High BY INCREMENT DO
    Low := 0;         (*Doesn't affect the counter variable.*)
    High := 1000;
    IO.StringOut ("Hello")
END
```

prints `Hello` only five times, and not 1,000 times. Even though the shaded assignments are valid, Low and High have already played their assigned roles in setting the bounds of the **FOR** loop. The initial and final values of the counter variable have already been calculated and stored.

A final **FOR** statement bug is subtle. What's wrong with this sequence?

the undefined control variable bug

```
IO.IntIn (Finish);
FOR Count := 1 TO Finish DO
    whatever the loop does
END;
IO.StringOut ("Number of repetitions was:    ");
IO.IntOut (Count)
```

The problem is that a **FOR** statement's counter variable is undefined *by definition* when the loop ends. It may have a value—in fact, it certainly has a value—but exactly what that value is isn't specified by the Modula-2 standard. (The reason the standard doesn't set a rule has to do with compiler implementation considerations, and is a little beyond our scope.)

Modula-2 Details

Syntax Summary

- **FOR** statement: causes a *counter variable* to assume a predetermined sequence of values; also, lets an action be repeated for a predetermined number of times:

  FOR *counter variable* := *initial value* **TO** *limit* **BY** *amount* **DO**
  sequence of statements
  END;

- The **BY** *amount* portion is optional.
- The initial value and limit must have the counter variable's type, and can't be REAL.
- The counter variable should be a local variable, and can't be a parameter.
- The initial value and limit are determined when the loop is first entered.
- The **BY** increment or decrement value can't be given with a variable.
- It's an error to make an assignment to the counter variable from within the loop.
- The counter variable is undefined on exit from the statement.
- nested statement: an 'inner' statement that takes place within the body of an 'outer' statement.

Modula-2 Details

Remember . . .

- **FOR** loops control the availability of data. They can generate a sequence of data values internally (via the counter variable), or by returning to particular variables again and again, or they can count values entered by the program user.
- 'Visual commenting' by consistently indenting nested statements is essential for keeping complex programs comprehensible. Commenting statement **END**s helps as well. Overall, a program's physical appearance should reflect its structure in Modula-2.
- Pseudocode is a pidgin Modula-2 that's used for program planning. Although pseudocode doesn't obey Modula's syntax rules, it illustrates the flow of information through a program, and the final code's basic construction.
- An *accumulator* is a variable that keeps a running sum. Remember to initialize it.
- Programs should provide information, and not just answers. Why? Because answers can be wrong—they reflect the programmer's understanding of the problem—even when the information they're based on is right. When possible, supply enough program output to let the reader draw her own conclusions.

- Label program output so that it makes sense, in addition to being right.
- The First Law of Debugging: 'It ain't what you don't know that hurts you. It's what you do know that ain't so.' Use confirmation comments to help verify what you think you've done.
- The Second Law of Debugging: 'Testing can show the presence of program bugs, but it says nothing about their absence.' Try to devise tests that will make a program fail, rather than succeed.

Zen Study Guide Big Ideas

The two Big Ideas in this chapter are:

1. *Most programs process a sequence, or stream, of data values.* Therefore, most programming languages include a statement that can generate basic streams of values—counting numbers, letters, odd numbers, and the like. *Most programs repeat actions.* Therefore, most programming languages include a statement that will repeat an action for a fixed number of times. In Modula-2, the **FOR** statement serves both purposes.
2. *You have to outline programs before you write them.* Pseudocode is a reasonable middle ground between the vagueness and imprecision of English, and the excessive precision of actual Modula-2 code. English is too informal to let you know what kind of problems you'll face when coding time comes around. In contrast, Modula-2 is too rigid to let you sketch out alternative ideas.

Zen Study Guide Questions

4-1 The first Self-Check Questions asked you how many times a **FOR** loop's action took place. Write a dozen more. Make them hard by

a) using CHAR counters,
b) decrementing rather than incrementing the counter,
c) having a negative initial value,
d) using bounds that cause the statement to be skipped, or
e) including syntax errors.

4-2 Nesting is inherently confusing. Therefore, one basic **FOR** loop test question is a variation on the following:

```
FOR Outer := A TO B DO
    FOR Inner := C TO D DO
        something that involves Outer or Inner
    END
END
```

Make up a few questions of this ilk. Make them harder by using the tricks suggested above, as well as by replacing C or D with Outer.

4-3 *Write a program that prints this pattern* . . . is always a good standby test question. To study for it by the Zen method, consider this sequence:

```
FOR Num := 1 TO A DO IO.CharOut (First) END;
FOR Num := 1 TO B DO IO.CharOut (Second) END;
FOR Num := 1 TO C DO IO.CharOut (Third) END;
IO.LineOut;
```

The sequence prints one line of a pattern—A instances of First, B instances of Second, and C instances of Third.

Practice creating values for A, B, and C. For example, you could do it like this:

```
FOR Count := 1 TO 5 DO
    A := Count - 1;
    B := Count;
    C := Count + 1;
    Call the procedures described above;
END
```

Print four different patterns, and sleep with them under your pillow. When you get up, see if you can *write a program that prints this pattern . . .*

4-4 *Look for the counter variable* doesn't quite have the ring of *Cherchez la femme*, but it's a basic tactic of **FOR** loop problem solving. Our first example program, Gauss's sequence, was solved by recognizing that the sequence 1, 2, 3 through 100 could be replaced by the counter itself. The sequence of odd numbers, in turn, was replaced by a counter incremented by 2.

We can use the counter to generate just about every regular sequence. For example, assume that Counter goes from 1 through N by ones. It's easy to recognize the sequences that these generate:

1/Counter *generates* 1/1, 1/2, 1/3, 1/4 ··· 1/N
Counter–1/Counter *generates* 0/1, 1/2, 2/3, 3/4 ··· N–1/N
Counter*Counter *generates* 1, 4, 9, 16 ··· N^2

Use the Zen Study Method to create a few *Write a loop that creates this sequence* problems. Use the lazy (e.g. my) method by first modifying the operation on Counter, then figuring out what sequence is created.

4-5 I hardly need to tell you that if you read a value, but don't use it, you've skipped it. **FOR** loops can be used to skip input values just as easily as to obtain them. For example, suppose that you have ten lines of ten values. You can read every value on every line with:

```
FOR Line := 1 TO 10 DO
    FOR Value := 1 TO 10 DO
        IO.IntIn (TheValue);
        DoSomethingWith (TheValue)
    END
END
```

However, you can either precede or follow the inner value-reading loop with a value-skipping loop. Make up three questions that require you to do so.

4-6 Table-making is a standard **FOR** loop application. Make up two questions for each of these kinds of tables:

a) *Conversion tables*, for instance, conversion of Fahrenheit to Centigrade.

b) *Arithmetic tables*, for instance, of sines or square roots.

c) *Comparison tables*, like the output of the income projection program.

4-7 *Here's the input, what's the output?* The basic question is built around a loop like:

```
Total := 0;
FOR Count := 1 TO 5 DO
    IO.IntIn (Value);
    Total := Total + Value
END;
IO.IntOut (Total, 1);
```

Write three *Here's the input, what's the output* questions that are complicated by *a*) reading the initial value and final value as part of program input, *b*) using a **BY** amount to fool the unwary into thinking that some input values will be skipped, *c*) changing the initial value and final value within the loop, *d*) having an extra input statement within the loop, or *e*) adding the counter variable to the running total.

4-8 *Here's the output, what was the input?* questions are my own personal favorite. The reverse sort (*Here's the input*) can be ground through eventually, but these have to be 'decoded' before they can be understood. Start with the basic:

```
IO.IntIn (Finish);
FOR Count := 1 TO Finish DO
    IO.IntIn (Number);
    IO.IntOut (Number)
END
```

This simply reads the number of values to be read and echoed, then does the reading and echoing.

You can make up more complicated variations on this theme by *a*) reading the start as well as the finish value, *b*) performing some arithmetic operation on the number before you print it, or *c*) using the previous input number as part of the arithmetic computation. Beware of looking for solutions that aren't unique, though. Make up three such problems.

4-9 Make up three variations on the Gauss summing problems.

4-10 Make up two variations on the multiplication table problem.

4-11 The basic outline of a program that uses a **FOR** statement to get data from the program user is:

> **FOR** *all available data*
>   *get a value;*
>   *process the value;*
> *print the results*

Some of the processes such a program might carry out include adding the input values or finding their average. Think of three other processes that might rely on the same basic program shell.

Exercises

Warmups and Test Programs

4-12 Write a procedure that prints all the even numbers between 3 and 43.

4-13 Write a procedure that prints every third capital letter, starting with "C".

4-14 Write a program that prints all numbers divisible by 5 from 50 down to zero.

4-15 Write a procedure that is passed two parameters—a character, and a length—then prints a line of that many characters.

4-16 Write a program that reads eleven numbers, and prints the difference between each successive pair. Thus, only ten numbers appear as output.

4-17 Write a procedure that averages a sequence of REAL numbers. The length of the sequence should be passed as a value parameter, while the answer should be returned with a variable parameter.

4-18 Error messages are often hard to understand, or even misleading. It can be very useful to make errors intentionally, just to see if the compiler makes a helpful diagnosis. So, try the forbidden! Write test programs that:

a) attempt to change the value of the counter variable;

b) use a counter variable of type REAL;

c) supply a variable as the **BY** amount.

4-19 Are there any limits on the size of the initial or final values? How about on the maximum **BY** amount size?

4-20 How many of the ASCII characters do interesting things? Write a program that attempts to print each character. To keep them from whizzing by, print each character's ordinal position, and ask the program user to hit a key (just read and ignore the character) between each output statement. If you're lucky, you may get to explore the existential question 'How can I debug a program that locks my keyboard?'

4-21 It is an unalterable fact that the value of the counter variable is undefined on exit from the FOR statement. However, it must have *some* value. On your system, what is it? Is the value consistent (i.e. equal to the final value) regardless of the counter variable's type? No matter what you find out, forget the answer.

4-22 How long does a statement take? Well, it depends what kind of statement it is. Most computer systems have a built-in command that lets you time the execution of a program. Find out what it is, then go exploring relative *time-costs* of different kinds of statements and operations. You'll find that a single statement's execution is far too fast to time meaningfully—you'll need at least a thousand (which you can get courtesy of the **FOR** loop). Some statements to time include:

a) Empty statements (in a loop that does nothing).

b) Assignments that don't involve any calculations: Test := 10;

c) Assignments that do involve calculation. Try three different variations: INTEGER arithmetic, REAL arithmetic, and arithmetic that requires a function call.

4-23 The word *increment* has been defined as 'the difference between two values of a variable,' while *decrement* is described as 'a negative increment.' In both cases the words are used as nouns. Nevertheless, we persist in using increment as a verb, as in 'Let's increment the counter variable.' Do you think this is appropriate? Why or why not? Can you extend your feelings to the noun *input* as well?

4-24 Two containers each hold exactly one liter of soda, although one container holds Coke, and the other Pepsi. Suppose that you pour precisely one tenth of the Coke into the Pepsi, and then pour one tenth of a liter of the disgusting mixture that results back into the Coke. Repeat the process exactly ten times, then answer this question: is there more Coke in the Pepsi, or Pepsi in the Coke?

Exercises

Programming Problems

4-25 Calculating a *weighted average* is like calculating an ordinary average, except that some values are given greater weight, or importance. As you know, weighted averages are very important in college; after all, examinations are given primarily in order to prepare you to take more examinations. Thus, it is essential that each test count more heavily in the final reckoning than its predecessor did.

Suppose that a set of course exams carries the following weights: 10%, 25%, 30%, 35%. Write a program that reads in ten sets of exam grades, then prints each student's weighted average, and the overall unweighted average *on each test* for the entire set.

4-26 Imagine, if you can, a set of Chinese boxes that are neatly nested one inside the next. The smallest box, which measures one inch on a side, fits inside a box with two inch sides, which is inside a box with three inch sides, and so on.

Next, imagine that you unpack the boxes and place them in a row. Thus, the length of the row is one inch, plus two inches, plus three inches, etc. Draw a square whose sides equal the length of this row.

Finally, take a great imaginary leap. Suppose that you unfold each of the boxes, and lay it down flat. Will it cover the square? If you can solve this in your head, you're done. If not, write a program that solves the problem for any number (say, up to ten or fifteen) of Chinese boxes.

4-27 A chain letter is the surest route to popularity with friends, neighbors, and the Federal Postal Inspector. In effect, *n* people sign the letter; then each one must send the letter to *n* more people. Calculate, and print a table of, the number of letters that will be generated in ten generations of a letter with 1, 4, 7, and 10 signers.

4-28 In the year 1202, a man named Leonardo of Pisa described a sequence that begins 0, 1, 1, 2 and continues indefinitely: each new number is the sum of the prior two numbers. Leonardo is better known as *Fibonacci* (from *Filius Bonaccii*, or son of Bonaccio), and his numbers are now called the *Fibonacci sequence*. The sequence has been studied for hundreds of years; there's even a journal (the *Fibonacci Quarterly*) devoted to articles on it.

Write a procedure that generates values in the Fibonacci sequence. Then, use it to answer these questions:

a) What is the largest Fibonacci number you can generate on your system?

b) Fibonacci devised his sequence to describe the growth of a rabbit population. If each pair produces a new pair monthly (starting at the age of one month), and no rabbits die, you will indeed find you have one pair, then two, then three, then five, etc. How many rabbits will there be in a year?

c) The ratio between successive Fibonacci numbers comes closer and closer to the *golden ratio*. What is it?

d) Write a subprogram that takes the golden ratio you just calculated, raises it to the *n*th power, then divides by the square root of five. Round the result, and print it. Can you compare this result, for increasing values of *n*, to the *n*th Fibonacci number?

4-29 As everybody knows, Californians hew toward intellectual pursuits. Their allegedly excessive interest in suntanning is a vile slander. In fact, they only lie in the sun in order to free their minds for their first love, conversation.

Recently, however, concern has been growing about the possible hazardous effects of oversunning. There are really two issues of risk: daily exposure, and cumulative exposure. Although suntanning progressively reduces the danger of burning, the hazardous effects of exposure to the sun's rays accumulate day after day.

Write a program that a Californian might use to keep track of her daily tanning time, as well as the total time she's spent in the sun. Have it print a table that is based on the assumption that the she starts out with fifteen minutes in the sun, and increases her time in the sun by one minute each day. Its output should begin:

```
Day          Time    Total Time
  1            15            15
  2            16            31
  3            17            48
 . . .
```

Options: Let the user choose the total number of days, change the first day's time, or increase each day's increment. Convert the cumulative total to hours and minutes.

4-30 Although the list is diminishing, there are still a few things that cannot be done with a hand-held electronic stopwatch. For example, suppose that a race requires four laps of a track. A stopwatch is capable of recording and reporting on the time of each lap, called a *split*, as well as the total elapsed time to and through that lap, called the *cumulative time*.

However, a stopwatch cannot let us know the relative difference, in percentage terms, between one lap and the next. Nor can it tell us the difference between the current lap's split, and the average split run in the race so far. Finally, a stopwatch is utterly incapable of projecting, at each stage of the race, what the final time will be.

Write a program that takes care of these tasks, and proves the eminent superiority of an $80,000 computer over a $19.95 wristwatch. The user must supply the number of splits that will be entered. You must choose a format for time entries. *Options*: *a*) Whatever the input format is, print times in both seconds, and minutes and seconds; *b*) get a goal time along with the number of splits, and calculate the current deviation from an absolutely evenly paced race; *c*) properly handle special-case races that don't have uniform splits (e.g. the 3 and 3/4 lap 1500 meter race).

'When you embed debugging code, think of the Gauguin painting Where do we come from? What are we? Where are we going?' (p. 179)

Chapter 5 The **IF** Statement

Data goes into a program, and information comes out. So far, we've managed this impressive transformation by doing things *to* data. Chapter 5 introduces *choice*, which will let us develop information by learning about data instead.

Making decisions has two components. First, we need a way to phrase the choice. In Modula-2, as in most programming languages, the choice is posed as a comparison that tells us something about data. Do two values equal each other? Is one greater? Is it less? The answer is always either yes or no—or, in programming terms, one of the BOOLEAN values FALSE or TRUE.

Second, we need a way to put our decision into action. The **IF** statement takes care of this part of the job by automating the choice between alternatives. Oddly enough, though, we'll find that the more Modula-2 code we learn to write, the less mechanical our programs will seem to be when they run.

This chapter's Program Engineering Notes are longer than most. That's because I want to pay extra attention to correct usage of BOOLEAN values—they'll be very important to Chapter 6 as well.

Barbers, so they say, are paid less for their ability to cut hair than for their skill at *not* cutting. Now that you've had some fun employing **FOR** loops to burn up impressive amounts of CPU time, try using **IF** statements and BOOLEAN expressions to learn the barber's restraint.

Section 5-1
Strictly Modula-2

LET'S BEGIN WITH AN EXAMPLE OF THE **IF** statement:

the **IF** statement in action

```
MODULE Guidance;
    (*Provides a guide for dating in today's increasingly complex world.*)
IMPORT IO;
VAR Age: INTEGER;
BEGIN
    IO.StringOut ("How old is your friend?   ");
    IO.IntIn (Age);
    IF (Age < 18) OR (Age > 34) THEN
        IO.StringOut ("Uh oh ... are you sure this is appropriate?")
    END;
    IO.LineOut
END Guidance.
```

```
How old is your friend?  17
Uh oh ... are you sure this is appropriate?
```

Guidance illustrates both of this chapter's concerns. The shaded portion is, of course, an **IF** statement. The **IF** lets us pass over and ignore the parts of a program that we don't want to use, or explicitly select those parts that we do. In the example, we select an output statement (`Uh oh ...`) if the value of Age falls within certain bounds, and skip the statement otherwise.

Guidance also demonstrates the construction of BOOLEAN (boo´-lee-an) expressions. This little segment:

a BOOLEAN expression

```
(Age < 18) OR (Age > 34)
```

actually contains two kinds of expressions. First, there are *relations* between INTEGER values—the comparisons between Age and 18, and between Age and 34—that are either true or false. Second, the **OR** operator creates an expression by joining the relations. If either one of them is true, then the entire segment has the BOOLEAN value TRUE.

IF Statement Syntax

Although BOOLEAN expressions will occupy most of our attention for the next few pages, let's look at the **IF** statement. The **IF** statement makes three kinds of decisions: *yes/no, either/or*, and *which alternative*? While the idea that form should follow function is more often applied to architecture than to programming languages, the **IF** statement has a trio of syntactic forms because each has a slightly different application. The different outcomes are easy to see if I outline samples of the three.

The basic *yes/no* choice has the simplest syntax. An action that consists of one or more statements is taken if the BOOLEAN expression is TRUE, or is skipped entirely if it's FALSE.

a yes/no choice

```
IF BOOLEAN expression THEN
    sequence of statements
END
```

If the expression and action are very short—and I'm sure I won't be adding extra statements to the action—I'll sometimes put the whole thing on one line, like this:

```
IF BOOLEAN expression THEN statement END
```

The **ELSE** part allows an automatic *either/or* choice. If the statement's BOOLEAN expression is FALSE, the **THEN** action is skipped, and the **ELSE** action is taken instead.

an either/or choice

```
IF BOOLEAN expression
    THEN sequence of
           statements
    ELSE alternative sequence
         of statements
END
```

indent to show grouping

> When an action includes several statements, always deeply indent the entire sequence. It doesn't affect the statement's syntax, but it makes your (and my) intention plain to a program reader.

The third option requires the choice of one, and only one, action out of several alternatives. Adding **ELSIF** parts lets additional conditions be considered if the first one fails. Note that the conditions aren't necessarily *mutually exclusive*, though. Although more than one condition might be met, only the first TRUE condition will have its action taken.

a which alternative choice

```
IF condition 1
      THEN sequence of statements 1
    ELSIF condition 2
      THEN sequence of statements 2
    ELSIF condition 3
      . . .
    ELSE alternative sequence of statements
END
```

The final, shaded, **ELSE** part is optional. It provides an alternative sequence of statements that's taken if none of the explicitly stated conditions are met. As before, regular indenting is important to help make the statement clear.

Modification Problems

1. Can you modify Guidance so that its advice is reversed?
2. Try using **ELSIF** parts to make Guidance give a more specific 'too young' or 'too old' message that depends on the value of Age.

Self-Check Questions

Q. This assignment gives Finished a value that depends on the outcome of a BOOLEAN expression—either TRUE or FALSE:

```
Finished := Response = "Q";
```

Rewrite it as an **IF** statement.

A. The expression is the same, but the assignments are more explicit:

```
IF Response = "Q"
    THEN Finished := TRUE
    ELSE Finished := FALSE
END
```

Q. Are either of these statements legal? If so, what value do you think will be assigned to Question?

a)
```
IF Response = "A" THEN   ELSE Question := "Z" END
```

b)
```
IF Answer = "A"
        THEN Question := "T"
    ELSIF Answer = "A"
        THEN Question := "U"
    ELSIF Answer = "A"
        THEN Question := "V"
    ELSE Question := "W"
END
```

A. Both statements are legal, although neither is very sensible. The first statement assigns "Z" to Question if Response doesn't equal "A", and does nothing otherwise. A missing action doesn't have any effect, but its syntax is legal.

In the second statement, Question is assigned "T" if Response is "A", and "W" otherwise. Thus, only the first TRUE branch of the statement is followed. As I pointed out earlier, Modula-2 syntax doesn't require the alternative conditions to be mutually exclusive. However, it's *very* confusing and usually a semantic error when they're not.

The Relational Operators

Like all expressions, BOOLEAN-valued expressions represent values. The simplest examples are the constants, or values, of the type BOOLEAN:

FALSE TRUE

the BOOLEAN constants

The standard function ODD(x) also represents a BOOLEAN value—FALSE if its INTEGER or CARDINAL argument x is even, and TRUE if it's odd. We can state longer BOOLEAN expressions by using the *relational* operators.*

relational operators

Math	*Modula-2*	*English*
=	=	*equal to*
<	<	*less than*
≤	<=	*less than or equal to*
>	>	*greater than*
≥	>=	*greater than or equal to*
≠	<> *or* #	*not equal to*

BOOLEANs are assertions

Anybody who's taken high school algebra has probably seen these relations referred to as *inequalities*. Programming languages use the fact that inequalities always make an assertion, or claim, that has to be either true or false. This makes them BOOLEAN-valued expressions. For example, these are either TRUE or FALSE:

```
LowerLimit > 5        ApplicantsAge <= 65
```

They assert that a variable named LowerLimit has an INTEGER or CARDINAL value greater than 5, and that the value of ApplicantsAge is less than or equal to 65. Since character sets are ordered, these BOOLEAN expressions, which confirm or deny the alphabetical ordering or equality of CHAR values, are just as plausible as their numerical counterparts:

```
Finished = "Y"      "X" <> CHR(63)      SecondLetter >= FirstLetter
```

relational operands

> In stating relations, it's essential that both operands—the values that are being compared—have appropriately comparable types.

On occasion, type conversion functions will have to be called to allow comparison between numerical values of different types. You should be aware, though, that the vagaries of computer arithmetic can make comparisons unreliable if they depend on precise REAL representations. As an extreme example, the REAL value 3.0*(10.0/3.0) might equal 9.99999999... instead of 10.0.

Since every relational operator expects two operands, and every two operands expect just one operator, some common mathematical notation can't be used in writing programs. In particular, multiple inequalities like 5<X<10 have no meaning in Modula-2. We'll soon see how to express such relations properly; this particular example is correctly written as (5<X) **AND** (X<10).

---

* One other relational operator, **IN**, is discussed with sets in Chapter 15.

BOOLEAN Constants, Variables, and Assignments

Constants and variables also represent BOOLEAN values. Note that these constant definitions are not BOOLEAN expressions, even though '=' is a relational operator under other circumstances.

BOOLEAN constants and variables

```
CONST THISprogramWorks = FALSE;
      TESTING = TRUE;
VAR Balance: REAL;
    Heir, Broke: BOOLEAN;
    Temperature: INTEGER;
    ActivateDebuggingProcedures, NotDivisibleBy2: BOOLEAN;
```

Assignments to BOOLEAN variables are like all other assignment statements: an expression is evaluated, and its result value is given to the variable. The constants of type BOOLEAN (FALSE and TRUE), user-defined BOOLEAN constants (like TESTING), and BOOLEAN variables may all appear on the right-hand side of the assignment.

assignment to BOOLEANs

```
Heir := TRUE;
ActivateDebuggingProcedures := TESTING;
Broke := (Balance<=0.0);
NotDivisibleBy2 := ODD(Temperature);
Broke := Heir;
```

In practice, the constants TRUE and FALSE are only occasionally found in relational expressions. If you look closely, you'll see that the two assignments below are identical. Assuming that Willy and Nilly are BOOLEAN variables:

these are the same

```
Willy := Nilly;          (*is the same as ... *)
Willy := ( Nilly = TRUE );
```

The first assignment has the same effect as the second because Nilly has the same value as Nilly = TRUE.

Self-Check Questions

Q. Suppose that these variables have been initialized as follows:

```
Alpha := TRUE;   Ch := "D";   Int := 3;   Beta := FALSE;
```

What are the values of these expressions? Are any of them illegal because of type clashes?

a) Beta	*b*) Alpha = Beta	*c*) Int <> "B"
d) Ch > "E"	*e*) Beta <> FALSE	*f*) Ch <= "D"
g) 3 >= Int	*h*) Ch = Int	*i*) Alpha = ODD(Int)

A. Expressions *a*, *b*, *d*, and *e* are all FALSE, while *f*, *g*, and *i* are TRUE. Expressions *c* and *h* are illegal because they attempt to compare CHAR and INTEGER values.

The BOOLEAN Operators **AND**, **OR**, and **NOT**

Three 'word-symbols'—**AND**, **OR**, and **NOT**—are the BOOLEAN operators. Just as the word-symbol **DIV** takes INTEGER operands to form an INTEGER-valued expression, the BOOLEAN operators use BOOLEAN operands to create BOOLEAN-valued expressions. BOOLEAN operators can combine tests that might otherwise require a series of **IF** statements. **AND** joins two conditions.

the **AND** operator

> If we have two BOOLEAN values (call them Maybe and Perhaps), the expression:
>
> Maybe **AND** Perhaps
>
> is evaluated as TRUE if both Maybe and Perhaps are TRUE. If either or both of them are FALSE, the entire expression is FALSE.

AND can appear in any BOOLEAN expression. It's frequently used to help set lower and upper boundaries on a value, e.g. as it does in this **IF** statement:

```
IF (Value>=5) AND (Value<=10) THEN
    IO.StringOut ("The value is between 5 and 10, inclusive.")
END
```

AND can also be part of an expression in an assignment:

```
Capital := (Letter>="A") AND (Letter<="Z");
```

The second operator, **OR**, yields a result of TRUE if one or the other (or both) of its operands has the value TRUE. This English sentence: *If I do well on the midterm, or ace the final, then I'll pass*, is a perfect example of '**OR**ing' two BOOLEAN values:

IF *(I do well on the midterm)* **OR** *(I ace the final)* **THEN**
    I'll pass
END

the **OR** operator

> The **OR** operator is less restrictive than **AND** is. The expression:
>
> Maybe **OR** Perhaps
>
> is TRUE if either Maybe or Perhaps, or both of them, are TRUE. It's only FALSE if Maybe and Perhaps are both FALSE.

For example, this program segment:

```
IF (PurchasePrice<=BankBalance) OR CreditIsGood THEN
    IO.StringOut ("Who should I make the check out to?");        etc.
```

lets a check be written if there's enough money in the bank to cover the purchase (PurchasePrice<=BankBalance), or if the value of the BOOLEAN variable

CreditIsGood is TRUE. Naturally, it's all right for both conditions to be TRUE, too.

The third BOOLEAN operator, **NOT**, reverses a BOOLEAN condition.

the **NOT** operator

> The **NOT** operator's result is the opposite of its operand. Thus,
>
> **NOT** Condition
>
> represents TRUE if Condition is FALSE, and FALSE otherwise.

For example:

```
IF NOT ODD(InputValue) THEN
   IO.IntOut (InputValue, 1);
   IO.StringOut ("  couldn't possibly be a prime number.")
END
```

Although we won't be using them, there are substitute characters for two of the BOOLEAN operators: & can take the place of **AND**, and ~ can be used instead of **NOT**.

Self-Check Questions

Q. Write each of these English statements as a BOOLEAN expression. Naturally, you'll have to make up some variable names as you go along.

a) I have more than enough money.
b) A equals 6, but B neither equals A nor 7.
c) 18 ≤ my last date ≤ 34
d) My age isn't *between* 20 and 30.
e) My IQ is not an odd number.
f) My middle initial is either "7" or "B", and I have fewer than five pimples.
g) A letter isn't a capital.

A. There are, of course, many ways to express the same conditions:

a) Money > Enough
b) (A = 6) **AND** (B <> A) **AND** (B <> 7)
c) (MyLastDate >= 18) **AND** (MyLastDate <= 34)
d) (Age <= 20) **OR** (Age >= 30)
e) **NOT** ODD(IQ)
f) ((Initial = "7") **OR** (Initial = "B")) **AND** (Pimples < 5)
g) (Letter <= "A") **OR** (Letter >= "Z")

Evaluating BOOLEAN Expressions

As you've seen in the examples, I'll generally use parentheses within BOOLEAN expressions. As a basic rule, equalities and inequalities must be parenthesized when **NOT**, **AND**, or **OR** appear in an expression. Parentheses *must* be used in these examples:.

parentheses are needed here

```
NOT (Key="T")
(Voltage=110) AND (Amperage<10)
(Limit<5) OR (Limit>=10)
(Temperature>80) AND Sunny
```

Operator precedence is the reason. All of the BOOLEAN operators have higher precedence than the relational operators, so A>B **OR** C>D is misinterpreted by the computer as A> (B **OR** C) >D, which is meaningless.

Parentheses may also be required to circumvent the regular precedence of the BOOLEAN operators.

BOOLEAN precedence

> The order of precedence in the BOOLEAN operator hierarchy is: **NOT**, **AND**, **OR**. **NOT** has the most precedence, while **OR** has least. Among terms of equal precedence, evaluation goes from left to right.

Suppose we want an action to take place if Maybe and Perhaps are both FALSE. This statement:

```
IF NOT Maybe AND NOT Perhaps THEN    etc.
```

does the job. These expressions:

```
NOT Maybe AND Perhaps
NOT Maybe OR Perhaps
```

might sound good in English, but their effect in Modula-2 is unexpected—only Maybe is being **NOT**ed. In fact, I'll give the unusual advice that you forget about the relative BOOLEAN operator precedences entirely. When **NOT**, **AND**, or **OR** appear in BOOLEAN expressions, it's good programming practice to use parentheses as internal documentation, even if they don't affect the expressions' value. This expression uses the smallest legal number of parentheses—zero:

always use parentheses

```
IF NOT Hot AND Humid OR Raining THEN    etc.
```

But this version is self-documenting and unambiguous:

```
IF ((NOT Hot) AND Humid) OR Raining THEN    etc.
```

partial evaluation

> The evaluation of a BOOLEAN expression ceases as soon as the expression's value is known.

Thus, an expression like P **AND** Q doesn't have to be fully inspected if the left operand—P—is FALSE. No matter what the value of Q is, the expression's value will be FALSE. Similarly, the evaluation of P **OR** Q will end prematurely if P is TRUE—the expression will be TRUE even if Q is FALSE.

The Program Engineering notes contain additional hints on using *identities* and *truth tables* to simplify complicated BOOLEAN expressions. Pitfalls of partial evaluation of BOOLEAN expressions are discussed there as well.

Self-Check Questions

Q. Assume that First, Second, Third, and Last are all BOOLEAN values. Which of these expressions state the same condition?

a) **NOT** First **AND NOT** Last
b) First <> Last
c) (First **OR** Second) **AND** Third
d) First = Last
e) **NOT** First **OR NOT** Last
f) (Third **OR** First) **AND** (Second **OR** Third)
g) **NOT** (Last **OR** First)
h) **NOT** (Last **AND** First)
i) (First **AND NOT** Last) **OR** (Last **AND NOT** First)
j) (First **AND** Second) **OR** Third
k) (First **AND** Last) **OR** (**NOT** Last **AND NOT** First)
l) (Third **AND** First) **OR** (Second **AND** Third)

A. Note that even when two expressions mean the same thing, one may be a whole lot clearer than the other. The answers are $a = g, b = i, c = l, d = k, e = h, f = j$.

Section 5-2 Standard Practices

IN CHAPTER 4 I SUGGESTED THAT YOU VIEW THE **FOR** statement as a means of controlling or generating *data*, and not simply as a mechanical repeater of *actions*.

It's useful to look at the **IF** statement in the same light—by considering its effect on values as well as statements. The **IF** statement uses a variety of techniques to control data:

IF as a data controller

a) *Selecting* a particular value or kind of value from a flow of data—a dollar sign, say, or numbers that can be divided by 42.

b) *Selectively updating* counters or other variables—counting the number of consonants versus vowels, or the storing the smallest value seen so far.

c) *Routing* data to the right part of a program—for instance, making procedure calls and passing along arguments in response to interactive user commands.

d) *Error-checking* data—making sure that it fits within certain boundaries, and fixing, replacing, or rejecting values that don't.

Keep this alternative view in mind as you read through a few examples of the **IF** statement in its basic forms: making yes/no *decisions*, either/or *choices*, or more complicated multiple-alternative *selections*. Later on, you might want to read through the chapter exercises just to get a feel for how the **IF** statement will be applied in each case.

Simple Checks: Yes or No

Our first problems poses a series of yes/no choices.

problem: dividing toys

> Ruth and Jim are modern parents who are determined to turn their home into Mr. Rogers' Neighborhood. To teach their Kristin and Heather to share, only toys that can be divided equally (marbles, toy soldiers, and the like) are allowed in the house.
>
> The problem is that Kristin and Heather have friends. Young Aphrodite from around the corner could show up, or Peace, Love, and their obnoxious little brother Happiness from next door, or they all might descend at once. Write a program that Jim and Ruth can use to find out if a particular toy that has many pieces might not be evenly divisible among some combination of kids.

For once I've managed to pose a problem that's going to seem much simpler if I phrase it mathematically! How many combinations of kids are there? There will be either two, three, five, or six. How can we decide if *n* can be evenly divided two, three, five, or six times? With the **MOD** operator. There's no remainder from, say, a two-way division if *n* **MOD** 2 equals zero. We can use the same approach to check out three, five, and six-way divisions.

Suppose that we pseudocode the steps of a program:

pseudocoding the program

get the number;
IF 2 *evenly divides the number, say so;*
IF 3 *evenly divides the number, say so;*
IF 5 *evenly divides the number, say so;*
IF 6 *evenly divides the number, say so;*

It's important to recognize that these statements are independent. The fact that 2 evenly divides a number doesn't disqualify 3 from working just as well. Similarly, even though any number divisible by 6 can also be divided by 2 and 3, we *still* have to check 2 and 3 separately. Each **IF** is considered in turn, regardless of what went before, or what will follow.

The completed program is shown below. As you read it, try to mentally develop some test data that will show up any shortcomings in MrRogers.

(*program follows*)

a yes/no example program

```
MODULE MrRogers;
   (*Determines if an input number can be divided by 2, 3, 5, and/or 6.*)
IMPORT IO;
VAR TheNumber: INTEGER;
BEGIN
   IO.StringOut ("How many toy pieces are there, Jim & Ruth?   ");
   IO.IntIn (TheNumber);
   IO.StringOut ("Socially acceptable combinations are:  ");
   IF (TheNumber MOD 2) = 0 THEN IO.IntOut (2,2) END;
   IF (TheNumber MOD 3) = 0 THEN IO.IntOut (3,2) END;
   IF (TheNumber MOD 5) = 0 THEN IO.IntOut (5,2) END;
   IF (TheNumber MOD 6) = 0 THEN IO.IntOut (7,2) END
END MrRogers.
```

```
How many toy pieces are there, Jim & Ruth?  18
Socially acceptable combinations are:  2 3 6
```

Modification Problems

1. Suppose that Jim and Ruth decide to become their children's best friends by including themselves in all games. What additional numbers have to be checked? Modify MrRogers to handle them.

2. The most obvious shortcoming of MrRogers is its lack of meaningful output if a number isn't divisible by any of the trials. After you read about **ELSE** parts, below, rewrite MrRogers so that it makes a special announcement if a toy is totally unacceptable.

3. MrRogers gets its data externally, from Jim or Ruth. Modify it to generate data internally (the INTEGERs 1 through 25 are sufficient), and print a table of its results.

Self-Check Questions

Q. The fact that **IF** statements are independent doesn't mean that they can't affect each other. For example, consider the two pairs of **IF** statements below. Rewrite each pair as a *single* **IF** statement without an **ELSE** part. Simplify the final assignments if you can.

a)
```
IF (Time >= 5) THEN Time := Time * 4 END;
IF (Time >= 20) THEN Time := Time DIV 2 END;
```

b)
```
IF (Age > 9) THEN Age := 7 + (3 * Age) END;
IF (Age > 36) THEN Age := 5 - (Age DIV 3) END;
```

A. Tricky, huh? A test case or two will show you that if the first statement's condition is met, the second one will also be met. If the first statement is skipped, so is the second.

a) `IF Time >= 5 THEN Time := 2 * Time;`

b) `IF Age > 9 THEN Age := 3 - Age;`

Making Choices: Either/Or

If the number supplied to MrRogers wasn't divisible by any of the checks, MrRogers' output would have looked strange, but the program would have still run. However, data will sometimes require a choice between two different courses of action. Consider this problem:

problem: maximum weights

> The United States Postal Service ships first class mail at a rate of $2.40 for the first two pounds, with roughly 42.4¢ for each additional pound. Unfortunately, there is a limit of seventy pounds per package. Write a program that asks for a weight in pounds, and returns either a postage price, or the bad news about overweight packages.

This time we'll use an **IF** statement with an **ELSE** part to create an either/or choice—either we calculate postage, or we print a message.

pseudocode

```
get the package's weight;
IF it is seventy pounds or less
    THEN calculate the postage rate;
         print it
    ELSE print an 'overweight' message
END
```

What if the program user enters a weight that's too light—less than the minimum, or even accidentally negative?

robustness

> Using **IF** statements to *error-check* input helps make programs *robust*—less sensitive to user errors or misuse. They end gracefully instead of crashing.

I've put a check for underweight packages in the final program.

As you can see, program Mailer, below, is slightly more complicated than the pseudocode draft. It uses an *auxiliary* or helping variable, Mailable, to record a bit of information—the package's can be mailed/can't be mailed status. If and when you try modifying Mailer, you'll find that auxiliary variables are a convenient defensive programming measure: checking the BOOLEAN condition is separated from acting on it.

> Clear formatting is crucial in this program. Although syntax determines semantics—spacing and indentation are for the benefit of human program readers, and not the computer—it's the programmer's duty to make her code readable. Format statements in a way that makes the alternative nature of the actions easy to see.

(*program follows*)

an either/or example program

```
MODULE Mailer;
  (*Calculates package postage. Checks for overweight packages.*)
IMPORT IO;
CONST EXTRApounds = 0.424; (*The per additional pound rate.*)
      BASErate = 2.40;     (*The two-pound minimum rate.*)
      MINIMUM = 2;         (*The minimum package weight.*)
VAR Mailable: BOOLEAN;
    Weight: INTEGER;
    Postage: REAL;
BEGIN
  IO.StringOut ("How much does the package weigh?"); IO.LineOut;
  IO.StringOut ("Round up to nearest pound.  ");
  IO.IntIn (Weight);
  IF Weight < MINIMUM THEN Weight := MINIMUM END;
  (*Regardless of input, we have a positive Weight value.*)
  Mailable := (Weight <= 70);
  IO.StringOut ("Your package  ");
  IF Mailable
    THEN Postage := (FLOAT (Weight-MINIMUM) * EXTRApounds)
                    + BASErate;
         IO.StringOut ("can be shipped for $");
         IO.FixOut (Postage, 3, 2)
    ELSE IO.StringOut ("is too heavy to ship.");
         IO.LineOut;
         IO.StringOut ("Better lug it yourself.")
  END; (*IF*)
  (*We've printed one of the messages.*)
  IO.LineOut;
  IO.StringOut ("Thanks for using this program.")
END Mailer.
```

```
How much does the package weigh?
Round up to nearest pound. 975
Your package is too heavy to ship.
Better lug it yourself.
Thanks for using this program.
```

Modification Problems

1. Transform program Mailer into program SquareRoot—have it ask for a number, then print its square root. Be sure to print an error message if the number is negative.

2. Suppose that Ruth and Jim (from the previous example) are using program Mailer to ship tightly sealed packages that they claim are toys. Since their packages can be subdivided into seventy-pound parcels—a two hundred pound shipment can be repacked as two 70 and one 60 pound packages—they're not worried about the packages being overweight. Modify Mailer so that it lets them know how many packages there will be, and what total postage is.

Self-Check Questions

Q. What's wrong with beginning the main **IF** statement in program Mailer like this?

```
IF (Mailable = TRUE) THEN . . .
```

A. As a BOOLEAN variable, Mailable represents either the value TRUE or FALSE. The rewritten statement is redundant; the computer is forced to go through the unenlightening exercise of determining that TRUE=TRUE is TRUE, or that FALSE=TRUE isn't.

Q. Rewrite this pair of **IF** statements as an **IF** statement with an ELSE part.

```
IF (Time < 37) THEN Time := Time DIV 4;
IF (Time > 36) THEN Time := 3 * Time;
```

A. Here, a careful inspection reveals that if the first statement is entered, the second will be skipped. The rewritten pair is:

```
IF (Time < 37)
   THEN Time := Time DIV 4
   ELSE Time := 3 * Time
END
```

Choosing Between Alternatives

alternative choices

The third **IF** statement form presents a choice between several alternatives. Since a single BOOLEAN expression can have only two different outcomes, each **ELSIF** part requires an additional BOOLEAN test. However, only one of the tests will be passed, and only one in the series of alternatives can be executed. A closing **ELSE** part, which is optional, can be used to provide a final alternative if *none* of the conditions are met.

It is useful to recognize that your knowledge about program data *increases* as a program works its way through a sequence of **ELSIF** parts. Think back to our introductory example, Guidance. Suppose that I wanted to issue distinct messages for each of three possibilities—a potential date is too young, too old, or just right. If our BOOLEAN checks are ordered correctly, only two tests are needed:

even failed tests increase knowledge

```
IF (Age < 18)
      THEN IO.StringOut ("Too young for your worldly ways.")
   ELSIF (Age > 34)
      THEN IO.StringOut ("Too old for your limited experience.")
   ELSE IO.StringOut ("Just right for a journey of mutual discovery.")
END
```

After the first two tests fail, we know that Age must be between 18 and 34, inclusive.

An application might require more individual tests, of course. Consider the problem of deciding if a single CHAR value is a letter, a digit, a punctuation mark, or something else. We find the answer by asking a sequence of questions, phrased as BOOLEAN expressions. In the example below, only one of the expressions can be TRUE, since the CHAR won't fall into more than one category.

choosing one alternative

```
read a CHAR value;
IF the value is a letter
      THEN count it as a letter
   ELSIF it is a digit
      THEN count it as a digit
   ELSIF it is a punctuation mark
      THEN count it as a punctuation mark
   ELSIF it is a space or tab
      THEN count it as a separator
   ELSE
      it didn't fall into any of the categories
END
```

The sequence will be left as soon as a test is passed, or a condition is met. The shaded **ELSE** part in the pseudocode sequence above provides an alternative to the entire sequence of **ELSIF** parts—a default that is met if all of the earlier tests fail. In this case, the only leftover values will be the unprintable control characters. We could count them, or leave the **ELSE** portion off entirely.

> The **ELSIF** form is efficient from the computer's point of view, because unnecessary checks can be skipped. More importantly, though, it gives the same message to a human program reader—that only one of several alternative actions is intended to be taken.

Modification Problems

1. Modify the first code segment to decide if two letters are in alphabetical order, reverse order, or are identical.

Self-Check Questions

Q. Rewrite this sequence using an **IF** statement with **ELSIF** parts. Your goal is to achieve the same effect with as little effort as possible—you may want to change the order of the checks. What range does this code assume the value of Grade falls into?

```
IF Grade > "F" THEN
    IO.StringOut ("You are in a league by yourself.")
END;
IF (Grade = "D") OR (Grade = "E") THEN
    IO.StringOut ("Not too good.")
END;
IF (Grade >= "F") THEN
    IO.StringOut ("Looks like you get to go again!")
END;
IF (Grade >= "A") AND (Grade <= "C") THEN
    IO.StringOut ("Not too bad.")
END;
```

A. Note that reordering was useful. Both schemes assume that no grade was 'less,' in terms of ASCII character ordering, than "A".

```
IF (Grade <= "C")
        THEN IO.StringOut ("Not too bad.")
    ELSIF Grade <= "E"
        THEN IO.StringOut ("Not too good.")
    ELSIF Grade = "F"
        THEN IO.StringOut ("Looks like you get to go again!")
    ELSE IO.StringOut ("Looks like you get to go again!");
          ("You are in a league by yourself.")
END;
```

Section 5-3
Extended Examples

LET'S TURN TO A FEW EXTENDED EXAMPLES OF THE **IF** statement. As usual, you will find that learning a new statement is a little like learning a new verb in a foreign language—it's sometimes easier to memorize conjugations than it is to drop the new word casually in a sentence.

Here, we'll consider examples that give us three different angles on problem solving and programming. Our first program, Position, relies on a BOOLEAN function to keep an **IF** statement readable. We'll take an unfortunately complicated expression out of the statement entirely, and replace it with a subprogram—an essential technique if you want to keep hard programs comprehensible.

Program two, Robbery, joins the **IF** statement to the **FOR** loop to create a problem solving engine of rather impressive power. We'll engage in an *exhaustive search* of a problem's *solution space*, which is to say that we'll create the programming equivalent of the Prince's search for Cinderella. Instead of trying on shoes, though, we'll generate and test every possible answer as we search for a fit.

Program three, Coins, is the longest of all. Although the problem it solves is easy in mathematical terms—it just makes change—it will be the hardest to program because I'll insist on *style*. We'll see the **IF** statement at work in every possible form.

Problem: Self-Describing Letters

Even if you hate numbers you can still love words. After reading our next example, you may wish to declare a temporary truce with the digits. The problem is:

problem: self-describing letters

> The word 'abode' is interesting because its first two letters, 'ab,' and its last two letters, 'de,' occupy the same position in the word as they do in the alphabet. The word 'cache' is in the same league, because it has two letters (the third and fifth) in identical word/alphabetical ordering position.
>
> Write a program that reads a word and lets you know how many letters are *self-describing*—i.e. how many letters are found in their alphabetical positions?

In the past, I've suggested that a good first step in any pseudocode is to think of, then outline, the solution's ultimate *goal*—its central procedure call, complete with parameters. Here, the central call is to a function. Suppose that we could rely on this BOOLEAN-valued function:

goal pseudocode

```
(*What I'd ultimately like to know is . . .*)
do they match? ( current position, current letter )
```

If I had such a function, I could use it in an **IF** statement:

expanding the pseudocode

```
IF they match ( current position, current letter ) THEN
    make the happy announcement
```

Indeed, if I had such an **IF** statement, I could use it in a loop that got the letters:

putting the pseudocode in context

```
FOR each letter in the word DO
    read a letter;
    IF they match ( current position, current letter ) THEN
        make the happy announcement
    END (*of the IF*)
    (*We have identified a self-describing letter.*)
END (*of the FOR*)
```

We seem to be doing remarkably well! As is usually the case, starting with a clear statement of our goal pointed the way to a program.

use BOOLEAN functions

> BOOLEAN functions expand our informal programming vocabulary. By making the flow of a program plain, they let Modula-2 retain some of the clarity of pseudocode. Use mock function calls as a design tool in pseudocoding, and program with them whenever complex BOOLEAN conditions have to be stated.

How can we implement the function? Well, it returns the result of a comparison that's either TRUE or FALSE:

function pseudocode

RETURN (*the current position in the word*) =
(*the relative position of the letter in the alphabet*)

Try to figure out how it's done before you read the final code of program Describe, below.

letter-position checker

```
MODULE Describe;
   (*Finds out how long a word is. Then, reads the word and reports on
    any letters whose alphabetical value equals their position in the word.*)
IMPORT IO;
VAR Length,                   (*of the word that's being checked*)
      Position: INTEGER;      (*in the word as we read it*)
      Current: CHAR;
PROCEDURE TheyMatch (Place: INTEGER;
                          Letter: CHAR) : BOOLEAN;
   (*TRUE if Letter is in alphabetical position Place.*)
VAR PositionInAlphabet: INTEGER;
   BEGIN
      PositionInAlphabet := 1 + ( ORD (Letter) - ORD ("a") );
      RETURN PositionInAlphabet = Place
   END TheyMatch;
BEGIN
   IO.StringOut ("How many letters are in your word?    ");
   IO.IntIn (Length);
   (*We know how much input data there will be.*)
   IO.StringOut ("Ok, type in the word.    ");
   FOR Position := 1 TO Length DO
      IO.CharIn (Current);
      IF TheyMatch (Position, Current)
         THEN IO.CharOut (Current)
         ELSE IO.CharOut ("*")
      END  (*IF*)
      (*We've indicated if the letter was self-describing or not.*)
   END; (*FOR*)
   IO.LineOut
END Describe.
```

(output follows)

```
How many letters are in your word? 10
Ok, type in the word. leadership
***de**hi*
```

Modification Problems

1. Give Describe the ability to check more than one word. Test cases you might want to try include **`predefinition`**, **`academician`**, and **`archetypical`**.
2. Turn Describe into program NonDescriptive—a program that finds input *digits* that *aren't* in the positions they describe. Thus, the input **`132476`** would produce as output `*32*7*`.

Problem: Stolen Gold

Program Describe obtained its data from the program user. However, as we well know, we can use a **FOR** statement to generate data internally. Let's turn to a problem that requires us to do so—the Mystery of the Stolen Gold Shipment.

problem: stolen gold

> Three desperados stole a shipment of gold bars late one night. After escaping to their hideout, they resolved to divide the booty in the morning and went to bed. However, as soon as one of the bandits heard the others snoring, he split the stolen gold into three equal piles, and found one bar left over. He buried one of the three piles under a tree, along with the extra bar. Then he went to sleep, sure that he had protected his interest in the treasure. Naturally, the other two outlaws were no more honest than the first. Each in turn crept to the cache of gold, divided it three ways, and found one bar left over, which he kept along with 'his' third.
>
> Soon came morning and the final three-way division. Oddly enough, this division also left one odd bar remaining. The highwaymen fought over this bar, and in an unprecedented three-way draw, shot each other dead.
>
> The problem I pose is this: Each of the four three-way divisions left exactly one bar. How many bars could have been in the entire shipment? Assume that the shipment contained no more than 500 bars.

We can see if any particular number solves the stolen gold mystery by mimicking the action of the desperados. If, after the repeated divisions and subtractions we have a remainder of 1, then it's a solution. However, we don't just want one solution—we want all of them.

solution space

A problem's *solution space* is its entire set of possible answers. An *exhaustive search* of the solution space means checking each and every potential solution.

I might pseudocode a solution to the stolen gold problem as:

an exhaustive search pseudocode

> **FOR** *every potential solution from* 1 *through* 500
> *test the solution to see if it solves the mystery*

Exhaustive searches are the opposite of the elegant solutions programmers usually pride themselves on finding. Instead, exhaustive search is the classic *brute force* method of problem solving—solving a problem by overpowering it.

brute force

Solving a problem by brute force isn't exactly what it sounds like—we don't write the problem down on a piece of paper, and then stomp on it until an answer crawls out. Instead, brute force implies that we're going to repeat an unsophisticated solution step many, many times. Think about the Count of Monte Cristo digging his way out of the Chateau d'If, one spoonful at a time for fourteen years. *That's* brute force.

Brute force sounds like a nasty way to attack problems, to be eschewed in favor of elegance whenever possible. Sure—on paper. But in the real world, brute force solutions are sometimes the best because they're the easiest to write quickly and correctly. They may waste a computer's time, but who cares? They save ours.

Just because you're going to rely on brute force doesn't mean that you can't be clever, though. Let's reconsider the stolen gold bars. Do we really have to work our way through all the divisions of every number in the solution space? No. We can make the search a little bit less exhausting by short-circuiting any failed test. A sequence of nested **IF** statements is our tool:

a less exhausting search

> **FOR** *every potential solution from* 1 *through* 500
>   **IF** *the first bandit's division leaves a remainder of* 1 **THEN**
>   **IF** *the second bandit's division leaves a remainder of* 1 **THEN**
>   **IF** *the third bandit's division leaves a remainder of* 1 **THEN**
>   **IF** *the final division leaves a remainder of* 1 **THEN**
>   *we've got a possible answer*

The **FOR** statement generates a series of candidate numbers. Nesting the **IF** statements as I have forms a connected sequence of questions. A wrong answer—a failed **IF** statement—acts as a short circuit, and halts further testing. This usage is common for minimizing the testing of large amounts of data.

Picture what will happen when a program based on this pseudocode runs. Suppose our trial number passes the first division. If it fails the second division, the third and final trials are not made. Thus, nesting the statements abbreviates the loop. Only numbers that pass each test will be considered in subsequent tests, and only numbers that pass every **IF** test are printed as solutions. You'll find that the confirmation comments are particularly helpful as you read program Robbery, below.

(*program follows*)

```
MODULE Robbery;
  (*Demonstrates nested IF statements in an exhaustive search.*)
IMPORT IO;
VAR TrialNumber, DividedNumber: INTEGER;
BEGIN
  FOR TrialNumber := 1 TO 500 DO     (*Generate a series of trials.*)
    IF (TrialNumber MOD 3) = 1 THEN  (*First desperado.*)
      DividedNumber := 2*(TrialNumber DIV 3);
      IF (DividedNumber MOD 3) = 1 THEN  (*Second desperado.*)
        DividedNumber := 2*(DividedNumber DIV 3);
        IF (DividedNumber MOD 3) = 1 THEN  (*Third desperado.*)
          DividedNumber := 2*(DividedNumber DIV 3);
          IF (DividedNumber MOD 3) = 1 THEN
            IO.IntOut (TrialNumber,3);
            IO.StringOut ("  is a solution.");   IO.LineOut
          END(*We've selected a solution.*)
        END  (*We've dropped out because the third desperado failed.*)
      END  (*We've dropped out because the second desperado failed.*)
    END  (*We've dropped out because the first desperado failed.*)
  END  (*FOR loop*)
END Robbery.
```

short-circuiting **IF** statements

```
 79 is a solution.
160 is a solution.
241 is a solution.
322 is a solution.
403 is a solution.
484 is a solution.
```

Problem: Making Change

The **IF** statement is a fundamental tool of programming that is liable to show up in every possible permutation of nesting and sequence. Let's work a final problem that requires all three **IF** statement applications—to make yes/no, either/or, and straight alternative choices.

problem: making change

> Write a change-making program. It should accept as input a price and amount of money tendered, then print the minimum number and type of coins required for change.

Who hasn't made change? First, you count out the dollars, then the half-dollars, and so on through the pennies. Aside from the detail of correctly ordering the coins, the change-making algorithm of program Coins isn't going to be a great advance in either computer science or mathematics. Nevertheless, a pseudocode outline makes it clear that **IF** statements will be employed extensively, and that no simple rule can govern their appearance:

pseudocode

```
find out the price and amount tendered;
decide if there's enough money;
IF there's no change
    THEN say thanks
    ELSE
        IF there are dollars in the change, return them;
        IF there are half-dollars in the change, return them;
            . . .
        IF there are pennies in the change, return them;
```

where *return them* implies a procedure call that contains a sequence like:

a coding detail

```
print the number of coins;
IF we're dealing with dollars
        THEN print the word 'dollar'
    ELSIF we're dealing with half dollars
        THEN print the words 'fifty-cent piece'
            . . .
    ELSIF we're dealing with pennies
        THEN print the word 'cent'
END;
deal with the possibility of a plural;
```

Making Change with Style

In *The Psychology of Computer Programming*, Gerald Weinberg warns that reading programs is a dying art. 'Just as television has turned the heads of the young from the old-fashioned joys of book reading,' he says, 'so have terminals and generally improved turnaround made the reading of programs the mark of a hopelessly old-fashioned programmer.'*

Unfortunately, writing programs, like writing literature, can sometimes only be learned by reading. A list of prescriptions for 'literate' programs I could set down might impress me and your teacher, but it's not nearly as helpful as a good clear piece of code.

Why does a program need to be written with style? Because programming is hard. Doing an arithmetic calculation for change is easy, but that's not our problem. Instead, we face the challenge of imitating the fine steps a human change-maker takes almost automatically. A real-life clerk would inform the customer if she were shortchanged, and so should our program. A person doesn't think about pluralizing words, but we have to teach the computer to say the

* Gerald Weinberg, *The Psychology of Computer Programming*, Van Nostrand Reihnhold 1971, pg. 6.

names of coins, and to add an 's' when it's needed for a plural. A human wouldn't bother announcing the coins that she *wasn't* returning as change, and neither should Coins.

How can style be incorporated into a program? Coins should also be as well-written as possible, following the program engineering guidelines that apply to every program:

engineering program Coins

- Its output should be clear.
- It should be robust, and able to deal with unexpected situations, like being shortchanged.
- It should be self-documenting where possible, but comments should be added to clarify less-than-obvious features.
- It should take advantage of procedures to minimize the length and complexity of its code.
- Indenting should be regular enough to help the eye automatically bracket the extent of nested statements.

It must also use statements—even the humble **IF** statement—correctly. As you read the code of program Coins, below, try to identify the different **IF** applications, as well as their relation to each other.

For example, does the order of statements, or parts of statements, ever matter? The **IF** statements in the main program of Coins are basically independent of each other. Since the statements are not nested, each statement's BOOLEAN expression is evaluated regardless of the previous statement's effect. Nevertheless, the statements must be correctly ordered, so that larger coins will be removed before smaller ones.

In contrast, the **IF** statements in procedure GetChange rely on **ELSIF** parts. Since the tests are alternatives, and only one can be passed, their order doesn't really matter in this application. Once one of the BOOLEAN conditions is met, the remaining tests will be skipped.

Try to subject Coins to a critical reading. What could you have done differently, and better? Is its division into procedures and functions clear, or would you have drawn the lines elsewhere? Are its names self-explanatory? Does it make good use of constants? Do its comments help you make the changes suggested in the Modification Problems?

Appraise the program. What are its shortcomings? Does it always work? Is it robust? Does it check for, and recover from, errors as well as it might? Could its output or interaction with the user be more attractive?

> Reading another person's program can be a difficult job. However, it can give you an insight into programming that can't be taught in class.

Learn to criticize your own programs by practicing on mine.

change-making program

```
MODULE Coins;
  (*Computes minimum coinage for making change.*)
IMPORT IO;
CONST DOLLAR = 100;
      HALFdollar = 50;
      QUARTER = 25;
      DIME = 10;
      NICKEL = 5;
      PENNY = 1;
VAR Price, Given: REAL; (*Input REAL dollar amounts ...*)
    Change: INTEGER; (*... but use INTEGER pennies internally.*)
    ChangeIsDue: BOOLEAN;
PROCEDURE ConvertToPennies (Price, Given: REAL) : INTEGER;
  (*Returns Price–Given expressed as a whole number of pennies.*)
  BEGIN
    RETURN TRUNC ( 0.5 + ( 100.0 * ABS ( (Price - Given) ) ) )
    (*Simulates the effect of a 'round' function by adding 0.5.*)
  END ConvertToPennies;
PROCEDURE GetChange (Unit: INTEGER; VAR Change: INTEGER);
  (*Prints number of coins. Reduces Change by that many Units.*)
  VAR Pieces: INTEGER;
  BEGIN
    Pieces := Change DIV Unit;
    Change := Change - (Pieces * Unit);
    (*We know how many coins to give, and what's still left over.*)
    IO.IntOut (Pieces, 1);
    IF Unit = DOLLAR (*Picks the proper action.*)
          THEN IO.StringOut ("  dollar")
       ELSIF Unit = HALFdollar
          THEN IO.StringOut ("  fifty-cent piece")
       ELSIF Unit = QUARTER
          THEN IO.StringOut ("  quarter")
       ELSIF Unit = DIME
          THEN IO.StringOut ("  dime")
       ELSIF Unit = NICKEL
          THEN IO.StringOut ("  nickel")
       ELSIF Unit = PENNY        (*Could this be a plain ELSE?*)
          THEN IO.StringOut ("  cent")
    END; (*Unit=DOLLAR IF*)
    IF Pieces>1 THEN         (*Takes care of multiple coins.*)
       IO.StringOut ("s")
    END;
    IO.LineOut
  END GetChange;
```

(program continues)

```
BEGIN (*Coins*)
  IO.StringOut ("What's the price?   ");
  IO.RealIn (Price);
  IO.StringOut ("How much did you get?   ");
  IO.RealIn (Given);
  (*We may, or may not, be given more than is actually due.*)
  ChangeIsDue := (Price > Given);
  Change := ConvertToPennies (Price, Given);
  (*Potential change is dealt with as INTEGER pennies.*)
  IF Change = 0
    THEN IO.StringOut ("Thanks!")
    ELSE          (*Could this be an ELSIF?*)
      IF ChangeIsDue
        THEN IO.StringOut ("Too little!  You're short by  ")
        ELSE IO.StringOut ("Your change is exactly  ")
      END; (*ChangeIsDue IF*)
      IF Change>=100 THEN GetChange (DOLLAR,Change) END;
      (*Remaining Change is less than one DOLLAR. The calls below are similar.*)
      IF Change>=50 THEN GetChange (HALFdollar,Change) END;
      IF Change>=25 THEN GetChange (QUARTER,Change) END;
      IF Change>=10 THEN GetChange (DIME,Change) END;
      IF Change>=5 THEN GetChange (NICKEL,Change) END;
      IF Change>=1 THEN GetChange (PENNY,Change) END
  END (*Change=0 IF*)
END Coins.
```

```
What's the price?  11.98
How much did you get?  3.79
Too little!  You're short by 8 dollars
1 dime
1 nickel
4 cents
```

Modification Problems

1. Modify Coins to make change in Italian lire. The basic banknote is the 1,000 lire bill, while coin denominations are 50, 100, 200, and 500 lire. What will you do about missing small change? Note: the singular of *lire* is *lira*.

2. The main program of Coins contains an **ELSE** part that is accompanied by the comment *Could this be an* **ELSIF**? Rewrite it that way.

3. Write a procedure MakeChange that contains all of the GetChange calls in Coins. Will the program require other changes in support of this modification?

Section 5-4

Program Engineering: Over-Engineering Software

IN CONTRAST TO THE STATEMENTS WE'VE USED previously, the **IF** statement lets us know something about what we're doing. Unfortunately, as we've seen before, bugs tend to enter at precisely the moment we start to believe that what we intended to do is what we've actually done.

From a program engineering point of view, we have two goals:

1. To make sure that what we *think* we know is really what we *do* know.
2. To make sure that the first goal holds for anybody else who happens to be reading our code.

In the past, we've used confirmation comments to help test our intentions against our accomplishments, and to let other program readers do the same. Now, let's look at some methods of building checks into the code itself. Our main tool will be *embedded debugging code*—test code that's built into the program *and left there*. First, though, I want to introduce a new tool—procedure HALT.

The HALT Procedure

The fact that a program is utterly dependent on its data doesn't mean that a programmer has to give up all control as well. She always has the opportunity to inspect data as it arrives, and to decide on an appropriate course of action—including halting the program.

For example, imagine a program that finds square roots. Nominally, it does just three things—gets a value, calls the MathLibØ.sqrt function, and prints a result. Strictly speaking, the program's job is to do something *to* its data.

However, we can rely on the **IF** statement to gain knowledge *about* program data, even when this isn't an explicitly stated part of the job of turning raw data into information and results. We have to do this before we can calculate a square root—we have to be sure that the program user gave us a positive number. Taking this step makes the program more *robust*—less likely to crash or, worse yet, produce incorrect results.

> Second-guessing the quality of program data is one of the key arts of program engineering. It's not that incorrect or improper data is hard to find. It is easy enough to spot negative numbers, or non-alphabetic input characters. Rather, it is hard to remember to be suspicious.

What can you do once you find flawed data? Let's consider three possibilities:

responses to flawed data

1. Note the error and proceed, possibly to an alternative process.
2. Fix the faulty data, or give the program user a chance to reenter it.
3. Announce the problem and quit.

We might use the first alternative in a square root program. In pseudocode, what we would do is:

1: take an alternative path

```
IF the number isn't negative
   THEN calculate its square root
   ELSE tell the user the facts of life as they relate to square roots
END
```

However, we could also follow the second alternative, perhaps like this:

2: fix the problem

```
IF the number is negative
   THEN Number := ABS(Number)
END;
calculate the square root;
```

In Chapter 6 we'll see how the **LOOP** statement can give the user a second chance to enter input.

The third alternative relies on a standard procedure called HALT.

> The standard procedure HALT immediately terminates program execution.

It's used like this:

3: terminate program execution

```
IF the number is negative
   THEN IO.StringOut ("Negative input.  Halting program.");
        HALT
   ELSE calculate the square root
END
```

HALT is only invoked in exceptional circumstances, when it becomes impossible for a program to continue correctly. Since the program stops immediately, it's usually a good idea to precede the HALT with a brief explanation of why the program is stopping. It may also be necessary to call a procedure like IO.LineOut to ensure that the explanation actually prints.

Embedded Debugging Code

If a program were an engineering project in the ordinary physical sense, the use of a fail-safe like HALT might be described as an intentional piece of *over-engineering*—making a structure strong enough to withstand unexpected loads or undetected weaknesses. But it is difficult to over-engineer software, and that has been one of the difficulties that programmers face. How can an extra-strength margin of safety be built into a program?

> Building debugging statements into your program before you need them, then leaving them there after you think you're done with them, is called *embedding* debugging code.

The classic bit of embedded debugging code relies on a constant declaration and an **IF** statement:

embedding debugging code

```
CONST DEBUGGING = TRUE;  (*Might be turned to FALSE later.*)
  . . .
IF DEBUGGING THEN
    IO.StringOut ("Debugging point 7.  Last input value was:    ");
    IO.CharOut (Letter);    etc.
```

When you embed debugging code, think of the Gauguin painting *Where do we come from? What are we? Where are we going?* The printed comment, whatever it is, must make two points: *Where am I*, and *What am I doing here?* Without these markers, the debugging output can be as difficult to decipher as the code itself.

A single BOOLEAN-valued constant like DEBUGGING doesn't give much control over the embedded code. An alternative approach is to number debugging points like this:

fine-tuning debugging

```
CONST DEBUGpoint = "A"   (*The point can be changed.*)
  . . .
IF "A" = DEBUGpoint THEN
    IO.StringOut ("Debugging point A.  Last input value was:    ");
    IO.CharOut (Letter);    etc.
```

Embedded debugging code may seem like an awful waste—it's put into a program before it is useful, and is left there after it isn't needed any more. Nor, frankly, do I set a good example for you in our sample programs—I don't embed debugging code here because it tends to make program volume blow up. Nevertheless, programming without planning for errors is (as Samuel Johnson said in reference to remarriage) an unfortunate triumph of hope over experience.

Boundary, Impossible, and Unavoidable Condition Bugs

Let's turn to some of the specific bugs that accompany the **IF** statement. In general, problems tend to arise when conditions that are easily stated in English must be translated into the much less forgiving world of BOOLEAN expressions.

Three kinds of errors are easy enough to check, even though they're not so easy to avoid entirely. A *boundary error*, in a BOOLEAN expression, is a bug that hinges right on the turning point between TRUE and FALSE. The expression is usually right, except for the close calls. An *impossible condition*, as the name implies, is an expression that can't be TRUE. An *unavoidable condition*, in contrast, can't be FALSE.

Although their effects can be very different, the root causes of these bugs can usually be traced to a single misplaced or misused symbol. Problems may be entirely accidental—a < slips in in place of a >—or they may reflect a flawed understanding of the difference between **AND** and **OR**. In any case, the moral for debugging is:

> Make a go/no go test of each **IF** statement. Be sure that the condition can be met, as well as failed. Take particular care to check the turning-point cases.

For example, consider the statements below.

```
IF (Age < 18) AND (Age > 34) THEN ...
    (*Impossible condition—how can Age meet both criteria?*)
IF (Age > 18) OR (Age < 34) THEN ...
    (*Unavoidable condition—Age can't help meeting one criterion.*)
```

In each case, the BOOLEAN operator is used incorrectly.

Boundary condition problems are usually due to a lack of care in choosing between < and <=, or > and >=. Very often, such problems arise when a statement is rewritten in order to reverse a condition. For example, this statement correctly specifies the range 18 through 34, including the ends:

```
IF (Age >= 18) AND (Age <= 34) THEN ...
```

This rewritten version, in contrast, introduces a problem:

```
IF (Age < 18) OR (Age >= 34) THEN ...
    (*Boundary condition bug—34 is unintentionally excluded.*)
```

As with the other problems, a little bench testing goes a long way. Here, the boundaries are the critical values around the test of Age—probably 17, 18 and 19, and 33, 34 and 35.

Sooner or later you will probably become convinced that your system's **IF** statement is broken. When that happens, it's time to check the values in BOOLEAN expression *before* you evaluate them:

```
(*Check the values before you use them.*)
IO.StringOut ('Age = ');    IO.IntOut (Age, 1);
IF (Age > ...
```

It often turns out, in programming as in life, that one of the things you're absolutely positively sure of is wrong. Frequently, the statement is failing because a variable wasn't properly initialized.

Simplifying BOOLEAN Evaluation

BOOLEAN expressions can seem hard to evaluate at first. Looking at them one term at a time helps bring them into perspective. The effect of the BOOLEAN operators can be summarized in the *truth table* shown below. It summarizes what happens when the operators are applied to various combinations of BOOLEAN operands:

NOT TRUE *is* FALSE
NOT FALSE *is* TRUE

truth tables

TRUE **AND** TRUE *is* TRUE
TRUE **AND** FALSE *is* FALSE
FALSE **AND** TRUE *is* FALSE
FALSE **AND** FALSE *is* FALSE

TRUE **OR** TRUE *is* TRUE
TRUE **OR** FALSE *is* TRUE
FALSE **OR** TRUE *is* TRUE
FALSE **OR** FALSE *is* FALSE

distributive laws

Sometimes BOOLEAN operator precedence makes an expression long enough to be unwieldy or confusing. Fortunately, the expression can usually be rewritten by following the *distributive* laws, as shown below. Assume that p, q, and r are BOOLEAN-valued expressions or variables:

(p **OR** r) **AND** (q **OR** r) = (p **AND** q) **OR** r
(p **AND** r) **OR** (q **AND** r) = (p **OR** q) **AND** r

De Morgan's laws

A similar set of relations is known as *De Morgan's* laws:

(**NOT** p) **AND** (**NOT** q) = **NOT** (p **OR** q)
(**NOT** p) **OR** (**NOT** q) = **NOT** (p **AND** q)

Partial Evaluation of Expressions

In the initial discussion of BOOLEANs, I pointed out that BOOLEAN expressions aren't always evaluated in full. Why not?

the basis of partial evaluation

By definition, Modula-2 requires that the BOOLEAN expressions (p **AND** q) and (p **OR** q) be evaluated from left to right like this:

p **AND** q *means* **IF** p **THEN** q **ELSE** FALSE **END**

p **OR** q *means* **IF** p **THEN** TRUE **ELSE** q **END**

The BOOLEAN outcome of these precise definitions is no different from the rules I gave earlier. However, their practical implication is that BOOLEAN expressions don't always have to be fully evaluated. In each case, it's conceivable that checking the value of the first operand alone will be enough to determine the value of the complete expression.

Making this distinction is important if fully evaluating a BOOLEAN expression could lead to a run-time error. For example, attempting to divide a number by zero is a common cause of crashes. An expression that checks the value of the denominator before carrying out the division avoids the problem:

relying on partial evaluation

```
(*Since evaluation stops after the first expression, this statement is correct.*)
IF (Denominator<>0.0) AND ((Numerator/Denominator)>Fraction) THEN
    IO.StringOut ("We have a lucky winner!")
END
```

By and large I won't rely on partial evaluation of BOOLEAN expressions to save the day. Here's why:

don't, in general, rely on it

> There are two reasons to shy away from reliance on partial evaluation of BOOLEAN expressions. First, partial evaluation may mask the presence of an inadvertent program error that *should* be exposed by a crash. Second, partial evaluation introduces an undocumented dependence on order.

Although they seem like incorrigibly bad news, crashes are sometimes useful diagnostic tests—perhaps Denominator was zero because somebody forgot to initialize it. The problem of unadvertised dependence on order is rather more serious. A seemingly cosmetic alteration introduced later while 'cleaning up' code may change the sense of the expression. Not all languages allow partial evaluation, and programmers sometimes forget about small details.

In my programs, I deal with these issues in one of two ways. First, any segment can be rewritten to make the order of the two evaluations explicit, and to allow for an error message:

explicit partial evaluation

```
IF (Denominator<>0.0)
   THEN IF (Numerator/Denominator)>Fraction
           THEN IO.StringOut ("We have a lucky winner!)
         END  (*inner IF*)
   ELSE IO.StringOut ("Denominator not initialized.")
END  (*outer IF*)
```

Second, if such elaborate checking isn't necessary, any statement that relies on partial evaluation will be explicitly documented like this:

documenting partial evaluation

```
IF (Denominator<>0.0) AND  (*then if*)
    (Numerator/Denominator)>Fraction THEN
   IO.StringOut ("We have a lucky winner!")
END
```

It may seem clunky, but wait until you've spent a day tracking down a bug that an eleven-character comment would have prevented.

Modula-2 Details

Syntax Summary

- **IF** statement: examines program data to decide what to do. Three forms are allowed: a *yes/no* choice:

IF *a condition is met* **THEN** *statements* **END**;

an *either/or* choice:

IF *a condition is met*
  THEN *statements*
  ELSE *alternative statements*
END;

or a *which alternative* choice:

IF *first condition is met*
        THEN *first alternative statements*
    ELSIF *second condition is met*
        THEN *second alternative statements*
    ELSIF *third condition is met*
        . . .
    ELSE *final alternative statements*
END;

- relational operators: compare two values of the same simple type:

Math	*Modula-2*	*English*
$=$	=	*equal to*
$<$	<	*less than*
$\leq$	<=	*less than or equal to*
$>$	>	*greater than*
$\geq$	>=	*greater than or equal to*
$\neq$	<> *or* #	*not equal to*

- BOOLEAN operators: build expressions from BOOLEAN-valued operands. & can substitute for **AND**, while ~ can substitute for **NOT**.

NOT A	*the opposite of its operand*
A **AND** B	TRUE *if, and only if, both operands are* TRUE
A **OR** B	TRUE *if either operand is* TRUE

- BOOLEAN expressions: are evaluated left-to-right. Evaluation stops as soon as an outcome is certain; in particular:

A **AND** B *means* **IF** A **THEN** B **ELSE** FALSE **END**
A **OR** B *means* **IF** A **THEN** TRUE **ELSE** B **END**

- operator precedence: relations—equalities and inequalities—have the lowest precedence, so as a rule relations are put in parentheses in BOOLEAN expressions. The exact precedence levels are:

NOT	*most precedence*
AND	
OR	
<, >, <=, >=, =, <>	*least precedence*

- HALT: a standard procedure that immediately stops program execution.

Modula-2 Details

- The constants of type BOOLEAN are FALSE, TRUE, in that order.

Remember . . .

- It is difficult to tell when BOOLEAN expressions rely on partial evaluation. For the sake of future program maintenance, always comment any expression that relies on partial evaluation to stay correct:

IF A **AND** (**then if**) B . . .

- Robustness—for now, the ability to withstand errors that can be anticipated—is a desirable program quality. **IF** statements should be used, when possible, to do error-checking on the 'reasonableness' of program values.
- Programs are meant to be read, as well as run. Use indenting and layout that clearly displays the grouping of statements.
- A problem's *solution space* is the entire range of potential answers. *Exhaustive search* programs explore the entire solution space. However, searches can often be short-circuited.
- DeMorgan's laws and the distributive laws can, and should, be used to simplify complicated BOOLEAN expressions.
- Over-engineered software tries to anticipate problems. *Embedded*, or built-in, debugging code is a particularly useful tool. Build it in before you need it, and maybe you won't need it at all.
- When possible, programs should detect and try to do something about flawed data. Always inform the user, fix the data when possible, and use HALT to stop program execution when necessary.

Zen Study Guide
Big Ideas

The Big Ideas in this chapter are:

1. *Decisions are made by comparing values.* No matter how complicated a choice may be, it generally starts with a comparison of two letters or numbers. The basic comparisons are made with relations; the BOOLEAN operators are needed to let decisions be more complex.

2. *Making a decision implies that part of a program might be skipped.* Normally, data passes from statement to statement. The **IF** statement, though, lets program actions be skipped—the flow of program data can be diverted.

3. *Computers may be built from glass and steel, but programs are made of eggshells.* When software works, it seems to be invulnerable, but the tiniest crack can lead to total collapse. Programs can have a margin of safety for error, but you have to put it there.

Zen Study Guide
Questions

5-1 The first point of attack on Chapter 5 is the BOOLEAN expression. Make up a few questions that ask you to:

a) *Evaluate this relation*, sticking to the relational operators.

b) *Evaluate this expression*, using the BOOLEAN operators.

5-2 *Write this condition as a* BOOLEAN *expression* comes next. For example, 'I've studied for at least three hours' might be Studied <= 3. Make up a half-dozen phrases of this sort.

5-3 BOOLEAN expressions can always be written in a variety of ways. For example, it was easy for me to turn this expression:

```
First <> Last    (*Assume that First and Last have BOOLEAN values.*)
```

into this one, which has exactly the same value:

```
(F rst AND NOT LAST) OR (Last AND NOT First)
```

Write, and then rewrite in a more complicated manner, a half-dozen BOOLEAN expressions. Then mix them up, and ask the question *Which pairs of expressions are the same?*

5-4 If you look at the warmup exercises, you'll see that it's easy to make up new BOOLEAN functions, like Even, Positive, and so on. Make up five *Write a* BOOLEAN *function that . . .* questions.

5-5 '*Rewrite a* this *as a* that' is the classic form of attack on a control statement. For example, I might ask you to rewrite this:

```
IF (A >= 3) AND (B < 6)
   THEN IO.StringOut ("Hello!")
   ELSE IO.StringOut ("Goodbye!")
END
```

as two separate **IF** statements. Make up three questions of this type—going from a single statement to a sequence of statements.

5-6 Now, try the same problem in reverse. See if you can come up with a short sequence of **IF** statements that can be replaced by a single statement with **ELSE** and **ELSIF** parts. You'll find that this isn't quite as easy as it seems.

5-7 The **FOR/IF** combination is frequently used to acquire knowledge about a stream of data that originates outside a program. Suppose that you can rely on an input sequence of fifty CHAR values. Make up three questions that require you to inspect the values:

a) as independent values; for instance, how many A's were there?

b) as paired values; for instance, how may A's are followed by B?

c) as sequences of values; for instance, how many words ended with ED?

5-8 As above, but look at a stream of INTEGERs. What is the second largest value? How about the smallest difference between successive values? How about the average of the positive and negative numbers? Three more questions, please.

5-9 The next step in our progression is to do something *to* the values in the stream. Naturally, you don't want to modify *every* value, because that would be too easy. Ask three questions that require you to learn something about each input value before you modify it (for example, double even numbers and halve odd numbers).

5-10 Finally, it's always interesting to generate a stream of data (as we did in program Robbery), and look for something interesting. For example, I might ask you to find every four-digit number with the following property: it is a perfect square, the first and second pairs of digits are both odd, and the sum of the two pairs of digits is also a perfect square. Make up two questions that would require a programmer to find numbers that have some crazy property.

5-11 Anybody can read the book and start programming. But only students who have done their programming homework can answer *Find the bug* questions. In other words, it's easier to know what's right than it is to find out what's wrong. For instance, *Find the bug:*

```
(*Inspects fifty characters, counts those that aren't "T" or "5"*)
Count := 0;
FOR i := 1 TO 50 DO
    IO.CharIn (Ch);
    IF (Ch <> "T") OR (Ch <> "5") THEN
        Count := Count + 1
    END
END
```

The **OR** should be an **AND**.

Make up three or four buggy program segments. In case you haven't been doing *your* homework, some sneaky bugs to throw in might be *a*) using uninitialized variables, *b*) initializing a variable inside the loop rather than before it, *c*) using the wrong BOOLEAN operator (as above), and *d*) including a boundary error by using the wrong relational operator.

5-12 Yes, it's time once again for *Here's the output, what was the input?* For example, you might provide a code segment that adds 3 to odd numbers and subtracts 2 from even ones, along with some sample output. It's possible to determine exactly what the input was. Make sure that it's possible to figure the input out, incidentally—it's very embarrassing when hands start to pop up in mid-test!

Exercises

Warmups and Test Programs

5-13 Being able to decipher error messages associated with **IF** statement syntax errors is an especially useful talent. What is the effect of a missing **END**? An extra **END** within the statement? An **IFELSE** (yes, that's right) part? An out-of-order **ELSE** part?

5-14 Writing incorrect programs is so much fun that I'll suggest a few more. What happens when you try to compile programs that include:

a) the expression A < B **OR** C > D?

b) the expression 1 **AND** 2?

c) the expression "a" <> 7?

5-15 Write a function Even that returns TRUE if its INTEGER parameter is an even number, and FALSE otherwise.

5-16 Write a function called Divisible that returns TRUE if its first INTEGER argument is evenly divisible by its second.

5-17 Write a function Positive that returns FALSE if its REAL argument is a negative number.

5-18 Write a function IsADigit that lets you know if its CHAR argument is a digit character.

5-19 Write a function WithinLimits that returns TRUE if its first CHAR argument is in the range given by its second and third CHAR arguments.

5-20 The Modula-2 **OR** operator is unable to differentiate between one TRUE operand and two. But suppose that we want an action to take place if A is TRUE, or if B is, but absolutely not if both of them are. This condition is called an *exclusive or*; some languages, but not Modula-2, provide an **XOR** operator to state it. Write a BOOLEAN function that takes two BOOLEAN arguments, and can take the place of **XOR**.

5-21 Suppose that you want an action to take place if two conditions are both TRUE, or both FALSE, but not otherwise. Write a BOOLEAN function that tests this condition.

5-22 Write a procedure that lets you know if a sequence of N input values is all (take your pick) positive, odd, or increasing.

5-23 Why must relations be put in parentheses when they're part of longer BOOLEAN expressions? Do you have any relations you like to see in parentheses?

5-24 Compare these two program segments. Is there any difference between them? Explain.

```
IF (a <= b) AND (a <= c)
      THEN Smallest := a
   ELSIF (b <= a) AND (b <= c)
      THEN Smallest := b
   ELSE Smallest := c
END;
```

```
Smallest := a;
IF b < Smallest THEN
   Smallest := b
END;
   IF c < Smallest THEN
   Smallest := c
END;
```

5-25 How long does an **IF** statement take? More to the point, how long does it take to evaluate different kinds of BOOLEAN expressions? Redo some of the timing studies you might have written in Chapter 4. Are BOOLEAN operators any more 'expensive' than relations? Are short-circuited evaluation sequences (i.e. an **IF** statement with **ELSIF** parts) really any 'cheaper' than the equivalent series of **IF** statements?

5-26 One cause of bugs associated with BOOLEAN expressions is the unexpected appearance of outcomes that were not explicitly ruled out. Here are a few everyday questions whose answers are similarly unexpected:

a) Two coins add up to thirty cents, yet one of them isn't a nickel. How can this be?

b) Some months have 30 days, and some have 31. How many months have 28 days?

c) Two people played seven games of chess, yet they each won the same number of games. What happened?

d) A boy broke his collarbone playing football. When he arrived at the hospital, his coach told the nurse 'My son is injured.' However, when the doctor set the bone, the boy complained 'Ouch, Dad! That hurts!' If the doctor isn't also the football coach, can everybody be telling the truth?

Exercises
Programming Problems

5-27 My local sporting goods store periodically makes the following offer: *Buy five items, and the sixth one is free!* Which is the free item? Is it the last one over the counter? The smallest? The lightest? No, of course not. It's the cheapest one, naturally.

Write a program that the store's electronic cash register might rely on. Have it read all six prices, find the lowest one, and print a total cost. Local tax is 6.5%, incidentally.

5-28 Genius, according to Thomas Edison, is a matter of 2% inspiration, and 98% perspiration. Nowadays, of course, perspiration is usually socially inappropriate, except for men and horses. Alternatives—sweating, glowing, and dripping—exist, and it is important to know when to use them, regardless of the consequences to genius.

According to my social secretary, the rules are:

Temperature	*Cooling Mechanism*
over 100°	*Dripping*
91° to 100°	*Sweating*
81° to 90°	*Perspiring*
71° to 80°	*Glowing*
70° or below	*Gleaming*

Write a program that can read a temperature, and let you know what to do.

5-29 For most of the past decades, power companies gave bulk discounts to their largest consumers. In effect, those companies that were most effective at squandering energy were given incentives to continue doing so.

Recently, matters have changed somewhat. For example, my own power bill last month set two levels of use:

Baseline Quantities	Gas — 34.8 Therms	Electric — 220.4 KWhrs
Baseline Price	$0.44181 / Therm	$0.08118 / KWhr
Over Baseline Price	$0.84204 / Therm	$0.12739 / KWhr

Write a program that takes four meter readings (starting and ending reading for both gas and electric meters), and calculates charges to the nearest cent.

5-30 A well-known author of computer science textbooks proposed the following grading policy in his course:

Homework 1 through 5	5% each
Midterm 1	15%
Midterm 2	25%
Final Exam	35%

His students pleaded for the following changes:

- Throw out the lowest homework grade.
- Throw out the lower of the first two test grades.
- Only count half of the final exam grade if it was lower than the average of the two midterm scores. Make up the difference by increasing the influence of the average of the first two exams.
- Let a homework score of 20 or less be replaced by a special project grade. However, don't permit more than two homeworks to be replaced like this, even if more special projects are done.

The well-known author agreed to institute these changes, but only if his students would write a program that could do the grading. Unfortunately, they could not, so he returned to his previous policy of passing only students who could prove that they had bought extra copies of his textbook for their relatives.

Can you handle the problem? Assume that the first line of grade data contains only the number of grades. Then, each successive line has the homework grades, followed by the exam grades. As noted, any homework score of 20 or below is followed by a project score, although only the first two special project scores are counted. Assume that all grades are on a scale of 1 to 100. Finally, in addition to printing each student's average grade, also print the average score on each of the exams.

5-31 To truly hip skiers, only one word has meaning: vertical. Fortunately, it is fairly easy for a skier to figure out how many vertical feet she has managed to ski in a day. Since, for all practical purposes, you can only ski downhill, all she has to do is to find the difference between the starting and finishing elevation of each run.

Runners who hope to be truly hip, however, face a much more difficult task. A run may begin and end at the same elevation, but have a considerable amount of up and down in between. The only way to calculate total vertical is to go along the trail with a topographical map, and add up all the increasing elevation changes.

For example, a typical run might be described in terms of these ten elevation figures:

```
500   530   550   550   575   540   500   480   465   500
```

The total change going up is 30 + 20 + 25 + 35, or 110 feet.

Write a program that will read a fixed number of elevations, and let you know how much vertical was covered—going up!

5-32 I'm sure that *Home on the Range College* ('where seldom is heard a discouraging word') was your first choice of schools. But since you didn't get in, you probably have to calculate your grade point average once in a while.

Write a program that calculates a grade point average. Assume that an A is worth 4.0, B is 3.0, and so on. Grades may be modified by + or – (worth plus or minus 0.3), except that A+ is still only 4.0. Out of respect for the students' self-esteem, F– counts as 0.0. To make the program a bit harder, assume that grades are separated by one *or more* blank spaces.

5-33 A number's divisors are the numbers that divide it evenly. For example, the divisors of 12 are 1, 2, 3, 4, 6, and 12. If a number is another number's divisor, an INTEGER division doesn't leave any remainder—A is a divisor of B if A **MOD** B equals 0.

It's easy enough to write a BOOLEAN-valued function that lets you know if a number is another number's divisor. Then, you can find a number's divisors through brute force: just test every value between 1 and the number.

a) Write a program that finds and prints a number's divisors.

b) Modify it to print a table of divisors for all numbers below 1003.

c) Modify that to find out what number below 100 has the largest number of divisors.

5-34 Yes, another question about divisors. We live in times that are reluctant to make value judgments; nevertheless, consider these traditional descriptions of a number, and the sum of its divisors (excluding the number itself):

- *Perfect* numbers equal the sum of their divisors (less themselves). 6 is perfect because it equals 1 + 2 + 3.
- *Abundant* numbers have divisors that add up to *more* than the number.
- *Deficient* numbers have divisors that don't add up to the number itself.

Write a program that answers these questions:

a) Which three numbers below 1000 are perfect?

b) What odd number below 1000 is abundant? What seven odd numbers below 5,000?

c) What are the relative proportions of deficient, abundant, and perfect numbers?

5-35 The word 'begin' is moderately interesting because it letters are strictly increasing—no letter alphabetically equals, or is less than, the letter before it. 'coop' also increases, but not strictly, since some of its letters are duplicated.

Write a program that can rely on, say, ten letters of input. Keep track of whether the letters are the same, increase, increase strictly, or decrease.

5-36 A *toggle switch* is just a two-position switch. Toggling usually refers to either/or switching, rather than on/off switching. For instance, a walkie-talkie might use a toggle switch to flip between sending and receiving. Naturally, you can build toggle switches into programs as well. For example, a program that reads and echoes characters might use the appearance of an exclamation point as a toggle—from that point on, letters should be echoed as capitals. A second appearance of the exclamation point switches back, so that letters echo in lower case.

Input: **`Here is ! a test ! of MY system.`**
Output: `here is A TEST of my system.`

The toggle switch itself can be implemented as a BOOLEAN variable. The **NOT** operator can be used to flip its current value.

Write a program that reads fifty values, and implements one of these toggles:

a) Switch between printing only odd and only even numbers, toggling at each zero.

b) Toggle between echoing characters in upper and lower case, switching at each period, question mark, or exclamation point. (Assume all input is lower-case.)

c) Read numbers, and switch between printing them as positive or negative. Start by printing odd numbers as negatives, and evens as positives, and toggle at zero.

'A loop does not end because it has achieved its goal. *Instead, it ends because the loop has reached its* bound.' (p. 195) Why the computer scientist never got out of the shower.

Chapter 6 The **LOOP** Statement

Can you do something hard, like finding every three-digit number that equals the square of the sum of its digits? Sure—you have the **FOR** loop and the **IF** statement. But can you do something easy, like counting the number of letters in a word? No way—you don't have a loop that stops when it reaches a blank. **LOOP** comes to the rescue.

Loops are important, so I'll present two classes of applications: text processing examples that involve words, and 'number crunching' examples that involve math. This chapter is long because I assume you'll cover just one of 6–3 or 6–4 in detail. Whichever path you choose, our focus will be on *design*, and reasoning about loops before they're written.

A special In Depth section discusses Modula-2's other loop statements, **WHILE** and **REPEAT**. Although the loop design method I like to use is most easily presented with **LOOP**, some programmers will feel more comfortable with these alternative loops.

Section 6-1

Strictly Modula-2

AS USUAL, WE'LL BEGIN WITH AN EXAMPLE. HERE'S a hard-working **LOOP**:

a **LOOP** example

```
MODULE AddLoop;
    (*Uses an $80,000 computer to simulate a broken $1.95 calculator.*)
IMPORT IO;
VAR Number, Sum: INTEGER;
BEGIN
    IO.StringOut ("How about giving me a series of numbers to add?");
    IO.LineOut;
    IO.StringOut ("Let a negative number mark the end of the list.");
    IO.LineOut;
    Sum := 0;
    LOOP   (*until we read a negative number*)
        IO.IntIn (Number);
        (**)IF Number < 0 THEN EXIT END;
        Sum := Sum + Number
    END;
    (*The last value read was negative. Sum is the non-negative total.*)
    IO.StringOut ("The grand total was:  ");
    IO.IntOut (Sum, 1);
    IO.StringOut (".   Oh no!")
END AddLoop.
```

```
How about giving me a series of numbers to add?
Let a negative number mark the end of the list.
83  197  214  100  3  69  -5
The grand total was: 666.  Oh no!
```

The **LOOP** statement supplies repetition at its most basic: a sequence of statements is repeated until an exit condition is met. Its syntax is simple:

LOOP
a sequence of statements that should include an **EXIT** *statement*
END

The loop's action includes a statement that controls the moment of departure from the loop. The reserved word **EXIT**, when encountered (usually like this):

IF *condition is met* **THEN EXIT END**;

causes a loop to be departed immediately. Program execution resumes following the **END** of the loop.*

Two ironclad rules of program engineering apply to **LOOP** statements:

* Technically speaking, the **EXIT** is not syntactically required, and some kinds of simultaneously running loops are set up without **EXIT** points. We'll talk about *coroutines* briefly in Chapter 20.

two ironclad rules

- There should only be a single statement that contains an **EXIT**.
- The **EXIT** point should be at the outermost level of the loop. It should not be nested, or difficult to find.

Together, these mean that each **LOOP** has just one exit point, and that the exit point is immediately apparent to the program reader. As a matter of legal Modula-2 syntax, these two rules can be broken. As a matter of programming practice, though, violations occur only under exceptional circumstances.

Modification Problems

1. Change AddLoop into a program that *multiplies* numbers until it reads a *positive* value.
2. Modify AddLoop so that instead of stopping at *any* negative number, it quits at a particular value supplied by the program user. Let this value (call it Sentinel) be one argument of a procedure that contains the loop.

LOOP and EXIT

In general, we'll find that understanding *why* a loop should exit lets us know *where* it should exit almost automatically. The **EXIT** point, which is usually guarded by an **IF** statement, might be at the beginning of the loop:

```
LOOP
   (**)IF condition is met THEN EXIT END;
   sequence of statements
END
```

or at the loop's end:

```
LOOP
   sequence of statements;
   (**)IF condition is met THEN EXIT END
END
```

or even somewhere in the middle of the loop:

```
LOOP
   sequence of statements;
   (**)IF condition is met THEN EXIT END;
   sequence of statements
END
```

Note that when the **EXIT** condition is met, any remaining steps in the loop are skipped. If you feel that they could be put into an **ELSE** part, you're absolutely correct—they're the alternative to leaving the loop. I generally don't use an **ELSE** part because it complicates the syntax and appearance of the code without yielding any offsetting benefit in clarity or correctness.

commenting practices

However, it's very important to clearly mark the **EXIT** point. In my programs, I'll always put empty comment brackets (like these: (**)) at the start of the line that contains the **EXIT** and its condition. This highlights the loop's exit point; making it easy to spot adds to the loop's self-documentation. When programs are more complicated, I'll also add the reason for loop exit as a comment on the first line.

the condition must change

> Every loop must eventually *terminate*, or end. To make sure this happens, the **IF** statement that controls the **EXIT** condition has to be possible to meet. Almost inevitably, this means that a variable, whose value is changed as part of the loop's action, helps state the **EXIT** condition.

A loop that does not obey this basic law will either fail to terminate (it will be an *infinite loop*), or it will always terminate on the first trip through (and not be a loop at all).

Self-Check Questions

Q. What's wrong with these **EXIT** conditions? Why are they probably in error?

a. **IF** TRUE **THEN EXIT END**
b. **IF** FALSE **THEN EXIT END**
c. **IF** ABS(Counter)<0 **THEN EXIT END**
d. **IF** (Value>10) **AND** (Value<=5) **THEN EXIT END**
e. **IF** (Value<=10) **OR** (Value>=11) **THEN EXIT END**

A. Try a 'go/no go' test—all the conditions are either unavoidable or impossible. Statements *a* and *e* can't help being TRUE, while *b*, *c*, and *d* are invariably FALSE. Since there's no way any of them can be modified, the 'TRUE' loops will exit immediately, and the 'FALSE' loops will repeat forever.

Loop Specification: Goals, Bounds, and Plans

It isn't hard to write a loop that will compile. However, it's always a good idea to lay out the loop's semantics—what it will do—in advance. For the purpose of planning, we can break every loop into three distinct parts:

parts of a specification

- A *goal* that states the overall purpose of the loop.
- A *bound* that gives the reason the loop will end.
- A *plan* that describes the action the loop is going to take.

In other words, the goal is what we hope to accomplish, the bound is our **EXIT** condition, and the plan explains how we'll approach the goal and bound.

Now it's time for an insight that is so obvious it's often overlooked.

bounds, not goals, end loops

> *A loop does not end because it has achieved its goal.* Instead, it ends because the loop has reached its *bound.*

The loop's goal describes the happy state of events we hope to attain. However, the bound condition supplies the strongest statement we can make with certainty when the loop is done. If the loop's plan was a good one—and program data was what it was supposed to be—we'll reach the bound just as we achieve our goal.

A *loop specification* is a pseudocode description of a loop's goal, bound, and plan. For example, the specification of the loop in our example program might have been:

loop specification

Goal: To know the sum of a series of input numbers.
Bound: We've read a negative number.
Plan: Read a number.
Add it to a running sum.

Note that the loop's goal and bound are quite distinct. I could set other goals with the same bound: find the numbers' product, or convert the numbers into base 9. I might also apply other bounds to the same goal: stop adding once the sum exceeds 500, or after we've read twenty numbers.

Stating the goal and bound separately help us to design the loop's plan. Most successful plans follow one simple rule—think about the bound first.

designing a loop plan

> Take the smallest step necessary to approach the bound. Then, and only if necessary, add additional steps that help reach the loop's goal. Finally, ask a question: should I exit *before* or *after* taking the extra steps?

The plan's first step, *read a number*, brings us toward the bound. An extra goal-oriented step, *add it to a running sum*, helps meet the goal of this particular loop. However, our **EXIT** point comes right in the middle. We want the loop to end as soon as it reaches its bound of spotting a negative number. In pseudocode, the example program's main loop is:

a pseudocoded loop

```
initialize an accumulator to hold the running sum;
LOOP
   approach the bound—read a number;
   (**)IF it was a negative number then EXIT;
   approach the goal—add the number to the running sum
END of the loop
```

If we don't leave the loop, we add the number to our running sum, and start the loop again. The Modula-2 version of the loop falls into place fairly effortlessly:

the Modula-2 version

```
Sum := 0;
LOOP
   IO.IntIn (Number);
   (**)IF Number < 0 THEN EXIT END;
   Sum := Sum + Number
END;
(*The last value read was a negative number. Sum is the non-negative total.*)
```

comment cautiously

> The confirmation comment that follows the loop is based on the loop's **EXIT** condition—the only statement we can make with certainty. Any additional claims have to be checked carefully against the action of the loop.

Self-Check Questions

Q. Come up with a bound and plan for each of these goals:

a) *Goal: Skip fifty input characters.*

b) *Goal: Find the position of the letter* F *in input that includes an* F.

c) *Goal: Divide* 35 *by* 6 *using subtraction.*

d) *Goal: Count the number of* F*'s in an input word.*

e) *Goal: Print an* INTEGER *in reverse.*

A. You might want to try drafting the pseudocode that applies to each of these loop specifications. Where will the **EXIT** point be?

a) *Bound: A counter equals* 50.
Plan: Read a value.
Add 1 *to a counter that started at* 0.

b) *Bound: Reading the letter* F.
Plan: Read a value.
Add one to a position counter.

c) *Bound: The remainder is less than* 6.
Plan: Subtract 6 *from the current remainder.*
Add 1 *to the quotient.*

d) *Bound: Reading a space character.*
Plan: Read a character.
Add 1 *to a counter if the character is an* F.

e) *Bound: The number has been reduced to zero.*
Plan: Print the one's digit by printing the number **MOD** 10.
'Strip off' the one's digit by recalculating the number **DIV** 10.

Loops and Their Bounds

Most of the attention we pay to loops involves bounds, so I'll organize loops according to our different motives for stopping them. Why would we want a loop to stop?

sentinel bounds

- A *sentinel* or marker value was encountered—usually, a number or character that marked the end of a certain kind of input. For example, a blank space is a sentinel that marks the end of a word, while 0 or –99 might mark the end of positive INTEGER input.

count bounds

- A *count* was attained—the loop's action was repeated enough times. You can recognize that the **FOR** statement is a special-purpose loop that's been customized for count bounds. However, **LOOP** is needed when the count won't necessarily increase on every trip though the loop, e.g. when we count certain kinds of input, rather than every input value.

limit bounds

- A *limit* was reached—one more iteration wouldn't get us closer to the loop's goal, and might have even led to error. Limits can be given as exact figures (reducing a number to zero), as boundaries (exceeding an upper size limit), or in relative terms (the difference between two successive calculations is less than 0.000001 percent).

data bounds

- The *data* was used up, or we reached an imaginary border within the data. Data bounds will be more common in later chapters, when we write programs that actively subdivide stored information into 'still relevant' and 'irrelevant' subgroups. However, text processing provides one major application here.

For example suppose our goal is to add numbers. Our bound can vary, depending on the task at hand. It's a:

sentinel, if we add numbers until we spot a negative one;

count, if we add just ten numbers;

limit, if we add numbers until the total exceeds 500;

data bound, if we add the contents of a file of numbers.

Now, the bounds I've described are terms of convenience. Nobody's going to test you on them, and you don't have to try to fit programming problems into one category or another.

Nevertheless, categories are useful because they help point out an important fact of program engineering: that loops often have *multiple bounds*, or more than one reason for terminating. Loops often have:

goal-oriented vs. error-avoiding bounds

- an *intentional* bound that is tied to achieving the loop's goal;
- one or more *necessary* bounds that guarantee loop termination, and are needed to prevent errors.

> When you establish a loop's bound condition, always ask: *Does this bound alone guarantee that the loop will end?* If the answer is no, add necessary bounds.

For example:

when necessary bounds are needed

- We may be looking for a sentinel value, but know that there are a limited number of input values—and that the sentinel may not appear.
- We may be repeating a computation that approaches a 'correct answer' or 'good enough' limit, but gets there so slowly that we want to put an upper count bound on the number of repetitions allowed.
- We may not be sure that a correct answer exists, or that our method of seeking it is correct, and wish to set a bound on the amount of time we spend trying.
- We may not be sure of the reliability of our data, and wish to provide a 'fail-safe' exit condition if it doesn't meet our expectations.

Usually, we'll find that buggy situations arise when the intentional bound is in one category, while the necessary bound is in another. Use the sentinel, limit, count, and data categories as a checklist for avoiding potential errors.

Self-Check Questions

Q. The code segments below all lack **EXIT** statements, but that doesn't mean that they don't have bounds. Read each segment's confirmation comment, then write an **EXIT** statement and show where it goes.

```
CharCount := 0;                                (*Segment A.*)
LOOP                                (*a*)
    IO.CharIn (Letter);             (*b*)
    CharCount := CharCount + 1;     (*c*)
END                                 (*d*)
(*CharCount represents the number of letters that came before a SENTINEL.*)

Guesses := 1;                                  (*Segment B.*)
LOOP                                (*a*)
    IO.IntIn (NewGuess);            (*b*)
    Guesses := Guesses +1;          (*c*)
END                                 (*d*)
(*NewGuess equals Answer, or five guesses were taken.*)

Tries := 5;                                    (*Segment C.*)
LOOP                                (*a*)
    IO.IntIn (NewTry);              (*b*)
    Tries := Tries -1;              (*c*)
END                                 (*d*)
(*NewTry equals Answer, or five guesses were taken.*)
```

A. It's important to read the initializing statement that precedes the loop, since it can affect the position of the **EXIT**.

A) **IF** (Letter = SENTINEL) **THEN EXIT END**; between *b* and *c*.
B) **IF** (NewGuess = Answer) **OR** (Guesses = 5) **THEN EXIT END**; between *b* and *c*.
C) **IF** (NewTry = Answer) **OR** (Tries = 0) **THEN EXIT END**; between *c* and *d*.

Section 6-2 Standard Practices

IN OUR PREVIOUS PROGRAMMING, CONFIRMATION COMMENTS have helped us check code. I'll start our discussion of loop programming practice by seeing how to expand the role of confirmation comments. They'll be more important than ever as *postconditions* that help establish loop correctness.

The rest of this section concentrates on examples of each kind of loop bound. The extended examples in sections 6–3 and 6–4 deal with more complicated programs that might require more than one kind of loop, or use loops that have more than one reason for exit.

Loop Postconditions

But first, let's look at design. The most important line of any program is the first line to follow a loop.

pre- and postconditions

A loop's *postcondition* describes our state of knowledge about program data after the loop. The postcondition should focus on facts that are relevant as the *precondition* of the next program segment.

problem: averaging numbers

We've already used postconditions informally—I called them confirmation comments. Let's look at an example that makes good use of its postcondition comments: a program that finds the average of a sequence of numbers.

Suppose that a number no larger than –1.0 is a sentinel that follows the last valid input. Here's how I might specify a loop that solves the first part of an averaging problem—adding and counting the numbers:

loop specification

Goal: Count and sum a sequence of input values.
Bound: We've read the sentinel.
Plan: Read a single value.
Add it to an accumulator that started at 0.
Increment (add 1 *to) a counter that started at* 0.

The specification is pseudocoded below. Notice the pains I've taken to limit my postcondition—the claim that's made on loop exit:

pseudocode

initialize the counter and accumulator to 0;
LOOP
  get a value;
  (**)**IF** *it's the sentinel* **EXIT**;
  add the value to the accumulator;
  increment the counter
END *of the loop;*
*(*We've read the sentinel. The counter represents the number of values read, and the accumulator represents their sum.*)*
calculate and report on the results;

> Most loop bugs derive from an incorrectly positioned **EXIT** point or an overstated postcondition. When you program, check one against the other. Does the postcondition follow from the **EXIT** condition? Does the **EXIT** condition support the postcondition?

Let me ask a question that will seem naive, if not downright dumb. Why do we leave the loop? Is it because we know the average of the input numbers? No! Is it because we are ready to compute the average—to divide the sum by the count? Again, an emphatic no! Instead, we leave the loop *because we have read the sentinel value.*

We can't reach any other sure conclusions—not even if there *were* any input values—without making additional checks. That's why it's important to recognize that the loop's postcondition is the next statement's *precondition.* It has to answer any questions the next statement will ask in order to work properly.

graceful degradation

As you read RealAvg, below, notice that it prints an explicit message if the data list is empty. This attention to detail could be quite important in a larger program, where an incorrect or unsupplied value could lead to wrong results, or even an inexplicable program crash. Explicit built-in error messages help a program to *degrade gracefully,* rather than failing catastrophically or invisibly.

average until sentinel program

```
MODULE RealAvg;
   (*Averages a series of numbers. -1.0 marks the end of data.*)
IMPORT IO;
CONST SENTINEL = -1.0;
VAR Value, Total: REAL;
    Counter: INTEGER;
BEGIN
   IO.StringOut ("Enter real numbers for averaging. End with ");
   IO.FixOut (SENTINEL, 1, 1);
   IO.StringOut (" or less"); IO.LineOut;
   Total := 0.0;          (*Initialization *)
   Counter := 0;
   LOOP  (*until we read the sentinel value*)
      IO.RealIn (Value);
      (**)IF Value <= SENTINEL THEN EXIT END;
      Total := Total+Value;      (*increase the accumulator*)
      Counter := Counter+1       (*increment the counter*)
   END; (*LOOP*)
   (*Count holds the number of values, and Total is their sum.*)
   IF Counter=0
      THEN IO.StringOut ("No data entered.")
      ELSE IO.StringOut ("The average of ");
           IO.IntOut (Counter, 1);
           IO.StringOut (" values is ");
           IO.FloatOut (Total/FLOAT(Counter), 8, 5)
   END (*IF*)
END RealAvg.
```

```
Enter real numbers for averaging.  End with -1.0 or less
23.9  85.68  227.0E02  0.00863  75.0 -1.0
The average of 5 values is 4.57692E+003
```

Self-Check Questions

Q. RealAvg checked the value of Counter to decide if any numbers had been entered for averaging. Could it have checked the value of Total instead?

A. Sure—but what it found out might be wrong. Suppose that the program was calculating the average temperature of a chilly icebox, and that every value entered was 0.0. The `No data entered` message would be wrong, since there would be a calculable average of zero.

Sentinel Bounds

Let's turn to specific examples of the different bounds, starting with sentinel bounds. Some typical applications that involve sentinels are:

- Finding the position of a particular value, or ignoring input until a particular value appears. The value itself serves as the sentinel.

where sentinels appear

- Doing calculations on a series of numbers. The sentinel follows the last 'legitimate' value in the series.
- Prompting a program user to provide a value that falls within a limited range or group of allowable entries ("Y" or "N", say, or a positive number). The correct value need not be any one particular value, but nevertheless it serves as the sentinel.
- Reading a line of characters. A control character (equal to the constant EOL from the IO library, which we'll discuss in more detail in 6–3) marks the end of each line.

In general (and experience will teach you this much better than memorizing) loops that obey sentinel boundaries are set up like this:

a typical sentinel-bound loop

LOOP
  approach the bound—get a value;
  (**)*check the bound*—**IF** *it's the sentinel* **THEN EXIT END**;
  approach the goal—do normal processing (if any) with the value
END

Let's consider a problem that illustrates the basic sentinel-bound loop; it uses a sentinel to mark the start of a certain kind of input. Here's the problem:

problem: skipping to a sentinel

> Dollar amounts appear in text that is interesting, but extraneous to our immediate interests. A line of input might begin:
>
> `Allocated to payoffs:  $100000.00 for favors . . .`
>
> Dollar signs only appear before dollar amounts. Write a procedure that skips past the irrelevant characters in this line, then reads and returns the number.

In other words, we want to read a number that's embedded in text. The dollar sign acts as a sentinel, because we assume that a sequence of characters that obeys the Modula-2 syntax of a REAL number will always follow a dollar sign. I can write the loop specification as:

loop specification

Goal: Find the REAL *number.*
Bound: We've read a dollar sign.
Plan: Read a character.

What an easy plan! As it turns out, taking the smallest step toward the bound also helps us meet the loop's goal. No additional action is necessary. Do we need a pseudocode version of the loop specification? I don't think so.

skip-to-sentinel procedure

```
PROCEDURE SkipToNumber (VAR TheNumber: REAL);
  (*Returns the REAL value that follows a dollar sign.*)
  VAR OneCharacter: CHAR;
  BEGIN
    LOOP
      IO.CharIn (OneCharacter);
      (**)IF OneCharacter="$" THEN EXIT END
    END;
    (*OneCharacter equals "$", so we're about to read a number.*)
    IO.RealIn (TheNumber);
  END SkipToNumber;
```

Thus, the loop takes us up to the number one character at a time. We exit the loop after reading the sentinel value, then read the number itself.

Let's turn to a slightly more difficult formulation of a sentinel problem.

problem: sentence length

> One definition of 'sentence' you won't learn in Composition 101 is 'a series of letters, spaces, and punctuation marks that ends with a period, question mark, or exclamation point.' Write a procedure that counts the number of characters in such a sentence.

As before, a sentinel marks the end of input. However, now the sentinel can have three different values: period, question mark, or exclamation point. I can outline the loop required by the sentence length counting problem as:

loop specification

Goal: Count the characters that precede a sentinel.
Bound: We've read the sentinel.
Plan: Read a character (approach the bound).
Add 1 *to a counter that began at* 0 *(approach the goal).*

A pseudocode expansion of the specification is accompanied by a confirmation comment—its postcondition:

pseudocode

```
initialize a counter variable to 0;
LOOP
  read a character;
  (**)IF it's a sentinel character EXIT;
  since we didn't leave, add 1 to the counter
END of the loop
(*The counter represents the number of non-sentinel characters read.*)
```

Order of statements becomes important when the plan has more than one step. We can't start with the **EXIT** check right off because we don't know if we've read a sentinel or not. Nor can we add one to the counter before making the **EXIT** check, since the sentinel might be the very first character. The final code is:

letter counting procedure

```
PROCEDURE LetterCount (VAR Count: INTEGER);
    (*Counts the number of letters in a sentence that ends with ., ?, or !.*)
    VAR Ch: CHAR;
    BEGIN
        Count := 0;
        LOOP
            IO.CharIn (Ch);
            (**)IF (Ch=".") OR (Ch="?") OR (Ch="!") THEN EXIT END;
            Count := Count + 1   (*If we didn't leave, increment the count.*)
        END
        (*Count represents the number of non-sentinel characters read.*)
    END LetterCount;
```

Modification Problems

1. Put procedure SkipToNumber inside a loop that finds the sum of the first five numbers embedded in the text I described.

2. Suppose that a number is a sequence of digit characters that ends with a space. Turn LetterCount into DigitCount. However, to make life harder, let the characters ".", "E", "+", and "−" appear in a number without being counted as digits.

3. What if I define a 'word' as a sequence of characters that ends with a space? Modify LetterCount so that it counts the number of words in a sentence. Don't forget to think about how you'll deal with a one-word sentence.

Self-Check Questions

Q. What is the purpose of this code segment? What do the shaded statements do?

```
IO.CharIn (Ch1);
Count := 0;
LOOP
    Ch0 := Ch1;
    IO.CharIn (Ch1);
    IF (Ch0 = "A") AND (Ch1 = "B") THEN EXIT END;
    Count := Count + 1
END
```

A. The segment counts the number of characters that come before the two-character sentinel **"AB"**. Ch0 and Ch1 act as a moving *window* that's two characters wide. The sequence below advances the window by sliding characters down:

```
Ch0 := Ch1;
IO.CharIn (Ch1);
```

Q. If the first initialization statement above were changed to:

```
Ch1 := "X"; (*instead of IO.CharIn (Ch1);*)
```

what would Count have to be initialized to?

A. If Ch1 were initialized to "X", then Count would have to begin at –1. Properly initializing the window would take an extra trip through the loop.

Count Bounds

Loops with count bounds are our second category of **EXIT** condition. Applications that call for count bounds include:

where count bounds are used

- Counting repetitions of an action. We're already intimately familiar with this application, thanks to our experience with the **FOR** statement.
- Counting data items—some number of input values. Again, this is precisely the sort of application we've employed **FOR** loops in.
- Counting events—input of vowels, or correct answers—rather than iterations. We are concerned with the results of a loop's plan, rather than how long it takes to achieve them.
- Providing an upper limit on iteration. The count serves as a necessary bound that can guarantee the termination of a runaway loop, or merely limit the number of incorrect answers in a guessing game.

In general (although once again, experience will be the best teacher), loops that have count bounds follow this basic pattern. The key statements initialize and increment the counter:

a typical count bound loop

initialize a counter;
LOOP
    take an action;
    if it's appropriate, increment (add one to) the count;
    (**)*check the bound*—**IF** *the count was reached* **EXIT**
END

How you initialize the counter will usually affect where you increment it, or how you state the *count was reached* test. For example, initializing the counter to 0 implies that you'll increment before the **EXIT** point, and test for a count that *equals* the bound. On the other hand, initializing the counter to 1 implies that the counter will be increased at the end of the loop, *or* that you'll check for a count that *exceeds* the bound. There's no easily stated rule to follow besides this:

> Always check count bound loops for off-by-one errors. Ask 'would it work if the bound were 1?'

when to use **FOR**

LOOP won't replace **FOR**, incidentally. The **FOR** loop initializes and updates the counter. It 'counts' by non-numeric types. Best of all, it cannot fall victim to either premature or laggardly exit. The virtues of the **FOR** loop are such that we will rely on it whenever simple problems of counting the number of actions or raw data present themselves.

However, the count that controls a loop won't always be tied to the number of actions the loop repeats—we may be interested in counting results, instead. Whereas a **FOR** loop is condemned to repeat a fixed number of times, a **LOOP** statement can rely on a count bound to let it repeat only as often as is necessary or practical; usually, until there have been a certain number of successes or failures at some task. For example:

problem: the computer tutor

> Computer-aided instruction has been the educational panacea of the future for some time now. Write a program that drills its user in multiplication, and won't quit until she makes three mistakes.

If we wanted to test a student on a particular 'times' table, a **FOR** loop would be ideal. It would take us through the entire table, and we could count errors as we went along. Our present problem is different, though. We want to count occurrences of a particular result—a mistakes. When enough mistakes have been made, we want to quit the loop.

Although this program will require some additional input and output, we can focus on the central drill loop.

loop specification

Goal: Quiz a user on her favorite multiplication table.
Bound: She made three mistakes.
Plan: Ask the question.
Get the answer.
Increase a Mistakes *count if it was wrong.*
Prepare the next question.

Note that most of the plan is dedicated to taking one step toward the bound. Only the last step, *prepare the next question*, is goal-related. In a sense, we reach the loop's goal—quizzing the user, as a side-effect of trying to attain the bound—getting those three wrong answers. I'll pseudocode the specification as:

pseudocode

initialize the count of Mistakes;
find out what table to test;
get the starting Times *multiplier*;
LOOP
    ask the question;
    read the answer;
    IF *the answer was wrong increase* Mistake *by one*;
    (**)**IF** Mistakes *equals the bound then* **EXIT**;
    increase Times *by one*
END *of the loop*

As usual, it's important to keep your eye on the bound. We don't leave the loop because the user made three mistakes. We leave it because a variable named Mistakes equals three. It's essential that the actual code back up our understanding of the exit condition. The final program is shown below.

a count-bounded loop program

```
MODULE Tutor;
   (*Mercilessly drills the user in multiplication.*)
IMPORT IO;
CONST ERRORlimit = 3;
VAR Quiz, Times,  (*Drills the Quiz times table, from Quiz * Times on up.*)
Answer, Mistakes: INTEGER;
BEGIN
   IO.StringOut ("Time to multiply.  What's a number you hate?  ");
   IO.IntIn (Quiz);
   IO.StringOut ("How high should I start the drill?  ");
   IO.IntIn (Times);
   Mistakes := 0;
   LOOP
      IO.StringOut ("What is  "); IO.IntOut (Quiz, 1);
      IO.StringOut ("  times  "); IO.IntOut (Times, 1); IO.StringOut ("?  ");
      IO.IntIn (Answer);
      IF Answer <> (Quiz * Times) THEN
         Mistakes := Mistakes + 1
      END;
      (**)IF (Mistakes = ERRORlimit) THEN EXIT END;
      Times := Times + 1
   END;
   (*The user made ERRORlimit mistakes.*)
   IO.StringOut ("I hope you appreciate the beauty of math now.")
END Tutor.
```

```
Time to multiply.  What's a number you hate? 3
How high should I start the drill? 7
What is 3 times 7? 21
What is 3 times 8? 25
What is 3 times 9? 27
What is 3 times 10? 30
What is 3 times 11? 32
What is 3 times 12? 36
What is 3 times 13? 42
I hope you appreciate the beauty of math now.
```

Modification Problems

1. Give Tutor whatever additional code it needs to be at least modestly helpful to its user. For instance, add a procedure that tells the user when she enters the wrong answer!
2. Suppose that I initialized Mistakes to 1. Modify Tutor so that it still works.

Self-Check Questions

Q. Rewrite these **FOR** statements as **LOOP**s. You can rely on the standard procedures INC and DEC. The call INC(x) increases x to the next higher letter or number, while DEC(x) decreases x.

a) **FOR** Ch := "A" **TO** "Z" **DO** action **END**;
b) **FOR** Count := 1 **TO** 9 **BY** 3 **DO** action **END**;
c) **FOR** Ch := "Z" **TO** "A" **BY** –2 **DO** action **END**;

A. Note that INC and DEC have two forms, and can be called with or without an increment or decrement distance.

a)
```
Ch := "A";
LOOP
    action;
    INC (Ch);
    (**)IF Ch > "Z" THEN EXIT END
END
```

b)
```
Count := 1;
LOOP
    action;
    Count := Count + 3;
    (**)IF Count > 9 THEN EXIT END
END
```

c)
```
Ch := "Z";
LOOP
    action;
    DEC (Ch, 2);
    (**)IF Ch < "A" THEN EXIT END
END
```

Limit Bounds

Let's turn to limit bounds. They are appropriate for:

- Loops that methodically reduce a number to zero, or to less than some known value. For instance, algorithms for **DIV** and **MOD** rely on repeated subtraction; eventually, we can tell that an additional subtraction would be useless, or cause an error.

limit bound applications

- *Successive approximation* methods. Here, each time a calculation is performed the result is compared to a previous calculation. The bound is reached when two successive calculations yield approximately equal results.
- *Bisection* methods that calculate mathematical approximations. The limit is set by approaching to within some tiny amount of an answer that is known to be correct.

- *Non-convergence* tests, which must often be applied to the iterative methods described above to ensure that calculations have not gone astray (producing meaningless results) or have grown uncontrollably (and threaten the legal limits of REAL or INTEGER).

I won't beat around the bush—loops that rely on limit bounds involve numbers. Fortunately, there is math and there is math. Although the example we'll work here deals with numbers, I've reserved more mathematically oriented examples for section 6-4. The problem I'll pose is:

problem: simulating **DIV** and **MOD**

Write a procedure that uses addition and subtraction to do long division. In other words, use + and – to get the effect of **DIV** and **MOD**.

You should remember this particular goal from second or third grade. A *dividend* is going to be divided by a *divisor*. We want to end up with the whole-number *quotient* and *remainder*.

It's not hard to attack the problem with a brute force algorithm—subtract as many 'whole' divisors from the dividend as we can. The count of 'wholes' equals what we'd get from calculating Dividend **DIV** Divisor. Whatever is left over, if anything, is the remainder; it equals Dividend **MOD** Divisor.

loop specification

Goal: Find a whole-number quotient and remainder.
Bound: No more 'whole' divisors can be removed from the dividend.
Plan: Reduce the dividend by subtracting the divisor once.
Add 1 *to a count of 'wholes' that began at* 0.

According to the specification, we're going to subtract divisors from the dividend. Whatever's ultimately left over can be assigned to the remainder. Personally, though, I find it less confusing to start out with a remainder that *equals* the dividend, and then subtract divisors from the remainder. That's what I've done in the long division pseudocode, below.

long division pseudocode

```
initialize a count of Wholes to 0;
initialize the Remainder to Dividend;
LOOP
   (**)IF the remainder is less than the divisor EXIT;
   subtract Divisor from the Remainder;
   add 1 to the count of Wholes
END of the loop
(*The Remainder is less than the Divisor. What else?*)
```

Try to complete the postcondition before you read on. What else do we want this comment to help confirm?

Notice the position of the **EXIT**. For the first time, it's at the very beginning of the loop.

> The bound condition stated in a loop specification may be satisfied without carrying out the loop's plan at all.

Thus, the exit point is put right up front. It has to be there, because the remainder might be smaller than the divisor to begin with.

The final Modula-2 version of our pseudocode is shown as procedure LongDivision, below. As you can see, I've taken particular pains on the postcondition that sums up the loop's goal. Simply restating the loop's **EXIT** condition isn't enough.

exiting at the loop beginning

```
PROCEDURE LongDivision (Dividend, Divisor: INTEGER;
                         VAR Wholes, Remainder: INTEGER);
  (*Simulates DIV and MOD using only addition and subtraction.*)
  BEGIN
    Wholes := 0;
    Remainder := Dividend;
    LOOP
      (**)IF Remainder < Divisor THEN EXIT END;
      Remainder := Remainder – Divisor;
      Wholes := Wholes + 1
    END
    (*We've reached the bound: Remainder < Divisor (it's really the
     remainder). Moreover, (Wholes*Divisor) + Remainder = Dividend
     (our arithmetic was correct).*)
  END LongDivision;
```

Modification Problems

1. Put LongDivision into a program that calls it until a Remainder of 3 is returned.
2. Suppose that you call LongDivision with a negative dividend and divisor. What will happen? Modify LongDivision so that it deals with this unfortunate circumstance properly.

Self-Check Questions

Q. Write a Modula-2 code segment that implements this loop specification. What do you think the loop's goal is? You can assume that Number starts out with some positive value.

Goal: For me to know, and you to find out.
Bound: Number *equals* 0.
Plan: Print the value (Number **MOD** 10).
Divide Number *by* 10 *using* **DIV**.

A. In Modula-2, the specification is:

```
LOOP
  IO.IntOut (Number MOD 10, 1);
  Number := Number DIV 10;
  (**)IF (Number = 0) THEN EXIT END
END;
```

The loop's goal is to print Number in reverse.

Q. That was so much fun we'll do it again. What does this do, and how does it do it?

> *Goal: Those who know don't tell, and those who tell don't know.*
> *Bound:* Number *equals* 0.
> *Plan: Add* 1 *to a counter that began at* 0.
> *Divide* Number *by* 10 *using* **DIV**.

A. The goal is given away by the variable identifiers:

```
Digits := 0;
LOOP
    Number := Number DIV 10;
    Digits := Digits + 1;
    (**)IF (Number = 0) THEN EXIT END
END;
```

The segment counts the number of digits in Number.

Data Bounds

The fourth kind of loop bound is called for when an explicit sentinel, count, or calculated limit can't be relied on to provide an exit condition. Instead, we leave a loop because we've run out of data to examine. In fact, the plan of a data-bound loop is generally devoted to arriving at this exhausted state as rapidly as possible! Data bounds appear in such applications as:

- Searching through organized collections of values. The position (e.g. in the 'middle') of data items helps to delimit the boundaries of a search.

where data bounds appear

- Reorganizing stored data according to bounds that aren't fixed, but are associated with particular values.
- Working with *streams*, or sequences of values (usually characters), whose end isn't marked by a sentinel. We'll discuss streams further in 6–3.

We won't use data bounds to any great extent until we start working with structured data types—in particular, with arrays and pointers. However, I'll describe an algorithm called *binary search* that will give you a feel for what a data bound consists of.

Suppose that you want to find a name in a telephone book. At first, unknown data includes the contents of the whole book, and is bounded by the first and last names. But instead of starting on the first page and patiently working your way to the end, though, why not follow this algorithm instead:

1. Look at the value that falls right in the middle of the boundaries; i.e. in the middle of the book.

divide and conquer

2. It might be the value you want. If it isn't, though, you've found a new lower bound (if it's less, alphabetically, than the name you want) or a new

upper bound (if it's greater, alphabetically, than the name you want) for continuing the search.

3. Start all over again at step 1 *unless* the lower and upper bounds have met, which means that there's no data left to split.

Binary search is a *divide and conquer* technique. Each time we inspect a name, we learn something about the data—we learn that half of it is useless because it's too big or too small. We conquer the data by making its unknown portion smaller and smaller. With binary search and ordered data, we can find one value out of thirty *million* by inspecting, at most, just 25 names.

Section 6-3 Extended Examples: Text Processing

WORK WITH CHARACTERS, AS OPPOSED TO NUMBERS, is called *text processing*. Procedure Echo, below, is an example of the basic text processing program. It *echoes*—reads, and immediately reprints—text without changing its appearance.

vanilla text-echoing procedure

```
PROCEDURE EchoText;
  (*Reads and echos text input while preserving its line structure.*)
  VAR Ch: CHAR;
  BEGIN
    LOOP  (*until we run out of characters to read*)
      IO.CharIn (Ch);
      (**)IF NOT IO.Done() THEN EXIT END;
      IF Ch = IO.EOL
        THEN IO.LineOut
        ELSE IO.CharOut (Ch)
      END  (*IF*)
      (*No more data could be read.*)
    END  (*LOOP*)
  END EchoText;
```

EchoText calls two identifiers from the IO library that we haven't used before.

IO.Done()

- IO.Done() is a BOOLEAN-valued procedure. It's TRUE if the CharIn call that preceded it actually read a character, and FALSE if there wasn't a character to read.

IO.EOL

- IO.EOL (most people spell it out, 'cause it can't be pronounced) is a constant that represents the carriage return. Because some systems use a two-character sequence as the return (one moves to the next line, one moves to the line's beginning) *all* Modula-2 systems pull a switch—if you read a carriage return, the system will give you IO.EOL instead.

Together, they let us read characters, one at a time, until we run out of data. If we happen to spot a character that equals IO.EOL, we can call

IO.LineOut to print a carriage return (simply echoing the IO.EOL character might not be enough). If Done() is ever FALSE, we leave the loop because there's nothing left to read.

The loop specification that underlies Echo is:

echoing text

Goal: Echo text while maintaining line structure.
Bound: There's no more input text.
Plan: Read a character.
If it equals IO.EOL, *call* IO.LineOut, *otherwise echo it.*

> The data bound **NOT** IO.Done() is nearly universal in text processing loops. It may appear as a loop's intentional bound—we want to exit because we've run out of data, or it may be required as a necessary bound—we *have* to exit because we've run out of input—but it almost always appears.

Now, if the computer waits for you to enter data when it executes the IO.CharIn procedure call, how can it ever know that there isn't any more? Well, if data is coming from a file (we'll see how, below) there's no problem—it gets to the end of the file, and that's that.

the end-of-file character

However, when you're typing in text yourself, a special control character will serve as the end-of-input marker. Since the character isn't standardized (on UNIX systems it's usually control-D, while MS-DOS systems use control-Z) I'll leave a space for you to write yours down here: __________.

Modification Problems

1. Put EchoText into a program, then run it and test it.
2. Change EchoText so that it counts, and returns via a variable parameter, the number of lines that are being echoed.
3. Modify EchoText so that it ignores carriage returns in input. However, you *should* print a carriage return after every period, question mark, or exclamation mark, instead.

Self-Check Questions

Q. Suppose that I had gotten dreadfully confused while writing this section, and had accidentally reversed Done() and EOL in procedure EchoText (as shown below). Will it work? What does it do?

```
LOOP
    IO.CharIn (Ch);
    (**)IF Ch = IO.EOL THEN EXIT END;
    IF NOT IO.Done()
        THEN IO.LineOut
        ELSE IO.CharOut (Ch)
    END (*IF*)
END (*LOOP*)
```

A. It works, but weirdly! The code segment ends successfully if it reads a carriage return; otherwise it just echoes characters. However, if there isn't any carriage return, it will print an endless stream of carriage returns.

Streams and Filters

Programs that process text are often called *filters*. That's because we usually think of the sequence of characters that text consists of as forming a *stream* of values. The stream might be temporarily frozen into a file, or it might be trickling out of your keyboard one character at a time. In either case, it's easy to view text processors as filters that change, use, or learn about the stream of values as it flows through the program.

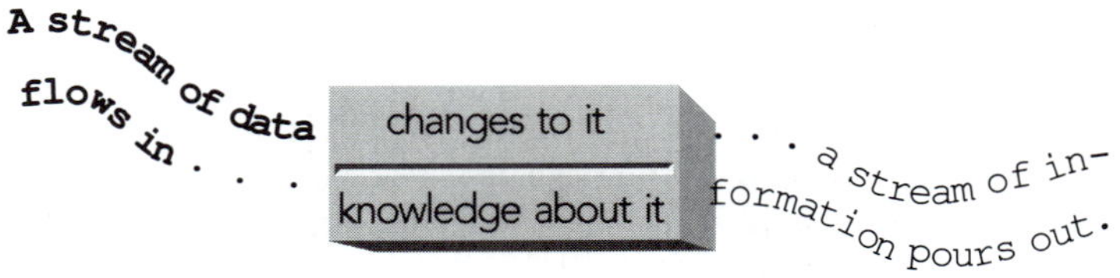

standard input and output

A program's *input stream* is its input—the characters we type into a keyboard. A program's *output stream* is the output it produces, and typically sends to a terminal screen. The keyboard and screen are usually referred to as the *standard* or *default* input and output streams.

The standard input and output streams can be *redirected* or rerouted. Redirecting input lets you process data that's currently stored *in* a file, while redirecting output lets you send your program's results *to* a file. Redirection can be arranged from within a Modula-2 program (we'll see how later), or at the operating system level (input or output are redirected when you issue an 'execute my program' command). On UNIX or MS-DOS systems, it's done like this:

run *program* **<** *InputFile* **>** *OutputFile*

Either '**<** *InputFile*' or '**>** *OutputFile*' is optional, and the exact **run** *program* command depends on your system. (If it occurs to you that there should be a way to let one filter program's *output* be used as the next filter program's *input*, you probably have a bright future in computer science.)

Regardless of the stream's origin or destination, it will pass through our program. A *process filter* does something *to* the stream of data:

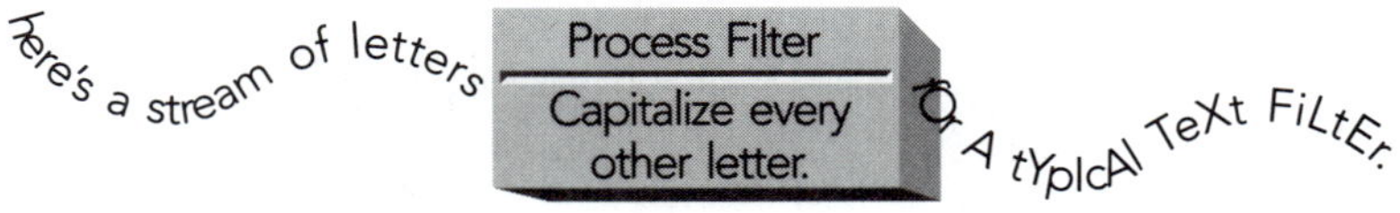

Text-handling problems that process filters typically solve include:

- Copying files, or searching for a particular value in a stream of data.

some process filters

- Capitalizing letters, transforming all letters to lower-case, or changing the order of values in a stream.
- *Stream editing*, or carrying out a stored sequence of editing commands on a stored file.
- Translating data from one form (say, binary) to another (like decimal).

In each case, the process filter applies some basic rule—the process—to the values in the stream. The simplest process filter does very little indeed—it just reads and echoes a sample of text without changing its line structure. The only processing that occurs involves translating an *input* carriage return into an *output* carriage return.

State filters, in contrast, produce information by learning something *about* the data stream.

What is *state*? Computer scientists talk about data's state as a shorthand way of saying 'What does this data mean? What is its status? What do we know about it besides its value?' For instance, the characters on this page have individual values, but they also occupy distinct and meaningful states: as words, as white space,* as illustrations, and so on.

also marginal notes

A *state transition* is a change from one state to another. The goal of most state filters is to spot or establish state transitions. Some of the problems we can solve by writing state filter programs include:

- Counting the number of words in the input stream (by counting in-a-word/not-in-a-word state transitions).

state transitions

- Printing one word or sentence per output line (by adding a carriage return whenever the in-a-word or in-a-sentence state changes).
- Compressing blanks in input (by turning off echoing of characters whenever we're in a blank-spaces state).
- Suppressing extraneous blanks at the beginning (*leading* blanks) or end (*trailing blanks*) of lines of text (by recognizing distinct before-the-line, during-the-line, and after-the-line states).

> Distinguishing between process and state filters is only useful if it suggests a method of problem solving—not for its own sake.

* Footnotes, too, of course.

Perhaps you can modify an existing process filter, or rephrase a complicated problem as a simpler state filter. As with loop bounds, you won't get tested on petty differences. You will find for yourself, though, that seeing the similarities between programs helps make your life that much easier.

Self-Check Questions

Q. The process at the heart of procedure EchoText was this statement:

```
(*Read a value for Ch, then . . . *)
IF Ch = IO.EOL
   THEN IO.LineOut
   ELSE IO.CharOut (Ch)
END
```

Write the central process statement or statements for each of these alternative goals:

a) Print every input letter as a capital, without worrying about layout.
b) Echo every other character, without worrying about layout.
c) Maintain line structure, but replace each "&" with two carriage returns.

A. It's not hard once you get the hang of it:

a) IO.CharOut (CAP(Ch));

b)
```
IO.CharOut (Ch);
IO.CharIn (JunkCh);  (*This character will just be ignored.*)
```

c)
```
IF Ch = IO.EOL
     THEN IO.LineOut
   ELSIF Ch = "&"
     THEN IO.LineOut;
          IO.LineOut
   ELSE IO.CharOut (Ch)
END
```

A Few Process Filters

Let's write a few process filters. These little programs are remarkably powerful, and they're easy to write.

The first step of most text processing loops will be *read a character*. This is the smallest step we can take toward the loop's bound—running out of data. The rest of the loop's plan will be a more or less complicated variation of procedure EchoText. I'll start with a variation that's about as complicated:

problem: counting lines and characters

Write a program that counts the number of lines and characters in its input stream.

Should carriage returns be counted as input characters? It's up to you—I personally don't. However, I do use the carriage returns, by assuming that they mark lines of input. A loop specification is:

specification

Goal: Know the number of lines and characters of input.
Bound: There's no more input.
Plan: Read a character.
Decide if the count of lines or characters should increase.

As usual, a pseudocode outline puts the specification in its context—we have to initialize variables and report on results as well as loop:

pseudocode

```
initialize LineCount and CharCount variables to 0;
LOOP
    read a character;
    (**)IF it couldn't be done then EXIT;
    IF it equaled IO.EOL then increment LineCount
                        otherwise increment CharCount
END of the loop;
report on the results
```

The final program is shown below. If you run this on your own, don't forget that your own system will have specific rules for *a*) adding an end-of-input character if you type in the program's data yourself, or *b*) letting you redirect input from an existing file if you don't want to wear out your fingers.

character and line counting program

```
MODULE Counter;
    (*Counts the number of characters and lines in input text.*)
IMPORT IO;
VAR Ch:  CHAR;
    ChCount, LineCount: INTEGER;
BEGIN
    ChCount := 0;
    LineCount := 0;
    IO.StringOut ("Give me a few lines of input.");  IO.LineOut;
    LOOP
        IO.CharIn (Ch);
        (**)IF NOT IO.Done() THEN EXIT END;
        IF (Ch = IO.EOL)
            THEN LineCount := LineCount + 1
            ELSE ChCount := ChCount + 1
        END
    END;
    (*The counters represent the number of returns and non-returns.*)
    IO.IntOut (ChCount, 1);  IO.StringOut ("  characters read,  ");
    IO.IntOut (LineCount, 1);  IO.StringOut ("  lines read.")
END Counter.
```

(*output follows*)

```
Give me a few lines of input.
Although I work, and seldom cease,
At Dumas pere and Dumas fils.
Alas, I cannot make me care
For Dumas fils and Dumas pere.
120 characters read, 4 lines read.
```

Modification Problems

1. Improve Counter so that it counts sentences as well as lines.
2. Convert Counter into a program that capitalizes and prints its input.
3. Change Counter into a program that lets you know only the number of lines that are *a*) shorter that ten characters, and *b*) longer than twenty.
4. 'Is there a carriage return at the end of the last line?' is one of the enduring existential questions of the computer age. Modify Counter so that its count will be correct either way. *Hint:* think about being 'in' a line as a state that changes briefly when you read a carriage return.

Text Searches and Multiple Bounds

One of the most common text-filtering problems is *searching*—sifting though a text stream for a particular value. Suppose that I ask the question simply:

problem: finding the letter s

> Write a program that finds the letter "s" in a sample of input, and lets you know its position.

If you think about it, this is really a sentinel bound problem: we want to find a value that equals the sentinel "s". On the face of it there's no difficulty; we just read characters until the sentinel shows up. But is the sentinel bound effective enough to *guarantee* termination? No. Suppose there isn't an "s" in input. Even though it's one of the most common letters in English, it doesn't have to appear.

necessary bounds guarantee termination

> Like most search loops, our search for an "s" requires a necessary bound—a bound that makes the loop stop if its intentional bound can't be met.

A variety of candidates might serve as the necessary bound. When should we call off the search? After some number of characters? At the end of the line? When there's no more input? It doesn't matter—as long as the necessary bound can *guarantee* that the loop will **EXIT**. Thus, the specification of a typical searching loop is:

loop specification for a text search

Goal: Find a sentinel.
Intentional bound: The sentinel was found.
Necessary bound: Data was somehow exhausted.
Plan: Read a single value.

In practice, the **EXIT** check begins with a check of the data bound. Was a character read at all? If not, the other checks aren't meaningful. The example below shows an appropriate occasion for relying on Modula-2's partial evaluation of BOOLEAN conditions. If **NOT** IO.Done() is TRUE, then the whole expression is TRUE. Evaluation quits before we have to look at Ch more closely:

checking the data bound first

```
(**)IF NOT IO.Done() OR (*then if*)
        (Ch = SENTINEL) OR (Ch = IO.EOL) OR
         (Count = LIMIT) THEN EXIT END;
```

Let's get back to our search for an "s". I'll specify the search loop as:

loop specification

Goal: Find the position of the first "s" *in input.*
Intentional bound: We read an "s".
Necessary bound: We ran out of input.
Plan: Read a character.
    Add 1 *to a position counter that began at* 1.

Pseudocode adds a few details. Read the loop's confirmation comment—its postcondition—carefully.

pseudocode

```
initialize the position counter to 1;
LOOP
   read a letter;
   (**)IF we hit a bound then EXIT;
   add 1 to the position counter
END of the loop;
(*We found the sentinel or ran out of input. If
 the sentinel was found, we know its position.*)
print the position of the "s" if one appeared
```

bounds, not goals, end loops

> Loops end because they reach their bounds, and not because they achieve their goals. When there are multiple bounds, *look after you loop* to decide exactly what caused termination.

A postcondition comment should help a program reader confirm her mental picture of what has happened. A postcondition that merely parrots the **EXIT** condition doesn't say enough—she can read the bounds for herself.

However, a postcondition that jumps to a conclusion about why the loop ended is even worse. We *hope* that we found the sentinel's position, but there's always the possibility that that's not why we left the loop. As you read program

FindS, below, you'll see that half its code is devoted to deciding exactly why we left the loop.

a search with multiple bounds

```
MODULE FindS;
   (*Reports the position of the first "s" it reads.*)
IMPORT IO;
CONST SENTINEL = "s";
VAR Position: INTEGER;
     Letter: CHAR;
BEGIN
   Position := 1;
   IO.StringOut ("Please enter your sample data."); IO.LineOut;
   LOOP
      IO.CharIn (Letter);
      (**)IF NOT IO.Done() OR (*then if*)
            (Letter = SENTINEL) THEN EXIT END;
      Position := Position + 1
   END;
   (*Now, we look after we loop.*)
   IF NOT IO.Done() (*Check this first, because Letter might not be defined.*)
      THEN IO.StringOut ("The sentinel wasn't found.")
      ELSE IO.StringOut ("The sentinel was found in position  ");
           IO.IntOut (Position, 1)
   END
END FindS.
```

```
Please enter your sample data.
Mary had a tiny lamb with fleece a pale white hue,
And everywhere that Mary went the lamb kept her in view.
To academe he went with her, illegal and quite rare.
It made the children laugh and play to view a lamb in there.
The sentinel wasn't found.
```

Modification Problems

1. Modify FindS so that it reports the line number, as well as the position within the line, of the sentinel.
2. Put the sentinel-searching code of FindS into a procedure that has two parameters—the value of the sentinel, and the position it's found in. What value might you give to the position parameter if the sentinel wasn't found?
3. Modify FindS so that it looks for, and prints the position of, any i before an e that follows a c.
4. The poem above is a *lipogram*—a composition that omits one letter—written by A. Ross Eckler. Can you rewrite it so that some other letter (say, 'e,') is missing?

A State Filter: Counting Words

Looking for changes in *state* is a useful approach to problem solving that finds many applications in text processing. Recall that state filters collect, and then use or act on information *about* the input stream. Counting words provides a good example:

problem: counting words

> We can define *words* as sequences of characters that are separated by blanks. Write a program that counts the number of words in a stream of text. There may be extra blanks between any two words, or at the beginning or end of the stream of characters.

The fact that there can be extra blanks prevents the kind of quick and easy solution that I prefer—count spaces, and subtract 1.

state transitions

But suppose that I rephrase the problem like this: look for, and count, state *transitions*—changes from one state to another. What states are we keeping an eye on? Well, as we read through the input stream one character at a time, we always know at least one fact about the data: either we're in a word, or we're not in a word. Those are our two states. It doesn't matter how long the word is, or how many blanks there are, because just one state will always prevail. If we can count state transitions, we can count words. Wow!

Rephrasing the problem of counting words as one of spotting state transitions is exactly the kind of elegant solution that makes life difficult for the people who give you examinations. The same solution can be applied to a variety of seemingly different problems:

some equivalent problems

Spot transitions from in-a-word to not-in-a-word, then
  Count words (increment a count after each transition)
  Compress blanks (print a single space after each transition)
  Print one word per line (print a carriage return after each transition)

Like most stream-processing loops, word-counting loops have data bounds. The smallest step a word counter can take toward the loop bound is trying to read a single character. This single step is also big enough to let us know if a state transition has occurred, since it just takes one character to enter or leave a word. A loop specification is:

loop specification

Goal: Count the number of words.
Bound: We ran out of data.
Plan: Read a single character.
  If a transition occurred, update the current state
    and add 1 *to the count of words.*

There are only two possibilities for the current state: either we're in a word, or we aren't. A BOOLEAN variable is well-suited for service as a *state variable* that represents the current state. We'll use the state variable like a switch, and count every time it goes from 'on' to 'off.'

pseudocode

```
initialize a state variable InAWord to FALSE;
initialize a counter variable to 0;
LOOP
    try to get a character;
    (**)IF NOT Done there's no more data, so EXIT;
    IF InAWord
        THEN IF the state changed, count it and update the state to FALSE
        ELSE IF the state changed, update the state to TRUE
    END the IF statement
END the loop;
report on the results;
```

The code could be a little simpler. After all, if the switch is already on, it doesn't matter if we turn it on again; just as if it's already off, we can turn it off again without causing problems. What's important is that we only count *transitions*, in one direction, from one state to the other.

checking the final state

Potential for a subtle error does exist, though. Suppose that we leave the loop while we're InAWord. Will the last word be counted? No, because as I've defined it, there hasn't been a state transition. One more check is needed before we report on the results. The completed program is shown opposite.

Modification Problems

1. Are there bugs in Words? You bet! Fix the bug that causes two words that are separated by a tab character to be counted as one word.
2. Change Words into CountSentences. In particular, deal with the problem of mistakenly counting '. . .' as three sentences.
3. Extend Words so that it finds out the average number of words per sentence in its input. Then get the average length of each word, too.

word counting program

```
MODULE Words;
   (*A state filter that counts the words in a sequence. One or more
    blank spaces separate words.*)
IMPORT IO;
CONST BLANK = "  ";
VAR InAWord, ItWasBlank: BOOLEAN;
     Count: INTEGER;
     Ch: CHAR;
BEGIN
   InAWord := FALSE;
   Count := 0;
   IO.StringOut ("Say something poetic."); IO.LineOut;
   LOOP   (*until there's no more input*)
      IO.CharIn (Ch);
      (**)IF NOT IO.Done() THEN EXIT END;
      ItWasBlank := (Ch = BLANK) OR (Ch = IO.EOL);
      IF InAWord
         THEN IF ItWasBlank THEN
                        InAWord := FALSE;
                       Count := Count + 1
                 END   (*We read a blank and there was a transition.*)
         ELSE IF NOT ItWasBlank THEN
                       InAWord := TRUE
                END   (*We read a letter and there was a transition.*)
      END  (*IF*)
   END; (*LOOP*)
   (*We know that we've reached the bound. However, if we're InAWord
    when we leave the loop, we still have to count the very last word.*)
   IF InAWord THEN Count := Count + 1 END;
   IO.StringOut ("The number of words was:    ");
   IO.IntOut (Count, 1)
END Words.
```

```
Say something poetic.
The forehead of Shelley was cluttered with curls,
And Keats never was a descendent of Earls,
And Byron walked out with a number of girls.
The number of words was:  25
```

Stream Handling Tools from IO

The IO library contains procedures that redirect I/O. They change the places where input comes from, and where output goes to. However, the exact details of how input and output files are changed is the least standard facet of Modula-2. Your system's versions of the open and close calls described below may differ in the way they obtain new file names. Program CopyFile, below, will give you an idea of how it's done.

text copying example program

```
MODULE CopyFile;
   (*Copies a file, after prompting for appropriate file names.*)
IMPORT IO;
VAR FromFile, ToFile: ARRAY [0..14] OF CHAR;
   (*These string variables will hold the file names.*)
PROCEDURE EchoText;
   (*Read and echo text input while preserving its line structure.*)
   VAR Ch: CHAR;
   BEGIN
      LOOP  (*until we run out of characters to read*)
         IO.CharIn (Ch);
         (**)IF NOT IO.Done() THEN EXIT END;
         IF Ch = IO.EOL
            THEN IO.LineOut
            ELSE IO.CharOut (Ch)
         END  (*IF*)
         (*No more data could be read.*)
      END  (*LOOP*)
   END EchoText;
BEGIN
   IO.StringOut ("Please enter the existing file's name:   ");
   IO.StringIn (FromFile);
   IO.StringOut ("Please enter the new file's name:   ");
   IO.StringIn (ToFile);
   IO.OpenInput (FromFile);
   IF NOT IO.Done() THEN (*the OpenInput call failed*)
      IO.StringOut ("Reading from keyboard -- sorry.");
      IO.LineOut
   END;
   IO.OpenOutput (ToFile);
   IF NOT IO.Done() THEN (*the OpenOutput call failed*)
      IO.StringOut ("Writing to screen -- sorry.");
      IO.LineOut
   END;
   EchoText;
```

(program continues)

```
    (*Return to the standard input and output.*)
    IO.CloseInput;
    IO.CloseOutput;
    IO.StringOut ("File copy complete.")
END CopyFile.
```

```
Please enter the existing file's name:  Suzy.wk
Please enter the new file's name:  My.wk
File copy complete.
```

Procedure OpenInput changes the standard input stream from the keyboard to somewhere else, usually a file. A companion, procedure CloseInput, restores the keyboard as the standard default. OpenInput takes just one argument—a string that names the file that's going to provide input. The string can be a variable, a constant, or literal text between quote marks:

redirecting input

```
OpenInput ("RawData");
(*Make file RawData the standard input source.*)
IO.CharIn (Ch);
(*All input, like Ch, will be read from file RawData.*)
CloseInput;
(*Go back to the default standard input source—the keyboard.*)
IO.CharIn (Ch);
(*From now on, Ch will be read from the keyboard as usual.*)
```

Another pair of procedures redirects output. Procedure OpenOutput changes the default for IO's output procedures. Instead of writing to the screen, they'll send output elsewhere—again, usually the file named in the call of OpenOutput. Its partner, CloseOutput, brings back the standard output—the screen.

redirecting output

```
OpenOutput ("Cooked");
(*Make file Cooked the standard output location.*)
IO.StringOut ("Here are the results.");
(*All new output will go to file Cooked.*)
CloseOutput;
(*Go back to the default standard output location—the screen.*)
IO.StringOut ("Back to the screen.");
(*Now, all out goes to the screen.*)
```

> Always take care of *all* communication with the program user before you redirect input or output. If you don't, she won't see your prompts, and your program won't read her keyboard input.

Now, there are no guarantees that any procedure associated with input or output will succeed. Fortunately, failure isn't catastrophic. In fact, sometimes the only way to find out if a file exists, or if there's a value to be read, is to try and fail.

how calls fail

- A call of OpenOutput might fail if the file couldn't be opened. Done() will be FALSE, and output will continue to go to the standard output.
- A call of OpenInput might fail if the file didn't exist. Done() will be FALSE, and input will continue to come from the standard output.
- A call of any input procedure would fail if there was nothing to read. Done() will be FALSE, and the input procedure's argument will be undefined.

Thus, function Done() is the IO library's all-purpose success or failure indicator. Done() always reflects the status of the most recently called IO procedure.

I've summarized the new features of the IO library below. The shaded function, termCH(), is included for compatibility with some systems, but we won't be using it in this text.

constant

Name	*Type*	
EOL	CHAR	*Equivalent to the end of line character.*

functions

Name	*Result type*	
Done ()	BOOLEAN	TRUE *if the last operation was successful.*
termCH ()	CHAR	*The character 'waiting' at the terminal.*

procedures

Name	
OpenInput (*StringValue*);	*Get future input from file 'StringValue'.*
OpenOutput (*StringValue*);	*Send future output to file 'StringValue'.*
CloseInput;	*Return to the standard input stream (the keyboard).*
CloseOutput;	*Return to the standard output stream (the screen).*

Background: What is termCH()?

termCH() represents the character that ended the most recently read string or number. In effect, this is the character that is about to be read. It is the *term*inator *ch*aracter that ended the most recently read string or number.

Huh? Let me try again. A call of IO.StringIn first skips leading blanks. Then, it reads (and saves) characters until it hits a blank or control character that marks the end of the string. This blank or control character is still part of the input stream, though—it could be read with IO.CharIn. However, IO.termCH() acts as a 'lookahead' function that lets us inspect without affecting the stream.

Wirth named and included both Done and termCH as variables in his original I/O library. I've retained his names (to avoid confusing you even more), but have implemented both as functions.

Modification Problems

1. CopyFile didn't try very hard to get a legal file name. Add procedures to the program that prompt the user for new file name—and repeat the prompts until the name of files that can actually be opened are supplied.
2. A file that is copied to itself is said to be *clobbered.* I'm sure you can figure out why. Modify CopyFile so that it can't accidentally clobber its input file.
3. Some Modula-2 systems will let alternative pieces of hardware, like printers, mice, musical keyboards, or speakers be substituted for the standard input or output. Find out if your system will let you make any of these substitutions. For example, can you make CopyFile 'write' to your printer?

Section 6-4
Extended Examples: Mathematical Processing

LET'S TURN OUR ATTENTION TO A FEW mathematically oriented programs. The first two examples, Euclid's algorithm and Newton's method, involve methods of calculation that predate computers by thousands of years. Nevertheless, they are so effective that they are among the most frequently programmed mathematical algorithms. The second pair of problems, finding pervasive roots and approximating π, use brute force methods that would never have been practical without computers.

Mathematical processing, like text processing, tends to follow typical models. All of our examples have limit bounds, and all but the first involve approximations. That's what calculation is like with REAL numbers—the only way to get exact figures is to be very clever about fudging and rounding! We'll try three different approaches to approximation:

three limits on approximation

- Using Newton's method to calculate square roots will bring us closer and closer to an exact answer; we'll quit when 'closer' gets arbitrarily tiny.
- Summing a series of pervasive roots will get us too close to distinguish between one calculation and the next; we'll quit when the computer isn't precise enough anymore.
- Approximating π will focus on getting closer to an exact answer than an existing approximation; we'll quit when we're good and ready.

Example: Euclid's Algorithm

Our first example implements an algorithm that is attributed to Euclid (about 330–270 B.C.E.). Although he's best known for his contribution to geometry (and for his crack—'There is no royal road to geometry'—that even kings have to study), Euclid also formalized a method for finding two numbers' greatest common divisor. It has come to be known universally as *Euclid's Algorithm.*

Euclid's algorithm is based on a fact, followed by a clever insight. The fact is that if a number divides both *Bigger* and *Smaller*, it also divides *Bigger–Smaller*. Why? Well, let's call the divisor *x*. Then:

the fact

Bigger equals some number *A* times *x*, and
Smaller equals some number *B* times *x*, so
Bigger–Smaller equals *A–B* times *x*.

which brings us back to where we started from—saying that if a number divides *Bigger* and *Smaller*, it also divides *Bigger–Smaller*.

Enough for the fact; let's get to Euclid's clever insight. First, though, take a moment to match wits with a guy who probably lived in a mud hut. You want to find the greatest common divisor (GCD) of two numbers. You know that the smaller number, and the numbers' difference, share the same GCD. Can you figure out a method of finding the GCD?

the insight

Give up? The insight is that replacing *Bigger* and *Smaller* by *Smaller* and the *Bigger–Smaller* difference is a step that can be repeated—it's an algorithm. How many times should the algorithm be repeated? Until the *Smaller* value equals zero. When it does, the *Bigger* value will equal the greatest common divisor, and we'll have the answer.

Let me plug in some numbers and grind through an example. Say the GCD is *x*, the bigger number starts as 9*x* and the smaller number begins as 7*x*. Then:

Euclid's algorithm in action

Bigger	*Smaller*	*Bigger – Smaller*
$9x$	$7x$	$9x - 7x = 2x$
$7x$	$2x$	$7x - 2x = 5x$
$5x$	$2x$	$5x - 2x = 3x$
$3x$	$2x$	$3x - 2x = x$
$2x$	x	$2x - x = x$
x	x	$x - x = 0$
x	0	*Bigger has been reduced to x—the greatest common divisor*

Now, although you may have never thought of it in this way, the **MOD** operator carries out precisely this kind of subtraction, but much more quickly. Note that the result of each **MOD** operation will always be smaller than the prior 'smaller' number, which make easy to keep *Bigger* and *Smaller* straight:

Euclid's algorithm using **MOD**

Bigger	*Smaller*	*Bigger* **MOD** *Smaller*
$9x$	$7x$	$9x$ **MOD** $7x = 2x$
$7x$	$2x$	$7x$ **MOD** $2x = x$
$2x$	x	$2x$ **MOD** $x = 0$
x	0	*Bigger has been reduced to x—the GCD*

Rephrasing Euclid's algorithm as a loop specification gives us:

loop specification

Goal: Find the greatest common divisor of two numbers.
Bound: One number has been reduced to zero.
Plan: Take the larger number **MOD** *the smaller.*
Save the new larger and smaller values.

Should we implement the algorithm as a procedure or function? Since it returns a single value and doesn't modify its arguments, most people calculate GCD with a function. A pseudocode restatement of the specification is:

function pseudocode

```
LOOP
    take Larger MOD Smaller;
    save the result and the smaller number;
    (**)IF Smaller is zero then EXIT
END of the loop;
RETURN the value of Larger
```

Finally, we have the Euclid's algorithm function itself:

Euclid's algorithm function

```
PROCEDURE GCD (Larger, Smaller: INTEGER) : INTEGER;
    (*Employs Euclid's algorithm to return the GCD of two numbers.*)
    VAR Remainder: INTEGER;
    BEGIN
        LOOP  (*until each instance of the GCD is removed from Smaller*)
            Remainder := Larger MOD Smaller;
            Larger := Smaller;
            Smaller := Remainder;
            (**)IF Smaller = 0 THEN EXIT END
        END;
        (*Larger represents the greatest common divisor.*)
        RETURN Larger
    END GCD;
```

Modification Problems

1. Rewrite GCD as a procedure that returns the number of iterations it took to find a GCD. Then, see how big a number you can get.
2. The chance that two numbers taken at random share a common divisor greater than 1 is $\pi^2/6$. Or is it? Put GCD in a program that tests this conjecture by calculating the GCD of every pair of numbers from 1,1 to *n*,*n*.

Example: Newton's Method

Our second historical example is a method of calculating square roots. It dates from the Hammurabi dynasty (ca. 1800–1600 B.C.E.)—the Old Babylonian era. Although the method is often attributed to Isaac Newton (hence its name) it was fully described on clay tablets, in lovely base 60 cuneiform, more than three millennia before Newton arrived. The algorithm is:

Take a guess (besides zero) at the number's square root. The assignment:

calculating square roots

```
Guess := ((Number / Guess) + Guess) / 2;
```

gives a new Guess value that is closer than the original to being correct, no matter how wild the original guess was.

> Newton's method is typical of iterative approaches to numerical approximation. The accuracy of the result is limited only by our patience, or by the number of places a computer provides for storing digits. There is no fundamental limit to the method itself.

approaching ε

A Modula-2 loop, however, must have a bound. In this case, the bound is provided by a mathematical limit: we continue to approximate the root until the difference between two subsequent approximations is quite small. In mathematical terms we'd say that the difference approaches ε, or *epsilon*, which is Greek for *itsy-bitsy*. A specification of the loop is:

Goal: Approximate a square root.
Bound: The difference between subsequent guesses is < EPSILON.
Plan: Take a new guess.
Save the old guess.

A common value for EPSILON is 1.0E–09, or one part in a billion. We can pseudocode a Newton's method function as:

Newton´s method pseudocode

Initialize an OldGuess *value;*
LOOP
  Calculate a value for NewGuess;
  (**)IF *it's within* EPSILON *of* OldGuess **EXIT**;
  Make NewGuess *the new* OldGuess
END *of the loop;*
RETURN *the most recent* NewGuess

The final function is shown below. I tried calculating the square root of 2 with OldGuess initialized to 1.414222—the Babylonians' best recorded estimate—and got to within 0.0000000001 of the correct root in just one iteration!

Newton's method procedure

```
PROCEDURE Newton (Number: REAL) : REAL;
  (*Use Newton's method to approximate a square root.*)
  CONST EPSILON = 1.0E-09;
  VAR OldGuess, NewGuess: REAL;
  BEGIN
    OldGuess := 1.0;  (*Take a lousy first guess.*)
    LOOP
      NewGuess := ((Number / OldGuess) + OldGuess) / 2.0;
      (**)IF ABS (NewGuess - OldGuess) < EPSILON THEN EXIT END;
      OldGuess := NewGuess
    END;
    (*The difference between approximations is less than EPSILON.*)
    RETURN NewGuess
  END Newton;
```

Example: Pervasive Roots

Our next loop problem also relies on a boundary set by a mathematical limit.

problem: pervasive roots

> The equation $y^2 - y = x$ isn't terribly exciting. Were we all unimaginative English majors, I'd probably solve it with the quadratic formula. Whole-number solutions are easy to find; for instance, $5^2 - 5 = 20$.
>
> However, as proto-computer scientists, I'm sure you'll appreciate the opportunity to see and solve the equation like this: as a *pervasive* square root:
>
> $$y = \sqrt{x + \sqrt{x + \sqrt{x + \sqrt{x + \sqrt{x + \cdots}}}}}$$
>
> Rather astonishingly (to me, at least), the long sequence of added square roots (and roots of roots) of 20 also converges to 5—the two equations really are equivalent. Write a program that calculates a pervasive root to the limit of your computer's accuracy.

Aren't computers wonderful! Now, not all solutions will converge on a whole number. But as we decided with Newton's method for finding square roots, computer arithmetic ultimately provides a limiting factor. Although we intend to add an infinite sequence, eventually the root of a root of a root, etc., becomes too small to influence the computer's REAL representation of the sum.

Solving the pervasive root problem obviously requires a brute force approach. Let's calculate the value of the pervasive square root of a given number to the limit of computer accuracy. At the same time, we'll keep track of the number of steps were required to reach that limit. Our bound, in this case, will be that two successive sums seem to be equal.

limits on REAL comparison

> Two REAL values that appear to be equal may not actually be equal. Comparisons between REALs are used to demonstrate the limits of REAL accuracy, rather than the accuracy of REAL calculations.

In other words, adding yet another tiny decimal to a REAL won't be an error, but it won't appear to change the REAL's value. This fact has caused nasty headaches for numerical programming over the years. It is only an infinitesimal exaggeration to say that it used to be common practice for computers to claim that the sine of 90° was very close to, but not quite, one.

Fortunately, we're not concerned with finding the *right* answer. Instead, we want an answer that will meet our bound:

loop specification

Goal: Count steps needed to approximate an infinite series of calculations.
Bound: The difference between successive calculations is undetectable.
Plan: Save the previous calculation.
Find the square root of (the number + the previous calculation).
Increment a step counter that began at 1.

In the pseudocode below, notice that the pervasive root is calculated from the inside out. If you will let me show it with mirrors:

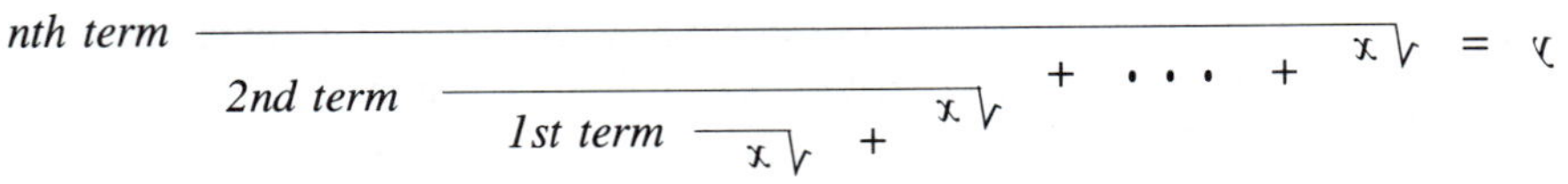

A pseudocode outline of the final program is:

pseudocode

initialize a counter to 1;
calculate the innermost root;
LOOP
    calculate the addition of another term;
    (**)**IF** *it 'equals' the prior calculation* **EXIT**;
    increment the counter
END *of the loop;*
print the results

The completed version of program Pervade is on the next page.

limit-bounded loop example

```
MODULE Pervade;
   (*Computes a pervasive square root to the limit of REAL accuracy.*)
IMPORT IO;
IMPORT MathLibØ;
VAR Number, Current, Last: REAL;
      Count: INTEGER;
BEGIN
   IO.StringOut ("Enter a number.    ");
   IO.RealIn (Number);
   Count := 1;
   Current := MathLibØ.sqrt(Number);
   LOOP   (*until two approximations are indistinguishable*)
      Last := Current;
      Current := MathLibØ.sqrt(Number + Last);
      (**)IF Current = Last THEN EXIT END;
      Count := Count + 1
   END;
   (*A more accurate guess wouldn't be stored more accurately.*)
   IO.StringOut ("The root took  ");
   IO.IntOut (Count, 1);
   IO.StringOut ("  steps to converge on  ");
   IO.FloatOut (Current, 12, 5)
END Pervade.
```

```
Enter a number.  20.0
The root took 8 steps to converge on 5.00000E+000
```

Example: Approximating π

Let's consider a final example of settling for an answer that is 'good enough'—finding a rational approximation to π.

The search for whole numbers that, when divided, equal π has provided mathematical employment for centuries. Famous approximations include 19/6 (the Ahmes Papyrus, circa 1,650 B.C.E.), 25/8 (a Babylonian guess), $223/71 < \pi < 220/70$ (Archimedes, around 150 B.C.E.), and 377/120 (Ptolemy of Alexandria, early first century). The most remarkable approximation comes from Tsu Ch'ung-chih (in the fifth century). His value—355/113, equal to 3.1415929—was not improved on for a thousand years.

Such searches are ultimately in vain, as π is irrational, and can't be expressed as the quotient of two integers. However, I can pose this problem:

problem: rationally approximating π

> On the assumption that you already know the value of π to an arbitrary number of decimal places, write a program that finds the smallest rational fraction that is closer to π than Tsu's approximation.

As we've seen, the customary way to state a 'good enough' limit is to say that:

ABS (π – *our approximation*) < EPSILON

EPSILON, as usual, represents a very small number. How small? For our purposes, the difference between π and Tsu's approximation is small enough:

choosing EPSILON

CONST EPSILON = ABS (π – 355/113);

A brute force approach to working up an approximation isn't very hard. Take any fraction. If it's too large, incrementing the denominator makes it a bit smaller, while increasing the numerator will make it a tad larger. I can outline the loop specification as:

loop specification

Goal: Find a rational approximation to π.
Bound: The approximation is closer to π *than* 355/113 *is*.
Plan: Make the current approximation slightly larger or smaller.

A first refinement, below, is typical of numerical approximation loops. If our first guess is close enough, we can exit. If not, we have to recalculate:

first pseudocode

```
make an initial approximation;
LOOP
   (**)IF the approximation is close enough EXIT;
   try to get closer
END the loop;
print the answer
```

The *initial approximation* can be any positive fraction. The *try to get closer* step is taken care of by adding one to the numerator or denominator.

second pseudocode

```
initialize a Numerator and a Denominator;
calculate the approximation, Numerator/Denominator;
LOOP
   IF the approximation is close enough EXIT;
   IF the approximation is too small
      THEN increase Numerator by one
      ELSE increase Denominator by one
   recalculate the approximation
END of the loop;
print the answer
```

We have to take one additional precaution. Suppose that we can't surpass the accuracy of Tsu's approximation for some time. Modula-2 supplies two additional types, LONGINT and LONGREAL, that can represent many more digits than ordinary INTEGER and REAL variables, so I've used them in program PiApprox. The call of LONGFLOAT is system dependent, though. You should check your own documentation to find out more about usage of the 'oversized' types.

The completed program is shown below. Its result is indeed a millionth or two closer to π. Before we congratulate ourselves on our calculative prowess too strenuously, though, I might point out that the accuracy of this answer was easily surpassed in the early fifteenth century by Al-Kashi, in Tamerlane's (of Mongol hordes fame) capital city of Samarkand. I've used his best result as my reference value for π in program PiApprox, below.

pi approximation program

```
MODULE PiApprox;
   (*Finds a rational approximation to PI that's closer than 355/113.*)
IMPORT IO;
IMPORT MathLib0;
CONST PI = 3.14159265358979;          (*From Al-Kashi.*)
      TSU = 355.000000000000/113.000000000000; (*From Tsu Ch'ung-chih.*)
      EPSILON = ABS(PI-TSU);
VAR Numerator, Denominator: LONGINT;
    Approximation: LONGREAL; (*Will hold Numerator/Denominator.*)
BEGIN
  IO.StringOut ("The next closer approximation to pi than 355/113 is");
  IO.LineOut;
  Numerator := 1; (*Let's start with a terrible guess.*)
  Denominator := 1;
  Approximation := LONGFLOAT(Numerator) / LONGFLOAT(Denominator);
  LOOP
    (**)IF ABS (Approximation-PI) < EPSILON THEN EXIT END;
    IF Approximation < PI
      THEN Numerator := Numerator + 1
      ELSE Denominator := Denominator + 1
    END; (*IF*)
    Approximation := LONGFLOAT(Numerator) / LONGFLOAT(Denominator)
  END (*LOOP*);
  (*Numerator/Denominator is closer to PI than TSU is.*)
  IO.IntOut (Numerator, 1);
  IO.CharOut ("/");
  IO.IntOut (Denominator, 1);
  IO.CharOut ("=");
  IO.FixOut (Approximation, 20, 14)
END PiApprox.
```

```
The next closer approximation to pi than 355/113 is
52163/16604=3.14159238737654
```

Modification Problems

1. Remove the assignment to Approximation from PiApprox. Are other changes required? What about making the assignment the first action the loop takes?

2. How long does it take for PiApprox to come up with its approximation? Add a counter to the program.

Section 6-5
In Depth: **REPEAT** and **WHILE**

MODULA-2 PROVIDES TWO OTHER CONDITIONAL LOOPING statements. Like **LOOP**, they're supplied so that an action can be repeated until a BOOLEAN condition is met. Like **LOOP**, they have distinct boundaries—the 'next' program statement is always the one that follows the loop. But unlike **LOOP**, their exit points are fixed. They are always entered at the beginning, and left at the end.

The **REPEAT** statement obeys this syntax:

the **REPEAT** statement

```
REPEAT
    sequence of statements
UNTIL a BOOLEAN expression is TRUE
```

The **WHILE** statement is similar, but the condition is found at the start of the statement:

the **WHILE** statement

```
WHILE a BOOLEAN condition is TRUE DO
    sequence of statements
END
```

Let's try a few examples. Suppose that we want to find a sentinel value in a sequence of input characters. Written as a **LOOP**, we'd say:

as a **LOOP**

```
LOOP
    IO.CharIn (OneCharacter);
    (**)IF OneCharacter = SENTINEL THEN EXIT END
END
```

As a **REPEAT** statement, the code is nearly identical:

as a **REPEAT**

```
REPEAT
    IO.CharIn (OneCharacter)
UNTIL OneCharacter = SENTINEL
```

Rewriting the segment as a **WHILE** statement requires a bit of thought, though. We'd like to say:

```
WHILE OneCharacter <> SENTINEL DO     etc.
```

but that won't work. Why not? Because OneCharacter isn't defined until we read a character. Writing the loop correctly requires that OneCharacter be initialized before the loop:

as a **WHILE**

```
IO.CharIn (OneCharacter);
WHILE OneCharacter <> SENTINEL DO
    IO.CharIn (OneCharacter)
END
```

If I pose a problem that is a bit more difficult, you can see why the three loops are not interchangeable alternatives. Suppose that instead of just spotting a sentinel, we read and add numbers that precede it. The **LOOP** solution is:

a summing **LOOP**

```
Sum := 0;
LOOP
    IO.IntIn (Number);
    (**)IF Number = SENTINEL THEN EXIT END;
    Sum := Sum + Number
END;
```

This **REPEAT** alternative seems easy, but it's incorrect:

a summing **REPEAT**, with bug

```
Sum := 0;
REPEAT
    IO.IntIn (Number);
    Sum := Sum + Number  (*Are you sure?*)
UNTIL Number = Sentinel
```

Can you see what the problem is? We can't avoid adding the sentinel to the sum. The **REPEAT** statement, which *always* takes the action, isn't suitable.

WHILE, in contrast, provides a correct solution. As before, though, we have to be sure that the variables that make up the BOOLEAN expression are defined before the statement:

a summing **WHILE**

```
IO.IntIn (Number);
Sum := 0;
WHILE Number <> SENTINEL DO
    IO.IntIn (Number);
    Sum := Sum + Number
END
```

By now we can make up a rule of thumb for choosing between **REPEAT** and **WHILE**:

- If the action will take place at least once, a **REPEAT** is appropriate.
- If the action might not take place at all, use a **WHILE** statement.

In effect, the **WHILE** statement sets up an *entry* condition. If it isn't met, the **WHILE** is skipped entirely.

A final example makes the contrast between entry and exit conditions clear. Suppose that we want to skip characters until we reach a sentinel *or* run out of data. The **LOOP** version, which has a positively stated exit condition, is:

LOOP with multiple bounds

```
LOOP
    IO.CharIn (Ch);
    (**)IF NOT IO.Done() OR (*then if*) (Ch = SENTINEL)
                                        THEN EXIT END
END
(*We ran out of data or read the sentinel.*)
```

As always, I use the *then if* comment to show that I'm relying on partial evaluation of the BOOLEAN expression.

Since the **WHILE** statement has an entry condition, the exit condition is implied, rather than being explicitly stated as above. The **WHILE** equivalent is:

WHILE with multiple bounds

```
IO.CharIn (Ch);
WHILE IO.Done() AND (*then if*) (Ch <> SENTINEL) DO
    IO.CharIn (Ch)
END
(*We ran out of data or read the sentinel.*)
```

I'll point out some of the problems this causes in a minute.

Loops and Language Design

Now that we've seen all of them in action, we can ask the big question: Why does Modula-2 have three statements that do essentially the same thing? The answer requires a short digression into the philosophy of programming language design. In the early days, philosophy was simple:

- Statements should *help the programmer*.

Help, in this case, meant relief from having to remember the codes of machine and assembly language. Statements were convenient, but they didn't really change the options presented by the computer.

In time, though, languages started to be designed for the convenience of programmers, rather than machines. The design philosophy became:

- Statements should help the programmer *accomplish her goals*.

Unfortunately, the statements that made one programmer's life simpler would, all too often, make programs extremely difficult to understand and fix or modify. The basic design philosophy had to be extended to:

- Statements should help the programmer accomplish her goals *clearly*.

Naturally, problems persisted. It turned out that clarity wasn't enough to deal with the inherent complexity of code. Programs that appeared to be open and understandable might actually contain bugs that were subtle, yet devastating. Philosophy got one more extension:

- Statements should help the programmer accomplish her goals clearly *and correctly*.

This approach to design is different from the others. It suggests that the best way to help programmers is, on occasion, to limit what they can do.

Statements like **REPEAT** and **WHILE** were designed to meet the dual goals of clarity and correctness. Restricting entry and exit to single, specific points doesn't prevent programmers from making mistakes or writing obscure code, but they do limit the opportunity for mischief. Formal methods that attempt to prove that code is correct rely on the fixed entry and exit points of statements like **WHILE**.

However, a restriction that is meant to make code more correct may have the unwanted side-effect of making it less clear. Reconsider this **WHILE** loop:

correct, but not clear

```
IO.CharIn (Ch);
WHILE IO.Done() AND (*then if*) (Ch <> SENTINEL) DO
   IO.CharIn (Ch)
END
```

If you find its BOOLEAN condition practically incomprehensible, you are not alone. Can you easily state the exit condition it implies? I can't. The problem can't really be avoided, since the condition must serve double-duty as an entry condition, and as an implied exit condition. Checking the correctness of the statement requires the programmer and reader to mentally reverse the sense of the BOOLEAN condition.

The **LOOP** statement is easier to work with than **WHILE** because it is less restrictive. **EXIT** conditions are stated in a positive sense, which makes them much easier to understand, and less error-prone to write. **LOOP** aspects that are potentially prone to abuse (for example, having multiple **EXIT** points) can be restricted by custom and practice (just as self-control manages to bar multiple **RETURN**s from procedures and functions).

the **goto**

In some ways, **LOOP** is an updated version of a statement that caused a great deal of havoc in the old days—the **goto**. The **goto** allowed unrestricted *jumps*—every statement in a program was numbered, and the **goto** could go to any of them. This led to such unbelievably complicated programs ('spaghetti' is the only printable term used to describe them) that **goto** was banned from polite society, and was suppressed (i.e. not taught) in languages, like Pascal, that kept it for historical reasons.

EXIT does 'jump' out of a loop, but it can only go to one place—the next program statement. In my opinion, the loop specification method we use for **LOOP** is so much easier and more reliable than **WHILE** loop planning methods that I rely on it in this text. However, **WHILE** and **REPEAT** are also in wide use, and you should know how to employ them.

Section 6-6
Program Engineering Notes

'GRACEFUL DEGRADATION' SOUNDS MORE LIKE THE name of a new wave band than a desirable program characteristic. But like other contributing factors to program robustness, the ability to recover from failure, even in part, is a key part of good program engineering. What does a robust program do when it encounters incorrect input? A robust program tries either to undo the error, or to recognize that it is in a state of error and act accordingly. In contrast, a non-robust program may crash, or produce incorrect output.

But it is not always necessary to give up—even gracefully—when confronted with problems. We can always use the **LOOP** statement to try again:

```
IO.StringOut ("What is your answer--yes or no?");
LOOP
    IO.StringOut ("Enter 'y' or 'n'.");
    IO.CharIn (Answer);
    (**)IF (Answer="y") OR (Answer="Y")
          OR (Answer="n") OR (Answer="N") THEN EXIT END
END;
(*Answer indicates yes or no.*)
```

a robust loop

For a more general-purpose example, consider procedure GetInput, below. It increases a program's robustness by letting the user correct improperly entered data. We don't know when an appropriate value will be entered, or what its exact value is. We *do* know, though, that the exit condition can conceivably be met, and that the crucial variable, Value, is reinitialized on each iteration.

input error check program

```
PROCEDURE GetInput (VAR Value: INTEGER;
                        Lower, Upper: INTEGER);
  (*Gets and returns a Value between Lower and Upper.*)
  BEGIN
    LOOP  (*until Value is in the proper range*)
      IO.IntIn (Value);
      (**)IF (Value>=Lower) AND (Value<=Upper) THEN EXIT END;
      (*If we're still here, Value is less than Lower, or greater than Upper.*)
      IF Value<Lower
        THEN IO.StringOut ("That's too small.  Try again.")
        ELSE IO.StringOut ("That's too large.  Try again.")
      END (*IF*)
    END (*LOOP*)
  END GetInput;
```

improving robustness

> Interactive programs should include such error-checking and mistake-correcting loops whenever possible. Nothing is more infuriating than a computer program that *can't* let a user change input she knows is wrong—or *won't* let her know what is right.

Reporting on problems with files is particularly important. A failed call of IO.OpenInput, say, is invisible to the program user. Eventually, she comes to realize that the program is waiting for *her* to type input, rather than reading it from a file. Redirecting input or output should always be checked for success.

reporting on failed redirection

```
IO.StringOut ("What file will input come from?  ");
IO.StringIn (FileName);
IO.OpenInput (FileName);
(*Now, let's check on the success of the change.*)
IF IO.Done()
   THEN IO.StringOut ("Now reading from file  ");
        IO.StringOut (FileName)
   ELSE IO.StringOut (FileName)
        IO.StringOut ("  could not be opened.   ")
        IO.StringOut ("Reading from standard input.)
END
```

```
What file will input come from? FactFile
Now reading from file FactFile
```

Whether you give the user another chance to enter a file name, or just let her know that the program will work with the default input or output, your program is robust—it's dealing with errors.

Loop Debugging

In my experience relatively few errors have to do with a loop's action. One can make mistakes, of course, but the sneaky bugs hide at the loop's *boundary conditions*—the circumstances of its first and last iterations. For example:

boundary condition problems

EXIT *condition bugs:* Is the **EXIT** condition doing what you intended it to do? Have you accidentally phrased an *unavoidable* condition that will always be met on the first trip through the loop, or an *impossible* condition that can never be met? Does the loop's **EXIT** condition guarantee that the loop will terminate? What if you run out of data, or don't arrive at an answer, or look for a sentinel that doesn't exist?

Off-by-one bugs: Is it possible for the loop to take place one time too often? One time too few? Should a < be a <=, or vice versa?

Synchronization bugs: Is the exact moment of **EXIT** properly synchronized with the loop's action? Is it possible that one last variable should have been updated before the loop was finished? Or that one last iteration of the loop was counted incorrectly?

> The KISS philosophy—*keep it simple, stupid!*—is excruciatingly important in loop programming. Always take the simplest possible step that approaches the loop bound; never add goal-oriented actions that might cause you to miss the bound.

For example, if a loop is bounded by the end of an input line, reading just one character could reach the bound. You can't read two characters and expect to avoid problems. Suppose, for instance, that we want to print every other character, starting with the first, on each line of input. This loop has a bug:

buggy character skipping

```
LOOP (*with bug*)
    IO.CharIn (Ch); (*Print this one.*)
    IO.CharOut (Ch);
    IO.CharIn (Ch); (*Skip this one.*)
    (**)IF (Ch = IO.EOL) THEN EXIT END
END
```

What if an input line has an odd number of characters? The end-of-line will be read and echoed, then the first character on the *next* line will be checked to see if it equals IO.EOL. One correct version is:

correct character skipping

```
LOOP
    IO.CharIn (Ch); (*Print this one.*)
    IF (Ch <> IO.EOL) THEN
        IO.CharOut (Ch);
        IO.CharIn (Ch) (*Skip this one.*)
    END;
    (**)IF (Ch = IO.EOL) THEN EXIT END
END
```

There's an extra line of code, but it makes the segment simpler.

Three Informal Methods

From the program engineering point of view, how can we approach boundary condition problems? I'll present three informal methods. The first tactic is the *termination checklist*:

first: termination checklist

> The common loop bound conditions—sentinel, count, limit, and data—provide a convenient checklist to help assure us that a loop will terminate.

I put loop bound conditions into categories mainly for this reason. However, there are two ways to use a termination checklist. The novice looks at what she *wants* to do:

a novice's termination checklist

1. *Sentinel:* Is there an **EXIT** sentinel? Is it the only one? Have you considered the possibility that the sentinel won't show up?

2. *Count:* Is there an upper limit on iterations? Should there be one? Are you counting the right thing? Can a partial result blow up before you reach the limit?

3. *Limit:* How long can calculations go on? Are additional iterations always meaningful, or can you get 'close enough?' Are you looking for a limit beyond your machine's accuracy? Is a fail-safe count bound needed?

4. *Data:* Is the loop limited by the availability of raw data? Should an additional check be added?

But the expert programmer follows up with these questions, which are aimed at what she's *not* looking for, rather than what she *is* seeking:

an expert's termination checklist

1. *Sentinel:* Should you be watching for a 'just in case' sentinel? For example, will the appearance of a sentinel like IO.EOL indicate that something has screwed up?
2. *Count:* Is a count bound on total number of iterations going to be a necessary precaution in a limit-bound loop?
3. *Limit:* Will a seemingly innocuous bit of arithmetic push the limits of computer accuracy? Might a partial result blow up (e.g. go beyond the legal limit for INTEGER)?
4. *Data:* Can you run out of data unexpectedly? Is a check for **NOT** IO.Done() going to be needed?

I call the second informal technique *debugging by triangulation*:

second: debugging by triangulation

> A loop has three sides: before, after, and during. To debug by triangulation, put debugging statements before the loop, immediately after the end of the loop, and within the loop (usually just before its final **END**).

The idea, of course, is to home in on the bug from all sides. For example:

```
IF BUGGY THEN IO.StringOut ("Loop 9a. Entering with . . .") END;
LOOP
    do stuff;
    (**)IF TimeToGo THEN EXIT END;
    do stuff;
    IF BUGGY THEN IO.StringOut ("Loop 9b. Inside, with . . .") END
END;
IF BUGGY THEN IO.StringOut ("Loop 9c. Exited with . . .") END;
```

Debugging by triangulation is based on an advertisement that an auto service company used to run: *You can pay me now* (i.e. for service), *or you can pay me later* (i.e. for repairs). Well, you can put all three debugging statements in at once, or you can put them in one at a time, but sooner or later you'll probably put them all in.

Why? Think back to the first law of debugging: *It ain't what you don't know that hurts you; it's what you do know that ain't so.* It's no shame to get confused, but it is foolish to think that it can't happen to you. When I debug my students' code, I throw in debugging statements right away because it saves me time in the long run.

Cooper's first law

Checking the values that established the **EXIT** condition is particularly important. The rule of thumb that I always follow is that when I'm sure about everything, but my program *still* doesn't work, one of the things that I'm sure about must be wrong. It usually turns out to be something dumb in a BOOLEAN expression; e.g. the variable I was sure had to be between twenty and thirty had never been initialized.

The third approach I take in loop debugging is the *execution profile*.

third: execution profile

> An *execution profile* is a count of the number of times each statement is executed, or the number of times each subprogram is called.

Execution profilers are commonly used to help improve the performance of complex programs—the profile supplies a quick pointer to where the program spends its time. However, the technique is useful in small programs too, especially if you have an idea of how many iterations you expect a loop to take. The profile code is easy to embed:

an embedded execution profiler

```
Iterations := 0;
LOOP   (*Loop 7*)
    Iterations := Iterations + 1;
    do stuff;
    (**)IF TimeToGo THEN EXIT END;
    do stuff
END;
IF PROFILING THEN
    IO.StringOut ("Loop 7, count:    ");
    IO.IntOut (Iterations, 3); IO.LineOut
END
```

Most Modula-2 systems have execution profiling as a compiling option, too.

Modula-2 Details

Syntax Summary

• **LOOP** statement: allows a sequence of statements to be repeated indefinitely. In practice, it contains an **EXIT** statement that halts the loop when its bound is reached.

```
LOOP
    sequence of statements;
    (**)IF bound has been reached THEN EXIT END;
    sequence of statements
END
```

Although Modula-2 syntax does not require it, as a general practice:

- Only one statement should contain the **EXIT**—there should be just one exit point.
- The **EXIT** point should be at the outermost level of the loop.

• the IO library: includes all procedures needed for text input and output, including:

Constant:
EOL = *the end-of-line character*;

Functions:

Done () : BOOLEAN;	TRUE *if the last operation was a success*
termCH () : CHAR;	*The input character about to be read*

Procedures:

OpenInput (*StringValue*);	*Get subsequent input from file StringValue*
OpenOutput (*StringValue*);	*Send subsequent output to file StringValue*
CloseInput;	*Return to the keyboard for input*
CloseOutput;	*Return to the screen for output*

Modula-2 Details

Remember . . .

• A loop's **EXIT** point should be easy to find. Have just one exit point, and comment it clearly.

• The **EXIT** condition must be attainable. This implies that part of the condition must be changed on each trip through the loop.

• A *loop specification* is a semi-formal loop outline. It splits a loop's description into three parts: the *goal*, or overall purpose of the loop; the *bound*, which gives the reason the loop will end; and the *plan*, which describes the actions the loop will take.

• Loop bound conditions can be put into distinct categories, but only for convenience. Most loop bounds involve counts, sentinels, limits, or data. These terms are intended only to provide a checklist for typical loop designs.

• An *intentional* bound is the reason we want a loop to end. However, since good intentions aren't always enough to guarantee loop termination, additional *necessary* bounds may be needed as well.

• A loop does *not* end because it has achieved its goal. It *only* ends when it reaches its bound. When a loop has multiple bounds, it must be followed by a check to determine which bound caused the loop to terminate.

• A loop's *postcondition* is a confirmation comment that describes a program's state on loop exit. It describes the loop's intended goal, of course, but it must also point out any possible deviations from that goal. It acts as the *precondition* of the next program statement.

• A loop's *boundary conditions* are the circumstances of its first and last iterations. The most common loop errors (in particular, *off-by-one* errors) occur there.

• A *stream* of data is a sequence of values. The *standard input* stream usually comes from the keyboard (but it can also come from a file or other input device). The *standard output* stream is usually directed to the terminal screen, but it too can be redirected (perhaps to a printer or file).

• IO.EOL represents the end-of-line or carriage return character—if the character you just read equals IO.EOL, you're at the end of a line. Carriage returns should be printed with IO.LineOut.

• IO.Done() is the general-purpose success/failure indicator for the IO library. If any operation (reading or writing a value, opening or closing a file, etc.) was a success, IO.Done is TRUE, otherwise it's FALSE.

• Text-processing programs are often called *filters*, because a stream of data flows through the program for modification or analysis. For convenience, we can describe filters that change the stream's contents as *process* filters—a process is invoked on each item. Filters that analyze a stream's contents can be called *state* filters—they learn about the stream by looking for *state transitions*, or the contents' meaning.

• A computer's representation of a REAL value is often a very close approximation, rather than an exact equivalent. Comparing REAL values says more about the limits of REAL storage than the correctness of calculations.

• *Graceful degradation* is one form of program robustness. If a program cannot recover from an error, it should still give the user an idea of what has caused the problem. Under no circumstances should a program fail abruptly or produce incorrect results.

• When possible, use loops to give interactive program users the opportunity to confirm and correct their input.

Zen Study Guide Big Ideas

The Big Ideas in this chapter are:

1. *Loops do many things, but they follow a few basic patterns.* If you can recognize the pattern before you begin to program, you'll have an idea of how to plan, and what bugs to watch out for.

2. *A loop's goal is not its bound, and its bound is not its goal.* The fact that a loop has terminated doesn't necessarily mean that it accomplished what you had in mind. Always keep a clear distinction between what a loop does and why it ends.

Zen Study Guide Questions

6-1 Difficulty in stating conditions makes for excellent test questions. Write a series of **EXIT** conditions that include impossible conditions (that can't be met) and unavoidable conditions (that will always be met), along with a few plausible conditions (that can be met correctly). The question, of course, will be *Which of these* **EXIT** *conditions are in error? Why?*

6-2 *Find the bug* questions are enjoyable to put on tests because they reward students who have done their homework—after all, they're the ones who have probably seen the bugs already. Take a half-dozen pieces of code from the text, and introduce the most devious bugs you can. Hints:

a) slightly change the **EXIT** condition;

b) change the position of the **EXIT** statement;

c) remove a statement from within the loop;

d) if there is an initializing statement before the loop, change any zero to one or vice versa;

d) put the initializing statement inside the loop instead of before it.

6-3 *Turn this* **FOR** *statement into a* **LOOP** is a good practice transformation. Most of these problems are, frankly, too easy to make good test questions. Nevertheless, try finding four examples from Chapters 4 or 5 that are relatively hard to turn into count bounded loops.

6-4 Stripped of its camouflage, the basic sentinel loop question is *Write a loop that spots the first . . .* A sentinel like *the first zero* or *the end of the line* is fine and dandy, but it isn't arbitrarily complicated, the way test questions are supposed to be. Think up a half-dozen sentinels that go from hard to harder (usually because they require your program to remember more and more current input states). Tricks to employ include requiring multiple sentinels (spot A or B or C), conditional sentinels (spot A followed by B), and compound sentinels (spot A followed by B, or C followed by D or E).

6-5 Procedure EchoText showed the basic text processing loop. It read and echoed each character, while printing a carriage return in place of each IO.EOL character. Thus, as a process filter, EchoText was pretty simpleminded.

Make up three *Write a program that . . .* problems that are really just outlandishly complicated variations of EchoText. For example, you might require the program to print a carriage return only before printing a dollar sign (instead of after reading IO.EOL), a question mark whenever you reach the end of an *input* line, and a plus sign at the end of every *output* line.

6-6 Let's try the same kind of variation with state filters. Recall that the basic state filter counted words by keeping an eye out for *in a word/not in a word* transitions. Use the Zen Study Method by thinking up three more counting problems that can be solved by spotting and counting state transitions.

6-7 I don't know if I would go so far as to claim that there are only three looping problems in the world. But if I did, *Write a program that does accounting for . . .* would be one of them. It might be financial accounting or scorekeeping, but the basic problem is addition and subtraction inside a loop. Make up a few variations on the accounting problem.

6-8 Pick a program from the text, say Tutor or LongDivision. Then, make up three sets of test data, two of which cleverly fail to check boundary conditions. Finally, ask the Zen Essay Question *Which set of data should you use to test this program, and why?*

6-9 *Rewrite this segment using that statement* is a particularly effective test question when loops are involved. The syntax of the **WHILE** and **REPEAT** statements can hardly be accused of posing much difficulty. But using them correctly requires a real understanding of the segment's semantics and intent.

Pick a few **LOOP** statements from the text, and present them as *Rewrite this using* **REPEAT** *or* **WHILE** problems. Can you come up with any general rules for making the transformations?

6-10 Time for *Here's the input, what's the output?* and *Here's the output, what's the input?* What are some of the things your example segment might do? How about capitalizing lower-case words, or doubling odd numbers, or counting words that begin with a capital letter, or only echoing increasing values in an input sequence of random numbers? As always, you will find that the interesting part is choosing the segments that *can* be run in reverse.

Exercises
Warmups and Test Problems

6-11 What error message will you get from compiling or executing a loop that doesn't have an **EXIT** statement? None. How long will such a loop go on for? Does your compiler impose a fail-safe upper limit on the number of statements a program can execute? If not, how can you halt a runaway program yourself?

6-12 Rewrite this segment without the first **IF** statement:

```
IO.CharIn (Ch);
IF Ch <> IO.EOL THEN BEGIN
    LOOP
        IO.CharOut (Ch);
        IO.CharIn (Ch);
        (**)IF Ch = IO.EOL THEN EXIT END
    END
END
```

6-13 Write a loop that reads a file of numbers, and stores only the odd numbers in a file named `odd.out`.

6-14 Computers are so fast that programmers who are trying to time their code will sometimes insert 'empty' loops. Such loops do nothing but approach their count bounds.

Run two experiments. First, see how high a loop has to count in order to provide a measurable delay in program execution (hint: use two separate loops—one to slow down your test harness, and one to test). How long does each iteration (assignment and test) take? Second, how does this compare to the speed of a **FOR** statement?

6-15 Write a loop that finds the first power of two that is equal to, or greater than, a number entered by a program user. Be sure that it will work correctly if the user enters zero.

6-16 Write a loop that reads and echoes input INTEGERs until the difference between any two successive numbers is greater than or equal to ten.

6-17 Suppose that you have a can of whipped cream. Every second the nozzle stays open, 2% of the remaining cream squirts out. How long, to the nearest second, will it take to empty half the can?

6-18 Write a procedure that reads and echoes characters, with the following proviso: turn echoing *on* whenever an "A" is read, and turn it *off* whenever a "T" appears.

6-19 Procedure GetInput, in section 6-6, showed how to make a program more robust by giving the user a chance to enter INTEGER input again. Write a similar procedure that calls either IO.OpenInput or IO.OpenOutput, and lets the program enter new names until a file is actually opened successfully.

6-20 Write a loop that reads a sequence of REAL values, and finds the number of values in the longest *decreasing* sequence. For example, in the sequence:

1.0 4.0 3.0 2.0 9.0 10.0 8.0 6.0 2.0 5.0

the longest decreasing sequence (10.0 through 2.0) has four values.

6-21 Write a procedure that reads and echoes text, preserving line structure. but only prints every other sentence starting with the first.

Exercises
Programming Problems

Note: Every Modula-2 system will include a library of functions and procedures for working with strings. Some of these exercises will be easier to do if you find, and use, your system's string-handling library.

6-22 Ifay ouyay ancay eadray histay, ouyay ancay robablypay igurefay outay hatway hetay roblempay isay.

6-23 Write a short program that performs the following short analysis of its input:

a) Print every letter that appears twice in a row.

b) Print every three-letter sequence that's in alphabetical order.

Some sample input for you is:

```
Miss Metteer, displaying calmness, deftly made crabcakes from
excess cottonseed.  A studious bookkeeper under a canopy, she
laughingly hijacked the corrupted terra cotta, too.
```

6-24 A *gerund* is the 'ing' form of a verb: *Seeing is deceiving, but eating is believing* contains four gerunds. Write a program that can translate the infinitive form of a verb (the 'to be' form) into the gerund form. Hint: you'll find that the last two letters of a verb usually tell you how the gerund is formed.

6-25 Think of a letter between "A" and "Z". With a little help (you have to tell me if my guess was too low or too high), I can figure out which letter it was in no more than five guesses. Your job is to *a*) figure out how I do it, and *b*) write a program that uses your method.

Then, write a program that picks the letter and lets a player guess. As above, the program should announce if a guess is high or low, and allow no more than five guesses.

6-26 An addict who likes cigarettes very much is willing to go to great lengths to continue her habit. She's even willing to roll new cigarettes from butts and newspaper. Suppose that she starts out with a carton (ten packs of twenty) of the filthy things. Assuming that she rolls a new cigarette from every three butts, how many smokes will she have in all? Will any butts be left over?

6-27 The least common multiple of two numbers is the smallest number that is evenly divisible by both of them. Write a program that finds a least common multiple. Hint: think about using procedure GCD.

6-28 One plus two is three. One plus two plus three is six. One plus two plus three plus four is ten. If we keep this up long enough, we'll find that the sum of the numbers 1 through n is a perfect square. Find the first three numbers for which this is true.

6-29 Did I ever mention that even most mathematical programming is really text processing? Well, where do numbers come from? Do you think that it's easy to read a REAL? Write a procedure that could replace IO.Realln. You can warm up for this by figuring out how to read an INTEGER one digit character at a time.

6-30 The January 1984 issue of *Scientific American* contains an article on an interesting sequence of numbers known as a *hailstone* series. The series is formed like this:

a) Pick a number.

b) If it's odd, triple the number and add one.

c) If it's even, divide the number by two.

Can you imagine why the sequence is likened to a hailstone? Although numbers bob up and down, eventually they reach a repeating 'ground' state: 4 2 1 4 2 1 . . . This has been shown for every number up to about 1.2E12.

Write a program that asks a user for a starting number, then answer these two questions:

a) How many steps does it take to get to the repeating 'ground' state?

b) What is the largest number the sequence reaches along the way?

If you're a go-getter, you might try generating a few numbers for inspection automatically. Which sequences, beginning at 100, have the greatest maximum value? Which are the longest? Have fun!

6-31 Suppose that a file contains the names of various sports:

```
Basketball
Weightlifting
Running
```

and so on. Write a program that reads and echoes the file one line at a time. Options: *a*) precede each output line with the words `Bo Knows`; *b*) if the input line is **`Diddley`** have the output line begin `Bo Don't Know`; as above, but if the input line is **`Hockey`**, just print `No`.

6-32 'Massaging the data,' when applied to the results of scientific tests, usually means 'fudging the answers.' Programs, however, frequently require that data be massaged into shape for processing. For example, some of the changes that might be needed are:

a) Rewrite a file of INTEGERs in REAL format; e.g. turn 1 into 1.0.

b) Turn lower-case letters into capitals.

c) Find, and get rid of, extraneous blank lines.

d) Make sure that alphanumerics (letters and digits) are the only characters in a file. (Spaces and carriage returns are usually acceptable exceptions.)

Can you manage to make all of these checks in a single trip through a file?

6-33 A *pretty printer* is a program that turns raggedy old piles of code into shiny new masterpieces. In other words, a pretty printer applies a consistent and readable standard for line breaks, indenting, and so on to programs. Pretty printing can't make a program work, but it can help you discover what's wrong with one.

Like so many programs, pretty printers can be as complicated as you'd like. Here are some features that a simple pretty printer might include:

a) Have all indenting be a multiple of five spaces.

b) Put each comment on a line by itself.

c) Don't let multiple blank spaces appear within a line.

d) Don't allow multiple blank lines.

e) Limit the length of any line, and indent by some fixed amount if you're forced onto a second line.

f) Number each program line.

Write a pretty printer that enforces these rules.

6-34 A *text formatter* is a program that applies various rules for indenting, line length, page length, and so on to text. Formatters can become arbitrarily ('unbelievably' might be a better word) complicated, but at heart they just read and echo characters.

Write a text formatter that follows these rules (using values for *a*, *b*, and so on entered by the program user):

a) Print exactly *a* lines per page.

b) Space *b* lines between each line of output.

c) Print *c* words on each line.

d) Doublespace between sentences.

e) Capitalize the first word of each sentence.

Add three more options to the list.

'Relying on run-time crashes to do error checking is like using telephone poles (instead of brakes) to stop your car. Still . . .' (p. 283)

Chapter 7 New Types and the **CASE** Statement

Chapter 7 begins with the last of Modula-2's control statements. **CASE** provides a choice of actions, or a choice of routes for data. The effect of **CASE** is a lot like the **IF** statement, but its syntax can make the flow of a program easier to follow.

Then, we'll turn to extensions of the simple value types. Although it took no great insight on Wirth's part to assume that you might find a type INTEGER, with values 1, 2, 3 and so on useful, it would have been rather extraordinary for him to build type FlavorTYPE, with values chocolate, vanilla, and strawberry into the language. Modula-2 lets us add new types ourselves.

Next, we'll turn to type *subranges*. If the *enumerations* described above giveth, subranging taketh away. A subrange is part of the full range of a type. It is a safety device—giving a variable a subrange type, rather than the entire type, limits it to representing a restricted sequence of values.

Finally, section 7–6 takes one of our periodic In Depth looks at a programming issue. We'll discuss the underlying issue—abstraction—and start to understand why enumerations, subranges, and statements like **CASE** are included in most high-level programming languages.

Section 7-1
The **CASE** Statement

AS ALWAYS, AN EXAMPLE. PROGRAM Gradient, below, uses a **CASE** statement to make a choice that depends on the value of a variable called Score. Although the program contains a variety of potential actions (including an empty statement), only one is executed.

CASE demonstration program

```
MODULE Gradient;
   (*Makes an appropriate comment about a user's grade.*)
IMPORT IO;
VAR Score: INTEGER;
BEGIN
   IO.StringOut ("What grade did you get?    ");
   IO.IntIn (Score);
   CASE Score OF
      0:  (*This empty statement is a null action.*)
     | 10: IO.StringOut ("Exceptionally Good!    ");
           IO.StringOut ("You are to be commended!")
     | 7 .. 9: IO.StringOut ("Good!")
     | 5, 6: IO.StringOut ("Barely Passing!")
     | 2 .. 4: IO.StringOut ("Flunking!")
     | 1: IO.StringOut ("Exceptionally Flunking!")
     ELSE IO.StringOut ("Your grade is in the forbidden zone.");
          IO.LineOut;
          IO.StringOut ("You'd better come to office hours.")
   END (*the CASE statement*)
END Gradient.
```

```
What grade did you get?  1
Exceptionally Flunking!
```

The first line of the **CASE** statement contains a *case expression*—here, Score—that can have any simple type except REAL. It's followed by a list of the **CASE** expression's potential values (the *case label list*), and the proper sequence of statements to take for each one. A vertical bar (|) separates each sequence. I'll always put the vertical bar before the next label to make it easy to spot.

Modification Problems

1. Modify Gradient so that it's more in line with current educational theory; i.e. accompany the `Flunking` message with additional lines of output that make your parents' sole responsibility for this clear. In other words, add at least one additional statement to each action.

2. Suppose that you wanted one message for grades below 0, and a separate message for grades greater than 10. How could you modify Gradient to do the job?

Strictly Modula-2

The vertical bar that separates one action from the next is the most complicated part of **CASE** syntax. Wirth could have chosen almost any unused character to serve as a separator; the vertical bar happens to be unobtrusive, yet clear.

empty statements

Suppose that a case label isn't supposed to have any action, as when Score equals 0. We show that nothing should take place by the simple expedient of not supplying any statements—the vertical bar follows the label and colon. By tradition, this 'nothing' is called an *empty statement*.

grouping actions

When a label requires two or more statements, as when Score equals 10, the vertical bar separates the '10' actions from the '7 .. 9' action. As always, I indent to show grouping, but the compiler only obeys the vertical bar.

An optional **ELSE** part supplies an alternative action that is taken if the expression's value doesn't match any of the label possibilities.

the **ELSE** part

> An **ELSE** part should be provided if there is a possibility that the **CASE** expression's value isn't listed. An **ELSE** part should appear even if its action is an empty statement.

Although some compilers will let an unmatched value 'fall through' the statement without ill effect, others may crash. Even though a crash might be desirable as a last-ditch check for errors, adding explicit error-handling statements is obviously a better way of dealing with unknown values.

The **CASE** expression is frequently a variable identifier, as in Gradient. However, it may be any sort of expression, as long as it, and the case labels, could legally appear in a BOOLEAN expression together. After all, the compiler will have to compare the **CASE** expression to the labels to decide what to do.

labels are constant expressions

The label values, in turn, have to be actual constant values (like 4 or "D"), or expressions that the compiler can figure out without executing the program. Thus, simple calculations are OK, but referring to variables is forbidden.

When the same action can be initiated by two or more possible values of the **CASE** expression, the constants are separated with commas. If a sequence of values is contiguous, the symbol ' .. ' means 'through and including.' Sometimes there's more than one way to show values:

5,6 *is the same as* 5 .. 6
"a", "b", "c" *is the same as* "a" .. "c"

As long as no particular constant appears more than once (since a call for two different actions would be illegal) the order in which the labels are specified doesn't matter.

Self-Check Questions

Q. Which of these values could appear as a **CASE** statement label? Assume that Hour is a variable, but that all capitalized identifiers (other than procedure calls) are constants.

a) 7 *b*) Hour *c*) RATE
d) RATE / 2.0 *e*) "0" *f*) –3
g) 0+1 *h*) ORD("T") *i*) TRUE
j) 5.0

A. The basic rule is that labels can't require program execution to be evaluated, and that they can't be REAL. Thus, *b* is out (if Hour is a variable), as are *d* (which must have a REAL result) and *j*. The others are legal. In particular, note that the BOOLEAN constants may appear as labels.

Q. Simplify this **CASE** statement:

```
CASE Value OF
     3: IO.StringOut ("Hi!");
   | 7: IO.StringOut ("Ho!");
   | 2, 5, 6: IO.StringOut ("Ho!");
   | 8: IO.StringOut ("Hi!")
    ELSE
END
```

A. I've left the **ELSE** part in even though no action is specified. The simplified statement is:

```
CASE Value OF
     3, 8: IO.StringOut ("Hi!");
   | 2, 5..7: IO.StringOut ("Ho!");
    ELSE
END
```

When to Use **CASE**

Why use **CASE** instead of **IF**? For two reasons:

- First, **CASE** labels are easier to use than BOOLEAN expressions for describing groups or ranges of values. The alternative nature of the actions is usually clearer to see.
- Second, the alternatives **CASE** presents are truly *mutually exclusive*. The order in which alternatives are considered doesn't matter, because each **CASE** expression can only call for one action.

This statement could be replaced with an **IF**, but it's hardly likely to be as clear:

categorizing characters

```
CASE Letter OF
     "a".."z": LowerCase := LowerCase + 1
   | "A".."Z": Capital := Capital + 1
   | "0".."9": Digit := Digit + 1
   | ".", ",", ";", ":", "?", "!": Punctuation := Punctuation + 1
   | " ": Space := Space + 1
    ELSE Special := Special + 1
END
```

The **CASE** advantage is even more obvious if nesting is involved. For example:

counting vowels and consonants

```
CASE InputCh OF
   "a".."z", "A".."Z":          (*only check letters*)
      CASE InputCh OF
         "a", "e", "i", "o", "u", "A", "E", "I", "O", "U": Vowels := Vowels + 1
         ELSE Consonants := Consonants + 1
      END (*inner statement*)
   ELSE Others := Others + 1
END (*outer statement*)
```

The **CASE** statement makes the programmer's intention clear. It's not that brevity is a virtue in itself, but the alternative of writing a long sequence of BOOLEAN expressions would lead to an unnecessarily complicated segment of code.

Relying on **CASE** for mutual exclusion is a more sophisticated application. As you may have found for yourself, alternatives that exclude each other can sometimes be more reliable than alternatives that don't. **ELSIF** parts may accidentally overlap or even contradict each other. **CASE**, in contrast, can use the compiler to help detect the bugs that last-minute addition can cause.

One other interesting feature of mutual exclusion is the possibility for parallel evaluation. Since the labeled alternatives really are mutually exclusive, and no more than one can be met, the computer can consider them in any order. More interesting still, several computers might work to consider the alternatives *at the same time*. We'll pursue this idea elsewhere.

Self-Check Questions

Q. Use **CASE** to write a loop that won't rest until it manages to read one of the characters a program user might reasonably enter in response to the question `Did you understand?`

A. Obviously, figuring out what a reasonable response is could be is half the problem. Can you add a carriage return to the list below?

```
LOOP  (*until a valid input value has been obtained*)
   IO.StringOut ("What's your answer?");
   IO.CharIn (Answer);
   CASE Answer OF
      "Y", "y", "N", "n", "H", "h", "?": Valid := TRUE
      ELSE Valid := FALSE
   END;
   (**)IF Valid THEN EXIT END
END
```

Standard Practices: Counting Digits

We'll be returning to the **CASE** statement after we discuss enumerated types, so let's just look at one long example now. The problem is:

problem: odd/even/zero counting

> Write a program that determines how many odd, even, and zero digits there are in a line of input numbers. Count zeros as both zeros and as even numbers. Count spaces and commas (but ignore all others), and stop when you reach the end of the current line.

The program inspects a stream of characters and collects information about its contents. We can call it a state filter, since we learn something about the stream rather than doing things to it. Like most filters, the digit counter requires a loop to step through the stream of characters. And, like most filters, it should probably include an extra check that will guarantee that the loop ends even if we don't spot the end of the line:

loop specification

Goal: Describe the contents of a stream of characters.
Intentional bound: We've read the IO.EOL *character.*
Necessary bound: We run out of text.
Plan: Read a single character.
Increment the appropriate count.

I'll rewrite the loop specification in pseudocode to help make sure that the **EXIT** condition goes into the right place:

pseudocode

```
initialize the counters;
LOOP
    get a character;
    (**)IF it's the sentinel OR the read wasn't Done() then EXIT;
    if appropriate, update a state variable that counts the character
END of the loop;
print the results
```

Why do I say that I'll *update a state variable*? Well, the variable that reads each character has a value, but it just describes one single data item. In contrast, the counter variables between them hold the state—what we know about so far—of the entire stream of input. Here's the pseudocode of the state update:

a pseudocode detail

```
CASE Current OF
    a zero:   add to the zeros count;
              add to the evens count
    | an odd number:   add to the odds count
    | an even number:   add to the evens count
    | a space or comma:   add to the others count
END the statement
```

Program Digits, below, puts the refinements together. Note the additional **CASE** check that precedes the call of the counting procedure, and ensures that its Current argument will have one of the values we're interested in counting.

digit-counting program

```
MODULE Digits;
   (*Count odd, even, and zero digits, and separators, in input.*)
IMPORT IO;
VAR Evens, Odds, Zeros, Others: INTEGER;
    Current: CHAR;
PROCEDURE StoreCount (ThisChar: CHAR;
               VAR Ecount, Ocount, Zcount, Other: INTEGER);
   (*Increments the proper total. ThisChar is a digit, comma, or space.*)
   BEGIN
      CASE ThisChar OF
         "0": Zcount := Zcount + 1;
              Ecount := Ecount + 1
       | "2", "4", "6", "8": Ecount := Ecount + 1
       | "1", "3", "5", "7", "9": Ocount := Ocount + 1
       ELSE Other := Other + 1
      END (*CASE*)
   END StoreCount;
BEGIN (*main program*)
   Evens := 0;
   Odds := 0;          (*Initialize the 'total' variables*)
   Zeros := 0;
   Others := 0;
   IO.StringOut ("How about giving me some stuff to count?");
   IO.LineOut;
   LOOP  (*for up to one line of data*)
      IO.CharIn (Current);
      (**)IF NOT IO.Done() OR (*then if*)
               (Current=IO.EOL) THEN EXIT END;
      CASE Current OF
         "0" .. "9", ",", " ":
                  StoreCount (Current, Evens, Odds, Zeros, Others)
         ELSE (*Do nothing. Why is this necessary?*)
      END (*END*)
   END; (*LOOP*)
   IO.StringOut ("Zeros: ");  IO.IntOut (Zeros, 1);  IO.LineOut;
   IO.StringOut ("Evens: ");  IO.IntOut (Evens, 1);  IO.LineOut;
   IO.StringOut ("Odds: ");  IO.IntOut (Odds, 1);  IO.LineOut;
   IO.StringOut ("Others: ");  IO.IntOut (Others, 1)
END Digits.
```

```
How about giving me some stuff to count?
2,597 18 528 637 9,083,002 6 47,100 319 10 590,093
Zeros: 8
Evens: 17
Odds: 19
Others: 14
```

Modification Problems

1. Take the loop out of the main program, and put it into procedure StoreCount.
2. Modify Digits so that it also lets you know which digits were divisible by *a*) 3; *b*) a one-digit number entered by the program user.
3. Modify Digits so that it reads text input, and reports on the number of characters among the first five letters in the alphabet, the last five, and the vowels.

Self-Check Questions

Q. What is the output of this code segment?

```
FOR Count := 2 TO 5 DO
    CASE Count MOD 3 OF
         0: IO.StringOut ("Heads, I win.    ");
       | 1: IO.StringOut ("Tails, you lose.    ");
       | 2: IO.IntOut (Count, 1)
    END
END
```

A. Note that when a number is taken **MOD** N, the answer must lie in the range 0 through N–1. The output is:

```
2Heads, I win.   Tails, you lose.   2
```

Section 7-2 Enumerated Types

IN EARLY COMPUTER LANGUAGES, EVERYTHING—commands, identifiers, values and all—had to be expressed in zeros and ones. As languages improved, the vocabulary of description expanded. Nevertheless, a fairly limited set of basic measures (usually equivalents of REAL, INTEGER, and CHAR) had to be employed to describe every value imaginable.

But the real world is filled with a number of things that already have perfectly good names. This month might be January or July; my favorite music might be rap, rock, reggae, ragtime, or retch (a local name for Muzak). In Modula-2, such groups of named values can be the basis of an *enumerated type* that's defined by the user. For example:

```
MODULE Menu;
    (*Demonstrates an enumerated type definition.*)
TYPE FruitTYPE = (banana, apple, orange, pear);
       VegetableTYPE = (cabbage, leeks, beets, okra);
VAR Appetizer, Dessert: FruitTYPE;
      Entree: VegetableTYPE;
BEGIN          etc.
```

enumerated type syntax

The first few lines of program Menue introduce two new types, FruitTYPE and VegetableTYPE. They are defined in the shaded *type definition part*, which always begins with reserved word **TYPE**. Each *type identifier* is followed by an equals sign and lists the type's value identifiers, or *constants*, within parentheses and separated by commas. The definitions and declarations from Menu allow assignments like:

defining enumerations

```
Appetizer := apple;        (*Appetizer gets the value apple.*)
Dessert := Appetizer;      (*Dessert is apple too.*)
Entree := leeks;           (*Entree is leeks.*)
Entree := cabbage;         (*Change Entree to cabbage.*)
```

Appetizer or Dessert could not be assigned values of type Vegetable, since that would cause a type clash. However, values or variables that have the same type can be compared:

```
IF Appetizer = apple THEN
   IO.StringOut ("An apple a day keeps the doctor away.")
END
```

ambiguous type membership is illegal

There are three things to remember when you use enumerations. First of all, no value can belong to more than one type, because being in two types would make type membership ambiguous. The definition below is illegal because the position of Friday, as well as its type, is unclear. Is it the last value of type WEEKday, or the first value of WEEKend?

```
TYPE WEEKday = (Monday, Tuesday, Wednesday, Thursday, Friday);
     WEEKend = (Friday, Saturday, Sunday); (*Illegal definition.*)
```

no external character representation

Second of all, enumerated type values can't be input or output. The constants of an enumerated type don't have any *external character representation*. This is a fancy way of saying that they cannot be read with IO.StringIn, nor printed using IO.StringOut. Later, we'll see how to use a **CASE** statement to help take care of the problem of input and output on a small scale.

scope of enumerated types

Finally, the same scope rules apply to type and constant identifiers as apply to variable and subprogram identifiers. Once a type has been defined, its name, and the names of its constants, are known in all subprograms—unless they are locally redefined. However, the identifiers of types (and their constants) are usually preserved globally. Since they are often used for communication between different parts of a program, they're seldom redefined.

Self-Check Questions

Q. Is this the beginning of a valid statement? Assume the definitions of program Menu.

```
IF Appetizer <> Entree THEN    etc.
```

A. No, because comparing values of different types causes a type clash. We can no more compare apple and leeks (even for inequality) than we could "S" and TRUE.

Q. Is this a valid type declaration? Why or why not?

```
TYPE LETTERS = ("A", "B", "C", "D", "E", "F");
```

A. It's illegal. "A", "B", etc., are not identifiers—they're constant values of type CHAR. The definition doesn't conform to Modula-2 syntax. We *could* use A, B, C, etc., as identifiers, but only if we omitted the quote marks.

Naming Values, Attributes, and States

Why complicate Modula-2 with enumerated types? They're appealing because they enhance program readability. Just as an individual constant definition relieves the program reader of remembering whether FLUNKING equals "D", "E", or "F", an enumerated type definition can name the values of any arbitrary group of values, expressions, or conditions.

enumerated types increase abstraction

> In a word, the benefit of enumerated types is abstraction—enumerated types let us deal with the *values*, *expressions*, and *states* that appear in a program in practical, human-oriented terms, rather than concrete, numerical, computer-directed ones.

Enumerating types to name elementary values is the most obvious application. For instance, the days of the week can be named:

```
TYPE DayTYPE = (Monday, Tuesday, Wednesday,
                        Thursday, Friday, Saturday, Sunday);
VAR AllDays, WorkDays: DayTYPE;
```

abstracting values

then used later on:

```
FOR WorkDays := Monday TO Friday DO
   ...        etc.
```

We can enumerate planets, colors, chess pieces, and the like just as easily. Although we could have used numbers—after all, we often refer to the months of the year with numbers—the association of numbers and values is often arbitrary and error-prone. Is day 1 Sunday or Monday? Meaningful names avoid the possibility of errors.

abstracting expressions

Enumerated values can also name *attributes*. Attributes are properties or characteristics of expressions that are distinct from literal values. For example, a BOOLEAN expression might have the literal value TRUE, but it could also carry some meaning: what is the significance of a TRUE expression?

Consider the type definitions below. RelationTYPE might be used to describe the outcome of comparing two values. CharGroupTYPE, in turn, might categorize a single character:

```
TYPE RelationTYPE = (LessThan, EqualTo, GreaterThan);
     CharGroupTYPE = (Letter, Digit, Punctuation, Spacer, Control);
```

In each case, the enumerated values name the attributes of an expression. The literal value of A = B might be TRUE, but we think of it in a different way—one value is EqualTo another. Procedure Compare, below, returns the attribute that describes a comparison between its arguments.

naming an A, B comparison

```
PROCEDURE Compare (A, B: CHAR) : RelationTYPE;
  (*Represents the relative alphabetical ordering of A and B.*)
  BEGIN
    IF A = B
          THEN RETURN EqualTo
        ELSIF A < B
          THEN RETURN LessThan
        ELSE RETURN GreaterThan
    END
  END Compare;
```

Finally, enumerated types are indispensable for recording and remembering the state of a process or computation:

abstracting states

- A search might be ongoing, successful, or failed.
- A mathematical algorithm might 'raise' an error because of overflow, underflow, or loss of precision.
- We might leave a loop in a known, but non-unique state—any of several potential bound conditions could have caused the **EXIT**.

I'll return to some examples of such state variables later.

Modification Problems

1. Add a fourth value to RelationTYPE called Undefined. Extend Compare so that it returns Undefined if either one of its arguments isn't a letter.
2. Put the RelationTYPE definition and procedure Compare into a program. Use them to let you know if characters in an input sequence always get bigger.

Self-Check Questions

Q. Suppose that program input is a series of numbers in numerical order. We wish to find a particular sentinel value. Define a type whose values indicate the status of the search. Then, declare a function that is passed the sentinel and current input value, and returns the status of the search.

A. Note the obvious similarity to procedure Compare, above.

(code follows)

```
TYPE StatusTYPE = (Continuing, Found, Failed);
PROCEDURE SearchState (SENTINEL, Current: INTEGER) : StatusTYPE;
   (*Returns the state of the program's search for SENTINEL.*)
   BEGIN
      IF Current < Sentinel
            THEN RETURN Continuing
         ELSIF Current > Sentinel
            THEN RETURN Failed
         ELSE RETURN Found
      END
   END SearchState;
```

Ordinal Procedures and Functions

Several functions and two procedures are defined in Modula-2 for working with the ordinal types, including enumerated types. They are:

type-handling functions

MIN(T), MAX(T) Given a type name T as an argument, MIN and MAX represent the least and greatest value of the type.

ORD(val) If val is a value of any ordinal type, ORD(val) represents the value's ordinal (counting) position within the type. Numbering begins with zero.

VAL(T, pos) Represents the value, belonging to type T, that occupies ordinal position pos.

Two procedures, each with two forms, are also provided for working with enumerated types.

type-handling procedures

INC(var), INC(var, n) var is a variable of any ordinal type. A call of procedure INC increments, or increases, var's value to the next value of the same type. If a second argument is supplied, var is increased to the n*th* next value.

DEC(var), DEC(var, n) Calling procedure DEC decrements, or decreases, the value of var to the next lower value. If a second argument is supplied, var is decreased to the n*th* lower value.

If some of these functions seem confusing at first, it might be because they break all the rules of ordinary functions. A type handling function can have an argument that's a type name or a variable of an unspecified ordinal type, and INC and DEC can even be given a varying number of arguments.

Let's look at a few quick examples. MIN and MAX have one common application—setting the bounds of a **FOR** loop. For example:

```
FOR Entree := MIN(VegetableTYPE) TO MAX(VegetableTYPE) DO
   IO.StringOut ("A vegetable a day keeps your colon OK.")
END
```

when to use **MIN** and **MAX**

> If your plan is to have a loop execute once for every value of a type, set its bounds in terms of MIN and MAX of the type name, rather than specific values. This clearly documents your intent to step through each value of the type.

It also contributes to the long-term health of the program, since changes in the exact definition of the type won't undo your goals in the loop. Even if more values are added to the type definition of VegetableTYPE, the loop above carries out your intentions without modification.

For a second brief example, consider the code segment below. It refers to my earlier definition of DayTYPE (the days of the week). Try to figure out whether the **FOR** statement could be rewritten as a **LOOP**.

```
FOR AllDays := MIN(DayTYPE) TO MAX(DayTYPE) BY N DO
    ...        (*do something with AllDays*)
END
```

The code segment does something with every N*th* day, starting with the first value of type DayTYPE. The long form of procedure INC can be used to step through the days, as long as we're careful not to try to assign AllDays a value beyond MAX(DayTYPE), the highest legal value. Here's one way to avoid the error:

simulating **FOR**

```
(*the rewritten FOR statement*)
AllDays := MIN(DayTYPE);
LOOP
    ...        (*do something with AllDays*)
    (**)IF (ORD(AllDays)+N) > ORD(MAX(DayTYPE)) THEN EXIT END;
    INC(AllDays, N)
END
```

Modification Problems

1. Use ORD and VAL to write your own versions of procedures INC and DEC.

Self-Check Questions

Q. Suppose that I make these definitions and declarations:

```
TYPE RockerTYPE = (Bruce, Al, David, Elvis, Harold);
VAR Singer: RockerTYPE;
```

What are the values of these expressions? Assume that Singer has been initialized to David. *Bonus:* what are the last names of the RockerTYPE values?

a) VAL(RockerTYPE, 3) *b*) MIN(RockerTYPE)
c) ORD(Singer) *d*) INC(Singer)
e) ORD(MAX(RockerTYPE)) *f*) DEC(Al)

A. I hope that you watched out for trick questions—not all of them are legal.

a) Elvis
b) Bruce
c) 2
d) This is a procedure call—not an expression. Singer becomes Elvis.
e) 4
f) Illegal procedure call—the argument must be a variable.

The bonus answers, in order, are Springsteen, Greene, Ruffin, Costello, and Melvin.

Section 7-3 Subrange Types

ONE FINAL TWIST TO MODULA-2'S BASIC TYPES allows the definition of named *subrange types*. A subrange definition gives a type identifier to a particular segment of any ordinal or enumerated type. This underlying type is then known as the subrange's *base type*, which is a term you must remember.

```
TYPE DayTYPE = (Monday, Tuesday, Wednesday,
                        Thursday, Friday, Saturday, Sunday);
     WEEKday = [Monday..Friday];          (*Subrange of DayTYPE.*)
     WEEKend = [Saturday..Sunday];        (*Subrange of DayTYPE.*)
     CAPITALletters = ["A".."Z"];         (*Subrange of CHAR.*)
     HOURSinADay = [0..24];               (*Subrange of CARDINAL.*)
VAR CardNight, SickDay: WEEKday;
    SailingDay, GameDay: WEEKend;
    HoursWorked: HOURSinADay;
    FirstInitial, MiddleInitial: CAPITALletters;
```

limits of subranges

In the example above, a DayTYPE variable might represent any of the days, but CardNight can only have a value in the WEEKday subrange. Similarly, HoursWorked can only represent a value from 0 through 24—the values included in the subrange HOURSinADay. Note that WEEKday and WEEKend have the same base type—DayTYPE. But any assignments from the wrong subrange, not to mention the wrong base type, will cause a program crash.

Default Subrange Types

A type name may appear just prior to the bounds of the subrange. This option helps sidestep a problem that has cropped up earlier—the distinction between INTEGER and CARDINAL types. Usually, the compiler figures out the base type of the subrange by looking at the type of the bounds.

> By default, a subrange defined with non-negative numerical bounds is assumed to be a CARDINAL subrange.

Unfortunately, this causes a problem in writing expressions that might involve negative numbers. I'll describe the fine points of such type clashes in a couple of pages. For now, just assume that we'd like to be able to restrict a variable to non-negative values for safety, but still treat it as a subrange of INTEGER for the purpose of arithmetic. Giving the subrange an explicit INTEGER base type solves the problem:

specifying a base type

```
TYPE TimeTYPE = INTEGER[0..60];
  (*Make TimeTYPE a subrange of INTEGER, instead of the default CARDINAL.*)
VAR Score: TimeTYPE;
    Adjustment: INTEGER;
BEGIN
   Score := 10;
   Adjustment := -5;              (*Score is still restricted to the subrange*)
   Score := Score + Adjustment;   (*[0..60], but the expression is now legal.*)
     ...
```

There is also a shorthand way to define subranges. The range of values a variable can represent may be specified when a variable is declared—but not when a parameter is declared:

shorthand subrange declarations

```
TYPE DayTYPE = (Monday, Tuesday, Wednesday,
                        Thursday, Friday, Saturday, Sunday);
VAR CardNight, SickDay: [Monday..Friday];
    SailingDay, GameDay: [Saturday..Sunday];
    HoursWorked: INTEGER[0..24];
    FirstInitial, MiddleInitial: ["A".."Z"];
```

subrange limits must be constants

The variables declared with this shorthand are like the variables in our last declaration, and represent the same limited range of values. User-defined constants can also be used to set the limits of a subrange. However, variables, function calls, or other expressions may *not* be. The effect of this restriction is to prohibit any attempt to actively set or alter the bounds of a subrange during program execution.

Self-Check Questions

Q. Which of these definitions are legal? If not, why not?

```
(*TYPE*)
a) SmallRealTYPE = [1.0 .. 5.0];
b) VOWELrange = ["a", "e", "i", "o", "u"];
c) VowelTYPE = ("a", "e", "i", "o", "u");
d) LetterTYPE = CHAR[1 .. 26];
e) DayNumbersTYPE = (1, 2, 3, 4, 5, 6, 7);
```

A. Unfortunately, not a single example is legal. *a* can't use REAL for a subrange. *b* can't declare a subrange of non-contiguous values. *c* can't have a new type that uses values already taken by CHAR. *d* can't be used to give a subrange an arbitrary type (here, CHAR), since named base types can only clear up ambiguity. Finally, *e* suffers from the same problem as *c*, although *e* could have been declared as a subrange.

Standard Practices

Is there any reason to define a subrange as a distinctly named type? Yes—we may have to use it to make declarations.

subrange parameters, function types

> When a value parameter, variable parameter, or function is declared, its type must have a name. The shorthand method of specifying subranges can't be used.

Variable parameters, in particular, are type-checked very strictly. A variable parameter and its argument should be defined with the exact same type name. In contrast, the argument of a value parameter (if it's a variable) can have a different type name, as long as its actual value falls in the value parameter's legal range.

A similar rule applies to **RETURN** statements—the actual value that's given to a function must be a legal value for the function to represent. I'll summarize the details of the *identical*, *expression compatible*, and *assignment compatible* type rules later in this chapter.

Function NextWorkingDay, below, returns a value of the subrange type WEEKday defined earlier. Since its value parameter is of type DayTYPE, its argument can be any value or variable whose base type is also DayTYPE, e.g. even a WEEKend-type variable. The function's result, though, belongs to the subrange type WEEKday.

a subrange-type function

```
PROCEDURE NextWorkingDay (Today: DayTYPE) : WEEKday;
  (*Represents the first working day to follow Today.*)
  CONST FIRSTworkDay = Monday;
        LASTworkDay = Friday;
  BEGIN
    IF (Today>=FIRSTworkDay) AND (Today<LASTworkDay)
      THEN INC(Today, 1);  (*INC is a procedure, not a function.*)
           RETURN Today
      ELSE RETURN FIRSTworkDay
    END
    (*The RETURN values have to be in the WEEKday subrange.*)
  END NextWorkingDay;
```

Modification Problems

1. Change NextWorkingDay into a procedure that returns the next school month. Naturally, you'll have to define a MonthTYPE to back it up.

Section 7-4 Extended Examples

FINALLY, LET'S LOOK AT A FEW LONGER PROGRAMS. OUR first example solves a problem you'll run into when you start to use enumerated types—they can't be written or read with ordinary I/O procedures.

> Suppose that you have enumerated a type that names several common housepets. Write TypeOut and TypeIn procedures that allow output and input of the values.

problem: enumerated type I/O

Let's make AnimalTYPE (a compendium of popular New Age housepets) our underlying type definition:

```
TYPE AnimalTYPE = (bat, ant, fish, rock, whale);
VAR Pet: AnimalTYPE;
```

Recall that the enumerated types have no external character representation. Even though we think of them as being strings—character sequences—the compiler is not concerned about their apparent English values. The statement IO.StringOut (Pet) won't compile, because Pet isn't a string.

Fortunately, Modula-2 provides the tools we'll need to carry out a translation. Output is taken care of by using a **CASE** statement to choose the proper argument for IO.StringOut. I've included an empty **ELSE** part out of habit, even though the alternatives I supply are intended to cover all the bases.

printing enumerated values procedure

```
PROCEDURE TypeOut (Value: AnimalTYPE);
   (*Prints the string that corresponds to a Value of AnimalTYPE.*)
   BEGIN
      CASE Value OF
          bat: IO.StringOut ("bat")
         | ant: IO.StringOut ("ant")
         | fish: IO.StringOut ("fish")
         | rock: IO.StringOut ("rock")
         | whale: IO.StringOut ("whale")
         ELSE
      END
   END TypeOut;
```

Input of enumerated type values is less convenient than output, because the **CASE** statement cannot be used in a symmetrical fashion—string values can't appear as **CASE** labels. The few lines of code below show an appealing, but illegal, example. They assume that TheString has been declared as a string-type variable.

an illegal try

```
(*Strings can't be CASE constants.*)
CASE TheString OF
     "bat":  Creature := bat
   | "ant":  Creature := ant
   | "fish": Creature := fish
       . . .     etc.
```

function Equal

Instead, we have to rely on the **IF** statement, as well as some understanding of string types and your system's string-handling library. It isn't standard, but every system will have one. Although names will vary, the library—call it STRINGS—will almost certainly let you import a BOOLEAN-valued function that compares two strings for equality. I'll call this function Equal, and assume it's TRUE if its arguments are the same, and FALSE otherwise.

Procedure TypeIn, below, relies on Equal to decide which string has been read. Since strings can't appear as **CASE** labels, we have to rely on an earlier method that uses the **IF** statement and **ELSIF** parts. In this situation, I actually use the final **ELSE** part, since the string that was entered might be entirely wrong.

input of enumerated values

```
(*Assume, among other imports, IMPORT STRINGS;*)
PROCEDURE TypeIn (VAR Creature: AnimalTYPE);
   (*Read a string and assign the proper value to Creature.*)
   VAR Name:  ARRAY [0..5] OF CHAR;
   BEGIN
      IO.StringIn (Name);
      IF STRINGS.Equal(Name, "bat") THEN Creature := bat
         ELSIF STRINGS.Equal(Name, "ant") THEN Creature := ant
         ELSIF STRINGS.Equal(Name, "fish") THEN Creature := fish
         ELSIF STRINGS.Equal(Name, "rock") THEN Creature := rock
         ELSIF STRINGS.Equal(Name, "whale") THEN Creature := whale
         ELSE IO.StringOut ("ERROR -- Creature is undefined.")
      END
   END TypeIn;
```

Modification Problems

1. Find out what your system's equivalent to STRINGS is, and test the TypeIn and TypeOut procedures. Add your favorite animal to the AnimalTYPE definition.

Example: Employee Payroll

A final example uses both **CASE** and enumerated and subrange types.

problem: employee payroll

Write a program that computes an employee's weekly pay, including benefits. Assume that we're paying time-and-a-half for Saturday, and double-time for Sunday work.

A problem like this is potentially very complex. Payrolls are a simple concept until you're actually on one. My last paycheck, for instance, contained deductions for state taxes, federal taxes, city taxes, disability insurance, life insurance, social security, contributions, and assorted minor items like my overdue faculty club bills. However, we will be satisfied to write an extremely simple version to begin with.

What does the most basic version of a payroll program have to do? Our flow of data consists of hours worked, given one day at a time. Our goal is to transform those raw hour figures into wages due. Since wage rates vary from day to day, we have to maintain a state variable that lets us know what day it is. A simple count of seven days' figures bounds the input data. Naturally, we look at the daily figures one at a time.

A **FOR** statement turns out to be particularly useful in this application. It can step through the days one at a time, and the counter variable can play double duty as a state variable. In pseudocode, we read in the hours day by day, and make adjustments for overtime as we go along:

first refinement

FOR *each day of the week*
    read in the number of hours worked;
    calculate the running pay total

We'll limit the variable that holds number of hours worked per day to a reasonable numerical subrange. A variable Workday, that take on values corresponding to the days of the week (call it DayTYPE), lets us refine the pseudocode:

second refinement

FOR Workday := MIN(DayTYPE) **TO** MAX(DayTYPE) **DO**
    read in the number of hours worked;
    calculate the running pay total

Within the *calculate the running pay total* section, how can we arrange for the Saturday and Sunday overtime rate? Why not use a **CASE** statement to select the proper pay scale? I've used actual labels rather than a closing **ELSE** part because I think it's a little clearer:

a coding detail

CASE Workday **OF**
    Saturday: *arrange for time-and-a-half*
    Sunday: *arrange for double-time*
    Monday .. Friday: *pay single-time*
END *the statement*

All that really remains is the definition of a new enumerated type:

data type definition

```
TYPE DayTYPE = (Monday, Tuesday, Wednesday,
                Thursday, Friday, Saturday, Sunday);
```

The addition of input, output, and a few calculations turn our pseudocode into program Payroll. Notice that Payroll is quite limited; it's a trial model that points the way to more sophisticated features, but doesn't really implement them. Payroll just handles one worker, at a single hourly wage and benefit rate. It only accepts full hours of work. Beyond the built-in subrange checks, it doesn't error-check input. Nonetheless, Payroll works, and can be upgraded later.

employee payroll program

```
MODULE Payroll;
   (*Computes one employee's weekly payroll using enumerated types.*)
IMPORT IO;
CONST BENEFITrate = 5.37;          (*Benefits add $5.37/hour.*)
   (*Benefits are only paid on actual hours worked, not overtime.*)
TYPE DayTYPE = (Monday, Tuesday, Wednesday,
                      Thursday, Friday, Saturday, Sunday);
VAR Workday: DayTYPE;
     HourlyRate, TotalWages, HoursCredited, Benefits: REAL;
     HoursWorked: INTEGER[0..10];  (*Avoid the CARDINAL default.*)
BEGIN
   IO.StringOut ("Please enter the hourly wage rate:   $");
   IO.RealIn (HourlyRate);
   TotalWages := 0.0;
   Benefits := 0.0;
   IO.StringOut ("Enter hours worked from Monday through Sunday.");
   IO.LineOut;
   FOR Workday := MIN(DayTYPE) TO MAX(DayTYPE) DO
      IO.IntIn (HoursWorked);
      CASE Workday OF
           Saturday: HoursCredited := 1.5*FLOAT(HoursWorked)
         | Sunday: HoursCredited := 2.0*FLOAT(HoursWorked)
         | Monday .. Friday: HoursCredited := FLOAT(HoursWorked)
      END; (*CASE*)
      Benefits := Benefits + (FLOAT(HoursWorked)*BENEFITrate);
      TotalWages := TotalWages+(HoursCredited*HourlyRate)
   END; (*FOR*)
   TotalWages := TotalWages + Benefits;
   IO.StringOut ("Wages for the week are $");
   IO.FixOut (TotalWages, 1, 2)
END Payroll.
```

```
Please enter the hourly wage rate:   $9.37
Enter hours worked from Monday through Sunday.
8  8  6  9  10  0  4
Wages for the week are $700.78
```

Modification Problems

1. Add a TypeOut procedure to Payroll that makes it easy to prompt for each day's wages by name.
2. Make Payroll more robust. For example, let the user correct a mistaken entry, or start over again if the total is not to her liking.

Section 7-5
Type Compatibility

JOIN ME NOW AS WE GRIND THROUGH A FEW details about types. Don't despair as you read—most of these details are here for the sake of future reference, rather than out of any present necessity. Most importantly, remember that the rules *reflect* some of the problems of storing values with a computer—they don't *cause* them.

Let's start at the heart of matters. On occasion, we may define subranges that are overlapping, similar, or even (apparently) the same. Understanding the differences between these subrange types is a confusing problem even for experienced programmers. For example:

a sample definition

```
TYPE CapitalLetterTYPE = ["A".."Z"];
VAR FirstInitial, MiddleInitial: CapitalLetterTYPE;
    LastInitial: ["A".."Z"];
```

Are FirstInitial and LastInitial variables of the same type? A reasonable person, seeing that the subrange definitions are equivalent, would say yes. However, Modula-2 draws three possible distinctions. I'll give them in decreasing order of strictness:

identical

- *Identical* types have to be defined with the same name. (Technically it's possible to change a name, but we won't ever do it.) FirstInitial and LastInitial, therefore, do *not* have identical types, even though they represent the same range of values. FirstInitial and MiddleInitial *do* have identical types.

expression compatible

- *Expression compatible* types can appear in an expression together. To be expression compatible, types must have identical base types, but one or both can be subranges. This is just common sense—values need a common characteristic if they're going to appear in an expression together. Unfortunately, INTEGER and CARDINAL are *not* expression compatible. Even if two such variables have the exact same value, they can't appear in an expression if their base types differ.

assignment compatible

- *Assignment compatible* types can be assigned to each other. To be assignment compatible, two types must have identical, expression compatible, or INTEGER or CARDINAL base types. Assignment compatibility is, frankly, a loophole intended to let INTEGER values (if they're nonnegative) be assigned to CARDINAL variables, and CARDINAL values (if they're small enough) be assigned to INTEGER variables.

Thus, FirstInitial and MiddleInitial have identical types, but neither is identical to LastInitial. However, all three variables are expression compatible, and are also assignment compatible.

Some Examples

Note that even the relatively lax rule of assignment compatibility doesn't always mean that an assignment is going to be legal. That depends on the actual values involved. For example:

```
TYPE LOWrange = [1..5]; (*Default base type CARDINAL.*)
     MIDrange = INTEGER[1..10];
     HIGHrange = [6..20];
VAR LowValue: LOWrange;
    MidValue: MIDrange;
    HighValue: HIGHrange;
    AnyValue: CARDINAL;
```

LowValue, MidValue, HighValue, and AnyValue are clearly not of identical types, because they're all declared with different type identifiers. Moreover, MidValue is not expression compatible with the others, since we've defined it as an INTEGER subrange. No matter what their values are:

LowValue + AnyValue
    is a legal expression: CARDINAL *subrange* + CARDINAL
MidValue + AnyValue
    is an illegal expression: INTEGER *subrange* + CARDINAL

All four variables, however, are assignment compatible. This may or may not mean that assignments are actually legal. These assignments are OK:

legal assignments

```
(*legal assignments*)
MidValue := LowValue;
AnyValue := MidValue;
```

MidValue, LowValue, and AnyValue are assignment compatible because their base types are all INTEGER or CARDINAL. They are legitimate assignments as well, because the actual value of LowValue falls within the subrange of MidValue, and the actual value of MidValue easily falls into CARDINAL range of AnyValue.

However, this won't always be the case. These assignments:

possibly legal assignments

```
(*potentially legal assignments*)
LowValue := MidValue;
MidValue := HighValue;
HighValue := AnyValue;
```

may or may not be valid, depending on the values of MidValue, HighValue, and AnyValue when the assignment actually takes place. If values are in the wrong subrange, a type clash will occur.

As I mentioned earlier, variable parameters and the variables that are their arguments must have identical types. This makes sense because a variable parameter merely renames its argument. But since value parameters treat their arguments as values, they need not have identical types—assignment compatibility is enough. Again, this makes sense because the argument is just used to initialize the value parameter. A run-time error occurs only for an 'out-of-range' argument.

type transfer functions

On occasion, programmers may want to get around the restrictions imposed by Modula's strict type checking. Under special circumstances, it is appropriate to use the name of a type as a *type transfer function*. This application is described in Chapter 20.

Self-Check Questions

Q. Suppose that we make the definitions and declarations shown below:

```
TYPE LowTYPE = INTEGER[1..7];
     MidTYPE = INTEGER[5..10];
     HighTYPE = INTEGER[11.20];
     BroadTYPE = INTEGER[1..20];
VAR LowValue: LowTYPE;
    MidValue: MidTYPE;
    HighValue: HighTYPE;
    BroadValue: BroadTYPE;
```

Which of these assignment are always in error, sometimes in error, or never in error?

a) LowValue := 2 * MidValue;
b) MidValue := LowValue;
c) BroadValue := HighValue;
d) HighValue := BroadValue;
e) MidValue := 3 * HighValue;

A. *a* and *e* are always going to be in error. *b* and *d* will sometimes be correct, and sometimes wrong, while only *c* is always going to be a legal assignment. For extra practice, you might want to try making up some procedure calls that incorporate the same sorts of errors.

Section 7-6 In Depth: Data Type Abstraction

TYPE ENUMERATIONS, SUBRANGES, AND THE **CASE** statement are particularly appropriate subjects for an in-depth discussion. None of them are absolutely necessary for carrying out any algorithmic goal, yet they have all found their way into modern programming languages.

The desirability of typing, in particular, has come to be an article of faith. But let's be contrary, and take a somewhat skeptical view. Are type requirements a virtue, the successful culmination of years of work in programming language design? Or are they, on other hand, a necessary vice, imposed on languages in a desperate attempt to maintain order? Does type represent progress, or defense?

Typing for Defense

Let's look at both sides. I'll consider the defensive position first, starting with Leopold Kronecker's century-old comment: 'God made the integers; all the rest is the work of man.' This was not a friendly statement for a mathematician to make; irrational numbers did not please Kronecker, and he felt that π was nonexistent! But his demand—that mathematics should be based on values that could be seen and felt, so to speak—was a defensive stand that was obeyed to the letter in the first generations of computer programming languages, which represented all values in bits and bytes.

Recent years have seen an explosion of alternatives to working exclusively with the binary digits early programmers used. In programming today, numbers have left the center stage, to be replaced by values whose distinction has been their type. As long as a programmer is willing to obey the ground rules for defining types, she can program in any terms she wants. Indeed, there are languages that go much further than Modula-2 in providing ways to create new types and operations.

In fact, defining values in terms of their types may be *the* distinguishing characteristic of modern high-level languages. But, in the words of Mary Shaw, a well-known language designer:

> 'Types are not God-given and type is not a central point of the universe. More than anything else, the notion of type is a means of organizing information about programs. It forces the program author to deal with certain issues. It is a means of summarizing information about values common to a subset of the program's variables. It is a means of expressing the kinds of operations that may be permitted on these values. The motivation for type arises from a need for discipline in programming. It arises from a need to maintain the integrity of data, and in recent times it is used to reduce the life-cycle costs of programs.'*

Shaw's position effectively presents the defensive view of types. Programs could be based entirely upon integers, just as, at the machine level, computation is based on binary arithmetic and comparisons. Nevertheless, we invent types because they provide a way of sensibly talking about and dealing with the practical problems—those 'certain issues'—that programmers face.

type is an artifact

We've already discussed a number of these issues at length. Built-in types like INTEGER, REAL, CHAR and so on prevent the intermingling of grossly unrelated values, and limit the use of operators (particularly mathematical operators) in meaningless circumstances. Enumerated types, as a class, introduce values

* SIGPLAN Notices, *Workshop on Data Abstraction, Databases, and Conceptual Modeling*, 16(1), 1981.

that can be compared—as long as both operands are from the same enumeration—but have no specifically defined operators.

enumeration

Thus, Monday isn't the same day as Friday, and we can use the relational operators to confirm this fact. On the other hand, Tuesday cannot be added to Thursday (unless we explicitly use the ORD function to get around type protections). Although we may think about the days as though they were numbered, and although the computer certainly represents them internally as numbers, the act of defining the days as an enumerated type tells other programmers 'These are NOT numbers, and if you want to provide operations you'll have to define them yourself.'

subranging

Subrange types make error avoidance even more practical and powerful. The conscientious language implementor or system programmer, recognizing that her disciples or subordinates may not adequately safeguard against implausible values, relies on subranges for two reasons:

1. to document the fact that a value is expected to fall within a particular range (in case the programmer has neglected to comment adequately);
2. to ensure that program execution halts if a variable takes on a value outside the range of reason (in case the programmer has neglected to error-check adequately).*

A perfect world would not require the use of subranges. In the real, practical world, though, comments cannot be peppered through a program as easily as a subrange type name like LOWERcase or THREEdigitNumber. Nor will any programmer willingly submit to the need to check every bit of data that is used by a program, particularly since the guideline 'within the range of reason' is so vague.

the defensive position

To summarize the defensive view, enumerations and subranges provide two types of protection. First, explicitly naming a type tells the program reader (or a future coder) that the programmer has some particular intention in mind for her variables and values. This helps the program to be understood in the present, and provides a guide if the program is modified in the future. However, there is a second line of defense, enforced by the computer with a crash, if the warning goes unheeded.

Typing for Extension

We need not look at type solely as a defensive programming device, though. Let's look at the other side of the coin, and see how the ability to deal in terms of

* A compiler may also rely on the limited number of values in a subrange to minimize the space it sets aside to hold a single value (e.g. 0..255 can be held in eight bits, but 0..256 will probably have sixteen bits set aside).

abstract values *extends* the programmer's options. In the initial discussion of enumerated types, I mentioned three kinds of type-related abstraction.

- New, elementary values can be given names: we can name colors, or the days of the week.

non-defensive applications

- Expressions and relations between values can be given names: Insignificant or Meaningful are useful, as are the more familiar LessThan and Equal.

- Program states can be named: searching, found, and missing describe the conditions that might prevail during the hunt for a particular input value.

Thus, enumerations can be used to expand, rather than restrict, the operations at our disposal. For example, values can be named with a simple type definition:

naming values

TYPE ColorTYPE = (red, blue, green, *etc.*);

Then, procedures or functions can be designed to carry out the arithmetic of mixing colors, so that Add (red, green) might return the value yellow.

(In case you're familiar with arrays, I'll mention that enumerated types can also supply array subscripts. This means that components can be allocated and identified in terms of abstract attributes: the 'big, red' element—and not just as the intersection of Cartesian-like coordinates: the 'row 2, column 9' element.)

Some languages (but not Modula-2) will even let standard operators like '+' be extended. Operator *overloading* means redefining operators so that they can be used with operands of nonstandard types like ColorTYPE.

Naming States

The third non-defensive application of enumerations I mentioned is for naming program states. Previously we've generally viewed state variables as a convenient, and sometimes necessary, means of describing the information we've collected about program data. However, we might use states to describe an algorithm—the entire execution of a program can be presented as a progression of states.

For example, consider the problem of writing a program to keep bowling scores. Suppose that I outline the rules. A bowled game consists of ten *frames* whose scores are added together. In each frame, the bowler attempts to knock down ten pins in one or two *trials*. The frame is over when all ten pins have been knocked down, or after two trials, whichever comes first.

- If fewer than ten pins have been knocked down in two trials, the current frame's score equals the number of pins toppled.

keeping bowling scores

- If exactly ten pins were knocked down in two trials (a *spare*), the current frame's score will equal 10 plus the number of pins downed in the first trial of the next frame.

- If exactly ten pins were felled in the first trial (a *strike*), the current frame's score will equal 10 plus the number of pins knocked down in her next two trials—whether those two trials require one or two additional frames.
- If the tenth frame ends in a spare, the bowler is granted one additional roll to complete tenth frame scoring.
- If the tenth frame ends in a strike, the bowler is granted two additional rolls to complete tenth frame scoring.
- If a foul was committed (for instance, by crossing the foul line at the top of the alley), that roll scores zero.

Mathematically speaking the problem is just arithmetic. However, the rules of scoring impose an elaborate set of state requirements. Although each individual piece of addition is simple, the rules of *what* is added *when* and *where* are not.

What are the states that describe the progress of a game? How about InASpare, InAStrike, LastFrame, and Foul? From the bowler's point of view, a game is a steady advance from the first frame to the last. From the scorer's box, though, 'finished' is just one of many states that may prevail during the course of a game. Numbers keep rolling in, but their significance is entirely dependent on the current state of a frame or prior frame.

Section 7-7 Program Engineering Notes

ENUMERATED TYPES AND **CASE** STATEMENTS MIGHT seem to be an unlikely pair. However, they prove to be an effective combination for keeping track of, and acting on, constantly changing program states.

State is usually misunderstood as referring to the condition of a program, as in 'My program is in a dismal state.' But it really means *knowledge* about *data*. As we know, a program turns data into information:

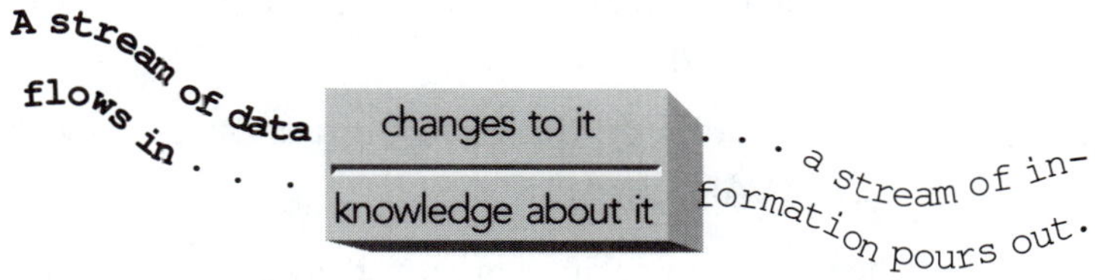

The simplest programs have just one focal point—they do something to data values (add them, or double them), or they learn something about data (is it all positive?). But more complex programs—the kind we'll be writing—usually have to do both. What we do will often depend on what we know, and that will change during program execution. And, when *determine the current state* becomes confused with *act on the current state*, problems result.

> Look before you leap. Be sure that you know the current state before committing yourself to a particular action.

For example, let's look at a concrete instance of loop design. Suppose that we want to read and somehow process numbers until we spot a sentinel or out-of-bounds value, or run out of input. Here's the solution we're most familiar with: a complicated **EXIT** condition:

establishing the state after the loop

```
LOOP
   IO.IntIn (Value);
   (**)IF NOT IO.Done OR (*then if*)
        (Value > UPPER) OR (Value < LOWER)
          OR (Value = SENTINEL) THEN EXIT END;
   . . .       rest of the loop
END;
(*Now, decide why we left the loop.*)
IF NOT IO.Done
      THEN value missing message and action
   ELSIF Value > UPPER
      THEN value too big message and action
   ELSIF Value < LOWER
      THEN value too small message and action
   ELSE (*Value = SENTINEL*)
      found the sentinel message and action
END
```

The shaded section is devoted to establishing our state after the loop. There is nothing incorrect about this approach. However, we are in an awkward position between the **LOOP** and the **IF** statement check that follows it: we don't know our state.

Now consider an alternative approach. It begins with the definition of an enumerated type:

a state type

```
TYPE ExitTYPE = (searching, under, over, found, missing);
VAR State: ExitTYPE;
```

Then, the loop is written in a way that establishes state *before* any action, including the **EXIT**, is taken. On a line-for-line basis we're using essentially the same code as we did before. But in the reordered code, we know our state at all times:

```
LOOP
   IO.IntIn (Value);
   IF NOT IO.Done
         THEN State := missing
      ELSIF Value > UPPER
         THEN State := over
      ELSIF Value < LOWER
         THEN State := under
      ELSIF Value = SENTINEL
         THEN State := found
      ELSE State := searching
   END
   (**)IF State <> searching THEN EXIT END;
   . . .        rest of the loop
END;
(*Now, take the appropriate post-loop action.*)
CASE State OF
     found: 'found the sentinel' message and action;
   | over: 'value too big message' and action;
   | under: 'value too small message' and action;
   | missing: 'value missing message' and action
END
```

The code is a bit longer, but it's simpler. We can tell what's going on.

Why Use Subranges?

Although they may not lead to the largest number of bugs in Modula-2 programs, the subrange types present the greatest confusion. It is important not to put the cart before the horse, though. As I pointed out, the rules reflect some realities of computing—they don't cause them.

- First, type separation is intended to make sure that values are never stored or retrieved incorrectly.
- Second, although the type of a variable is known at compile-time, the type of an expression *once you allow subranges* might not be known until run-time.
- Third, at the machine level, different machine language commands may be used to carry out the same operations—for example, the exact addition command used depends on the type of its operands.

In effect, types attach an invisible set of rules to every value and variable. They make sure we never add apples and oranges.

Why bother to make the effort of learning to deal with subranges? They're desirable for three reasons—self-documentation, program efficiency, and anti-bugging.

documentation

1. Self-documentation. Knowing the range of values that a variable is going to represent, and saying so at the time of variable declaration, helps demonstrate that you have a firm grasp of what your program does.

It also helps another person who may be working on your program *get* an idea of appropriate values within a program. The declaration:

```
KilnTemperature: INTEGER;
```

says nothing, whereas:

```
KilnTemperature: [400..1200];            (*or, better yet . . .*)
KilnTemperature: [400..MAXtemperature];          (*or, still better . . .*)
KilnTemperature: TempRangeTYPE;
```

is informative.

efficiency

2. Program efficiency. A variable declared to represent only a limited range of values can be dealt with (by the compiler) in a more economical manner than a variable that can represent *any* value of its type.

This is really the least important reason. Under certain circumstances, though, a program may require so many variables that limiting the storage they require is a valid programming consideration.

antibugging

3. Antibugging. Restricting the values that a variable may represent to a range of values we know it *should* have extends the protection. We help assure ourselves (and the computer) that any operations we'll try to carry out will make sense.

Real life often places limits on the values that a variable can reasonably represent. A payroll program may 'work', but allow 37 deductions, or –2. A checker-playing program might devote a considerable amount of time looking for the ninth row of a checkerboard. A computer croupier could spin a computer roulette wheel, and decide the ball has landed on number 39—which doesn't exist.

These are all obviously bugs that should, and could, be spotted or prevented by the programmer. The most annoying aspect of bugs, though, is that you don't see them until it's too late. When a program announces that the sum of two and two is five, the programmer knows that something has gone wrong, and takes another look at her code. The results of a more complex program, however, are more likely to be taken for granted—even if some input datum or partial result hidden within the program is totally absurd.

the moral

> Subrange types provide a constant check on variables and assure us that they have values appropriate to their application. They help prevent a very dangerous kind of program—one which appears to be reliable, but is not.

Using subranges doesn't absolve the programmer of responsibility for error checking and keeping track of data within a program. The subrange philosophy is rather nihilistic—if a variable takes on an inappropriate value, the program stops! Relying on run-time crashes to do error checking is like using telephone poles (instead of brakes) to stop your car. Still, strong type checking makes Modula-2 sympathetic to a programmer's woes—data inevitably gets screwed up for reasons beyond the control of programmer, program user, or computer. Subrange types won't make programs work, but they will make it easier to debug programs and to keep them running.

Modula-2 Details

Syntax Summary

• **CASE** statement: makes a choice between two or more alternatives, based on the value of a *case expression* that can't be REAL. The alternatives are indicated by *case labels*, which have to be constants that can be evaluated by reading, rather than running, the program. An optional **ELSE** portion provides an action for any unlabeled value of the case expression.

```
CASE Expression OF
    CONST1 : action;
   | CONST2, CONST4 : action;        (*a list of labels*)
   | CONST5 .. CONST8 : action       (*a range of labels*)
   ELSE action                       (*this would include CONST3*)
END
```

• enumerated type: a type declaration that names, and provides the constants of, a sequence of values.

```
TYPE MotownTYPE = (Temptations, Smokey, Marvin, Supremes);
     NoWaveTYPE = (Prince, RickeyLee, Elvis);
```

• subrange type: a type declaration that names a subsequence of an existing *base type*. If the base type is ambiguous, a particular type name can be supplied:

```
TYPE LowerTYPE = ["a" .. "z"];
     SMALLcardinals = [1..10];  (*Base type is CARDINAL, the default.*)
     SMALLintegers = INTEGER[1 .. 10];  (*Makes the base type INTEGER.*)
```

• ordinal functions and procedures: are built into Modula-2. The standard functions include:

MAX(T), MIN(T)	*maximum and minimum values of type* T
ORD(value)	*ordinal (counting) position of* value *in its type*
VAL(T, position)	*value of type* T *in ordinal position* position

Standard procedures include:

INC(variable)	*change* variable *to its successor*
DEC(variable)	*change* variable *to its predecessor*
INC(variable, n)	*change* variable *to its* n*th successor*
DEC(variable, n)	*change* variable *to its* n*th predecessor*

Modula-2 Details

Remember . . .

- The effect of a **CASE** expression that can't be found in the **CASE** label list isn't defined. To avoid trouble, you should include an **ELSE** part with every **CASE** statement you write.
- The constants of enumerated types must be distinct. Two enumerated types can't share values, although subranges of a single base type can.
- Enumerated type constants are ordered by their definitions. Their ordinal numbering always starts with zero.
- The constants of an enumerated type can't be input or output because they have no external character representation. Instead, they must be translated to and from strings.
- Use MIN(T) and MAX(T) to set the bounds of a **FOR** loop that's supposed to cycle through all of a type's values.
- By default, a numerical subrange with non-negative bounds is assumed to have base type CARDINAL. This can be overridden, as shown above.
- Two variables have the same, identical type only if they're declared with the same type identifier. A variable parameter and its argument must have identical types. Assignments between variables of identical types will always be valid.
- Two values are expression compatible only if they have the same base type.
- A value and a variable are assignment compatible (the value can be assigned to the variable) if they have identical types, expression compatible types, *or* if they're INTEGER and CARDINAL values in an appropriate range (i.e. a positive INTEGER can be assigned to a CARDINAL, or a small enough CARDINAL can be assigned to an INTEGER. A value parameter and its argument must be assignment compatible.
- Subrange types are useful because they help make programs self-documenting, and enforce automatic run-time checks on program values.

Zen Study Guide

Big Ideas

The Big Ideas in this chapter are:

1. *Type membership is meant to ensure that values will always be stored and retrieved correctly.* Enumerated and subrange types are natural extensions to the built-in standard types, because they increase the programmer's vocabulary without endangering the correctness of programs.

2. *The type of variables can be determined at compile-time, but expressions are not evaluated until run-time.* The compatibility rules reflect this fact—they don't cause it.

3. *The* **CASE** *statement is meant to show mutually-exclusive alternatives.* It's an alternative to, but not a replacement for, the **IF** statement.

Zen Study Guide
Questions

7-1 The syntax of enumerations and subranges is entirely arbitrary and sometimes confusing. Write a half-dozen definitions that ask the question *Enumeration, subrange, or incorrect?* Some of the syntactic details you can confuse include *a)* using parentheses instead of square brackets, or vice versa; *b)* using a non-contiguous subrange of values (e.g. the vowels); *c)* including the same constant in more than one type definition.

7-2 Ask three prose questions that require the definition of new types (e.g. the months) and subranges (particular seasons).

7-3 A multiple-choice question that's easy to pose (and grade) is put together by declaring a few variables, then listing a series of values and expressions. Which values can be assigned to which variables? The legal values fall within appropriate subranges, while illegal values have the wrong type, or are outside of a particular subrange. Write the question.

7-4 Asking whether a procedure call has legal arguments is a little more complicated than checking assignments. Recall that variable parameters have to have the same types as their arguments, while value parameters only have to be assignment compatible. Write three headings with at least two parameters each, then make up a half-dozen legal and illegal calls.

7-5 The **CASE** statement is prone to causing syntax problems at first. Take a legal example from the text, then create a few *Where's the bug?* problems by introducing syntax bugs. Rely on *a)* missing vertical bars; *b)* a misplaced **ELSE** part; *c)* an illegal expression in the list of labels; *d)* a label with the wrong type; *e)* a label that appears more than once.

7-6 When can an **IF** statement with **ELSIF** parts be replaced by a **CASE** statement, and vice versa? Describe four problems (e.g. scoring exams, recording income) and ask which statement is most appropriate to use.

7-7 An alternative way to approach the same problem is to ask that an **IF** be rewritten as a **CASE**, if possible. Write three **IF** statements for rewriting as **CASE** statements. Naturally, at least one of them should be impossible to rewrite.

7-8 State filtering—collecting information about a stream of data—is a standard **CASE** application, as shown in program Digits. Make up three counting questions that can be solved with a **CASE** statement. Each one should require at least three different categories.

7-9 Program Digits actually contains a slight twist—zeros are counted as both zeros and even numbers. Make up three more counting questions that have some values that fall into more than one category, e.g. given an input stream of digits, ignore zeros, but count odds, evens, threes and sixes, and fives.

7-10 Writing conversion programs is a standard problem for enumerated types and **CASE** statements. For example, converting to and from foreign currencies requires enumerating a type (for the currency names), writing input and output procedures (to get and print the names), as well as a **CASE** statement that chooses the right conversion formula. Make up two conversion program questions, and draft (but don't write) their solutions.

Exercises
Warmups and Test Programs

7-11 Is there a limit on the length of new type names? On the length of the constants of enumerated types? Does your system allow any non-standard characters (like underlines) to appear in identifiers?

7-12 What error messages will you get from an incorrectly placed **ELSE** part? Overlapping labels? What happens if the value of a case expression is not given as a label—and the **CASE** statement doesn't have an **ELSE** part?

7-13 A **CASE** label can indicate a range of values, e.g. "A".."Z". But is there a limit on the size of the range? In other words, suppose that the case expression is an INTEGER. Can a label's range go from 1..1000? 1..MAX(INTEGER)?

7-14 There is some inconsistency in Modula-2 implementations on the use and availability of functions INTEGER(x), CARDINAL(x), FLOAT(x), and TRUNC(x). The three questions to ask are:

a) Is the function available?

b) What is the type of its argument?

c) What is the type of its result?

Write test programs that document the type transfer functions on your system.

7-15 What happens when you try to use INC or DEC to increment or decrement a variable to a value that will fall outside of its legal range?

7-16 What is the purpose of this **CASE** statement? Could the range of the last label be extended beyond 10?

```
IO.IntOut (N, 1);
CASE N OF
     1: IO.StringOut ("st");
   | 2: IO.StringOut ("nd");
   | 3: IO.StringOut ("rd");
   | 4..10: IO.StringOut ("th");
END
```

7-17 Where's the bug?

```
(*Read an input sequence that is supposed to contain only a through f,
 and echo the characters b and d.*)
LOOP
   IO.CharIn (Letter);
   (**)IF Letter = IO.EOL THEN EXIT END;
   CASE Letter OF
         "b", "d": IO.CharOut (Letter);
      | "a", "c", "e", "f": IO.CharIn (Letter)
      ELSE IO.StringOut ("Illegal input character.")
   END
END
```

7-18 Consider this program segment:

```
First := 1;
FOR Last := 7 TO First BY -1 DO
   FOR LetterNumber := First TO Last DO
      Print (LetterNumber)
   END
END
```

Its output is:

```
whereas wherea where wher whe we w
```

Write procedure Print (which should, of course, contain a **CASE**).

7-19 What is an ordinal type? What simple type isn't an ordinal type? What does it mean to say that a type is enumerable?

7-20 Suppose that we make this type definition:

```
TYPE NicknameTYPE = (Boss, BigBopper, Killer, MainMan);
```

Write a procedure that takes a parameter of type NicknameTYPE and prints its value.

Exercises

Programming Problems

7-21 Write a program that uses the state approach described in section 7-7 to keep bowling scores.

7-22 Write a program whose input is a date supplied as a month, day, and year, e.g. **7/29/53**.

a) Print the day's number within the year, i.e. if the date supplied is **2/3/85**, the output should be be `34`. Don't forget that a leap year occurs every fourth year except if the year is divisible by 100, but not 400.

b) Read a second date, and then calculate and print the number of days from one to the next (that is, including one but not the other).

7-23 Assuming input as above, write a program that prints the month and day in words. To save some trouble, assume that 1/1/87 (which falls on a Thursday) is the earliest possible day. Proper output for **1/2/87** would be: `Friday, January 2, 1987.`

7-24 Write a program that reads an INTEGER value, then prints the 'word' equivalent. For example, for input **23567**, output should be:

```
Twenty three thousand five hundred sixty seven
```

Hint: As an easier warmup, write a program that might drive the electronic voice-box the phone company uses for automated directory assistance. For numbers of arbitrary length, start by by reading the number as a sequence of digits so you'll know its length.

7-25 Write a little stack calculator. In stack notation, input is given as:

x y operation

where *x* and *y* are values, and *operation*, for our purposes, will be a single character. The more familiar order of evaluation is, of course:

x operation y

Your calculator should accept two numbers and an operand, e.g.:

2.34 5.92 * or **-1.03 16 /**

and perform the correct operation. Have your calculator correctly handle +, –, *, /, and the INTEGER operations **D** and **M** (for **DIV** and **MOD**).

7-26 Two biology students are carrying out important scientific research on fleas. One student's job is to motivate the flea to jump (through use of a little electric flea prod), while the other student keeps tabs on its travel. The second student, unable to bear the sight of the suffering flea, resolves to write a program that will take over her job.

Write a program that tracks the flea as it hops around. Start at the origin *x, y* = (0,0) and print the current position after each hop. Each input should consist of a distance, and a direction. For example, input **2.3 nw** would mean 'move 2.3 units northwest.' Legal directions should be:

n nw w sw s se e ne

7-27 Write a program that accepts an input INTEGER and prints the correct representation in Roman numeral form. Can you do the reverse as well?

7-28 Suppose that our universe contained just four fruits—say, lemons, oranges, watermelons, and bananas. Aside from making fruit salad rather dull, we would find that just two different questions—Is it round? Is it yellow?—could be used to determine a fruit uniquely.

However, not all questions have just two answers. Consider the questions that might describe an animal:

diet = carnivorous, vegetarian, omnivorous
size = humungous, regular, tiny
hair = long, short, none

and so on. We will find that twenty-seven different animals can be uniquely identified through just three different questions. Write a program that does so. Please send your answer to the author.

7-29 A chessboard is empty except for a solitary white piece defined by a letter:

`p` = pawn
`n` = knight
`r` = rook
`b` = bishop
`q` = queen
`k` = king

Assume we number the board from 1 to 64 *lexicographically*, so that square 1 is at upper left and square 64 is at lower right, with numbers running as you read a printed page.

Write a program that asks for a piece's name and location, then prints a list of squares the piece may legally occupy on its next move. Assume, incidentally, that a pawn starts out on the row numbered 9 through 16.

7-30 A mischievous bartender likes to fool her more tipsy customers by pouring drinks containing two different liquors, in equal quantities, into the serving glass.

Write a program that asks for two liquors from the list below, then prints the alcoholic content of the mixed drink. Assume that:

beer	4% alcohol
ale	7% alcohol
wine	12% alcohol
bourbon	40% alcohol
scotch	60% alcohol
rum	77% alcohol

The bartender is not exactly sober herself, so allow for the possibility that the two liquors are the same.

7-31 Consider the scene in a closed-room, secret design session for the 1989 model of a big-name Detroit automobile. All around the conference table are engineers and executives trying to strike a balance between greed (which leads them to build the cheapest car possible) and fear (which pulls in the other direction, toward more reliable machines).

Now, the relation between the mean time between failures (MTBF) of a system and its subsystems can be described as:

$$\frac{1}{MTBF_{Total}} = \frac{1}{MTBF_{Sub\ 1}} + \frac{1}{MTBF_{Sub\ 2}} + \cdots + \frac{1}{MTBF_{Subn}}$$

and is correct if the failure of any subsystem dooms the entire car.

Suppose that the costs and MTBF's of various subsystems are:

Subsystem		MTBF	Manufacturing Cost
brakes	disc	4 years	$15.00
	drum	5 years	$25.00
engine	Wankel	3 years	$1,067.00
	conventional	6 years	$1,850.00
suspension	air	9 years	$430.00
	oil	7 years	$320.00
electrical	computer	2 years	$130.00
	standard	4 years	$40.00

Write a program that models each combination of subsystems (a quadruply nested **FOR** loop will do nicely), and answers these questions:

a) Which system is the cheapest?

b) Which has the longest mean time between failures?

c) Which system has the lowest cost per failure-free year?

Hint: Since an enumerated type value can't be printed directly, use a **CASE** statement to print its character equivalent.

7-32 The word *therein* is interesting because at least eleven words can be 'cut' from it without rearranging any letters—*the, here, in*, and so on.

Suppose we label the letters from 1 through 7. It turns out that a triply nested **FOR** loop (calling on a **CASE** statement for actual printing) can be used to generate every possible subsequence of the letters in the word. How? What are the words?

7-33 There are twenty amino acids, and each one has one-letter and three-letter abbreviations. They are:

Alanine	Ala	A	Lysine	Lys	K
Arginine	Arg	R	Methionine	Met	M
Asparagine	Asn	N	Phenylalanine	Phe	F
Aspartic acid	Asp	D	Proline	Pro	P
Cysteine	Cys	C	Serine	Ser	S
Glutamine	Gln	Q	Threonine	Thr	T
Glutamic acid	Glu	E	Tryptophan	Trp	W
Glycene	Gly	G	Tyrosine	Tyr	Y
Histidine	His	H	Valine	Val	V
Isoleucine	Ile	I	Asn and/or Asp	Asx	B
Leucine	Leu	L	Gln and/or Glu	Glx	Z

1. Write a program that translates a sequence of proteins (supplied as a series of one-letter abbreviations) into the proper amino acids. Let the space character indicate a missing acid.

2. Write a program that uses nested **CASE** statements to translate three-letter abbreviations into either one-letter abbreviations, or the amino acids' full names.

As data, you might want to use these sequences, which give the first forty amino acids for human alpha hemoglobin and myoglobin:

alpha hemoglobin = `V LSPADKTNVKAAWGKVGAHAGEYGAEALERMFLSFPTT`
myoglobin = `G LSNGZWE VLNVWGKVEPNIAGHGEEVLIRLFKGHPET`

'The most common labor of all is visiting each array component.' (p. 304)

Chapter 8 The **ARRAY** Type

Arrays bring to data what loops bring to action—power. We can hardly imagine programming without loops; a few reserved words multiply our efforts beyond comprehension. The array is an enormously powerful programming device for controlling large amounts of data. Ultimately, we're only going to be limited by our own ingenuity when it comes to devising new schemes for storing information.

We'll begin with a look at applications. Arrays encourage *visual thinking*—if you can mentally picture an arrangement of information, you can probably build it out of an array. Since most array programs use loops, we'll also spend some time looking at array-specific looping techniques.

Using structured types effectively requires some high-powered approaches to program design. Our introduction to *data abstraction* and the use of *primitives* begins here, and will be continued in the next few chapters.

Section 8-1

Strictly Modula-2

LET'S BEGIN WITH AN EXAMPLE. Program Reversal, below, stores a sentence in an array, then prints it in reverse.

array demonstration program

```
MODULE Reversal;
   (*Reads a sentence, then prints it in reverse.*)
IMPORT IO;
CONST LIMIT = 80;
TYPE ArrayTYPE = ARRAY [1 .. LIMIT] OF CHAR;
VAR Length,           (*the current sentence length*)
    i: INTEGER;       (*the FOR loop's counter variable*)
    Hold: ArrayTYPE;  (*the array variable*)
BEGIN
   IO.StringOut ("Type in a sentence that's up to 80 letters long.");
   IO.LineOut;
   Length := 1;
   LOOP
      IO.CharIn ( Hold[Length] );
      (**)IF (Hold[Length] = IO.EOL) THEN EXIT END;
      Length := Length + 1;
   END;
   (*Postcondition: Hold[1] through Hold[Length–1] holds the sentence.*)
   FOR i := Length–1 TO 1 BY –1 DO
      IO.CharOut (Hold [i])
   END
   (*We've printed them in reverse order.*)
END Reversal.
```

```
Type in a sentence that's up to 80 letters long.
evil star war nuts
stun raw rats live
```

arrays are structured types

Arrays let us change the *structure* of variables. This is different from changing their *contents*, as we did when we enumerated new simple types. The definition **TYPE** ColorTYPE = (red, blue, green) lets a variable hold a new kind of value, but just one instance of it. A ColorTYPE variable isn't all that much different from one of type INTEGER.

But being able to define new *structured types*, like arrays, radically alters the scale of our work. The definition:

```
TYPE BigTYPE = ARRAY [1..100] OF INTEGER;
```

creates a type name—BigTYPE—that can be used to declare a variable that holds a hundred *component* values. An extra zero in the right place:

TYPE VeryBigTYPE = **ARRAY** [1..1000] **OF** INTEGER;

makes VeryBigTYPE—a structure of one thousand components.

You can picture a BigTYPE or VeryBigTYPE variable as a very long row of distinct INTEGER values. When the row just extends in one direction, it's called a *one-dimensional* array. But if we want, we can create a structure with more than one row of components. Here is a *two-dimensional* array definition:

a two-dimensional array

TYPE HugeTYPE = **ARRAY** [1..1000], [1..1000] **OF** INTEGER;

HugeTYPE is two-dimensional because it has rows *and* columns, as though it were a giant checkerboard. In all, it has one thousand rows of one thousand components each—a one *million* component total!

Modification Problems

1. Feel up to modifying Reversal already? Then have it read, then echo in reverse, a line of INTEGERs. Let the end of the line *or* a negative number end the input.

2. Add a procedure to Reversal that initializes an argument of type ArrayTYPE to all blanks.

What Arrays Can Do

Let's take a quick look at some of the things arrays let us do. The first we've already seen—arrays can hold lots of data easily. Better yet, we don't have to name the components individually. Each component's position has a *subscript* that uniquely identifies it.

> Since we can select components at will, the array is a *random access* structure. Its stored values don't have to be visited in any particular order.

subscripts, not names

Simple variables must be named and declared one at a time.

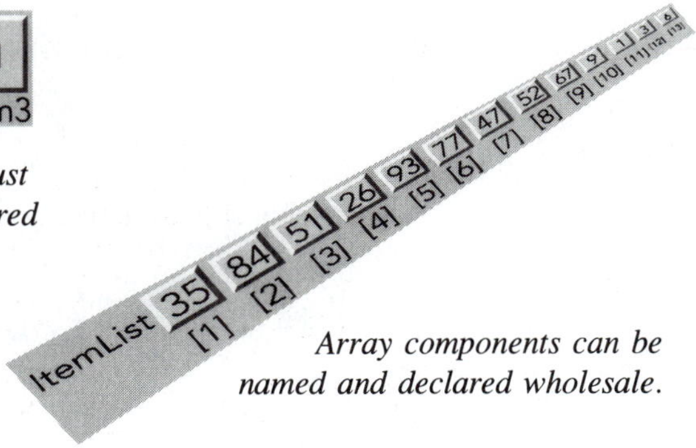

Array components can be named and declared wholesale.

Because an array has a beginning and an end, the *order* in which we store values contains information too:

storing in order

If we want, we can treat the subscripts that name each component as a *coordinate* system. Every component of a one-dimensional array has one coordinate, those in two-dimensional arrays have two, and so on:

by numerical coordinates

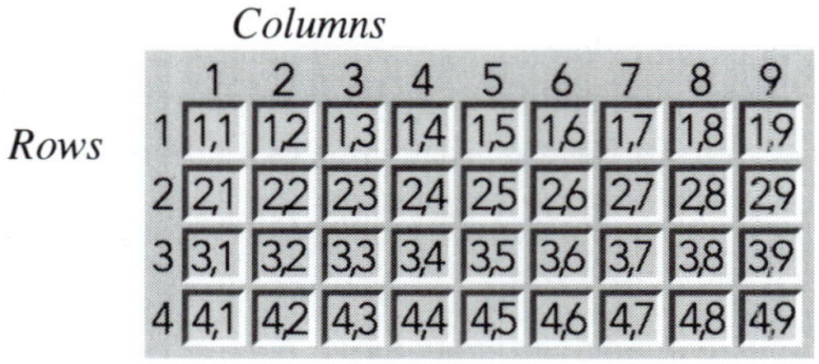

The row, column coordinates act as subscripts for the array components.

Why limit coordinate subscripts to numerical values? We can use the new values of enumerated types instead. For example, a two-dimensional array might use the days of the week to subscript one dimension, and hours of the day to subscript the other:

by non-numerical coordinates

A typical work schedule for California writers.

If an array's subscripts are the values of an enumerated type, we can describe a component in terms of its *characteristics*:

by characteristic

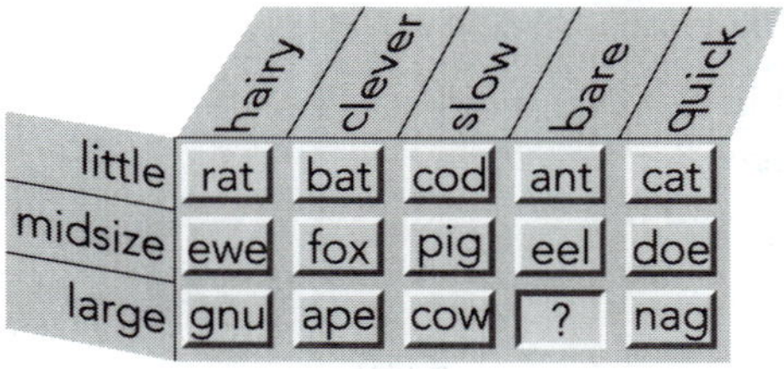

What kind of animal is large and bare?

We can also work the other way around—if we find a component that is interesting, we can use its position to find its *attributes*:

by attribute

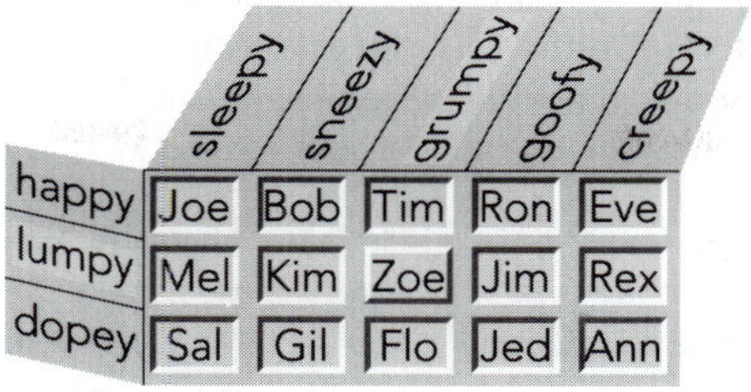

What sort of person is Zoe?

An array might also be used to associate pairs of values that are related to each other, but possibly have different types or uses. The array acts as a function, storing the correspondence between the values of two different groups:

to associate values

A few useful words in case you ever find seven Italian snakes under your chair.

Sound exciting? It is! But before we can start to program, we have to get acquainted with basic array syntax and usage.

Array Definitions

Type definitions in general tell the compiler how to set aside storage for a variable. The definition of an array type in particular describes three things:

1. The type of the array's components, or stored values. Components can be simple (like CHAR or REAL), or can have structured types themselves.

components, subscripts, size

2. The individual 'names' of the components, which are supplied by their *subscripts*. All of the subscripts aren't given—just the lower and upper *bounds* of each dimension. Subscripts can have any ordinal type; they can't be REAL.

3. The size of the array—how many components it holds—as determined by its bounds and the number of dimensions it has.

The reserved word **TYPE** always starts the definition.

The array's dimensions can be given in two different ways. Usually, we'll supply the first and last value of each dimension as though we were defining a subrange. However, the name of a 'small' ordinal type like BOOLEAN or CHAR, or of an enumerated or subrange type can be used instead. If a name is given, the compiler looks up the first and last values of that type, and sets the array bounds with them. Some typical definitions are:

specifying the dimensions

```
TYPE
  TideLevelTYPE = ARRAY [0..11] OF REAL;
    (*one dimension, twelve components*)
  LowerCaseCountTYPE = ARRAY ["a".."z"] OF INTEGER;
    (*one dimension, twenty-six components*)
  BoardTYPE = ARRAY [1..8],[1..8] OF BOOLEAN;
    (*two dimensions, sixty-four components*)
  CodeTableTYPE = ARRAY ["a".."z"],BOOLEAN OF CHAR;
    (*two dimensions, fifty-two components*)
  TolerancesTYPE = ARRAY ["A".."C"],[1..3],[-2..0] OF REAL;
    (*three dimensions, twenty-seven components*)
  CharCountTYPE = ARRAY CHAR OF INTEGER;
    (*one dimension, as many components as there are CHAR values*)
```

Arrays subscripted by CHAR are defined like this so that they'll be portable. The bounds of a CHAR array may vary on some systems, but a program that uses one will always compile.

> With the exception of open array parameters, described later, the size of an array has to be known at compile-time.

restrictions on bounds

This means that bound values have to be type names, constants, or simple arithmetic expressions. They can't be supplied with variable identifiers, or by expressions that can't be calculated without actually running the program.

Self-Check Questions

Q. What is the size of each of these arrays? Are any definitions illegal?

a) **TYPE** StorageTYPE = **ARRAY** [1..5], [1..10] **OF** CHAR;
b) **TYPE** MeasureTYPE = **ARRAY** [–5..5] **OF** REAL;
c) **TYPE** TruthTYPE = **ARRAY** BOOLEAN **OF** INTEGER;
d) **TYPE** ScienceTYPE = **ARRAY** [1.0 .. 5.0] **OF** REAL;
e) **TYPE** DigitTYPE = **ARRAY** ["1", "2", "3"] **OF** CHAR;
f) **TYPE** GraphTYPE = **ARRAY** ["0".."9"], [0..9] **OF** INTEGER;

A. The answers are:

a) 50 values of type CHAR.
b) 11 (don't forget the 0th value) of type REAL.
c) 2 INTEGER values, subscripted FALSE, TRUE.
d) Illegal—the dimensions can't be given with REAL.
e) Illegal—the dimension has to be a range, not just values.
f) 100 values of type INTEGER (each dimension has ten values).

Assignment

Once we can define array types we can declare array variables. Let's start by looking at a one-dimensional array. LowerCaseCountTYPE has INTEGER components, but uses the lower-case CHAR values as subscripts.

```
TYPE LowerCaseCountTYPE = ARRAY ["a".."z"] OF INTEGER;
VAR Paper, Program: LowerCaseCountTYPE;
```

> There are two ways to make assignments to array variables: all at once, or one component at a time.

assigning to arrays

When two variables have the exact same type (as Paper and Program do, above), we can assign one's value to the other. Naturally, the variable on the right-hand side of the assignment had better be initialized first itself.

assigning to components

To refer to an individual component, put its subscript between square brackets. Even though bounds have to be given with constants, subscripts can include variables or arithmetic expressions. These are all legal assignments to the arrays we just declared:

using subscripts

```
Paper["a"] := 5;
Program["a"] := Paper["a"];
Paper["b"] := Paper["a"];
Program["a"] := Program["a"] + 1;
Program[ CHR(98) ] := Program[ CHR (ORD ("z") -25 ) ];
Paper := Program;  (*an array-to-array assignment*)
```

Program and Paper were easy to deal with because they had just one subscript apiece. But suppose that an array has two or more dimensions? No problem—the array variable will just have two or more subscripts:

```
(*We change this component's value.*)
ThreeDimensions[Sub1, Sub2, Sub3] := TwoDimensions[Pos1, Pos2];
                                        (*We inspect this component's value.*)
```

It *is* important to remember that no matter how long and complicated subscripts might make an identifier, they're just pieces of a name. ThreeDimensions[Sub1, Sub2, Sub3] is certainly more complicated than, say, Foobar, but the component it represents might be identical to Foobar in type and value. Similarly, TwoDimensions[Pos1, Pos2] might seem to be more complex than, say, 16, but it could easily represent that very INTEGER.

Frequently, it will turn out that an array's components are also arrays. For convenience, multiple subscripts can all be placed within a single pair of square brackets. For example, given these definitions and declarations:

```
TYPE OneWordTYPE = ARRAY [0..10] OF CHAR;
     ManyWordsTYPE = ARRAY [1..20] OF OneWordTYPE;
VAR OneWord: OneWordTYPE;
    List: ManyWordsTYPE;
```

arrays of arrays

we can make the assignments below.

variations on array assignment

```
OneWord[0] := "a";      (*since OneWord components have type CHAR*)
List[1] := OneWord;     (*since List components are OneWordTYPE*)
List[2][0] := "b";      (*since the smallest component of List is a CHAR*)
List[2,0] := "c";       (*this is a legal shorthand for the previous notation*)
```

We might even take advantage of the array-to-array assignment to initialize the entire list:

initializing arrays

```
(*Initialize a single word.*)
FOR Count := 0 TO 9 DO
   OneWord[Count] := " "          (*Fill the word with blank characters.*)
END;
(*Initialize the entire list of words.*)
FOR Count := 1 TO 20 DO
   List[Count] := OneWord         (*Fill the list with blank words.*)
END
```

> As you know, a string constant, between quote marks, can be assigned to a string array variable if it's small enough to fit. The array variable must have at least one extra component.

Modula-2 automatically stores the NUL character 0C at the end of any string we assign. The first character goes in OneWord[0].

```
OneWord := "Fits_Right";     (*OneWord[10] equals 0C.*)
OneWord := "Short";          (*OneWord[5] equals 0C.*)
```

Now, finally, the rules. The three important things to remember about subscripts are:

subscript rules

1. Every subscript must have the proper type. If an array has, say, INTEGER-valued bounds, the subscripts have to be INTEGER as well.
2. Every subscript has to be within the proper subrange. If an array's bounds are ["a".."d"], then ["A"] and ["e"] are illegal subscripts. They'd cause a *subscript range* error and program crash.
3. If an array uses more than one subscript, they have to be in the proper order. At best, misordered subscripts will cause compile-time or run-time crashes. At worst, though, the misordered subscripts might still have the proper types and ranges, and cause bugs that are extremely difficult to find.

Self-Check Questions

Q. Suppose that I make these definitions and declarations. Which of the subsequent statements are legal? Why not?

```
TYPE TenTYPE = ARRAY [1..10] OF INTEGER;
       FiveTYPE = ARRAY [1..5] OF CHAR;
VAR TenArray: TenTYPE;
     FiveArray: FiveTYPE;
     Letter: CHAR;
  ...
```

a) TenArray := 10;
b) FiveArray [6] := "a";
c) IO.IntIn (TenArray[1]);
d) TenArray [TenArray[1]] := 3;
e) Letter := FiveArray;
f) Letter := FiveArray ["a"];

A. Not too many are legal, I'm afraid.

a) Illegal—we can't assign a single value to the whole array.
b) Illegal—FiveArray's highest subscript is 5.
c) Legal.
d) Legal if TenArray[1] holds a value in the range 1..10.
e) Illegal—we can't assign an array to a single character.
f) Illegal—FiveArray is subscripted by numbers, not letters.

A Few Definition Exercises

Data type definitions are the word problems of computer science—solutions aren't hard, but getting to them from the problem statement is.

Let's consider a few quick examples. First, suppose that we want an array that will hold ten numbers, like the grades of a series of quizzes:

quiz scores

```
TYPE QuizTYPE = ARRAY [1..10] OF INTEGER;
```

If we want to record percentages instead of scores, we have to define an array with REAL components:

quiz rankings

```
TYPE PercentTYPE = ARRAY [1..10] OF REAL;
```

QuizTYPE and PercentTYPE were both subscripted by the numbers 1 through 10. We can define an array that is subscripted by single digit characters instead:

counting digits

```
TYPE DigitCountTYPE = ARRAY ["0".."9"] OF INTEGER;
```

It might be used to record the number of times each digit had appeared.

Let's get a little more adventurous, and define some two-dimensional arrays. The screen of a computer terminal usually has 24 rows of 80 characters each. An equivalent array type is:

```
TYPE ScreenTYPE = ARRAY [1..24], [1..80] OF CHAR;
```

terminal screens

How about a building that has seven floors, where each floor contains offices numbered 100 through 199? We can hold the status of each office (which ones are occupied?) in a BOOLEAN-valued component:

a building plan

```
TYPE OccupiedTYPE = ARRAY [1..7], [100..199] OF BOOLEAN;
```

Can we define a type that might score the game of Tic-Tac-Toe? It's time to enumerate a new type for the array's components:

tic tac toe

```
TYPE MarkTYPE = (X, O, Empty);
     BoardTYPE = ARRAY [1..3], [1..3] OF MarkTYPE;
```

Let's also begin to enumerate types to serve as subscripts. GradeTYPE, below, has a separate CHAR component for each value of SubjectTYPE:

course grades

```
TYPE SubjectTYPE = (English, Math, CS, Gym, Italian);
     GradeTYPE = ARRAY SubjectTYPE OF CHAR;
```

Subranges can also provide subscript values. Suppose that you're interested in storing the names of telephone subscribers in three exchanges: 525, 526, and 527. How can you hold the name associated with each of the ten-thousand subscribers (e.g. 525-0000 through 525-9999) in each exchange? How about:

phone listings

```
TYPE ExchangeTYPE = [525..527];
     NameTYPE = ARRAY [0..50] OF CHAR;
     ListingTYPE = ARRAY ExchangeTYPE, [0..9999] OF NameTYPE;
     (*Frankly, this is likely to overpower your Modula-2 system!*)
```

Finally, we can define array types that have arbitrarily many dimensions. This collection of, say, temperature readings:

extensible arrays

```
TYPE HourlyTYPE = [1..24];
     DailyTYPE = [1..31];
     MonthlyTYPE = [1..12];
     ReadingsTYPE = ARRAY HourlyTYPE, DailyTYPE,
                              MonthlyTYPE OF REAL;
```

can easily be extended to include shorter or longer time units, perhaps by the simple expedient of defining an array of ReadingsTYPE values:

```
TYPE DecadeTYPE = [1..10];
     ExtraReadingsTYPE = ARRAY DecadeTYPE OF ReadingsTYPE;
```

Self-Check Questions

Q. Suppose that I declare these variables, based on the types defined above. Show a legal assignment for each one:

```
VAR Quiz: QuizTYPE;                    Digit: DigitCountTYPE;
    Occupied: OccupiedTYPE;            Board: BoardTYPE;
    Grade: GradeTYPE;                  Listing: ListingTYPE;
    Reading: ReadingsTYPE;             Extra: ExtraReadingsTYPE;
```

A. Note the alternative ways (as in *c* and *d*) to list multiple subscripts.

a) Quiz[3] := 7;
b) Digit["3"] := 7;
c) Occupied[1,100] := FALSE;
d) Board[1],[1] := X;
e) Grade[Italian] := "A";
f) Listing[525][7636] := "Cooper";
g) Reading [24,31,12] := 53.5;
h) Extra[5][24,31,12] := 57.9;

Open Array Parameters and HIGH

variables vs. parameters

When it comes to arrays, variables and parameters play by slightly different rules. When an array *variable* type is defined in a program, the type of its components and the size of the array have to be known. In contrast, when a one-dimensional array *parameter* is declared in a subprogram heading, its size doesn't have to be specified. It can be an *open array parameter*.

What's the purpose of open array parameters? Well, programmers frequently want to write general-purpose procedures or functions that work just as well whether their array arguments have ten components, or ten thousand. We'll be willing to restrict the arrays to one kind of component—all INTEGER or REAL—but not to one length.

open array parameters

An open array parameter is an **ARRAY OF** *its-components'-type*. It can be declared either as a value parameter or as a variable parameter, but it can only be one-dimensional. Open array parameters can used as arguments. They cannot appear in whole, array-to-array assignments.

For example, the heading:

a value parameter

```
PROCEDURE OpenValue (Data: ARRAY OF CHAR);
```

declares Data as an open array value parameter. Its argument can be any one-dimensional array of CHAR. This heading:

a variable parameter

```
PROCEDURE OpenVariable (VAR Info: ARRAY OF CHAR);
```

declares Info as an open array variable parameter. Its argument must be a one-dimensional array of CHAR. As with all variable parameters, the argument will be permanently changed by any assignments we make to Info.

Are we ready to use open array parameters yet? No, because I've left out one important fact. How will we know what subscripts to use?

function HIGH

The standard function HIGH takes an open array parameter name as its argument, and returns the number of components it has, *minus one*. All open array parameters have bounds of [0 .. HIGH (*the-parameter-name*)].

Thus, open array parameters always have INTEGER subscripts that begin at zero, and go up to the value returned by HIGH.* The argument of an open array parameter can have *any* type of subscript with any bounds, as long as it's one-dimensional.

Let's look at one brief example. Procedure Multiply, below, uses an open array variable parameter. Multiply takes an array that holds REAL components, then multiplies every current component value by the value of its second argument, Times. Since a variable parameter literally renames its argument, changes to the variable parameter are permanent.

open array variable parameter example

```
PROCEDURE Multiply (VAR Info: ARRAY OF REAL; Times: REAL);
    (*Multiply every component of Info by Times.*)
    VAR i: INTEGER;
    BEGIN
        FOR i := 0 TO HIGH(Info) DO
            Info[i] := Times * Info[i]
        END
    END Multiply;
```

Modification Problems

1. Change Multiply so that it takes an array of CHAR values as its argument, and capitalizes every letter it contains. Use the standard CAP function.
2. Although it isn't standard, some Modula-2 systems allow open array parameters with more than one dimension. Their syntax is something like this:

```
PROCEDURE Multiple (Argument: ARRAY OF ARRAY OF components );
```

Does your system allow them? How is procedure HIGH modified to let you know its dimensions?

Self-Check Questions

Q. Could a variable of the array type we defined earlier:

```
TYPE LowerCaseCountTYPE = ARRAY ["a".."z"] OF INTEGER;
VAR Paper, Program: LowerCaseCountTYPE;
```

be used as the argument of an open array parameter? Show a heading. What would its bounds be?

* On some older systems, HIGH may return a CARDINAL value. You can treat it as an INTEGER with the call INTEGER (HIGH (*the-parameter-name*)) *as long as the* CARDINAL *argument is less than* MAX(INTEGER).

A. Any one-dimensional array can be an open array parameter, even if it was declared with twenty-six CHAR subscripts.

```
PROCEDURE Example (Results: ARRAY OF INTEGER);
  (*Can be given Paper or Program as an argument.*)
```

However, open array parameter subscripts are always INTEGER, and the lower bound is always zero. It's as though Results were defined as an **ARRAY** [0..25] **OF** INTEGER.

Q. What does function WhoKnows, below, do? Why add 1 to HIGH(Data) in the **RETURN** statement?

```
PROCEDURE WhoKnows (Data: ARRAY OF REAL) : REAL;
   VAR i: INTEGER;
       Sum: REAL;
   BEGIN
      Sum := 0.0;
      FOR i := 0 TO (HIGH(Data) DO
         Sum := Sum + Data[i]
      END;
      RETURN Sum / FLOAT(HIGH(Data)+1)
   END WhoKnows;
```

A. WhoKnows calculates and returns the average of the REAL values stored in Data. We have to add 1 to the number returned by HIGH to get the true length of the array, since HIGH returns the length minus one.

Section 8-2
Standard Practices: The Basic Loops

I SEEM TO RECALL PROMISING YOU THAT ARRAYS would let you soar with the eagles. First, though, let's plod with the pigeons for a bit, and refresh the loop reasoning skills that almost all array applications will require. You can hope, with some reason, that learning to solve these basic examples will prepare you for the innumerable variations that life (and the exercises) will throw your way. We'll look at the four basic categories of loops:

- Count bounded loops are mainly used to *traverse* (travel through) an entire array.
- Sentinel bounds are needed for array *searching*.
- Data bounds are useful in *divide and conquer* methods that work by subdividing arrays.
- Limit bounds are required by many kinds of *simulations*. I'll describe an application, but won't do any coding.

Suppose that I create a couple of simple arrays to work with. Sequence, of type ListTYPE, will be a one-dimensional array of INTEGER values. Table, of type MatrixTYPE, is a two-dimensional array of CHAR values.

```
CONST MAXrow = 5;
      MAXcol = 10;
TYPE  ListTYPE = ARRAY [1..10] OF INTEGER;
      MatrixTYPE = ARRAY [1..MAXrow], [1..MAXcol] OF CHAR;
VAR Sequence: ListTYPE;
    Table: MatrixTYPE;
```

our practice arrays

Notice, incidentally, how the constants MAXrow and MAXcol help define MatrixTYPE. Although the practice of using defined constants instead of actual numbers isn't universal, we'll see below that it is helpful when array inspection loops are nested.

Count Bounds

The most common labor of all is visiting each array component. This is sometimes called *traversing* the array; the bound is set by the number of components. Arrays have to be traversed to:

arrays and count boundaries

- initialize the array, or modify every stored value;
- print the array's contents, or retrieve them for some other purpose;
- find the largest or smallest stored value, or count the number of components that share some characteristic;
- shift the location of a sequence of components within the array;
- do vector and matrix arithmetic.

The loop specification that underlies most array traversals is:

a general loop specification

Goal: Do something to or with the array's components.
Bound: We've reached the end of the array.
Plan: Add one to a counter.
Do something with the current component.

FOR is usually the loop of choice. The basic array traversal pseudocode is:

one-dimensional pseudocode

FOR *each component or column in the array*
inspect or process the current component;
report on the results;

For example, we can add the contents of the Sequence array with:

adding components

```
Sum := 0;
FOR Subscript := 1 TO 10 DO
   Sum := Sum + Sequence[Subscript]
END;
IO.StringOut ("The sum of the components is:    ");
IO.CharOut (Sum, 4);
```

What do you think the next snippet of code does?

comparing components to subscripts

```
Count := 0;
FOR Subscript := 1 TO 10 DO
   IF Sequence[Subscript] = Subscript THEN Count := Count + 1 END
END;
(*What does Count represent now?*)
```

It counts components whose subscripts and stored values are the same.

Traveling through a two-dimensional array takes a nested loop. Think of the problem as repeating our *column-by-column* pseudocode once for each row:

two-dimensional pseudocode

```
FOR each row in the array
   FOR each component or column in that row
      inspect or process the current component;
report on the results;
```

For example, suppose that we wanted to print the subscripts of every "X" stored in the two-dimensional Table array. Expanding and slightly modifying the outline above into Modula-2 gives us the code shown below.

a row-first search

```
(*Search for and find at least one hidden "X".*)
IO.StringOut ("The value 'X' is stored in components:    ");
FOR i := 1 TO MAXrow DO
   FOR j := 1 TO MAXcol DO
      IF Table[i,j] = "X" THEN
         IO.IntOut (i, 3);
         IO.IntOut (j, 3);          (*Print the coordinates.*)
         IO.LineOut
      END  (*IF statement*)
   END  (*inner FOR*)
END; (*outer FOR*)
```

> It's important to keep the subscripts straight. Table is a two-dimensional array that has a [MAXrow, MAXcol] component. However, it doesn't have a [MAXcol, MAXrow] component: [5, 10] exists, [10, 5] doesn't.

Bear this in mind when you start to follow the near-universal practice of using the convenient, but meaningless, names i and j as subscript values!

Self-Check Questions

Q. Modula-2 stores two-dimensional arrays in *row-major order*—a fancy way to say that the first subscript names the row, while the second subscript identifies a column in a row. So, what does this segment of code do? Notice that the loop bounds, MAXcol and MAX-row, seem to be reversed.

```
FOR col := 1 TO MAXcol DO
    FOR row := 1 TO MAXrow DO
        inspect Table[row,col]
    END (*inner, row-by-row FOR*)
END; (*outer, column-by-column, FOR*)
report on the results
```

A. The code traverses a two-dimensional array column-by-column, rather than row-by-row. I've reversed the regular loop order so that row, the row counter, increases more rapidly than col, the column counter.

Q. A checkerboard can be seen as an eight by eight array. However, only every other square is colored, starting with a colored square on the top, left-hand border. Write a loop that traverses a checkerboard, but only stops on the colored squares.

A. Since we want to visit each row, the outer *row-by-row* code is a typical **FOR** loop. The inner, *column-by-column* code changes slightly, since we want to alternate between starting with the first and second square of each row.

```
FOR row := 1 TO 8 DO
    IF ODD(row) THEN Start := 1 ELSE Start := 2 END;
    FOR col := Start TO 8 BY 2 DO
        visit Board[row, col]
    END
END
```

Sentinel Bounds

Sentinel bounds are the second staple of array processing. Although we start out by traversing an array, we're liable to cut the trip short, so **FOR** loops won't do. Typical sentinel problems include:

- finding the location of a particular stored value;

arrays and sentinel boundaries

- processing words or sentences stored in an array (spaces or punctuation marks act as sentinels);
- locating a subsequence with a particular characteristic (e.g. it is ascending or descending).

a warning

> We will usually find that when we're searching an array for a sentinel, a count bound will be needed as well. If the sentinel isn't found, we have to guarantee loop exit lest we sail past the end of the array.

For example, suppose we want to find the first number greater than 0 in the one-dimensional Sequence array (which we defined as an **ARRAY** [1..10] **OF** INTEGER). The simple solution is to inspect components, one by one, until we see one that fits the bill:

a simple one-dimensional search

```
Current := 1;  (*initialize the subscript counter*)
LOOP
   (**)IF Sequence[Current]>0 THEN EXIT END;
   Current := Current + 1  (*move to the next component*)
END  (*Postcondition — Sequence[Current]>0 — we think!*)
```

> Stating a postcondition doesn't guarantee that a program will survive to reach it. If no stored value is greater than 0, we'll run past the end of the array. The *range error* this causes will crash the program.

Thus, the loop specification typically has a necessary bound that guarantees termination (the count) in addition to the intentional bound that helps get us to our goal (the sentinel). But if we do add a second possibility for exit, we can't be sure why we left the loop. In pseudocode:

pseudocode

```
initialize the subscript counter;
LOOP
   (**)IF we reach the sentinel OR the end of the array EXIT;
   add 1 to the subscript counter
END of the loop;
decide why we left the loop
```

The Modula-2 version, below, has a postcondition and post-loop check we'll see again and again:

a safer one-dimensional search

```
Current := 1;  (*the subscript also acts as a counter*)
LOOP
   (**)IF (Sequence[Current]>0) OR (Current=10) THEN EXIT END;
   Current := Current + 1
END;
(*Postcondition: If Sequence[Current] isn't greater than zero, no component is.*)
IF Sequence[Current] <= 0 THEN
   IO.StringOut ("No value exceeded zero.")
END
```

Now, suppose that we're searching for a sentinel—say, an "X"—in a two-dimensional array of CHAR. Will anything change? Yes. We can use a problem solving method called *case analysis*—just a common-sense look at the alternatives that might arise—to help us move through the array. A specification of the array-searching loop is:

loop specification

Goal: Find the position of the letter "X".
Intentional bound: Spotting the sentinel.
Necessary bound: Reaching the last component of the array.
Plan: Advance to the next component.

case analysis

The case analysis helps decide what *advance to the next component* means. This wasn't really a problem for a one-dimensional array search, since *advance to the next component* always meant *add one to the subscript*. But the 'next' component in a two-dimensional array is sometimes on the next row:

advancing to the next component

```
initialize the subscript counters;
LOOP
   (**)IF we reach the sentinel OR the end of the array EXIT;
   IF we haven't reached the last column in the current row
      THEN advance to the next component in this row
      ELSE advance to the first component in the next row
END of the loop;
decide why we left the loop
```

The program segment below implements the loop pseudocode. The **EXIT** condition parallels the bound of the one-dimensional array search, but of necessity, it is more complicated. Nevertheless, the loop postcondition—and the requirement for a post-loop check—remain the same.

safely searching a two-dimensional array

```
(*Search the two-dimensional Table array for an "X".*)
Col := 1;
Row := 1;          (*Prepare to inspect the first component.*)
LOOP
   (**)IF (Table[Row, Col] = "X") OR
        ( (Col = MAXcol) AND (Row = MAXrow) ) THEN EXIT END;
   IF Col < MAXcol
      THEN Col := Col + 1
      ELSE Row := Row + 1;
           Col:= 1
   END (*IF*)
END; (*LOOP*)
(*Postcondition: If Table[Row, Col] isn't an "X", no component is.*)
IF Table[Row, Col] <> "X" THEN
   IO.StringOut ("No value equaled 'X'.")
END
```

Modification Problems

1. Suppose that Table were an array of INTEGER values. Modify the code segment above so that it locates the component that equals the product of its subscripts.

Self-Check Questions

Q. This program segment is intended to search a Sequence array (that is, an **ARRAY** [1..10] **OF** INTEGER) and let us know if it holds a 3. Assuming that Found is a BOOLEAN variable, what's wrong with the code? It isn't a syntax problem.

```
FOR Current := 1 TO 10 DO
    Found := (Sequence[Current] = 3)
END;
IF Found THEN IO.StringOut ("Found it.") END
```

A. In effect, Found acts like a switch. To spot the bug, imagine that there isn't a 3 in Sequence[10]. Found will become FALSE even if there's a 3 elsewhere in the array. If, for some reason, we want to use a **FOR** statement to find a sentinel, we have to invent a one-way switch that can be turned on, but not off.

```
Found := FALSE;
FOR Current := 1 TO 10 DO
    IF Sequence[Current] = 3 THEN
        Found := TRUE
    END
END;
IF Found THEN IO.StringOut ("Found it.") END;
```

Data Bounds

Data bounds appear frequently in array programming. They're exceptionally useful for *divide and conquer* problem solving methods that:

arrays and data boundaries

- describe sections of the array in terms of relative borders, rather than by fixed subscripts or stored data values;
- divide a large array into smaller pieces that can be dealt with separately;
- place barriers between the potentially 'useful' information stored in an array, and a steadily growing 'useless' portion;
- gradually shrink the unexplored portion of an array until it either holds the solution we're looking for or is empty.

auxiliary subscripts are pointers

> Data bounds are generally set with *auxiliary* subscript variables. They serve as *pointers* to the beginning and end of the array segment we're interested in.

Let's look at an example that highlights the use of these pointers. Consider the problem of identifying *palindromes*—words or phrases that are the same forward and backward. According to the rules of the game, capitalization, punctuation, and spaces are ignored (so I'll assume that any phrase we try to check has no capitals, spaces, or punctuation). These are all palindromes:

a few palindromes

Harass sensuousness, Sarah.
Dennis and Edna sinned.
Sore was I ere I saw Eros.
Satan, oscillate my metallic sonatas.

Let me describe a way to identify palindromes that makes it easy to see why subscript pointers are handy. A phrase is a palindrome if:

how to identify a palindrome

- its first and last letters are the same;
- the remaining phrase is a palindrome.

Our algorithm will use subscript pointers to mark the first and last letters of the steadily shrinking *remaining phrase*. Get two pencils, and use them as pointers to help you check one of the phrases above. Eventually you'll run out of letters, or arrive at a mismatch.

Now, suppose that the phrase is held in an array. Our pointers start out at the beginning and end, so we can compare the first and last letters. Each time we have a match, we can move the pointers over by one component. If there's ever a mismatch between the current 'first' and 'last' components, the phrase is automatically disqualified; it's not a palindrome. But as long as letters keep matching, we can continue to move the pointers, and shrink the unexplored portion of the phrase. A specification of the heart of the palindrome spotter is:

loop specification

Goal: Decide if an array contains a palindrome.
Bound: The unknown portion of the array is empty **OR**
two components are different.
Plan: Narrow the array by one component at each end.

Pseudocode includes the all-important *look after the loop*:

pseudocode

get the lower-case, spaceless, depunctuated phrase;
initialize pointers to the first and last components;
LOOP
(**)**IF** *the unknown portion of the array is empty* **OR**
two components don't match then **EXIT**;
move the pointers one component closer
END *of the loop;*
decide if the phrase was a palindrome

How can we tell when *the unknown portion of the array is empty*? Well, suppose that First and Last mark the ends of the array. It's easy to move them a step closer together:

narrowing the unknown

```
First := First + 1;
Last := Last – 1
```

But when is the unknown part of the array empty? It's safe to say:

```
(First + 1) >= Last
```

when the unknown can't be shrunk any further (try it out).

Function Palindrome, below, makes the check. Be sure that you understand why a phrase is a palindrome if, and only if, Phrase[First] equals Phrase[Last].

data bound example

```
PROCEDURE Palindrome (Phrase: ARRAY OF CHAR) : BOOLEAN;
  (*Returns TRUE if Phrase holds a palindrome.*)
  VAR First, Last: INTEGER;
  BEGIN
    First := 0;
    Last := HIGH(Phrase)
    LOOP
      (**)IF ( (First + 1) >= Last) OR  (*then if*)
              (Phrase[First] <> Phrase[Last]) THEN EXIT END;
      Last := Last + 1;
      First := First – 1
    END;
    (*Postcondition: We've inspected the whole phrase OR found a mismatch.*)
    RETURN Phrase[First] = Phrase[Last]
  END Palindrome;
```

Modification Problems

1. Suppose that the quality control of data isn't up to par, and phrases with capitals, spaces, and punctuation are sneaking in. Make Palindrome robust enough to deal with the situation.

Limit Bounds

Finally, we arrive at limit bounds. When arrays are used to *simulate* or *model* real-world processes, an entire arrayful of data will often be subjected to some action again and again. For example, each component might be averaged with the values of its nearest neighbors. As with mathematical limits, we can reach a point of diminishing returns—a point at which one more trip through the array might not improve its contents.

Limit bounds are used to spot that point. Some typical applications are:

arrays and limit bounds

- Modeling *diffusion* or *flow*; for instance, tracking the spread of pollution through an environment, or of heat through a conductive surface.
- Modeling population *growth*; simulating conditions that might lead to eventual survival or extinction.
- Modeling *cycles*; trying to detect rhythmic patterns that change, but are stable (like the cyclic, yet predictable, population of locusts).

Problems like this are a little beyond us because they have three equally difficult parts: figuring out what starting data is, figuring out what to do with it, and figuring out how to recognize that we've done enough!

One classic that's pretty straightforward, though, is modeling heat flow across a metal surface. We assume that one or more spots are inexhaustible modeling heat flow

heat sources, while other areas (perhaps the edges) are insatiable heat consumers. An array holds the starting temperatures at regular intervals across the metal surface.

The surface's temperature will, naturally, reflect a flow from hot to cold areas. We can simulate the process by averaging adjacent array components—really just a counting bound, since we visit each component in turn—until the change between one sweep and the next is very small. For all practical purposes, the array's contents have managed to find a state of equilibrium. In pseudocode:

a limit bound application

```
LOOP (*until the temperature is relatively stable*)
   LOOP (*reset each component's temperature*)
      average a component with its neighbors
      (**)IF we're at the last component EXIT — the count bound
   END; (*inner loop*)
   (**)IF there wasn't much change EXIT — the limit bound
END (*outer loop*)
```

You'll get a chance to write a program like this in the exercises.

Section 8-3 Extended Programming Examples

AS WE TRY TO SOLVE HARDER PROBLEMS, WE'LL SEE a shift in our approach to program planning: away from *actions*, and toward *data*. We'll be less worried about choosing a course of action, and more concerned with deciding how to store information.

Why? Well, there is a close partnership between a program's algorithm and its data. As the equation that names one of Niklaus Wirth's textbooks puts it:

Algorithms + Data Structures = Programs

This title made a strong statement in 1976—that teaching programming, which had previously focused on algorithms and methods, was incomplete unless it also paid plenty of attention to the data structures that algorithms used. But why *structures* rather than *types*?

structures are abstract

> Data types are *concrete*; like statements, they're built-in features of a programming language. Data structures, in contrast, are *abstract*. Like algorithms, they are language-independent. They can be constructed in any language, and they can be studied apart from any particular application.

there is a tradeoff

A program's algorithm might be likened to its plot—what happens? Its data structures, in contrast, provide the cast of characters—who does what, and to whom? In programs, as in literature, there can be a tradeoff between the two: a programmer who puts time into designing a good data structure may be rewarded with a less complicated algorithm. Conversely, an algorithm that has to compensate for a quick and dirty data structure may end up being unnecessarily complex and detailed.

A programming language can help or hinder algorithms by the data types it provides as building blocks. The general lack of built-in types is one reason that True Programming Gods don't like languages like BASIC and FORTRAN. They have control statements like Modula-2's, but their limited stock of data types (they only have arrays) forces programmers to waste time explaining data in terms the language can handle.

Data Abstraction

Data structures are typically described in terms of their *primitive operations.* Primitives carry out basic operations; for instance, the primitives of type INTEGER are addition, subtraction, and so on. Indeed, numbers are usually meaningful only because of what we can do with them. We can describe type INTEGER by describing exactly what its primitives do.

primitives define abstractions

> A data type or structure that is defined in terms of its primitives, rather than by the way it's implemented, is called a *data abstraction.*

Data abstraction isn't all that different from the *procedural abstraction* we've been using all along. We haven't had any trouble with talking about imaginary pseudocode procedures as though they really existed. All we needed was to agree on what the procedures *would* do *if* we had them. Using abstractions has been quite convenient, because it let us avoid nitty-gritty details of implementation until we were ready to deal with them.*

Data abstraction gives us the same advantage. We're not tied down to any particular implementation of some method of storing or manipulating data. Data abstractions, like procedural abstractions, are language-independent: we decide what we want to do before we choose the best way of doing it.

A data abstraction's primitives tie it to reality. We can *talk* about primitives all we want. But in general, *writing* a primitive will require that we know some of the details of how a data abstraction is implemented. Some of the primitives that most data abstractions will require are:

common primitives

- An *initialization* procedure that sets up the value or structure to begin with.
- An *assignment* procedure that lets values be stored or changed.
- An *inspection* procedure that retrieves or prints the stored values.

* Indeed, if you think about it for a moment, you'll recognize that when the procedures come from library modules, like IO, that day of reckoning is put off indefinitely. *I* did the work; *you* just know how the primitive input and output operations are called!

Object-Oriented vs. Goal-Oriented Design

Now, we've used one basic approach to programming since the beginning of this book—goal-oriented design. All of our pseudocode and specifications have been aimed directly at solving particular problems.

Data abstraction won't always fit into goal-oriented design, though. Indeed, we'll often have the feeling that worrying about highly specialized goals is a *distraction*—we want to be generalists, rather than specialists. Looking at the goal too soon tends to make programmers settle on a final implementation before they really have to. Fortunately, we can follow an alternative approach.

object-oriented design

> In technical jargon, the term 'data object' is often used synonymously with 'value' or 'structure.' Programming by looking at data abstractions first, before considering goal-oriented procedural abstractions, is called *object-oriented design.*

Data abstraction is one of the tools of object-oriented design, just as procedural abstraction is a tool of goal-oriented design. Generally, we'll find that as programs get larger, we'll rely more on object-oriented design tools. More sophisticated programming applications, such as a Modula-2 system's library modules, tend to rely almost exclusively on data abstraction:

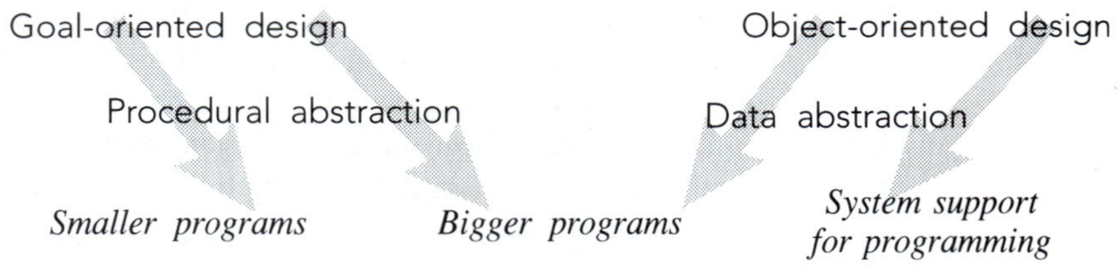

Object-oriented design is rich with possibilities. For instance, we can distinguish between different kinds of data abstractions. Informally, an abstract type is defined by its values—numbers, say, or characters—and by the primitives that manipulate them. An abstract structure, in contrast, is defined by relations *between* values—perhaps they're in some kind of order—and by the set of primitives that maintain the structure. The structure is the abstraction; the values themselves don't really matter.

But for the next few chapters, we won't worry too much about drawing refined distinctions between abstracted types and structures. That won't become a concern until we start to package our abstractions in modules that might be reused again and again. For the time being, we can just talk about designing data abstractions, plain and simple.

Example: Counting Characters

Let's start with a program that depends more on its data structure than its algorithm. Here's the problem:

problem: character counting

The ambition of every person of our era is to go on *Wheel of Fortune* and play for the car. In case you have been in a coma for the past decade, Vanna exposes a vowel and five other letters in a hidden phrase, and you try to guess the rest.

Assume that you are preparing for your appearance, and would like to know which letters are most likely to appear. Count the total number of characters in a sample of input, then print a table of the relative number of appearances of lower-case letters.

The grunt work of counting and tabulating values is an entirely typical array application. The analysis itself has other, more traditional applications:

a few applications

- Decoding an encrypted message, on the assumption that it will have the same letter frequencies as plain text.
- Seeing if different languages have unique letter-frequency 'signatures.' These might be used to help produce nonsense text that, despite being meaningless, can instantly be identified as being in a particular language.
- Determining if Morse code is as efficient as it might be—do common letters have brief codes and vice versa?

We can start to sketch a character-counting data structure in abstract terms. I'll say that:

multiple-character primitives

> Initialize *will initialize the character-count data structure*;
> CountData *will add* 1 *to the count of a character*;
> CountIs *will let us know the count stored for any single value*;

Between them, these three primitives hit all the bases: initialization, assignment, and inspection. Suppose I gave you these three procedures. Could you 'solve' the problem of counting characters? Sure you could. A rough pseudocode might be:

first pseudocode

> *use* Initialize *to initialize variables as necessary*;
> *get and process the input data with* CountData;
> *use* CountIs *to calculate and print the program's results*;

primitives define the abstraction

> Note that all of these are primitive operations that can't be written without knowing how the underlying data structure is going to be implemented. However, they fully define the abstraction for our purposes. We don't *have* to know anything else about it.

In fact, I can take a step that's pretty amazing: I can write the headings of the primitive subprograms, as I have below. The only commitment I'm making here is that Ch is a CHAR variable. *How* the primitives are going to work is an issue for another time; what's important now is *what* they do.

describing the data abstraction

TYPE CountTYPE *underlies the structure that holds the counts;*
VAR Count: CountTYPE;
    Ch: CHAR;
PROCEDURE Initialize (**VAR** Count: CountTYPE);
  (**Initializes the* CountTYPE *structure.**)
PROCEDURE CountData (**VAR** Count: CountTYPE; Ch: CHAR);
  (**Counts the appearance of a character.**)
PROCEDURE CountIs (Count: CountTYPE;
                    Ch: CHAR) : INTEGER;
  (**Returns the number of times* Ch *was counted.**)

This little collection doesn't define a data *type*—you can see that I haven't even tried to describe what the CountTYPE definition is going to be. However, it does define a data *abstraction*—we know exactly how to use the structure even though we don't know what it's made of. A more refined version of the pseudocode takes advantage of the extra detail:

second pseudocode

```
Initialize (Count);
initialize a Total counter;
LOOP
   read Ch;
   (**)IF NOT IO.Done() THEN EXIT END;
   CountData (Count, Ch);
   add 1 to the Total
END of the loop;
FOR Ch := each of the lower-case letters
   print CountIs(Count, Ch) / Total as a percentage
END of the for
```

Stop for a moment to recognize how the object-oriented approach we've followed differs from our old methods:

the object-oriented approach

- We began by getting an idea of what kind of information a solution to the character-counting problem would require.
- Then, we outlined some data handling tools—the primitives Initialize, CountData, and CountIs.
- Finally, we returned to the traditional goal-oriented approach and outlined a program that used the new tools.
- Our next step will be to define the underlying data type and write the bodies of the primitives.

In contrast, the old method would have had us begin by getting an idea of what steps the program would take. We might have decided exactly how data would have been stored, then outlined procedures for getting data and calculating results. We would have considered *first* what we now put off until *last*—deciding how to hold the character counts.

Implementing the Primitives

How are we going to hold character counts? Well, we could write the data abstraction primitives using twenty-six different counters: one for each letter. But from our present learned vantage point this approach is too tedious to bother with. Why declare 26 different variables (and have procedure calls with 26 parameters), when just one array can hold all the counts at once? Consider this definition:

the CountTYPE definition

```
TYPE CountTYPE = ARRAY CHAR OF INTEGER;
VAR Count: CountTYPE;
```

Count is an array whose bounds are MIN(CHAR) and MAX(CHAR). After it has been initialized, I can add one to any character's count with the assignment:

```
Count[Ch] := Count[Ch] + 1;  (*Increment the proper counter.*)
```

When we ultimately PrintResults, we can ignore all components outside of the "a".."z" sequence.

Now that we know the definition of the underlying data type, we can think about implementing the abstraction's primitives. Procedure Initialize can traverse the entire array via a **FOR** loop:

```
(*Initialize the array.*)
FOR Ch := MIN(CHAR) TO MAX(CHAR) DO
   initialize the current component to 0
```

the primitives

Procedure CountData's job was to add 1 to one character's total. As we've seen, its central assignment is:

```
Count[Ch] := Count[Ch] + 1;  (*Increment the proper counter.*)
```

Finally, CountIs returns the count that's been collected for a single character. Its main statement is:

```
RETURN Count[Ch]
```

Let's turn to our program. If I wanted, I could use procedural abstractions to describe what it's going to do:

initialize variables as necessary;
get and process the input data;
calculate and print the program's results;

If I did, these goal-oriented procedures would call Initialize, CountData, and CountIs just as they might call IO.CharIn or IO.FixOut. However, I think that writing them out as procedures will make our program overly complicated.

A final version of program ChCounts is shown below. Shading shows how it might be broken into goal-oriented procedures. The output is an analysis of the program text, contained in my file ChCount.MOD. I also added an extra touch—can you figure out how the table of output was produced?

character counting program

```
MODULE ChCounts;
   (*Count input characters, print relative frequency of lower-case letters.*)
IMPORT IO;
TYPE CountTYPE = ARRAY CHAR OF INTEGER;
VAR Count: CountTYPE;          (*Count holds individual character counts.*)
    Ch: CHAR;                  (*Ch holds the current input character.*)
    Total,                     (*Total counts the total number of letters.*)
    LineCount: INTEGER;        (*LineCount helps produce neat output.*)
PROCEDURE Initialize (VAR Count: CountTYPE);
   (*Initializes the CountTYPE structure.*)
   VAR Position: CHAR;
   BEGIN
      FOR Position := MIN(CHAR) TO MAX(CHAR) DO
         Count[Position] := 0
      END
   END Initialize;
PROCEDURE CountData (VAR Count: CountTYPE; Ch: CHAR);
   (*Counts the appearance of a character.*)
   BEGIN
      Count[Ch] := Count[Ch] + 1
   END CountData;
PROCEDURE CountIs (Count: CountTYPE; Ch: CHAR) : INTEGER;
   (*Returns the number of times Ch was counted.*)
   BEGIN
      RETURN Count[Ch]
   END CountIs;
BEGIN
   (*the initialization segment*)
   IO.StringOut ("Reading from ChCounts.mod . . .");
   IO.LineOut;
   IO.OpenInput ("ChCounts.mod");
   IF NOT IO.Done() THEN
      IO.StringOut ("Could not open input file.  Exiting.");
      HALT
   END;
   Total := 0;
   Initialize (Count);

   (*the processing segment*)
   LOOP   (*Count appearances of all available characters.*)
      IO.CharIn (Ch);
      (**)IF NOT IO.Done() THEN EXIT END;
      CountData (Count, Ch);
      Total := Total + 1
   END; (*LOOP*)
   (*We know individual and overall character counts.*)
```

```
    (*the results segment*)
    IO.StringOut ("Lower-case frequencies (among all input characters):");
    IO.LineOut;
    LineCount := 1;
    FOR Ch := "a" TO "z" DO  (*Print the output table.*)
      IO.CharOut (Ch);
      IO.StringOut (" = ");
      IO.FixOut ( 100.0 * (FLOAT(CountIs(Count, Ch))/FLOAT(Total)), 4, 2);
      IO.StringOut ("%    ");
      IF (LineCount MOD 5) = 0 THEN IO.LineOut END;
      LineCount := LineCount + 1  (*New-line after every fifth character.*)
    END; (*FOR*)
    IO.CloseInput
  END ChCounts.
```

```
Reading from ChCounts.mod . . .
Lower-case frequencies (among all input characters):
a = 2.90%  b = 0.18%  c = 1.38%  d = 0.60%  e = 3.81%
f = 0.55%  g = 0.55%  h = 1.84%  i = 2.80%  j = 0.00%
k = 0.05%  l = 1.61%  m = 0.46%  n = 4.55%  o = 3.68%
p = 0.83%  q = 0.09%  r = 2.21%  s = 1.79%  t = 5.74%
u = 3.40%  v = 0.28%  w = 0.23%  x = 0.09%  y = 0.09%
z = 0.28%
```

Modification Problems

1. Reconstruct ChCounts with goal-oriented procedures. You can have them call the primitives that ChCounts uses now, or you can 'decompose' the program entirely—in effect, rewrite it from scratch.
2. Modify ChCounts so that *a*) it *folds* upper-case letters—counts them as though they were lower-case, and *b*) it only prints each letter's appearance as a proportion of all *letters*, rather than of all *characters*. Are you tempted to modify the definition of CountTYPE? If you do, how will it affect the primitives?
3. The data abstraction we defined didn't have a 'print' procedure per se. Instead, it gave us a primitive that would return a single stored value. However, a Dump procedure that prints an array's contents can be incredibly useful during debugging. I'll talk about this more in the Program Engineering Notes; for now, write a Dump procedure for a variable of type CountTYPE.

Arrays and Textfiles

Let's solve another problem with the object-oriented approach. It involves one of the classic programming problems—transferring stored data from one form to another. The problem is easy enough to understand:

problem: file reversal

> A group of urban archeologists has thrown away their filecards. Instead, they're using laptop computers to log discoveries. As each layer of debris is unearthed, its contents are carefully entered, one description per line, from the uppermost layers to the oldest. The archeologists are mighty pleased with themselves when they get back to the lab, diskettes in hand.
>
> Unfortunately, their homebound colleagues are less enthusiastic. They don't care about all the bottles of Dippity-Doo and assorted pimple creams that are described at the beginning of each file. Instead, they want the files to be in chronological order. They want the descriptions to be stored in reverse, from the oldest to the newest.
>
> Write a program that will reverse the order of entries, line by line, in the survey team's files. You can assume that no file has more than five hundred lines, and that no line is more than 80 characters long.

Don't get unnerved by the mention of files. We're just using them the way we did in Chapter 6—as the source of an input stream of text, or the place where the stream of output might be sent.* Our problem is that computer files are sometimes less convenient than paper filecards. If the entry lines were written on filecards, we could reverse their order by taking filecards from the bottom of the pile instead of from the top. However, a computer file won't let us deal from the bottom.

what our abstraction needs

> The file reversal problem requires us to come up with a data abstraction that has one characteristic like a file: it can hold many lines of characters. However, it needs a property that's different—we want to be able to 'read' the lines in reverse.

It doesn't take too much imagination to see that an array will probably do just fine. We can certainly store characters in an array, and an array *row* matches up pretty well conceptually with a file *line*. Most importantly, we can easily inspect the array in reverse, from the last line to the first. If we really wanted to show off, we could print the individual lines in reverse.

Let's keep the array type in the back of our mind, but try to define the new abstraction in terms of primitive operations for initialization, assignment, and inspection. However, we can expand our notions of what these primitive operations do. We're used to assigning or printing just one INTEGER or CHAR at a time. But for our new abstraction, 'assignment' can mean 'read and store all available input.' Similarly, 'inspection' can mean 'print all stored data in reverse.'

---

* Recall that there are two ways to change the default (standard) input and output locations: at the operating system level (by giving a special version of your 'execute my program' command), or within a program (by using procedures IO.OpenInput and IO.OpenOutput). You might want to reread the discussion near the end of section 6-3.

primitives, first draft

Initialize *does any necessary initialization of the abstraction;*
GetData *will fill the data abstraction;*
RetrieveInReverse *will print the abstraction's contents;*

We can assume that the assignment procedure, GetData, will read from the standard input, and that the inspection procedure, RetrieveInReverse, will send its data to the standard output.

Next, let's try a more formal outline of the primitives. What parameters will they require? All three of them will obviously need the variable that actually holds the data. It will also be useful to know how many lines of data there are. If GetData reads 100 lines of input (thus filling lines 1 through 100 of the data-storing variable), then we'll want RetrieveInReverse to do the opposite—to print the contents of lines 100 down to 1.

describing the data abstraction

```
TYPE StoreTYPE defines a two-dimensional array of CHAR;
PROCEDURE Initialize (VAR Data: StoreTYPE);
  (*Does any initialization needed for Data.*)
PROCEDURE GetData (VAR Data: StoreTYPE;
                   VAR Rows: INTEGER);
  (*Fills Data (preserving line structure) and returns a line count.*)
PROCEDURE RetrieveInReverse (Data: StoreTYPE;
                             Limit: INTEGER);
  (*Prints the contents of Data from line Limit down to line 1.*)
```

data abstraction supports object-oriented design

> Once again, we've used the object-oriented approach, and described our tools as the primitives of a data abstraction. This lets us put off decisions about the exact definition of the underlying data type, and the implementation of the primitive operations.

As before, we can outline a final program pretty easily by using the new data abstraction:

program pseudocode

```
prompt for input, or prepare to read from a file;
Initialize (Data);
GetData (Data, Length);
announce the output, or prepare to write to a file;
RetrieveInReverse (Data, Length)
```

Implementing the Primitives

As always, the moment of truth arrives: we have to implement the primitives. What choices do we have for the type definition of StoreTYPE? Well, we could create a two-dimensional grid of single letters:

```
TYPE StoreTYPE = ARRAY [1..rows], [1..columns] OF CHAR;
```

alternatives for the underlying type

Or we could have a more complicated definition of an array of lines:

```
TYPE OneLineTYPE = ARRAY [1..columns] OF CHAR;
     ManyLinesTYPE = ARRAY [1..rows] OF OneLineTYPE;
```

The primitives can be implemented with either one, since we never pass just one line at a time as a parameter. As far as our *use* of the primitives is concerned, it doesn't matter. Why don't I use the simpler version now, and let you do the more complicated one as a modification problem later?

> Neither definition is 'right' or 'wrong.' Rather, the alternatives both have advantages and disadvantages that should be considered.

Let's outline the primitives. Each one uses a loop to traverse the StoreTYPE array; as luck would have it, one loop has a count bound, one has a data bound, and one has a sentinel. Procedure Initialize has to visit every component of the array.

the NUL character

> It's common to store the ASCII NUL character for initialization, since it won't print and never appears in ordinary text. The value 0C (zero, followed by C) can be assigned to any CHAR variable or array component.

I can outline a loop specification for Initialize as:

loop specification for Initialize

Goal: Initialize an array.
Bound: Reaching the last column of the last row.
Plan: Store 0C*, the null character.*
Increment a row or column counter.

Since the whole array is being traversed, two nested **FOR** statements—the inner statement handles one row, the outer statement goes from row to row—will be easier to deal with than a **LOOP**.

Procedure GetData needs a loop that reads all available input, one character at a time. As each character is read, it has to be stored in the Data array. Reading the end-of-line character is, of course, a signal that subsequent input should be put on the next row of the array. Except for the fact that it saves each value it reads instead of immediately printing it, GetData is a lot like the basic process filter Echo.

loop specification for procedure GetData

Goal: Copy a file's contents into an array.
Bound: The end of the input file.
Plan: Read a character into the current array component.
If it's not IO.EOL
then advance to the next array column.
else, change the IO.EOL *character to a* NUL
advance to the first column on the next array row.

I'll assume that there actually is an IO.EOL character at the end of each line.

Finally, procedure RetrieveInReverse goes through the array, from the last line to the first, printing characters. However, since not every line of input will necessarily be 80 characters long, RetrieveInReverse won't have to visit every component on every line. As soon as we get to a component that holds a NUL character, we're done with the current line, and can move on. We need a loop that performs a case analysis: what it does depends on the current component's value.

loop specification for procedure PrintInReverse

Goal: Copy an array's contents into a file, from last line to first.
Bound: We've printed the first line.
Plan: If we're at a printing character, print it, and
advance to the next array column,
Otherwise, advance to the first column of the prior array line,
and print a carriage return.

Does the file reversal program need additional pseudocode planning? Maybe. You might want to pseudocode the primitives before you read the final version of FileRev, on the next couple of pages. But as it happens—and as will often happen with well-designed data abstractions—the main body of the final program consists of little more than calls of our tools for initialization, assignment, and inspection.

Modification Problems

1. Modify FileRev so that it uses the alternative definition of StoreTYPE (as an array of single lines). How can you simplify the contents of the primitives? Note that the main program calls of the primitives don't require any change.
2. FileRev isn't particularly robust. Fix it up so that it *a*) can deal with a missing input or output file, *b*) can handle an input file that has more than FILEsize lines or LINEsize characters on each line, *c*) a file that contains the ASCII null character. What you do is up to you.
3. Modify FileRev so that it reverses each line, but keeps the sequence of lines as they began. Should you change the data abstraction, change the inspection primitive, or add an additional inspection primitive? Make an argument in favor of each alternative.
4. Can you figure out how to base the data abstraction on a file rather than an array? Hint: use a couple of auxiliary files, plan to do a lot of copying, and store the file in reverse instead of retrieving it that way.

(*program follows*)

file reversal program

```
MODULE FileRev;
  (*Prints the contents of a file in reverse line-order. Note size limits on input.*)
IMPORT IO;
CONST LINEsize = 80;          (*Use constants for easy modification.*)
      FILEsize = 50;          (*These are referred to in all primitives.*)
      NULL = 0C;              (*This is the ASCII NUL character.*)
TYPE StoreTYPE = ARRAY [1..FILEsize], [1..LINEsize] OF CHAR;
VAR Data: StoreTYPE;
    Length: INTEGER;
PROCEDURE Initialize (VAR Data: StoreTYPE);
  (*Fills the array with null (ASCII 0) characters.*)
  VAR i, j: INTEGER;
  BEGIN
    FOR i := 1 TO FILEsize DO
      FOR j := 1 TO LINEsize DO
        Data[i,j] := NULL
      END
    END
  END Initialize;
PROCEDURE GetData (VAR Data: StoreTYPE;
                          VAR Row: INTEGER);
  (*Fills Data (preserving line structure) and returns a line count.*)
  VAR Column: INTEGER;
  BEGIN
    Row := 1;
    Column := 1;
    (*Row and Column are always one ahead.*)
    LOOP  (*until there's no more data*)
      IO.CharIn (Data[Row, Column]);
      (**)IF NOT IO.Done() THEN EXIT END;
      IF Data[Row, Column] <> IO.EOL
        THEN Column := Column + 1  (*handle ordinary characters*)
        ELSE Data[Row, Column] := NULL;  (*handle carriage returns*)
             Column := 1;
             Row := Row + 1
      END  (*IF*)
    END; (*LOOP*)
    (*Postcondition: Row equals the number of lines of input in Store, plus 1. If
     an input line has fewer than LINEsize characters, NULL marks the end-of-line.*)
    Row := Row - 1
  END GetData;
```

```
PROCEDURE RetrieveInReverse (Data: StoreTYPE; Last: INTEGER);
  (*Prints the contents of Data backward, starting at line Last.*)
  VAR Row, Col: INTEGER;
  BEGIN
    Row := Last;  (*I could just use Last, but this is clearer.*)
    Col := 1;
    LOOP
      (**)IF Row = 0 THEN EXIT END;
      IF (Col <= LINEsize) AND (*then if*) (Data[Row,Col] <> NULL)
        THEN IO.CharOut (Data [Row, Col]);  (*handle ordinary characters*)
             Col := Col + 1
        ELSE Row := Row - 1;  (*handle carriage returns*)
             Col := 1;
             IO.LineOut
      END (*IF*)
    END (*LOOP*)
  END RetrieveInReverse;
BEGIN (*main program*)
  Initialize (Data);
  IO.OpenInput ("FileRev.dat");
  IF NOT IO.Done() THEN
    IO.StringOut ("Couldn't open input file 'FileRev.dat'.  Exiting.");
    HALT
  END;
  GetData (Data, Length);
  IO.CloseInput;
  IO.OpenOutput ("Results");
  RetrieveInReverse (Data, Length);
  IO.CloseOutput
END FileRev.
```

```
Couldn't open input file 'FileRev.dat'.  Exiting.
```

Section 8-4 Two Small Projects

HOW WILL YOU LEARN TO WRITE BIGGER, MORE complicated programs? By writing programs that are bigger, and perhaps more complicated. This section has one of each.

Our first example focuses on *bigger*. It should be big, because we'll be creating a new world—the habitat and breeding ground of the computerized life form known as a *cellular automaton*. To keep complication within reason, I'll use the object-oriented approach, and think about the program's data structures before getting to its algorithm. In addition, I'll try to design the structures as data abstractions. I'll define them in terms of what primitive operations do, and put off worrying about how the types are defined, or how the primitives will be written.

The second example hits *harder*. It's one of the classic programming problems: Edsgar Dijkstra's problem of the Dutch National Flag. In contrast to many of our array examples, the Dutch National Flag is not particularly concerned with data abstraction. In fact, the data type is supplied as part of the problem statement. Instead, it requires an extremely interesting algorithm development. Read through it if you can, because it demonstrates techniques that are applicable to much more sophisticated programs.

Cellular Automata

> In Greek mythology, the machinery of the universe was the gods themselves. They personally tugged the sun across the sky, delivered rain and thunder, and fed appropriate thoughts into human minds. In more recent conceptions, the universe is created complete with its operating mechanism: once set into motion, it runs by itself. God sits outside of it and can take delight in watching it.*

Computers are often used to simulate real-life events. But instead of imitating a simplified version of some real-life phenomenon, let's start with the full-fledged embodiment of a computer life form: a *one-dimensional cellular automaton*.

A one-cell automaton is a simple machine that blindly obeys a set of built-in rules. Rules are usually pretty basic, like 'turn a light on.' A *one-dimensional* cellular automaton has all of its cells in a row. The automaton as a whole consists of single machines—the cells—that individually follow rules. Automata can have more dimensions; a very famous one known as Life (it's described in the exercises) is two-dimensional.

Creating a one-dimensional automaton's rules is very easy. Suppose that we want each cell to blink on and off. The rule 'if you're off, turn on, and if you're on, turn off' does the job.

transition rules

Current value:	on	off
New value:	off	on

Since the automaton's state changes, such rules are generally known as *transition rules*. Each change of state, when applied to every cell, is called a *generation* of the automaton.

More interesting transition rules take the cell's next-door neighbors into account. What if we say that an 'on' cell has value 1, while an 'off' cell has value 0. There are eight possible combinations for a cell and its two neighbors:

---

* Tommaso Toffoli, and Norman Margolus, *Cellular Automata Machines*, MIT Press, 1987. This is one of those rare books that's fascinating whether you understand it or not.

values of three neighbors

Now, let's say that we want to make up a transition rule for the cell in the middle to obey. Since there are eight possible values for the three-cell neighborhood, any rule we make up has have eight parts. Here are some ways that I could phrase it:

- With explicit statements: 'If the neighborhood value is 000 turn off, if the neighborhood value is 001 turn on' and so on.
- In terse English, assuming that my listener knows that I'm talking about successive neighborhoods: 'off on off on off on off on.'
- Using a mathematical shorthand for words: 0 1 0 1 0 1 0 1 0 1.

As you might guess, we'll go with the third alternative. Here's a before-and-after view of each of the eight neighborhoods, under rule 0 1 0 1 0 1 0 1:

a table of transition rules

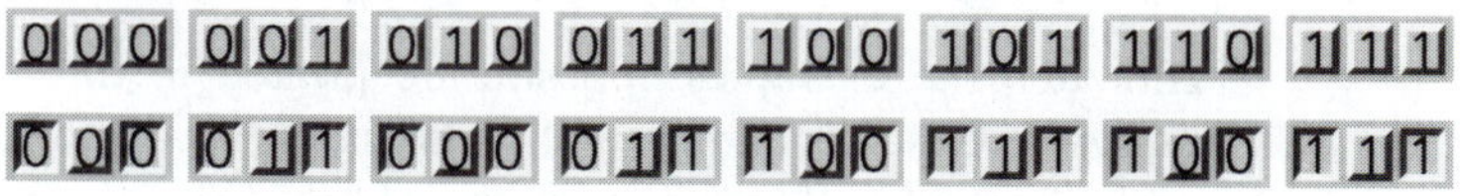

choosing a new value for the middle cell

If I want to, I can number the individual transitions 0 through 7. Why 0 through 7? Because if I read the values of the neighbors (000, 001, and so on) as three-digit binary numbers, they equal the decimals 0 through 7. Here's the before-and-after picture, with each transition numbered:

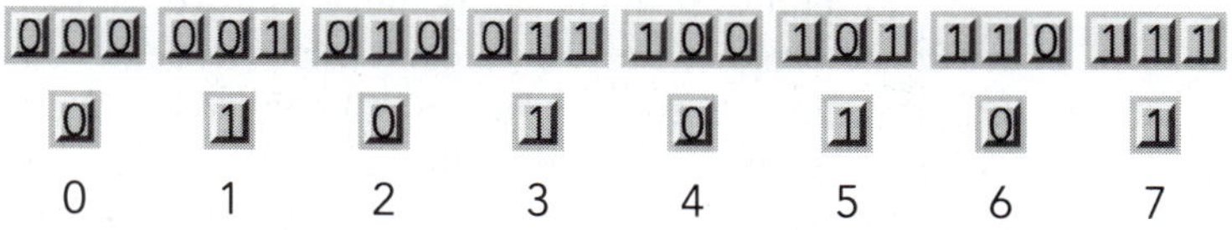

numbering the eight possible transitions

How many different rules are there? Pretend that the rule itself is a binary number (e.g. 00000000, 00000001, through 11111111). Translated into decimal notation, 256 different rules are possible. Later, I'll refer to them as rules 0 through 255. To summarize:

- Each of the eight possible combinations of neighbors is described by a decimal *transition number* 0 through 7.
- Each transition, 0 through 7, has a new value of 0 or 1 associated with it. Together, the zeros and ones can be read as an eight-digit *transition rule*.
- There are 256 different transition rules possible, numbered 0 through 255.

Now, describing a single transition of a single cell isn't especially interesting. Suppose, though, that we define an entire row of cells—a one-dimensional cellular automaton—and apply the transition rule to every cell in the row simultaneously. If we continue this process for a number of generations, we'll end up with a screen that looks like one of the examples shown on the next page. Note that some images start with just a single 'on' cell, while in others individual cells are initialized to 'on' or 'off' at random.

The action of an automaton can be changed drastically by altering just one number in the transition rule, or by changing the starting value of just one cell. The transition rules eventually fall into four categories:

the automaton's life cycle

1. those that evolve into a single state (i.e. they die) no matter what the starting row is;
2. those that develop a simple, regularly repeating cycle;
3. those that develop chaotic, constantly changing patterns that don't seem to have any regular structure;
4. those that generate long-lived, unpredictable patterns of great complexity.

Although the examples opposite are printed in tiny type, they were created using the Cellular program we'll write in this section, with DARKcharacter = "A", each line's LENGTH = 140 cells, and a DISPLAYlength = 100 generations.

A Data Abstraction for Cellular Automata

How can we approach the problem of programming a one-dimensional cellular automaton? Can we even use the object-oriented approach when there's so much emphasis on following rules?

Certainly. It might not be obvious to you at first glance, but the automaton stores data—the on/off state of each of its cells. Like most data structures, it has to be initialized (say, to turn all cells 'off'). We have to be able to make assignments so that we can turn individual cells 'on.' There has to be an inspection procedure if we hope to make pretty pictures like the ones above. Finally, the automaton requires one very unusual primitive—a way to make it obey a transition rule, and advance from one generation to the next.

Suppose that I describe a data abstraction called GenerationTYPE. As usual, I'll just lay out a rough sketch of its primitives; detailed enough to define the abstraction, but not so detailed that we have to decide how to write it:

the automaton data abstraction

TYPE GenerationTYPE *probably defines a one-dimensional array of cells;*
PROCEDURE Initialize *initializes the cells to zeros—'off';*
PROCEDURE TurnOn *changes a cell's value to one—'on';*
PROCEDURE Generate *obeys the transition rule;*
PROCEDURE Print *displays the current generation of the automaton;*

GenerationTYPE and its primitives only solve part of our problem, though. A one-dimensional automaton program will also have to read a transition rule. Another abstraction will come in handy:

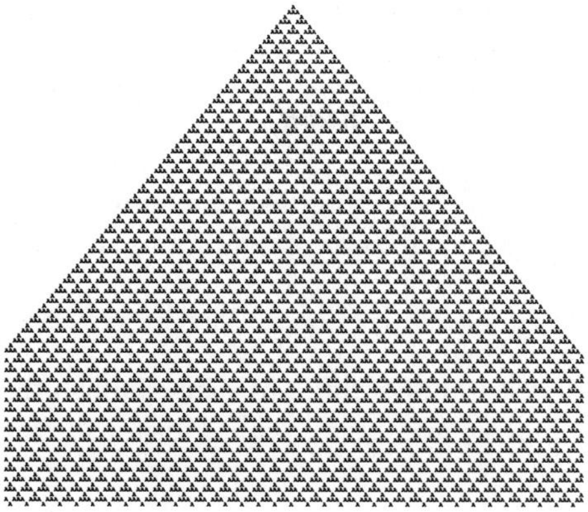

Rule 54 (00110110) *from a single cell.*

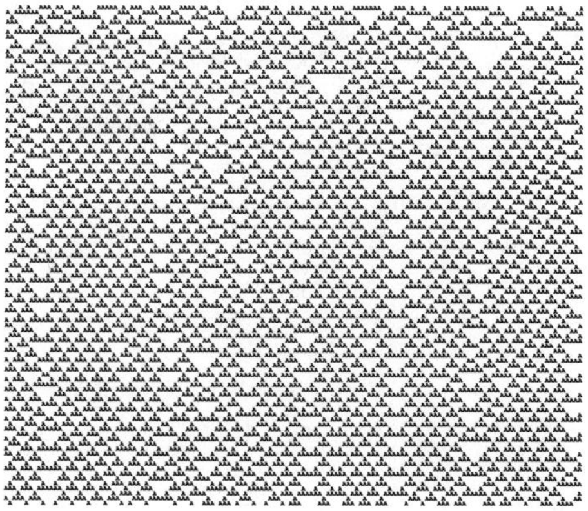

Rule 54 *from a random start.*

Rule 150 (10010110) *from a single cell.*

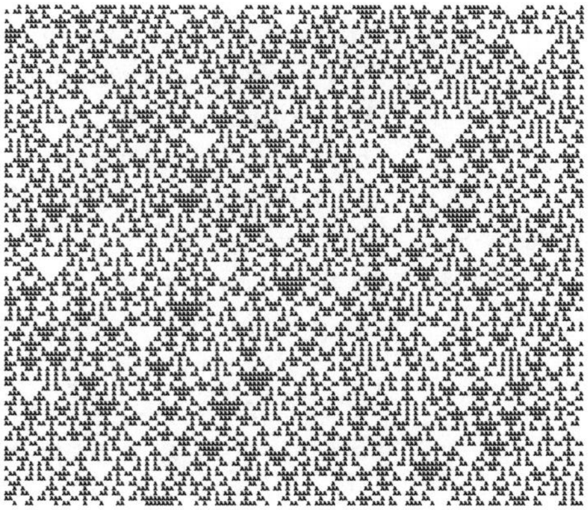

Rule 150 *from a random start.*

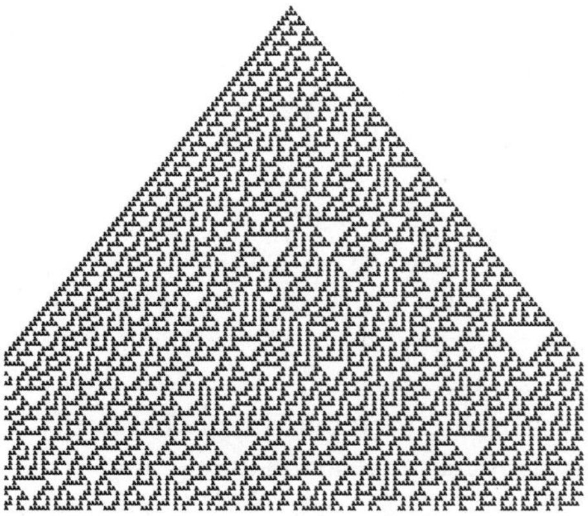

Rule 30 (00011110) *from a single cell.*

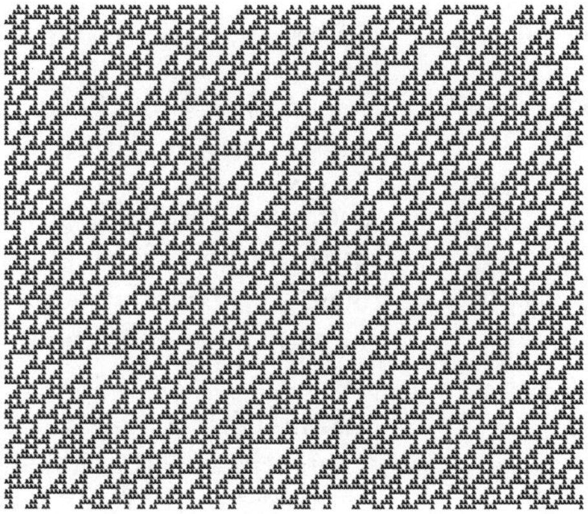

Rule 110 (01101110) *from a random start.*

For additional information on one-dimensional cellular automata, see Scientific American, March, 1984 (pg. 12) and September, 1984 (pg. 188), or Byte magazine, December, 1986 (pg. 181).

the rule table abstraction

> **TYPE** RuleTYPE *is also likely to be an array, since it stores* 8 *values*;
> **PROCEDURE** InitializeRule *gets the transition rule from the program user*;
> **PROCEDURE** PrintRule *makes sure that we obtained the rule correctly!*;

clusters

> It is not uncommon for a program to rely on more than one underlying data abstraction. Each abstraction consists of a *cluster* of definitions and declarations.

Let's take a closer look at the cluster of definitions that establish the automaton. I'll make the automaton a bit narrower than a typical terminal screen:

the definition underlying GenerationTYPE

```
CONST LENGTH = 75;
TYPE GenerationTYPE = ARRAY [1..LENGTH] OF INTEGER;
VAR Last, Next: GenerationTYPE;
```

Note that I'm drawing the line on abstraction by having the array store INTEGERs in its components. Even though I probably could create a new type that consists of zero and one, or on and off, it would make my life more complicated than I need it to be right now!

GenerationTYPE primitives

Procedure Initialize doesn't worry me—it's easy enough to make all components equal 0. Nor does procedure TurnOn inspire panic—it just assigns a 1 to an array component. Even Print is straightforward. When printout time comes, I can print an asterisk for every stored 1, and a blank space for each 0.

But how will Generate work? Obeying a transition rule implies that the value of a cell is going to change. But each cell's new value depends on the *current* value of its neighbors. If cell values change as we carry out transitions, how can we simultaneously evaluate every cell in the array? Let's delay consideration of Generate for a bit.

We can approach the RuleTYPE cluster the same way we outlined GenerationTYPE. Remember the suggestion I made about numbering the eight different transitions? Why don't we store them that way in an array called Rule:

a closer look at RuleTYPE

```
TYPE RuleTYPE = ARRAY [0..7] OF INTEGER;
VAR Rule: RuleTYPE;
```

RuleTYPE primitives

The other RuleTYPE primitives I suggested will be easy for us. InitializeRule reads and stores the eight-digit sequence of zeros and ones that set the transition rules. Procedure PrintRule, in turn, helps verify that we've stored the rule correctly by printing the contents of the Rule array.

> Although PrintRule isn't required by the problem statement, nearly every data structure should have a 'print' operation to aid in debugging.

What do we have at this point? The outlines of two clusters: GenerationTYPE and its primitives, and RuleTYPE and its primitives.

A Closer Look at the Primitives

Let's take a closer look at the primitives, especially Generate, the hard one. Recall that the problem was to follow the transition rules for an entire generation at once. Since the current generation helps establish values for the next generation, we have to use two GenerationTYPE arrays:

> Last *will hold the current, soon-to-be previous generation.*
> Next *will hold the upcoming, just-being-born generation.*

the Generate primitive

Each component of Next will get its value from three components of Last. For instance, Next[5] will have its value determined by Last[4], Last[5], and Last[6]. In general, the value of *any* Next[i] (besides the first and last) will be found by applying the proper transition rule to Last[i–1], Last[i], and Last[i+1].

Now comes the hard part: how do we apply a transition rule? Will Next[i] be 0 or 1, off or on? It would sure be nice to be able to say:

breeding the next generation

```
FOR i := 1 TO LENGTH DO    (*i.e. for each component*)
   Next[i] := whatever is right, given Last[i–1], Last[i], Last[i+1],
                 and the rule we're supposed to be following
                 (*i.e. then a miracle happens!*)
END;
Last := Next;  (*Replace the old generation with the new one.*)
Print (Last);  (*Print the next line of a pretty picture.*)
```

Just what is this *whatever is right*? Well, recall three facts:

- The new cell makes a transition to 0 or 1 that depends on which of eight possible values the old three-cell neighborhood has.
- If we treat the three neighbor values as digits of a binary number (000, 001, 010 and so on) their value will be the single decimal value 0 through 7.
- The Rule variable is an **ARRAY** [0..7] **OF** INTEGER.

where miracles come from

Suppose that in our imaginations, we number the transitions 0 through 7. We can store the 0 or 1 that each transition requires in components 0 through 7, of Rule. Here's how we'd store the rule 01010101:

components of a stored rule

So, how can we take a neighborhood (say, 101) and decide which component of the stored rule should apply? Why not 'read' the neighborhood as though it were a three-digit binary number?

the automaton reproduces!

> Next[i] *gets the value stored in* Rule [*the neighborhood's decimal equivalent*];

The arithmetic isn't hard. Suppose that the neighborhood is 1 0 1. The decimal equivalent of 101 is $1*2^2 + 0*2^1 + 1*2^0$, or 5. The rule we want to obey would be found in Rule[5], above, which happens to be 1.

This system is pretty neat, but we're not going to use it exactly as stated. As it happens, the world's expert on one-dimensional cellular automata, Stephen Wolfram, stores his rules the other way around. In effect, he puts the 000 rule in Rule[7], the 001 rule in Rule[6], and so on, down to the 111 rule in Rule[0].

Although Wolfram's system isn't the most obvious, it is the standard rule-numbering scheme, so I'll obey it. I could store the rule backward, as Wolfram does, but I find that confusing. Instead, I'll look it up backward, as I've done below. Suppose that Last[i] is the center of a three-component neighborhood. We can find the new value, for Next[i], like this:

```
Next[i] := Rule [7 – ( (4*Last[i–1]) + (2*Last[i]) + (1*Last[i+1]) ) ]
```

Subtracting the decimal equivalent of the neighborhood from seven does the trick—it reads the rule in reverse. The problem of breeding a new generation is essentially solved.

None of the other primitives are as hard as Generate. Let's look at the GenerationTYPE cluster. First, there's Initialize:

procedure Initialize

> **FOR** *every component of* Last
>   *initialize the component to* 0

Procedure TurnOn stores a 1 in any particular component. It's just an assignment statement:

procedure TurnOn

```
Last [WhichOne] := 1;  (*Turn on a component.*)
```

Next, Generate obeys the transition rule. We just saw how it works. But what about the first and last components, which are each missing a neighbor? By convention, people usually assume that the array 'wraps around,' so that the last component is followed by the first. Can you figure out how it's done?

procedure Generate

> *take care of* Next[1] *and* Next[LENGTH];
> **FOR** *all the in-between components*
>   Next[i] *gets the proper value, using* Rule *and* Last

Finally, procedure Print shows the current value of one generation of the automaton. Since each new generation prints on the bottom of the screen, and is followed by a carriage return, we automatically maintain a record of the automaton's development.

procedure Print

> **FOR** *every component of the current generation*
>   **IF** *it's a* 1
>   **THEN** *print a dark character*
>   **ELSE** *print a space*

Next, let's turn to the RuleTYPE cluster. Procedure InitializeRule reads a rule that's typed in by the program user. A simple version would be:

procedure InitializeRule

> **FOR** *each of the eight components of the* Rule
>   *read a value*

A more complicated version of InitializeRule might let the user enter a number between 0 and 255, then figure out which of the rules from 00000000 to 11111111 it corresponded to. (I've written it that way in the final program so that I could sneak in an example of base conversion.)

Finally, procedure PrintRule is our dump routine. It makes sure that I handled the conversion correctly:

procedure Print

FOR *each of the eight components of the* Rule
    print its value

The Cellular Solution

Let's stop for a minute to assess our position. We've taken three steps:

- We roughly outlined the primitives of two data abstractions.
- We stuck our necks out by suggesting specific data types for the abstractions to be based on.
- We took a closer look at the implementation of each of the primitives.

As it happens I was lucky—the specific data types I suggested worked out. That's partly because I'm experienced, and partly because the definitions weren't especially complicated. However, I might have gotten smarter in the course of drafting the primitive implementations, and had to go back and revise the type definitions. In fact, I can still do it—maybe to store off and on values instead of having arrays of INTEGER.

> All program design involves some backtracking and revision. The approach we've taken limits damage, though. We can change the way primitives are implemented, or the underlying types they're based on, without affecting the way that they're used.

Here is a final, formal outline of the GenerationTYPE and RuleTYPE data abstractions we've been working on:

the GenerationTYPE cluster

```
CONST LENGTH = 75;          (*About the width of a screen.*)
TYPE GenerationTYPE = ARRAY [1..LENGTH] OF INTEGER;
PROCEDURE Initialize (VAR Line: GenerationTYPE);
   (*Initializes the first generation to all zero values.*)
PROCEDURE TurnOn (VAR Cells: GenerationTYPE;
                          WhichOne: INTEGER);
   (*Sets one particular component to 1.*)
PROCEDURE Generate (Last: GenerationTYPE;
             VAR Next: GenerationTYPE; Rule: RuleTYPE);
   (*Applies the transition rules contained in Rule to the last generation.*)
PROCEDURE Print (Current: GenerationTYPE);
   (*Prints the current generation of cells.*)
```

the RuleTYPE cluster

```
TYPE RuleTYPE = ARRAY [0..7] OF INTEGER;
PROCEDURE InitializeRule (VAR Rule: RuleTYPE);
    (*Gets a generation rule from the program user.*)
PROCEDURE PrintRule (Rule: RuleTYPE);
    (*Prints the rule as a sequence of zeros and ones.*)
```

Let's take these tools and develop a pseudocode of the final program. A very first outline is:

draft pseudocode

get a rule from the program user;
do any other necessary initialization;
decide what the first generation starting conditions are;
FOR *some number of generations*
    display the current generation
END *of the loop*

Now I'll say the exact same thing using the tools we've just built:

pseudocode refinement

InitializeRule (Rule) *to a generating rule;*
PrintRule (Rule) *to confirm the rule that was entered;*
Initialize (Last) *to all zeros;*
TurnOn (Last, WhichOne) *as desired;*
Print (Last) *to display the first generation;*
FOR *a couple of dozen generations*
    Generate (Last, Next) *to create the next generation;*
    Print (Next) *to display it;*
    Last := Next *to make way for the next generation*
END *of the loop*

The final program, Cellular, adds a few extra refinements. Its format is somewhat unusual, because Cellular takes advantage of a formatting option that I've ignored in the past: there is more than one type definition part.

clustering definitions and declarations

> If it adds to program clarity, it is not necessary to maintain the usual order of single constant, type, and subprogram definition parts. Instead, each type definition can be accompanied by its cluster of constants and subprograms. Main program variables can be declared just before the start of the main program.

Clustering definitions makes it easier to keep track of a program's main structures. The feature shouldn't be used, however, as an excuse to scatter definitions and declarations willy-nilly. It's a good idea to use an easily spotted bracket (I use (*\/\/\/\/\/\/\/ *)) to set off each cluster.

one-dimensional cellular automaton program

```
MODULE Cellular;
  (*Displays several generations of a one-dimensional cellular automaton.*)
IMPORT IO;
CONST DARKcharacter = "#";
      LIGHTcharacter = " ";  (*i.e. a space*)
      DISPLAYlength = 25;  (*number of generations to print*)
(*Defining the Rule Table cluster \/\/\/\/\/\/\/\/\/\/\/\/\/\/\/\/\/\/\/\/\/\/  *)
TYPE RuleTYPE = ARRAY [0..7] OF INTEGER;
PROCEDURE InitializeRule (VAR Rule: RuleTYPE);
  (*Gets a decimal-valued generation rule from the program user,
   then stores it as the appropriate binary sequence in Rule.*)
  VAR TheRule, Divisor, i: INTEGER;
  BEGIN
    LOOP  (*Gets the rule value as a decimal.*)
      IO.StringOut ("Choose any rule from 0 through 255:   ");
      IO.IntIn (TheRule);
      IO.LineOut;
      (**)IF (TheRule >= 0) AND (TheRule <= 255) THEN EXIT END
    END; (*LOOP*)
    (*Postcondition: TheRule is in the range 0..255.*)
    (*Next, convert the decimal rule into its binary equivalent.*)
    Divisor := 128;
    FOR i := 0 TO 7 DO
      Rule[i] := TheRule DIV Divisor;
      TheRule := TheRule MOD Divisor;
      Divisor := Divisor DIV 2
    END
  END InitializeRule;
PROCEDURE PrintRule (Rule: RuleTYPE);
  (*Prints the rule as a sequence of zeros and ones.*)
  VAR i: INTEGER;
  BEGIN
    FOR i := 0 TO 7 DO
      IO.IntOut (Rule[i], 1)
    END
  END PrintRule;
```

(program continues)

```
(*Defining the Cellular cluster \/\/\/\/\/\/\/\/\/\/\/\/\/\/\/\/\/\/\/\/\/ *)
CONST LENGTH = 75;          (*About the width of a screen.*)
TYPE GenerationTYPE = ARRAY [1..LENGTH] OF INTEGER;
PROCEDURE Initialize (VAR Line: GenerationTYPE);
  (*Initializes the first generation to all zero values.*)
  VAR i: INTEGER;
  BEGIN
    FOR i := 1 TO LENGTH DO
      Line[i] := 0
    END
  END Initialize;
PROCEDURE TurnOn (VAR Cells: GenerationTYPE;
                              WhichOne: INTEGER);
  (*Sets this component to 1.*)
  BEGIN
    Cells[WhichOne] := 1
  END TurnOn;
PROCEDURE Generate (Last: GenerationTYPE;
                 VAR Next: GenerationTYPE; Rule: RuleTYPE);
  (*Applies the transition rules contained in Rule to the last generation.*)
  VAR i: INTEGER;
  BEGIN
    FOR i := 2 TO LENGTH – 1 DO
      Next[i] := Rule [ 7 – ( 4*Last[i–1] + 2*Last[i] + 1*Last[i+1] ) ]
    END;
    (*Postcondition: except for the first and last components, Next is defined.*)
    (*Can you use MOD in the loop to get rid of these special-case assignments?*)
    Next[1] := Rule [ 7 – ( 4*Last[LENGTH] + 2*Last[1] + 1*Last[2] ) ];
    Next[LENGTH] := Rule [ 7 – ( 4*Last[LENGTH–1] +
                               2*Last[LENGTH] + 1*Last[1] ) ]
  END Generate;
PROCEDURE Print (Current: GenerationTYPE);
  (*Print the current generation of cells.*)
  VAR i: INTEGER;
  BEGIN
    FOR i := 1 TO LENGTH DO
      IF Current[i] = 1
        THEN IO.CharOut (DARKcharacter)
        ELSE IO.CharOut (LIGHTcharacter)
      END
    END;
    IO.LineOut
  END Print;
```

```
VAR Rule: RuleTYPE;
    Last, Next: GenerationTYPE;
    Ch: CHAR;              (*used to initialize Last*)
    Counter,               (*also used in initializing Last*)
    GenerationCount: INTEGER;
BEGIN (*main program*)
   InitializeRule (Rule);
   IO.StringOut ("Choosing rule:    ");
   PrintRule (Rule);
   IO.LineOut;
   Initialize (Last);
   IO.StringOut ("Initialize the first generation by typing a line of characters.");
   IO.LineOut;
   IO.StringOut ("Any non-blank will be read as an ´on´ cell.");
   IO.LineOut;
   Counter := 1;
   LOOP
      IO.CharIn (Ch);
      (**)IF (Ch = IO.EOL) OR (Counter > LENGTH) THEN EXIT END;
      IF (Ch <> "  ") THEN TurnOn (Last, Counter) END;
      Counter := Counter + 1;
   END;
   IO.StringOut ("FirstGeneration initialized to:"); IO.LineOut;
   Print (Last);
   FOR GenerationCount := 1 TO DISPLAYlength DO
      Generate (Last, Next, Rule);
      Print (Next);
      Last := Next
   END
END Cellular.
```

```
Choose any rule from 0 through 255:  150
Choosing rule:  10010110
Initialize the first generation by typing a line of characters.
Any non-blank will be read as an on cell.
w uf qwoi faoij dj   dqfew  edw  fd fjjfj  feq[j fqj  fqejj   fej foj frq j
FirstGeneration initialized to:
```

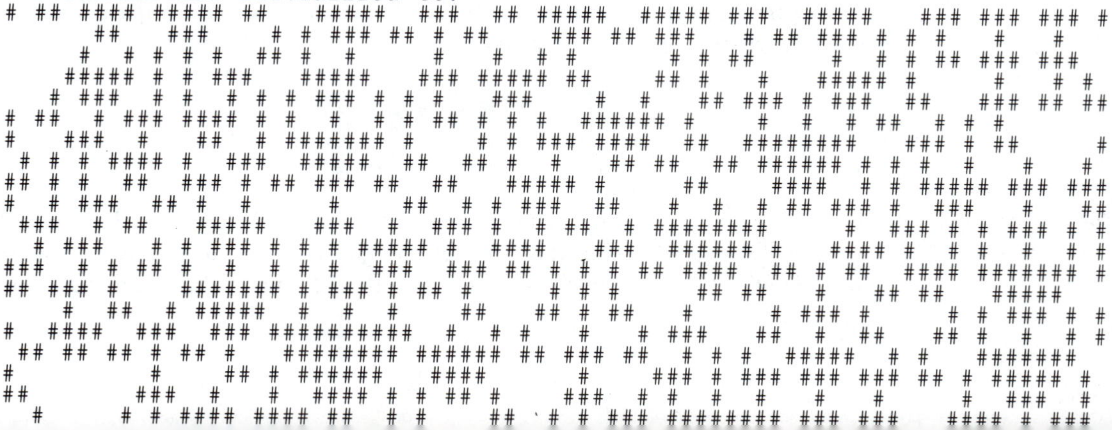

Modification Problems

1. Cellular prints DISPLAYlength generations, then quits. Improve the program so that printing is interactively controlled by the user. Let her *a*) print one screenful at a time; *b*) print *n* generations at a time; *c*) skip *m* generations before printing again.

2. The output of a one-dimensional automaton is sometimes described as having a *space* dimension (the individual cells on each output line) and a *time* dimension (the full line). Modify Cellular so that it compresses space (by only printing every other cell) and time (by only printing every other generation). Note that the program still needs the in-between values to calculate new generations—they're just not being printed. You might want to double the LENGTH and DISPLAYlength figures.

3. Change the underlying data type definitions so that the arrays have components of type ShadeTYPE, with values off, on. Are any changes required outside the primitives?

4. If you did 3, above, try extending ShadeTYPE to four different values (e.g. off, light, dark, on). Modify procedure InitializeRule to allow input of a four-outcome rule.

Dijkstra´s Dutch National Flag

Our final problem has acquired a great deal of renown because it illustrates methods that apply to much more sophisticated jobs. E.W. Dijkstra named his *Dutch National Flag* problem in honor of his country's flag, which has three horizontal stripes: one red, one white, and one blue. The problem is:

problem: Dutch National Flag

> Suppose that an array of length N holds three different values; say, red, white, and blue. Write a program that puts all the red values at the left end of the array, the blue ones at the right end, and the white values in the middle.
>
> The trivial solution—go through the array to count all the pebbles first—is forbidden, naturally. You've got to do it all in one pass, or traversal, of the array.

Solving the Dutch Flag problem relies on some techniques I touched on earlier: using subscripts as pointers, developing a case analysis to guide decision making, and exploiting data bounds to determine the right time for loop exit.

But how will we get started? Suppose I employ the time-honored problem solving method of developing a simpler example first. Indonesia's flag, a relic of its colonial past as the Dutch East Indies, has only two stripes: one red, and one white. We can imagine, then, that a Flag array starts out as:

R	R	W	R	W	W	R	W	R

mixed red and white components

the ´Indonesian Flag´ problem, algorithm one

Let me suggest the following algorithm for sorting the mixed array. I'll work my way through the array from left to right. Since we only have two colors to sort out, I'll keep the red components on the left end of the array, and throw the white components to the right end. Since I'll have to remember where the last red and first white components are, I'll need a couple of auxiliary variables. As you can see below, they're used as *subscript pointers*; they separate the sorted part of the array from the unsorted unknown.

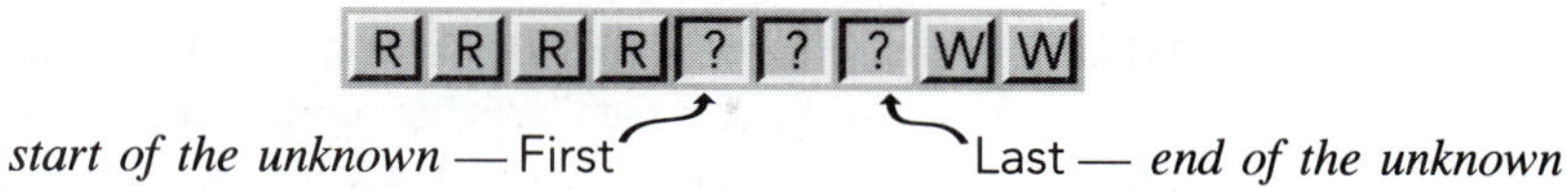

How can I tell when the array has been unjumbled? Well, let's imagine the situation a few steps in the future. Once the array has been neatly divided into red and white, the two sorted parts will pass each other. Thus, we can use the subscript pointers as data boundaries. They let us know that the array is sorted by letting us know that the unknown portion is empty.

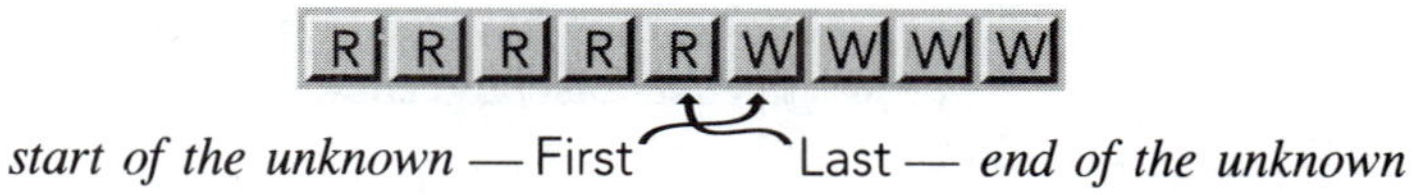

algorithm one in detail

Let's walk through the algorithm in more detail. At first, our boundary pointers (call them First and Last) are situated at the Flag array's first and last components. They will always subscript the first and last unknown components. Suppose that Flag[First] is red. We leave it alone, and increase First by one, heading toward Last. We might even go all the way through the array like this, if every component is red.

At the same time, the component subscripted by Last pointer might be white. In this case, we'd decrease Last by one, heading back toward First. As before, we'd travel the entire array like this were every component white.

the switch

One possibility remains. Suppose the array isn't perfectly ordered in advance, and doesn't just hold components of one color. If we follow the rules suggested above, we'll eventually find that Flag[First] is white, while Flag[Last] is red. When this happens, we swap, or exchange, the two components, and advance *both* subscripts.

In case you haven't noticed, we've developed a case analysis. We have an appropriate action to take regardless of what situation is encountered. In pseudocode:

move the bound´ case analysis

```
IF the current component Flag[First] is red
      THEN increment (advance) the First pointer
   ELSIF the current component Flag[Last] is white
      THEN decrement (decrease) the Last pointer
   ELSE swap Flag[First] with Flag[Last]
         increment Current;
         decrement Last;
END the if statement
```

Now, unless the Flag array happens to be just one or two components long, this pseudocode plan is going to have to be put into a loop. What will the **EXIT** condition be? Well, there's no sentinel or mathematical limit bound, and even a count is not at all obvious.

> What is clear, though, is that the First and Last pointers, which mark the borders of the 'unknown data' area, get closer and closer together. Thus, we can use them to provide a data bound condition—we leave the loop when we run out of unknown data.

In pseudocode:

two-color flag pseudocode, version one

initialize First *and* Last *to the array's beginning and end;*
LOOP
    carry out the case analysis described above;
    (**)**IF** First *and* Last *pass each other* **EXIT**
END *of the loop*
(**Postcondition: there are no more unknown components to check.**)

An Alternative Case Analysis

That was a great algorithm, wasn't it? But it's not the only one I came up with. Before we move on to the full three-color version of the Dutch National Flag problem, let me present a different two-color solution.

Superficially, the second algorithm is like the first: it uses a case analysis, and it's inside a loop. However, both the analysis and the loop bounds are quite different. Instead of tossing the white components to the end of the array, I'll toss them to the end of the red section. I can get the effect of sliding *all* the white components to the right just by swapping a single component.

The sorted array segments are built like this. Suppose that this is our current position. I'll call the pointers RedBorder and WhiteBorder. WhiteBorder advances to the right, and it always points to an unknown.

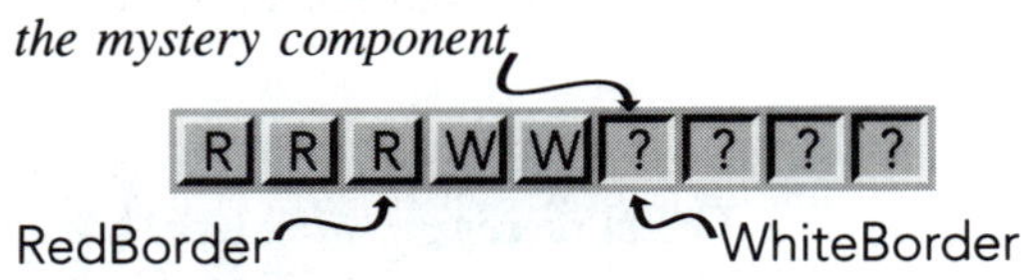

The rule we'll follow is:

If the next unknown component is white, advance the white border only.
If it's red, swap it with the leftmost white component and advance both borders.

Suppose the mystery component is red. The array and border pointers end up like this:

the new one is red

Now imagine instead that the mystery component was white. We'd wind up this way:

the new one is white

Now for a more detailed explanation. I'll begin with the RedBorder set to 0—just before the first array component—and the WhiteBorder set to 1—the first array component. For all intents and purposes, the RedBorder pointer is always one component behind the first known white value. If there aren't any red components, it doesn't even point to the array.

The WhiteBorder, in contrast, always points to the first of the remaining 'unknown' components. It is always one component *ahead* of the *last* known white value. If there aren't any white components, of course, it's one step ahead of the last red component.

Let's consider all the possibilities. Suppose that the current WhiteBorder component's value is white. Great! I can advance the white border by one, and start again. If advancing takes me past the end of the array, I'm done.

But what if the WhiteBorder component is red? Now it gets tricky. I advance the RedBorder by one. This takes me one step into white territory (if there are stored white values) or to the WhiteBorder (if there are no white components). Then, I must exchange the white RedBorder component with the red WhiteBorder component, and advance the WhiteBorder component by one.

Why does the exchange and advance work? Because afterward, the starting conditions hold again: RedBorder points to the last known red component, and WhiteBorder points to the first of the remaining 'unknowns,' if there are any.

Stated as a loop specification, the second algorithm is:

loop specification, algorithm two

Goal: Segregate an array's red *and* white *components.*
Bound: Reaching the last array component.
Plan: If the current WhiteBorder *component is* white, *then*
advance the WhiteBorder.
If the current component is red, *then*
advance the RedBorder;
swap the value at RedBorder *for the value at* WhiteBorder;
advance the WhiteBorder.

Read the version two pseudocode carefully, because it will be useful in solving the final three-color flag problem.

a two-color pseudocode, version two

```
initialize RedBorder to 0;
initialize WhiteBorder to 1;
LOOP
   CASE the current WhiteBorder component is
      white: increment WhiteBorder
      red: increment RedBorder;
           swap Flag[RedBorder] with Flag[WhiteBorder];
           increment WhiteBorder
   END; (*CASE*)
   (**)IF WhiteBorder goes past the end of the array EXIT
END (*LOOP*)
```

What is the loop's **EXIT** condition? WhiteBorder, the white border pointer, is increased by 1 on each trip through the loop. In effect, it is a counter variable—the loop ends when inspecting the counter itself shows that we're past the end of the array.

> The version two pseudocode has one other notable feature. Since we're keeping the known red and white components on the left-hand side of the array, the right-hand end is free. This is in contrast to the version one pseudocode, in which the right-hand end was needed to store red components.

Solving the Three-Color Flag Problem

Now, in algorithm two, the right end of the array spent most of its time languishing as an unknown region. However, we can make good use of it to solve the full Dutch National Flag problem. We'll use the right end of the array for storage as we did in algorithm one—but store blue components. In effect, we combine versions one and two of the two-color flag solutions.

three-color pseudocode

```
initialize RedBorder to 0; (*i.e. before the first known red*)
initialize WhiteBorder to 1; (*i.e. the first of the remaining unknowns*)
initialize BlueBorder to N, the array limit plus 1; (*i.e. the last unknown*)
LOOP
   CASE the current WhiteBorder component is
      white: increment WhiteBorder
      red: increment RedBorder;
           swap Flag[RedBorder] with Flag[WhiteBorder];
           increment WhiteBorder
      blue: swap Flag[BlueBorder] with Flag[WhiteBorder];
            decrement BlueBorder
   END; (*CASE*)
   (**)IF WhiteBorder passes BlueBorder EXIT
END (*LOOP*)
```

Take a close look at the case analysis. Either the WhiteBorder will advance by one component to the right (if the current component is either red or white), or the BlueBorder will 'advance,' or actually decline, by one component to the left (if the current component is blue). Since the unknown area is always between the WhiteBorder and the BlueBorder, the loop's EXIT condition is a data boundary—the unknown area shrinks to zero when these two borders pass.

My aim in this section was to explore an algorithm for solving the Dutch National Flag problem. I'll give you a program that tests our solution *gratis*. It's hardly more than a test harness; look at how I've initialized the flag. But don't worry—you'll get to improve it however you want.

Dutch National Flag program

```
MODULE Dutch;
   (*A test harness for a procedure that solves the Dutch Flag problem.*)
IMPORT IO;
CONST LIMIT = 15;          (*Size of the flag.*)
      DEBUGGING = TRUE;  (*Controls a call of Print in FlagSort*)
TYPE ColorTYPE = (red, white, blue);
     FlagTYPE = ARRAY [1..LIMIT] OF ColorTYPE;
     BoundTYPE = [0..LIMIT+1];
VAR Flag: FlagTYPE;
PROCEDURE Initialize (VAR Flag: FlagTYPE);
   (*A truly simple-minded procedure that initializes the flag.*)
   BEGIN
      Flag[1] := red;    Flag[2] := white;  Flag[3] := blue;   Flag[4] := red;
      Flag[5] := white;  Flag[6] := blue;   Flag[7] := red;    Flag[8] := white;
      Flag[9] := blue;   Flag[10] := red;   Flag[11] := white; Flag[12] := blue;
      Flag[13] := red;   Flag[14] := white; Flag[15] := blue
   END Initialize;
PROCEDURE Print (Flag: FlagTYPE);
   (*Displays the contents of the Flag array.*)
   VAR Count: BoundTYPE;
   BEGIN
      FOR Count := 1 TO LIMIT DO
         CASE Flag[Count] OF
              red:  IO.StringOut ("R  ")
            | white:  IO.StringOut ("W  ")
            | blue:  IO.StringOut ("B  ")
         END;
      END;
      IO.LineOut
   END Print;
```

(program continues)

```
PROCEDURE FlagSort (VAR Flag:  FlagTYPE);
   (*Sorts an array of mixed red, white, and blue values.*)
   VAR RedBorder, WhiteBorder, BlueBorder:  BoundTYPE;
   PROCEDURE Swap (VAR First, Second:  ColorTYPE);
      (*Exchanges two component values.*)
      VAR Temp:  ColorTYPE;
      BEGIN
         Temp := First;  First := Second;  Second := Temp
      END Swap;
   BEGIN
      RedBorder := 0;
      WhiteBorder := 1;
      BlueBorder := LIMIT + 1;
      LOOP
         CASE Flag[WhiteBorder] OF
               white:  WhiteBorder := WhiteBorder + 1;
             | red: RedBorder := RedBorder + 1;
                   Swap (Flag[RedBorder], Flag[WhiteBorder]);
                   WhiteBorder := WhiteBorder + 1
             | blue:  BlueBorder := BlueBorder – 1;
                    Swap (Flag[WhiteBorder], Flag[BlueBorder])
         END;  (*CASE*)
         IF DEBUGGING THEN Print (Flag) END;
         (**)IF WhiteBorder = BlueBorder THEN EXIT END;
      END  (*LOOP*)
   END FlagSort;
BEGIN   (*main program*)
   IO.StringOut ("Initializing the flag to:");   IO.LineOut;
   Initialize (Flag);
   Print (Flag);
   FlagSort (Flag);
   IO.StringOut ("Flag sort complete.")
END Dutch.
```

```
Initializing the flag to:
R W B R W B R W B R W B R W B
R W B R W B R W B R W B R W B
R W B R W B R W B R W B R W B
R W B R W B R W B R W B R W B
R W W R W B R W B R W B R B B
R W W R W B R W B R W B R B B
R R W W W B R W B R W B R B B
R R W W W B R W B R W B R B B
R R W W W R R W B R W B B B B
R R R W W W R W B R W B B B B
R R R R W W W W B R W B B B B
R R R R W W W W B R W B B B B
R R R R W W W W B R W B B B B
R R R R W W W W W R B B B B B
R R R R W W W W W R B B B B B
R R R R R W W W W W B B B B B
Flag sort complete.
```

Modification Problems

1. If *my* grade depended on it, I wouldn't be too happy about taking credit for the user interface of procedure Initialize. *a*) Improve it, then *b*) describe the data you would use to test program Dutch.
2. Find a country whose flag has *four* colored stripes. Then, modify FlagSort so that it can handle the four-color flag problem.

Section 8-5 Program Engineering Notes

IMAGINE THE DIFFICULTY OF CONSTRUCTING A BUILDING from invisible bricks and mortar. Beams fly into thin air, girders sail into space, and, most likely, rubble collects at the foundation.

Building a program is not at all a dissimilar proposition. As long as unseen parts fit together, construction proceeds apace. However, when invisible components aren't where they're supposed to be, the error messages start to pile up and programming grinds to a halt. The most useful tool a program engineer can have is sight—the ability to look inside her program's components. Translated into precept form, we have a basic rule of working with structured types:

> *Build tools before you build programs.* As you define new types, write primitives that will let you look inside them.

Visibility is more of an issue for structured types (like arrays) than for simple types (like plain INTEGER or CHAR variables) because the structure itself may hold information. In simple variables, *what* is the value is the only real question. Arrays, in contrast, require answers to *where* is the value as well—what is its subscript, and how does it relate to its neighbors?

Two Tools — Load and Dump

The two most useful tools go by rather unceremonious names:

- Load initializes an array for testing.
- Dump prints the contents of an array.

Specialized load and dump procedures aren't necessary for simple variables. After all, we have assignment statements and procedures for input and output from the IO library. But for structured variables, like arrays, they're absolutely essential. How can an array-manipulation procedure be tested if you aren't certain about the original contents of the array? And how can you be certain until you've taken a look inside?

Load procedures create a level playing field for program testing. After all, what you intend to do *with* data is only part of the programming problem.

know your test data

> *Getting there is half the fun.* Before a program can modify data it must obtain it. Before you can debug a modification procedure, you have to be able to give it data.

For example, testing the searching procedures we'll use in the next chapter requires a load procedure that can initialize an array in four different ways: to just one value, to ascending and descending sequences of values, and to a set of values I stored in a file. I certainly didn't want to enter all of this test data by hand, so I wrote a general-purpose procedure.

You probably won't need to be convinced about the usefulness of dump procedures. However, you should take a little time to write them well. Two pieces of advice are:

- *Label the output.* Print the subscripts along with the stored values.
- *Format the output.* Take an extra five seconds to make sure that subscripts aren't confused with values.

An array usually holds a large amount of information. A dump of its contents should be helpful rather than confusing. A typically unhelpful array dump is:

an unhelpful dump

```
(*Print the contents of a ten-component array.*)
FOR i := 1 TO 10 DO IO.IntOut (Data[i], 3) END
```

```
089023089042809230893226721007
```

Just adding labels doesn't necessarily help, either:

another useless version

```
(*Print and label the contents of a ten-component array.*)
FOR i := 1 TO 10 DO
    IO.IntOut (i, 1);
    IO.IntOut (Data[i], 3)
END
```

```
10892023308940425809623078938226972110007
```

Instead, try adding brackets, white space, and an occasional carriage return to make output clear:

help at last

```
(*Print, label, and format the contents of a ten-component array.*)
FOR i := 1 TO 10 DO   (*Fix the output of these.*)
    IO.StringOut ("   [");
    IO.IntOut (i, 1);
    IO.StringOut ("]:  ");
    IO.IntOut (Data[i], 3)
    IF (i MOD 5) = 0 THEN IO.LineOut END;
END
```

```
[1]: 089   [2]: 023   [3]: 089   [4]: 042   [5]: 809
[6]: 230   [7]: 893   [8]: 226   [9]: 721   [10]: 007
```

Note that load and dump procedures stand apart from the primitives that help define a data abstraction. They aren't necessarily meant to be well-refined additions to a textbook process of program development. Instead, they're the hammer and pickaxe of programming—nitty-gritty tools that do a simple job quickly.

Test Harnesses and Incremental Development

The load and dump procedures described above could, and probably should, be built into any program that uses a structured data type. However, it is not uncommon to build a testing tool that also stands apart from the final product.

> A *test harness* is a program shell that is used to test procedures in isolation, before they are integrated into a more complex final program.

Way back in Chapter 1 I described a program as a delivery system for procedure calls. A test harness is precisely that—a delivery system for procedure tests that contains:

- the type definition;
- load and dump procedures;
- the new procedure that is undergoing testing or modification.

The new procedure can be tested without having to deal with a main program that is more complex and finicky than the harness is. Once it works, the new procedure can be transferred to the main program.

A library module is, in a sense, the ultimate test harness. The collection of primitive routines that manipulate some hidden data type are written and tested independently of their final application. For all we know, test procedures like load and dump are still available (albeit unadvertised) parts of the library!

Using test harnesses is part of a more general approach to programming called *incremental program development*. An increment is a small amount; incremental development means programming by improvement. There is one basic reason for its appeal.

incremental program development

> A working program that only does 80% of the job but works is better than a program that does 99% of the job but crashes.

As the saying goes, writing the first 80% of a program takes 80% of the time, while polishing up the last 20% takes the other 80% of the time. Part of the reason for this weird arithmetic is that as programs become more complex, changes have a tendency to introduce unexpected effects.

Incremental programming tries to isolate the effect of changes. We add new features in preference to adding new functions, and add new functions rather than writing new programs. The short-program approach of *write / compile / test* becomes:

incremental additions

define types / compile / fix;
add load and dump procedures / compile / test;
add first processing function / compile / test / fix;
add features / compile / test / fix;
add second processing function / compile / test / fix;
keep adding features / and compiling / and testing / and fixing.

When a program is large enough, or when more than one person is working on it, the preliminary work on each *add functions and features* segment can take place in a test harness rather than in the final program.

Dealing with Data Deluge

As you have probably gathered by now, each chapter promises a simpler life as long as you learn umpteen new things to worry about. What's happening, of course, is that new material really does simplify the solution of problems—but only of problems you wouldn't have attempted otherwise. Overall, we're solving problems that are progressively more difficult and complicated.

Work with arrays is inherently more complex than work with individual variables because we have more balls in the air at once. There is a deluge of data in contrast to the few piddly drops we're used to. Here are three ideas that will help make life easier:

build scale models

1. *Build scale models.* Reduce the size of each array to the smallest practical size for testing your approach.

Consider program FileRev, which reversed the contents of a file of 500 lines, with 80 characters on each line. In principle, is there any difference in reversing a file of just ten 5-character lines? No. Is multiplying two 4-digit numbers any different from multiplying two 100-digit numbers? As long as the same data structure is used for both, again the answer is no.

What *is* different? Testing. As difficult as it is to spot a bug in a few lines of output, it is impossible to spot a bug when output runs longer than one screen. And what about input? Do you really *want* to type in two 100-digit numbers over and over again? Worse, do you really want to check a 101-digit answer? In short, scale models reduce testing to reasonable terms.

check intermediate results

2. *Avoid combining intermediate steps.* Try to check the results of every action.

Suppose that you are writing a text editing program. If an array contains 20 lines of text, the command *delete lines 6 through 10* might be implemented as a single Delete command. However, I would much rather present the sequence of actions like this:

checks between steps

```
Dump (TheArray);
erase the contents of 6 through 10;
Dump (TheArray);
move lines 11 through 20 to 6 through 15;
Dump (TheArray);
erase the contents of 16 through 20;
Dump (TheArray);
change TotalLength from 20 to 15;
```

Why? So that I can confirm that the array's physical contents match my mental image of what has happened. Even though I 'know' that copying one line over another erases it, I would still rather take the extra step of explicitly erasing the vanishing line first. The extra step doesn't make a correct program any better, but it helps me spot the unexpected bug in my own frequently incorrect code. Perhaps I read the wrong numbers for deleting, or perhaps an earlier command has messed up the array's contents.

make data self-explanatory

3. *Embed debugging information in data.* During testing, try to make test data self-explanatory.

This simple suggestion is particularly important for text processing programs. There is nothing visibly wrong with the next few lines as test data:

poor test data

```
I am failing computer science.
Sunshine beckons.
Printout hits the floor.
```

However, they do not shine a spotlight on bugs. Suppose that my test command is *move line 3 to follow 1.* The output:

```
Sunshine beckons.
Printout hits the floor.
I am failing computer science.
```

is wrong. Line 1 now follows line 3, rather than vice versa.

In contrast, the test data below may be of more modest interest as haiku, but it is a little more useful for debugging programs. Its order is obvious:

better test data

```
1 Life to Live.
2 pants suits.
3 piece power bands.
```

A Collection of Subscript Bugs

Technique or no, there will still be bugs. Syntax problems are a plague at first, but you should have the experience interpreting compile-time error messages by now to work through them. In fact, you can use the compiler to check your definitions as soon as you make them; after all, nobody ever claimed that a program had to do something. This program can be compiled in an attempt to find the bug it contains:

compiling as a syntax check

```
MODULE Test;
TYPE TwoTYPE = ARRAY [1..5, 1..10] OF INTEGER;
    (*Can you see the syntax bug yet?*)
VAR TwoArray: TwoTYPE;
BEGIN
END Test.
```

The syntax bug may even be easy enough for the compiler to make a temporary fix:

```
TYPE TwoTYPE = ARRAY [1..5], [1..10] OF INTEGER;
  - - Replaced array bound [1..5, 1..5] with [1..5], [1..10]
```

Run-time array errors generally derive from subscripts whose type is correct (so they pass the compile-time check), but whose value during program execution falls outside the legal range of the array. For example, suppose that you have been assured that the letter "T" can be found in a twenty-component array. This loop searches for it:

the ´off-the-end´ bug

```
i := 1;
LOOP
    (**)IF (TheArray[i] = "T") THEN EXIT END;
    i := i + 1
END;
IO.StringOut ("The T is in component  "); IO.IntOut (i, 1)
```

This is all well and good if the "T" is really there. But if it's not, we eventually try to inspect TheArray[21]—a component that doesn't exist.

test subscripts

> Always, but always, make sure that your intended loop bound will guarantee timely termination. A count bound on the subscript value is almost always appropriate.

The corrected segment below still contains a bug. To spot it, jot down the loop's postcondition:

the ´spurious location´ bug

```
i := 1;
LOOP
    IF (TheArray[i] = "T") OR (i = 20) THEN EXIT END;
    i := i + 1
END;
IO.StringOut ("The T is in component  "); IO.IntOut (i, 1)
```

Despite the fact that we've prepared to leave the search if the "T" wasn't there, we blithely announce that it is in component 20 if and when we *do* bail out. The shaded line must be replaced with a post-loop check:

look after you loop

```
(*Postcondition: If TheArray[i] doesn't hold a "T", no component does.*)
IF TheArray[i] = "T"
    THEN IO.StringOut ("The T is in component  "); IO.IntOut (i, 1)
    ELSE IO.StringOut ("The T wasn't there.")
END
```

Out-of-range subscripts also come up in two-dimensional arrays. If both subscripts have the same type, say, INTEGER, then misordered subscripts will pass muster at compile-time. They can even work at run-time as long as a search doesn't delve too deeply. For example, suppose that an array has five rows and ten columns. Normally, we'd refer to a component as:

```
(*A correct application.*)
AnArrayVariable [Row, Column] where Row ≤ 5 and Column ≤ 10
```

However, as long as the current value of Column stays at five or less, this reference is going to be correct, too:

```
(*This isn't necessarily incorrect.*)
AnArrayVariable [Column, Row]
```

This kind of bug is tough because it's intermittent—the code works until you really push the array to its limits. A useful debugging technique is to embed *subscript printers* in your code just before an array reference:

embedded subscript printers

```
IF DEBUGGING THEN
    IO.StringOut ("At point 9. Subscripts (row, column) are:");
    IO.IntOut (Row, 3); IO.IntOut (Column, 3); IO.LineOut
END;
AnArrayVariable [Row, Column] has something done to it
```

> Fixing bugs is easy—it's finding them that's hard. Before you decide where a bug *isn't*, find out where you are.

Subscript printing code is exceptionally useful when subscripts are being calculated on the fly. Such calculations are susceptible to off-by-one errors of the simplest sort. For example, what are the eight components that surround SomeArray[i, j]? Are they:

neighbors of i,j?

above:	[i–1, j–1]	[i–1, j]	[i–1, j+1]
alongside:	[i, j–1]		[i, j+1]
below:	[i+1, j–1]	[i+1, j]	[i+1, j+1]

Not necessarily! What if the component is on an edge of the array? Not every component *has* eight neighbors; in fact, a corner component has only three.

Open Array Parameters

Open array parameters are exceptionally useful for writing general-purpose code, but you must watch out for two all-purpose bugs. They arise when you forget either of two facts:

1. An open array parameter's lower subscript bound begins with zero.
2. The HIGH function represents the number of components in the array *minus one*.

Suppose that I define an array type like this:

```
TYPE DataTYPE = ARRAY [1..LIMIT] OF CHAR;
```

then pass it to an open array parameter:

```
PROCEDURE DoStuff (Data: ARRAY OF CHAR);
```

Here are the two bugs in their native habitat. I'm trying to initialize the array to hold only "Z"s.

more off-by-one errors

```
FOR i := 1 TO HIGH(Data) DO Data[i] := "Z" END;
  (*Oops—we've ignored Data[0].*)
FOR i := 0 TO LIMIT DO Data[i] := "Z" END;
  (*Oops—we've gone past the end of the array.*)
```

The corrected code segments (they're identical in effect) are:

```
FOR i := 0 TO HIGH(Data) DO Data[i] := "Z" END;
FOR i := 0 TO LIMIT–1 DO Data[i] := "Z" END;
```

Don't forget that on some older Modula-2 systems, HIGH may return a CARDINAL value. Make the call like this: INTEGER(HIGH (*argument*)) to temporarily treat it as an INTEGER (and don't forget that the argument should be in the legal range of positive INTEGER).

Semantic Bugs

You will find, as time goes on, that the kind of bugs described above are the ones that you're happy to have, because the compiler helps you find them. Semantic bugs are harder to deal with, because you may not even be aware that they're occurring.

By definition, you are afflicted with semantic bugs when a program runs, but arrives at the wrong conclusion. Testing is the only way to find semantic bugs—the choice is up to you whether you run the test, or a grader does! Recall that since testing can only show the presence of bugs, and remains silent about their absence, you have to be pretty aggressive in ferreting out potential errors.

Since array manipulation usually involves loops, semantic problems with arrays tend to arise from improper loop formulations. For example, off-by-one errors are common in shifting data within an array, or between two arrays. Suppose that we have this array:

what we have

T	e	s	t	–	–	–	–	–

and want to copy the first Length components (1, 2, and 3 if Length equals 3) into the last Length components.

what we want

T	e	s	t	–	–	T	e	s

Here's an incorrect attempt at making the shift:

```
(*An incorrect shift.*)
FOR i := 1 TO Length DO
    AnArray [LIMIT – i] + 1 := AnArray[i]
END
```

Can you see the problem? Suppose that Length is 3 and LIMIT is 9. Then, we're moving components 1, 2, and 3 into components 9, 8, and 7.

what we get!

T	e	s	t	–	–	s	e	T

The corrected assignment would be:

```
(*The corrected shift.*)
FOR i := 1 TO Length DO
    AnArray [(LIMIT – Length) + i] := AnArray[i]
END
```

My advice is to remember that, first and foremost, you're writing a loop to generate subscript values. Any array manipulations that take place are, in a way, secondary applications of the stream of subscripts. Testing the loop can be best simplified by leaving the array out of the picture entirely—test the loop by printing the subscripts it generates:

testing subscript streams

```
(*Testing subscripts in isolation.*)
FOR i := 1 TO Length DO
    IO.StringOut ("To:  ");    IO.IntOut ((LIMIT – Length) + i , 1);
    IO.StringOut ("    From:  ");    IO.IntOut (i , 1);
    IO.LineOut
END
```

```
To: 7   From: 1
To: 8   From: 2
To: 9   From: 3
```

This is a very common sort of problem, so I'll debug a second faulty example. Suppose that we want to reverse the contents of an array, so that the last component is moved to the first and so on. This code segment is incorrect:

a faulty reversal

```
(*Tries to reverse an array, but really leaves it in its starting configuration.*)
FOR i := 1 TO LIMIT DO
    Temp := TheArray[i];
    TheArray[i] := TheArray[LIMIT];
    TheArray[ (LIMIT –i) + 1 ] := Temp
END
```

Why? Well, let's imagine once again that the array manipulation loop's purpose is to generate a stream of subscript values. The goal of the incorrect loop above is really:

Incorrect goal: Generate every subscript from 1 *through* LIMIT.

But what we actually want is a little bit different:

Correct goal: Generate all LIMIT **DIV** 2 *pairs of subscripts.*

As before, we can try to debug the code by printing the stream:

another subscript printer

```
(*Assume that LIMIT = 10.*)
IO.StringOut ("Generating pairs of subscripts for an array reversal");
IO.LineOut;
FOR i := 1 TO LIMIT DIV 2 DO
    IO.IntOut (i, 3);
    IO.IntOut ( (LIMIT – i) + 1, 3); IO.StringOut (",      ")
END;
```

```
 1 10,      2  9,      3  8,      4  7,      5  6,
```

If we're satisfied with it, the shaded parts can come out, and the actual shift of data in the array can go in.

Modula-2 Details

Syntax Summary

- array: a structured type that contains *components* of any one type. The type must be defined before variables can be declared:

```
TYPE AnArrayTYPE = ARRAY [1..N] OF INTEGER;
      TwoDimensionTYPE = ARRAY [1..N], ["A".."Z"] OF CHAR;
VAR OneArray: AnArrayTYPE;
     TwoArray: TwoDimensionTYPE;
```

- bounds: in the array type definition, they establish the array's size and the form of its subscripts.

[1..10]	*bounds given with literal values*
[LOWER..LIMIT]	*bounds given with constants*
CHAR	*bounds given by an ordinal type name*
[*bound*], [*bound*]	*bounds of a multi-dimensional array*

- subscripts: help construct the name of an array variable by identifying a particular component:

  OneArray[1] *is an* INTEGER *variable*
  TwoArray [1], ["A"] *or* TwoArray [1, "A"] is a CHAR *variable*

- components: can be of any type, including an array type.
- open array parameters: are one-dimensional array parameters whose bounds are not fixed:

  PROCEDURE OpenArray (Parameter: **ARRAY OF** CHAR);
  (*Parameter'*s implied bounds are* [0..HIGH(Parameter)]*)

- Open arrays can be either value or variable parameters.
- HIGH(*open array parameter*) is a standard function that returns the length (number of components) of an open array parameter, minus one. On some systems HIGH may return a CARDINAL rather than INTEGER result.

Modula-2 Details

Remember . . .

- If it makes a program clearer, definitions and declarations may put into separate clusters, as long as all definitions and declarations are made before the program's statement part.
- Arrays are random-access types because components can be stored or inspected in any order.
- The values that set array bounds can't be variables, but they can be constants, or other simple expressions that can be figured out without running the program. If an ordinal type name is used to set bounds, they are automatically set to the first and last value of the type.
- An array-to-array assignment gives values to all of an array's components at once. Such assignments can only be made between arrays that were declared with the same type identifier. Open array parameters cannot be used in array-to-array assignments.
- An array component's identifier may be complicated by brackets and subscripts, but ultimately, it's just the name of a variable or value. Subscripts can be expressions of any kind; they have to be properly ordered, and within the legal range of the array's bounds, though.
- Auxiliary variables can be used as *pointers* to particular array components. They're useful in *divide and conquer* problem solving methods that split arrays into 'interesting' and 'uninteresting' segments.
- *Object-oriented design* is an alternative to the *goal-oriented* design we've used thus far. It begins by asking how a program uses data, rather than starting

by trying to break the problem down into goal-oriented procedures. We'll generally use object-oriented design to build the tools that make goal-oriented programming easier.

- *Data abstraction* is one technique of object-oriented design. A data abstraction outlines a type or structure in terms of the *primitive* operations it must allow, rather than the Modula-2 type definitions it should be based on. The primitives usually include only operations that can't be implemented without knowing details of the underlying definition.
- The ASCII NUL character is 0C, which has type CHAR. It's often used to initialize arrays of CHAR. Unlike the space character, 0C has no external character representation at all—it doesn't print anything.
- Some advice on arrays: use Load and Dump procedures to initialize and print their contents for testing purposes, check intermediate results when operations are complicated, and embed *subscript printers* to check subscript values before you rely on them.
- Some advice on dealing with larger programs: build them in incremental stages, use *test harnesses* to try out individual procedures, test with scale models, and embed debugging information in test data.

Zen Study Guide Big Ideas

The Big Ideas in this chapter are:

1. *Every structured type requires rules for specifying the types, and constructing the names, of stored values.* The specific rules of each type differ, but their purpose is the same: to let values be stored and retrieved without confusing the compiler.
2. *Arrays have as much information in their structure as they do in their stored values.* Information is built into arrays in various ways: by the kinds of subscripts we use to define them, by the components we choose to store values in, and by the position of stored values relative to each other. A well-designed program can use this built-in information to simplify other parts of its algorithm.
3. *Data abstraction separates the use of a data-handling tool from its implementation.* A data abstraction's primitives act as a performance specification (of what the abstraction can do), and as a requirements specification (of what its eventual implementation must allow).
4. *Object-oriented design is an alternative to, and not a replacement for, goal-oriented design.* But as new types are introduced, it's not surprising that many of the problems we'll see are most easily solved by focusing on data handling before we worry about refining complicated algorithms.

Zen Study Guide Questions

8-1 *How many components does this array have?* Fortunately, if you can answer one such question, you can probably answer them all. Make up one inordinately complex definition, then ask how many components it contains.

8-2 To get anywhere with arrays, you have to know how to access stored values. To test somebody's understanding of access, you have to give her the chance to answer one these questions incorrectly:

a) Do the subscripts have the right type?

b) Are they in the proper range of values?

c) Are they in the right order?

d) Is the type of a component correct? What is it?

e) Has a type name accidentally appeared in place of a variable identifier?

Take one of the groups of array type definitions from early in the chapter, and use it to create a series of *Is this statement legal?* questions.

8-3 Understanding the standard function HIGH is absolutely essential. The two sometimes confusing details of HIGH are *a*) that it treats the array's range as INTEGER, regardless of its actual type, and *b*) that the lower limit is considered to be zero, regardless of its actual value. Write two *Where's the bug?* problems that present procedures that use HIGH incorrectly. (As always, on some systems HIGH may return CARDINAL rather than INTEGER.)

8-4 Traversal is an array topic that you can and must have down cold. There are several ways to pose traversal problems:

a) *Initialize this array*, perhaps so that its contents match its subscripts.

b) *Print an array's contents in this pattern*, perhaps in reverse, or in rows twice as long.

c) *Move this array's contents to that array*, again, perhaps changing relative ordering, or perhaps working from arrays with different dimensions.

d) *Merge these arrays*, perhaps interleaving two short arrays into one long array.

Make up three traversal problems you might encounter on an exam.

8-5 Looking for a sentinel within an array is a pretty cut and dried problem. It's easy to make the same question much more difficult by looking for a sequence of values within an array. Naturally, the sought-for sequence will be held in an array as well.

The simplest version of the problem is: *Are the components in these two arrays identical?* A slightly more complicated version is *Does one array match the other array in reverse?* Come up with two more variations on the problem.

8-6 A harder version of the above problem assumes that one array is shorter than the other. Two ways of phrasing questions are *Can this short array's contents be found within this long array?* and *Can this short array's contents be found in this two-dimensional array?* Again, come up with two variations on this theme.

8-7 If a practice question has been *Do the values stored in this array increase?* as you work from left to right, it's a reasonable bet that the exam question will be *Do the values in this array decrease or stay the same?* as you travel the same route. Make up three more questions of this ilk.

8-8 Some problems can't be solved without arrays, while others needn't be solved with them. For example:

a) Find the (alphabetically) least word in a series of words.

b) Calculate the sum of a series of numbers.

c) Find the second largest number in a sequence of input numbers.

d) Read exactly two hundred numbers and print them in decreasing order.

e) Read exactly two hundred numbers and print the ones that fall within a specific range.

Problems *a* and *d* need arrays for solution, while the rest don't.

Make up another half-dozen questions of this sort. Try to choose examples that appear to require arrays but really don't, and vice versa.

8-9 The section titled *A Few Definition Exercises* should give you an idea of how a type definition can be made progressively more complicated. Make up five word problems that require a type definition for solution; for instance, *Define a type that might be used to represent the results of a track meet.*

8-10 It's not hard to transfer problems from one realm (say, the line by line processing of text) to another (say, the row by row processing of an array of CHAR). Nearly any problem that's drawn from one topic can be transferred to the other.

Go back to Chapter 6. Find your favorite text problems: three process filters, and three state filters. Rephrase them so that they involve arrays. (If you liked this problem, you'll love the next one!)

8-11 Transforming data from one representation to another is an especially confusing proposition. One example we worked in the text (program FileRev), was built around the problem of moving lines of text into rows of an array, and then back into lines of text again.

Working from the problems you developed in the last question, make up two problems that are confusing because they require you to get values from one kind of representation (like an array), and transfer them to another kind (like the output stream). You can complicate the questions by requiring that a process filter of some sort change the values as they jump the bridge.

8-12 To be really quick with arrays, it's good to know at least two ways to attack a problem. Suppose I state this question:

> A two-dimensional array of CHAR has one row that contains only X's. Which row is it?

There are a number of ways to approach a solution. One not especially good version is:

```
FOR every row in the array
    set a FoundIt variable to TRUE;
    FOR every column in a row
        IF the component stored there isn't an X
            THEN set FoundIt to FALSE
    IF FoundIt is still TRUE, print that row number
```

My pseudocode is poor because it goes through the whole array—even after the "X"-filled row has been found.

Make up three *Rewrite this code segment* problems. An easy way to get by on this is to start out with the good version!

8-13 Any easy array problem can be made more difficult by the simple expedient of enclosing it within a bigger array. For example, knowing that you have mastered *Are all the values in this one-dimensional array identical?* the crafty instructor instead asks *Find the row of this two-dimensional array whose values are all identical*, while the truly wicked teacher asks *Find the column whose components increase in value*.

I'm sure that you can be at least as crafty, if not as truly wicked, as a teacher. Dress up five one-dimensional array problems in two-dimensional clothing.

Exercises

Warmups and Test Programs

8-14 What is the largest array you can define? Can you define an **ARRAY** INTEGER **OF** anything? Does the type of the array's components (BOOLEAN, INTEGER, or REAL) change the maximum legal array size? What if the components are structured themselves?

8-15 What error messages do you garner for:

a) trying to inspect a component that's beyond the end of an array;

b) misordering the subscripts of a two-dimensional array;

c) leaving out one subscript of a two-dimensional array;

d) trying to assign a simple value to an array variable;

e) defining an array **FO** CHAR;

f) using parentheses instead of brackets in a definition?

8-16 Saying that arrays are random-access implies that the time-cost of access to every array component is equal. Is this true? Is a multi-dimensional array more expensive than a one-dimensional array? Can you compare the cost of reading one thousand items from an array to that of reading them from an input file?

8-17 Recall that the difference between value parameters and variable parameters is that a value parameter is an independent copy of its argument, while a variable parameter simply renames its argument.

How long does the copying process take? Define a few very large arrays, then compare the elapsed time of a procedure call that involves a value parameter, one that involves a variable parameter, and an initializing assignment to the array without any procedure call at all.

8-18 Assuming that you have suitably defined arrays, write code segments that initialize one-dimensional arrays to:

a) The numbers 1 through 10 (and in reverse).

b) The letters "a" through "z" (and in reverse).

c) The numbers 1.0 through 10.0 (and in reverse).

d) The odd numbers from 3 through 17 (and in reverse).

8-19 Assuming that you have a suitable two-dimensional array, initialize it in these ways:

```
 1  2  3  4        1  5  9 13
 5  6  7  8        2  6 10 14
 9 10 11 12        3  7 11 15
13 14 15 16        4  8 12 16

 4  3  2  1        4  8 12 16
 8  7  6  5        3  7 11 15
12 11 10  9        2  6 10 14
16 15 14 13        1  5  9 13
```

8-20 Outline procedures that find a value in *a)* a one-dimensional array, *b)* the same array working backward, *c)* a two-dimensional array, *d)* the same array but searching columns first, *e)* the same array but searching columns backward first.

8-21 A spelling checker requires that you store up to one hundred 10-character words. Define a suitable type for storing them.

8-22 Write a program that echoes text, but only prints every other instance of each letter.

8-23 Write a program that interleaves two arrays, component by component, into one array that's twice as long.

8-24 Write a program that finds the *n* largest values, in order, in an array of INTEGERs. Hint: as you find and print each 'largest' value, figure out a way to get rid of it.

Exercises

Programming Problems

8-25 The **Hum** package is a collection of programs, written by Bill Tuthill, that are useful for work in the *hum*anities. They count and rearrange text in ways that make it easier to study.

One interesting command that had never occurred to me was **`wheel`**. It 'rolls' through the text, printing *n* words at a time. For example:

```
Here is a
is a three
a three word
three word example          etc.
```

Write a program that wheels through a file in groups that can range, at the user's option, from one to ten words per line.

8-26 An *anagram* is produced by rearranging the letters of one word or phrase to produce another. Write a program that determines if two sequences (of up to ten letters each) are anagrams. Test it on *cat/act*, and *scare/races/cares/acres*.

8-27 A *palindrome* is a series of letters that spell the same words both forward and backward, as Napolean's lament 'Able was I ere I saw Elba.' A *numerical palindrome* is a bit simpler—it's just a number that's the same front to back and back to front. 1234321 is a numerical palindrome, but 123431 isn't.

Numerical palindromes can usually be generated from arbitrary INTEGERs with an easy algorithm: reverse the number, then sum the original number and its reversal. It may take many such steps, but a palindrome generally results: 561+165=726, 726+627=1,353, 1,353+3,531=4,884, which is palindromic. There are some notable exceptions to this—196 is the smallest—from which a palindrome *never* results, however.

Write a program that generates palindromic numbers up to fifty digits long. Announce the number of steps required, and be sure to check for, and halt before, overflow (just in case you try one of the 5,996 numbers below 100,000 that do not appear to be palindromic no matter how long you persist).

8-28 'I didn't know whether I should laugh or cry' is, in a way, a colloquial description of the mathematical phenomenon known as a *saddle point*. A saddle point is simultaneously a high point and a low point; its name comes from the kind of saddles that horses wear. The saddle point is found right where you sit: at the high point of the side-to-side, stirrup to stirrup curve, and the low point of the front-to-back, pommel (or horn) to cantle curve.

We can spot a saddle point in some two-dimensional arrays, as well. For example, component 3,3 of this array is a saddle point, because the smallest value in a row and the largest component in a column occupy the same position.

```
3 4 1 6 5 9
1 7 2 4 2 1
8 9 4 5 6 8
5 3 3 3 9 5
6 2 1 1 1 6
4 2 2 8 7 4
```

Write a program that is able to locate a saddle point in an $n \times n$ square array. Can you search for a largest column component/smallest row component with the same code? (Hint: rotate the array.) Assume that there is only one saddle point in the array. Happy trails!

8-29 Computers are terrific at doing arithmetic. Unfortunately, the numerical data types that are built into most programming languages strictly limit the size of numbers. It's a rare Modula-2 system that even allows dozen-digit INTEGER values.

Fortunately, it isn't very hard to build tools that will allow operations on numbers that are hundreds, or even thousands, of digits long. Design and implement a data abstraction that supports 100-digit whole numbers. At a minimum, you should supply primitives for input, output, and addition. Naturally, you can add additional primitives (subtraction, multiplication, and so on) if you like.

8-30 Suppose we create an array that holds a relatively random sequence of characters, say:

```
A  F  K  Z  M  L  B  Q  R  C  J
```

We can test an individual's short-term memory by printing the characters, then blanking the screen. The test subject is asked to try to reenter the correct sequence.

Have your program test and report on these measures of recall:

a) The number of letters entered that appeared somewhere in the list.

b) The number of letters entered in their proper positions.

c) An 'analog' score that smoothly penalizes out-of-position guesses. For each letter that is in the list, add 1 to the score if it's in the correct position, 1/2 if it's off by one, 1/3 if off by two, and so on.

Do your results agree with published studies that indicate that most people's short-term memories can hold seven, plus or minus two, values? Does presenting sequences of random length (rather than progressively longer) have any effect on the subject's ability to recall?

8-31 It's not hard to imagine using an array to represent a checkerboard. But can you imagine playing a game of checkers on it? Or, more to the point, can you imagine a checkerboard as a data abstraction, complete with primitive operations?

Plan and implement a set of procedures that support operations on a checkerboard variable. Include primitives for:

a) *initialization*, to position checkers at the start of play;

b) *inspection*, to print the current checkerboard; and

c) *input* and *assignment*, to make moves, but only if they're legal.

Naturally, you'll have to specify an input format for moves. Some bonus enhancements to consider are:

d) Detect a move that causes the moved piece to become endangered.

e) Detect a move that causes *any* piece to become endangered.

f) Take a piece automatically.

What additional primitives will this automatic playmaking require?

8-32 DNA molecules are comprised of triples that consist of four different organic bases—adenine, cytosine, guanine, and thymine, usually abbreviated as A, C, G, and T. Although the entire molecule can be incredibly long, a single missing base can have catastrophic effects. For example, one particular rat gene associated with hereditary diabetes differs from the normal 1,000-base sequence by just one missing G residue:

normal sequence	GGA	AGC	GGA	GGC	CGC
diabetic sequence	GGA	AGC	GAG	GCC	GCT

The resulting *frame shift mutation* is apparent in the third triple.

The problem of spotting off-by-one errors of this sort has interesting applications in text processing as well; e.g. in comparing two files that differ by only one line. For now, though, just write a program that is able to find the location of a frame shift mutation in a DNA base sequence. Be sure to confirm that the molecules differ only by this one base, and do not diverge entirely at the point of departure.

8-33 Many maps include a table of distances between points. The table is basically a square two-dimensional array, subscripted on each axis by a list of cities, with each component showing a distance. The distance between a city and itself, of course, is zero.

Write a program that implements a table of distances between cities. Let a program user employ it to:

a) Find the distance between any two cities.

b) Find the cities closest to, and furthest from, any city.

c) Find the total distance on a trip between any sequence of cities.

d) Find out whether or not a journey between two cities can include a detour, to a third city, that will not extend the total trip length by more than *x*%. The user should only enter the first and final locations.

8-34 Computer modeling can show that events that appear catastrophic at close range can be quite stable, and recur naturally in long-term cycles. For example, forest fires are usually viewed as disastrous, yet periodic, naturally occurring (e.g. by lightning) fires are an essential part of healthy forest ecology.

An interesting variation on the theme involves competition between species through *infestation*. For example, the spruce budworm is an unpleasant little beastie that defoliates balsam and spruce trees. Unfortunately for the budworm, these trees are replaced by beeches, which they find indigestible. The beeches, in turn, are eventually invaded and displaced by spruces and balsams, and the cycle starts again.

A simpler model of infestation can be set up with just three rules:

a) An infested area becomes defoliated next year.

b) A defoliated area becomes green next year.

c) An infested site infects its neighbors to the north, south, east, and west next year, if they are currently green.

Write a program that tracks, and displays, a sequence of generations. Assuming a green forest to begin with, use a variety of patterns to initialize the model (at the center of the forest), including:

```
iiii          gig          dgd
dddd          gdg          igi
```

where i means infested, g means green, and d means defoliated. Note that when a focus, or center-point, of infestation is known, intentional defoliation can be used to break the cycle. Perhaps you should try to make your program run backwards as well!

8-35 You've probably seen electron micrographs of incredibly twisted and tangled strands of deoxyribonucleic acid, better known as DNA. But have you ever paused to wonder about the statistical properties of the path a typical strand takes? In some sense, the end of a strand simply takes a random walk, much like the random Brownian motion of a gas molecule. However, the strand is limited by *steric hindrance*. In English, this means (at the very least) that an effective model of a DNA molecule's path should not be allowed to pass through itself.

The result of this restriction is that, while differential equations can be used to analyze and mathematically simulate Brownian motion and other random walks, a *self-avoiding* random walk must be simulated by computer.

Write a program that simulates 50 steps of *a*) a random walk, and *b*) a self-avoiding random walk. For convenience, assume that up to eight moves can be made from any point—a diagonal step takes the same amount of time as a vertical or horizontal one. Start in the center of a 101×101 (i.e. –50..50) array each trial. Run at least 100 trials of each walk, and plot a histogram of the final *x*-axis positions.

8-36 Write a program that lets two players play the game of *Nim*. One version of this game starts out with three rows of pebbles:

```
o o o o o o o
  o o o o o
    o o o
```

On each turn a player may remove one or more pebbles, but can only take them from a single row. The winner is the player who removes the last pebble. Can you use binary arithmetic to figure out a winning strategy? The secret is in the columns!

For a harder problem, expand your program to let the computer play one side. Under what conditions can it always win? Can changing the number of rows or pebbles lead to a disadvantage?

8-37 Topographical maps are easily represented with two-dimensional arrays of INTEGER or REAL-valued heights. However, the bird's-eye view they provide is sometimes difficult to appreciate after a day of lugging a 70-pound backpack.

a) Write a program that locates *peaks*. We define a peak as any point that is higher than its eight neighbors.

b) Write a program that locates *valleys*. A valley is any series of three points along a row or column that are lower than their twelve neighbors. (Can you deal with longer valleys as well?)

'As always, multiple bound possibilities lead to an ambiguous postcondition.'
(p. 367)

Chapter 9 Searching and Sorting: An Excursion into Algorithms

Programs carry out algorithms, but developing algorithms is something we've paid relatively little attention to. We've spent most of our time learning the general ideas of programming, and the specific facts of Modula-2. We haven't worried much about looking at alternative methods of carrying out any single task.

But this book isn't just about programming in Modula-2; it's about computer science, too. The study of algorithms, quite independent of any language they might be written down in, or even any specific purpose they might serve, is one of the foundations of the field. And within the arena of algorithms, two particular applications stand out: searching and sorting.

Searching is looking for things. After all, what good is a computer that can hold a million pages of data if you can't find what you need? The basic problem is simple enough to state—find a value that's stored in an array. But we'll find ourselves writing five different procedures to do the job.

Sorting is arranging things. We don't just sort for the sake of sorting, of course—this is computer science, not flower arranging. Instead, an awful lot of sorting gets done today in order to make searching easier tomorrow. We'll attack the problem in three different ways, and try to see why we might choose one approach over another.

(The primitives developed in this chapter can be found in the sorts and searches libraries on the accompanying disk.)

Section 9-1 Searching

IF COUNTING THINGS IS A MAJOR THEME OF programming, then searching for them is one of the primary minor chords. There so much to say about something that seems so simple because nothing can be taken for granted. No decision can be made—How will the data be stored? How will additional items be saved? How will we look for values?—without affecting the others.

Searching *algorithms* are usually described in terms of a pretty basic formulation of the problem: trying to find a number that's supposed to be stored in an array of numbers. Although the data that has to be searched might start out somewhere else (in a file, perhaps), it can usually be put into an array. And even though we might not be looking for a number per se, the problem of trying to find a name, or an identification code, or a value of some other type usually turns out to be equivalent to the hunt for an ordinary INTEGER.

Nothing could be more goal-oriented than an algorithm, but I'll continue using some techniques of object-oriented design to help focus our discussion. The purpose of a search is to find the position of a particular stored value. Why don't I make a search procedure one of the primitive operators of a new data abstraction? I'll change the way it's implemented, but not the way it's used. This will let us compare different implementations within a constant context. Some of the other primitives the data abstraction might need include:

the data abstraction

```
PROCEDURE Initialize (VAR Data: ARRAY OF INTEGER;
                          Init: INTEGER);
  (*Initializes Data to Init.*)
PROCEDURE GetNumbers (VAR Data: ARRAY OF INTEGER);
  (*Loads Data with test data.*)
PROCEDURE Sort (VAR Data: ARRAY OF INTEGER);
  (*Sorts the array, if necessary. We'll be writing this one later.*)
PROCEDURE SearchMethod (Data: ARRAY OF INTEGER;
                  Value: INTEGER; VAR Position: INTEGER);
  (*Sets Position to the location of Value in Data, or –1 if absent.*)
```

We'll write five (!) versions of *SearchMethod*:

alternative search implementations

Search *is the basic component-by-component search of an unordered array.*
StateSearch *is a state-oriented version of the* Position *algorithm.*
Linear *makes a component-by-component search through an ordered array.*
Quadratic *is like* Linear, *but it takes bigger jumps.*
Binary *uses a divide and conquer algorithm for the fastest search of all.*

I'll use open array parameters with all these procedures to help emphasize their general-purpose nature. In section 2 of this chapter, we'll be looking at three alternative algorithms for an implementation of procedure Sort.

Searching Unordered Arrays

The first variation of the searching problem assumes that an array—call it Numbers—isn't in any special order. Our task is to find the position of a particular value, say, 7, or to report that the value wasn't found. We set the stage with:

```
Initialize (Numbers, MAX(INTEGER));
  (*Postcondition: Numbers only holds a number we'll never look for.*)
GetNumbers (Numbers);
  (*Postcondition: Numbers is defined, but isn't necessarily ordered.*)
SearchMethod (Numbers, 7, Position);
  (*Postcondition: If Position isn't -1, it holds the 7's position in Numbers.*)
```

the ground rules

using postconditions

> Note the use of postconditions. Each one clearly documents any assumptions that subsequent statements may rely on.

How will the search proceed? Well, we hope to find a sentinel—the stored number 7. However, since the sentinel won't always be found, the sentinel alone won't guarantee loop termination. A necessary bound—we've inspected every component—has to be part of any searching loop. In outline, the search loop is:

an unordered search specification

Goal: Locate a value, or decide it isn't there.
Intentional bound: Spotting the value.
Necessary bound: We've reached the last component.
Plan: Advance to the next component.

As always, the multiple bound possibilities lead to an ambiguous postcondition. The search loop is followed by an **IF** statement that decides which of the bound conditions caused the loop to terminate.

first approach to searching

```
PROCEDURE Search (Data: ARRAY OF INTEGER;
                        Value: INTEGER; VAR Position: INTEGER);
  (*Search procedure for an unordered array. Position is -1 for absent Value.*)
  VAR i: INTEGER;  (*the current array component*)
  BEGIN
    i := 0;  (*Open array parameter bounds are [0..HIGH(Data)].*)
    LOOP
      (**)IF (Data[i]=Value) OR (i=HIGH(Data)) THEN EXIT END;
      i := i + 1
    END;  (*LOOP*)
    (*Postcondition: If Data[i] isn't Value, then no component is.*)
    IF Data[i] = Value
      THEN Position := i
      ELSE Position := -1
    END (*IF*)
  END Search;
```

A State-Oriented Version

I pointed out that Position's central loop leaves us in an ambiguous situation on **EXIT**. We have to check the contents of Data[i] to decide whether or not we found Value. However, we're perfectly capable of writing the loop unambiguously—if you're willing to not be frightened by the prospect of using a *state-oriented* approach.

In the past, we've written state-oriented programs as *filters* that processed input data. You'll recall that a state filter spots and records or acts upon *state transitions*—changes in our knowledge about a program's data.

However, we can spot state transitions when we look at data stored in an array, too. What states can exist during an array search? How about:

states can name bound conditions

- found—the sentinel bound was satisfied because the value was found.
- absent—only the count bound was satisfied.
- searching—neither of the bounds have been satisfied yet.

In effect, the state values name the loop's bound conditions. The possibility of a state transition looms every time we look at a new component:

loop specification

Goal: Locate a value, or decide it isn't there.
Bound: Our state is either absent *or* found.
Plan: Advance to the next component.
Update the state.

I've rewritten the search code using the state-oriented approach below.

a state-oriented search

```
PROCEDURE StateSearch (Data: ARRAY OF INTEGER;
                              Value: INTEGER; VAR Position: INTEGER);
  (*State-oriented search of an unordered array. Position is –1 for absent Value.*)
  TYPE StateTYPE = (found, absent, searching);
  VAR State: StateTYPE;
      i: INTEGER;
  BEGIN
    i := 0;  (*Open array parameter bounds are [0..HIGH(Data)].*)
    LOOP  (*until we're not searching anymore*)
      IF i > HIGH(Data)  (*Why is this state set first?*)
            THEN State := absent
        ELSIF Data[i] = Value
            THEN State := found
        ELSE i := i + 1;
              State := searching
      END; (*IF*)
      (**)IF State <> searching THEN EXIT END
    END; (*LOOP*)
    (*Postcondition: if Value was there, State = Found and Numbers[i] = Value.*)
    CASE State OF
        found: Position := i
      | absent: Position := –1
    END (*CASE*)
  END StateSearch;
```

In a simple application like this, naming states is simply a curiosity—in fact, it makes the program more complicated. But loops sometimes have **EXIT** conditions that are difficult to phrase, or which may be added to as time goes on. 'State bounds' clarify reasons for **EXIT**, and help ensure that the bound conditions are evaluated in the proper order.

Modification Problems

1. Write simple versions of procedures Initialize and GetNumbers. What is suitable test data for Position and StateSearch? Remember—you're trying to make the search procedures *fail*, not show that they can work.

2. Add another possibility to the state-oriented search—that the array was initialized to MAX(INTEGER), but it was never filled with data. If that's the case, have Position return the value MAX(INTEGER).

Self-Check Questions

Q. In procedure StateSearch, why is the absent state the first one updated? Why wouldn't this sequence work as well?

```
LOOP (*A version with a bug.*)
    IF Data[i] = Value
            THEN State := found
        ELSIF i > HIGH(Data)
            THEN State := absent
        ELSE i := i + 1;
              State := Searching
    END; (*IF*)
    (**)IF State <> searching THEN EXIT END
END; (*LOOP*)
```

A. The rewritten version above contains a potential bug. If Value isn't there, then we'll eventually try to inspect a component that's beyond the end of the array. The rewritten version *could* be made to work by changing the i > HIGH(Data) expression to i = HIGH(Data) instead.

Q. The *size* of a searching problem is usually described by the number of values that are actually inspected. This number is referred to as *n*. If *n* is big, then the search is a big problem; if *n* is small, the problem is small.

When an array isn't ordered, we might have to look at every single value to find the one we want. Another way of phrasing this is to say that *worst-case size* of the search is *n*. What are the *best-case* and *average-case* sizes of the problem—searching for a number in an unordered array? In other words, at best, and on average, how many of the *n* components must we look at before we find the number?

A. At best, the size of the problem is just 1. That's because the value we're seeking might be the very first one we inspect. In the average case, the number might be anywhere. On average, though, we'll find it after looking through about half, or *n*/2, of the components.

Note that if the number we want has a good chance of not being there at all, the average case gets worse and worse. Can you figure out why?

Searching Ordered Arrays

When the values in an array aren't ordered, inspecting each component until either the proper one is found, or none are left to inspect, is the only sure strategy for finding a value.

But when the component values *are* ordered, many approaches are more effective. We can explore three of them: linear, quadratic, and binary searches. Our ground rules will be:

the ground rules

```
Initialize (Numbers, MAX(INTEGER));
  (*Postcondition: Numbers only holds very big numbers we'll never look for.*)
GetNumbers (Numbers);
  (*Postcondition: Numbers is defined, but not necessarily ordered.*)
Sort (Numbers);
  (*Postcondition: Numbers is defined and ordered, smallest to largest.*)
SearchMethod (Numbers, 7, Position);
  (*Postcondition: if Position isn't -1, it holds the 7's position in Numbers.*)
```

As before, we'll always return a negative Position value if the number we're searching for isn't found.

Linear Search

I'll begin with a *linear search*. Since we're looking for a sentinel, the search of an ordered array is quite similar to the search of an unordered array. But can you guess why I had Initialize fill the array with very large values, rather than filling it with zeros, or very small numbers? I did it to guarantee that a value *that equals or exceeds* the sentinel will always be found. This little trick makes it easier to state the bound condition:

loop specification

Goal: Locate a value, or decide that it isn't there.
Bound: The current component equals or exceeds the sought value.
Plan: Advance to the next component.

Is there much difference between the algorithms of procedure Linear, below, and procedure Search, from a few pages back? No. There's just a tiny change in one condition:

```
(*from procedure Search*)
IF (Data[i] = Value) OR (i = HIGH(Data)) THEN EXIT END;
```

little changes have big effects

has been changed to:

```
(*in procedure Linear*)
IF (Data[i] >= Value) OR (i = HIGH(Data)) THEN EXIT END;
```

But however minuscule the change from = to >= might be, it's essential to remember the two key *preconditions* that make Linear work:

- any array components that aren't being used were initialized to very big numbers;
- the array was sorted before the search began.

They make the linear search of a *sorted* array more effective than the search of an *unsorted* array. The increased effectiveness shows up when the value isn't there:

advantage of Linear over Search

> On average, a linear search through an unsorted array will either find the sought value, or decide that it's not there, after looking at half the components. In contrast, an unsorted array that doesn't hold the value will always have to be searched all the way through.

Searching an unordered array can't be improved beyond Search, but Linear is just the first step in improving ordered array searches.

Note a small simplification in procedure Linear, below. Instead of declaring a local variable as the subscript counter, I'm using the variable parameter Position.

a linear search procedure

```
PROCEDURE Linear (Data: ARRAY OF INTEGER;
                  Value: INTEGER; VAR Position: INTEGER);
  (*Linear search for Value. Return negative Position if not found.*)
  BEGIN
    Position := 0;
    LOOP
      (**)IF (Data[Position] >= Value) OR
            (Position = HIGH(Data)) THEN EXIT END;
      Position := Position + 1
    END;
    (*Postcondition: If Data[Position] isn't Value, then no component is.*)
    IF Data[Position] <> Value THEN
      Position := -1
    END
  END Linear;
```

Modification Problems

1. Write a state-oriented version of Linear.
2. Modify Linear so that it searches from the high end of the array, instead of from the low end.

Quadratic Search

The linear search patiently trudges toward the sought value one component at a time. It's not hard to imagine that we would find the value a bit faster if we could take larger jumps:

> *take big jumps to get close to the sought value;*
> *take single steps to locate it exactly*

What's a reasonable size for the *big jumps*? Well, we'd like to strike a balance between the amount of ground covered by the big steps, and the number of single steps that might be left over. On one hand, a relatively small jump size wouldn't travel through the array quickly enough. On the other, a very large jump size could, in the worst case, leave a lot of single steps left over.

ramifications of some jump sizes

	Array size (n = . . .)			
Jump size	10	1,000	1,000,000	1,000,000
(1)	10	1,000	1,000,000	999,999
(*n/10*)	10	10	10	99,999
(*log_2 n*)	3	10	20	49,999
(*sqrt n*)	3	32	1,000	999
	Number of big jumps at worst for n = 10, 1,000, and 1,000,000			*Single steps at worst for n = 1,000,000*

Having the jump size remain constant, at 1, merely duplicates the linear search. Surprisingly, perhaps, making the jump very large—either a tenth of the total array size, or the base 2 logarithm of it—doesn't leave us in much better shape. Whatever we save by making big jumps at first is lost by the having to take many, many single steps in the final approach to the number.

However, if we make the jump size equal the square root of the array's length, the number of big jumps and single steps will be balanced—and, when they're balanced, the very worst case won't be so bad. At worst (as well as at best, *and* on average) the number of single steps will just about equal the number of big jumps. That's what happens for a million-component array in the table above. At worst, the value we're looking for is hidden in component 999,999. It will take us 999 big jumps, plus 999 single steps to get there—a much better performance than any of the other alternatives in the table.

balancing jumps and steps

> The *quadratic search* algorithm—quadratic means *square*—is based on a jump size that equals the square root of the number of components that must be searched.

I've used the state-oriented approach to write procedure Quadratic, below. Recall that the very rough outline of the search was:

take big jumps to get close to the sought value;
take single steps to locate it exactly.

A more detailed pseudocode outline is:

a state-oriented refinement

```
calculate the step size;
LOOP
    update states and Position;
    (**)IF State = CloseEnough THEN EXIT END
END of the big jump loop;
(*Postcondition: the value in Data[Position] is <= the number we want.*)
LOOP
    update states and position;
    (**)IF State <> searching THEN EXIT END
END of the single step loop;
(*Postcondition: If State = found then Data[Position] = Value.*)
set Position to –1 if State equals absent
```

Because of type mismatches, the *calculate the step size* isn't quite as easy as it might be. Calculating and rounding the square root requires that we import the MathLibØ library, then make a horrible assignment along the lines of:

a horrible assignment

```
(*Approximate the rounded square root of the number of components in Data.*)
JumpSize := TRUNC (MathLibØ.sqrt (FLOAT (HIGH(Data)+1)));
```

The completed procedure is shown on the next page.

quadratic search procedure

```
PROCEDURE Quadratic (Data: ARRAY OF INTEGER;
                     Value: INTEGER; VAR Position: INTEGER);
  (*Quadratic search for Value. Return negative Position if not found.*)
  TYPE STATUS = (searching, found, absent, CloseEnough);
  VAR State: STATUS;
      JumpSize: INTEGER;
  BEGIN
    (*Assume that MathLibØ has been imported.*)
    JumpSize := TRUNC (MathLibØ.sqrt (FLOAT (HIGH(Data)+1))) ;
    Position := 0;
    State := searching;  (*Initialize the current State.*)
    LOOP  (*by big jumps until we're CloseEnough*)
      IF (Position + JumpSize) > HIGH(Data)
          THEN State := CloseEnough  (*Or we'd pass the end of the array.*)
        ELSIF Data[Position+JumpSize] > Value
          THEN State := CloseEnough  (*Or we'd pass the sought value.*)
        ELSE Position := Position + JumpSize  (*State remains unchanged.*)
      END; (*IF*)
      (**)IF State = CloseEnough THEN EXIT END
    END; (*the big jump LOOP*)
    (*Postcondition:  Data[Position] <= Value.*)
    State := searching;  (*Reset the current State.*)
    LOOP  (*by single steps until we're not searching*)
      IF Position > HIGH(Data) THEN State := absent
        ELSIF Data[Position] > Value THEN State := absent
        ELSIF Data[Position] = Value THEN State := found
        ELSE Position := Position + 1  (*State is unchanged.*)
      END; (*IF*)
      (**)IF State <> searching THEN EXIT END
    END; (*the single-step LOOP*)
    (*Postcondition:  If Value is there, State = found and Data[Position] = Value.*)
    IF State = absent THEN
      Position := -1
    END
  END Quadratic;
```

Modification Problems

1. Build a test harness for Quadratic, then modify the procedure so that it prints the number of big and small steps it takes while searching.

2. As it's written, Quadratic's first loop doesn't distinguish between getting CloseEnough to the number we want, and actually finding it. Rewrite Quadratic so that found is one of the states that's set in the first loop.

3. As long as the size of the big jumps is fixed, we can't improve on the size (the square root of n) we've been using. However, you can search more quickly by changing the jump size—starting with big steps, then reducing them. Can you figure out a way of modifying Quadratic to do it? Hint: add a third loop.

Self-Check Questions

Q. You guessed it—just how good is the quadratic search algorithm in terms of n, the number of values there are to be searched? What's the best, worst, and average case?

A. As usual, the best case is easy to figure out—we spot the number we want on the first try. The best case is the constant 1.

The worst case starts to improve on the linear search. At worst, the value we want is in the next-to-last component. How many big jumps will it take to get 'close enough?' The square root of n, less one. How many little steps can there be left? Again, about the square root of n steps. The worst case size of the problem, then, is twice the square root of n.

By the same reasoning, we can see that the average case equals the square root of n. After all, on average there will be half that many big jumps, plus half that many single steps.

Binary Search

Our third and final method of searching an ordered array is the *binary search* algorithm. It's the fastest method we'll see. Binary search is quick because the jump size changes—it starts big, but gets smaller on each jump. The first jump is half the size of the array, the second jump is one-fourth of the array's length, the third jump is one-eighth, and so on.

Binary search is the basic *divide and conquer* algorithm. Here's how it works. Assuming, as usual, that the array is ordered, start by looking at the middle component. If the value it holds is too high, go to the middle of the bottom half of the array, and look again. If the middle component's stored value was too low, go to the middle of the top half instead. Then, repeat this 'split the remainder' step until we find the value we want, or decide that it doesn't exist.

How would you characterize the **EXIT** condition of a binary search loop? Instead of thinking about jumps, imagine what happens to the boundaries of the unknown portion of the array. At first, the entire array is unknown—its boundaries are the first and last components.

?	?	?	?	?	?	?	?	?	?	?	?	?

all of the stored values are unknown

After the first guess, though, we're able to set new lower and upper bounds on the unknown part that we're still concerned with:

moving the data bound

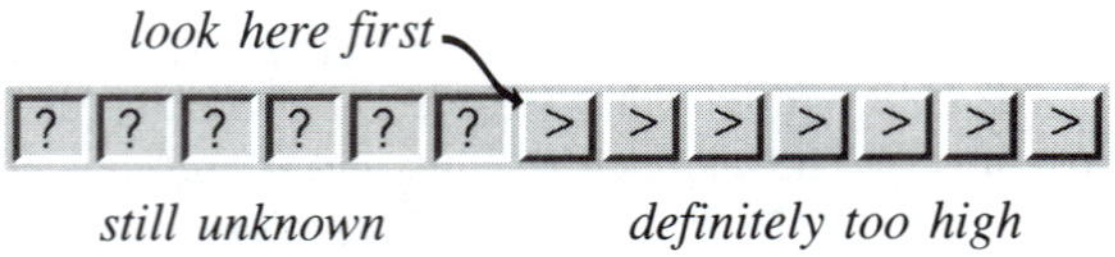

A second look, in the middle of the unknown area, shrinks the bounds even further:

Eventually—in fact, pretty quickly—we'll run out of unknown data. If we happen to find the value, the loop has an ordinary sentinel bound. If we don't find it the searching loop requires a data bound that guarantees termination—the lower and upper borders of the unknown area collide (just as they did when we checked palindromes in Chapter 8.

A loop specification for binary search would be:

loop specification

Goal: Locate a value, or decide it isn't there.
Intentional bound: We've found the value.
Necessary bound: The lower and upper bounds of our search coincide.
Plan: Pick a component midway between the lower and upper bounds.
Reset the lower or upper bound, as appropriate.

As always in the case of multiple loop bounds, we have to decide why the loop terminated. In pseudocode, a first refinement of the binary search algorithm is:

first pseudocode

initialize the starting upper and lower bounds to the array's end;
LOOP
  inspect the component in the middle of the unknown area;
  (**)**IF** *we find the value or run out of unknown data* **EXIT**;
  reset the boundaries of the unknown area
END *of the loop*;
decide why we left the loop

How do we locate the middle component? Well, no matter what the lower and upper bounds of the unknown portion are, we can always find the middle by adding the bounds and dividing by two. After each wrong guess, the old middle becomes the new lower or upper bound. In practice, we'll add or subtract 1 each time, since we've already checked the actual middle value. A second refinement would be:

pseudocode refinement

```
initialize Lower and Upper to the lower and upper array bounds;
LOOP
  let Middle equal (Lower plus Upper) DIV 2;
  (**)IF we find the value or run out of unknown data EXIT;
  IF the value of Data[Middle] is low
    THEN make Middle (plus 1) the new lower bound
    ELSE make Middle (minus 1) the new upper bound
END of the loop;
decide why we left the loop, and return an appropriate Position
```

The binary search is implemented below. As in the earlier searching examples, I return a negative Position if the sought Value isn't found. This time, though, I've dispensed with the state variable, and written the searching code as tersely as possible.

binary search procedure

```
PROCEDURE Binary (Data: ARRAY OF INTEGER;
                        Value: INTEGER; VAR Position: INTEGER);
  (*Binary search for Value. Returns Position = –1 if not found.*)
  VAR Middle, Lower, Upper: INTEGER;
  BEGIN
    Lower := 0;
    Upper := HIGH(Data);
    LOOP
      Middle := (Lower + Upper) DIV 2;
      (**)IF (Data[Middle] = Value) OR
                  (Lower > Upper) THEN EXIT END;
      IF Value < Data[Middle]
        THEN Upper := Middle – 1
        ELSE Lower := Middle + 1
      END (*IF*)
    END; (*LOOP*)
    (*Postcondition: If Data[Middle] isn't Value, no component is.*)
    IF Data[Middle] = Value
      THEN Position := Middle
      ELSE Position := –1
    END
  END Binary;
```

Modification Problems

1. Rewrite Binary as a function.
2. Off-by-one errors are common in binary search procedures. What test data would you use to find them? Embed enough debugging code in procedure Binary to let you spot and fix any reasonable problem.

Self-Check Questions

Q. As usual, what are the best, worst, and average cases in a problem of size n using procedure Binary?

A. As always, the best case takes just one step—the number we want is the first one we look at. In the worst and average cases, binary search is even better than quadratic search. Consider the worst case. Here are four different ways to state the problem (and, implicitly, the answer):

- How many times can we divide an array of length n in half before we're left with an array of length one?
- If a jump size starts at ½ of n, then goes to ¼, ⅛, and so on, how many steps will it take to work our way down to $1/n$?
- Starting with one, how many times can you double a number before it's larger than n?
- To what power should you raise 2 if you want it to be bigger than n?

In each case, the answer is the same—it's the logarithm, base 2, of n. This figure, written $log_2\, n$, is the worst-case size of the search.

Finally, what's the average case? On average, we have to look at half of the components. Surprisingly, perhaps, this takes just one less step than looking at them *all*. The average seach is $(log_2\, n) - 1$.

Section 9-2 Sorting

Computers end up spending a lot of time sorting—it's been estimated that up to a quarter of the time of centralized computers is devoted to sorting things. Why? Well, as our discussion of searching makes clear, it's an awful lot easier to find a value in an array that has been sorted than to search an array that's in no special order.

why sort?

> Computers sort data for the same reason the telephone company alphabetizes the phone book. Nobody needs to have the names in alphabetical order—it just makes it an awful lot easier to look one up. Wanting to have the names in alphabetical order is relatively unimportant; being able to look up one name very quickly is crucial.

Searching an unordered array might require that we look at every single component. Searching through an ordered array, as we've just seen, can be cut down to just $log_2\, n$ inspections.

We'll look at sorting algorithms as though we were considering alternative implementations of a Sort procedure. Recall that the other primitives of the 'sorted array' data abstraction were:

the data abstraction

```
PROCEDURE Initialize (VAR Data: ARRAY OF INTEGER;
  (*Initializes Data to Init.*)      Init: INTEGER);
PROCEDURE GetNumbers (VAR Data: ARRAY OF INTEGER);
  (*Loads Data with test data.*)
PROCEDURE SearchMethod (Data: ARRAY OF INTEGER;
                        Value: INTEGER; VAR Position: INTEGER);
  (*We just looked at several alternatives for this one.*)
PROCEDURE SortMethod (VAR Data: ARRAY OF INTEGER);
  (*Sorts the array into descending numerical order.*)
```

The three versions of *SortMethod* we'll write are:

versions of *SortMethod*

> Select *is based on the selection sort algorithm.*
> Insert *is based on the insertion sort algorithm.*
> Bubble *is based on the bubblesort algorithm.*

I'll start with a look at the different algorithms, then compare them briefly. For variety, I'll assume that the order we want is always largest to smallest—the opposite of our searching order. Finally, we'll implement the three sorting procedures.

Selection Sort

The first approach, *selection sort*, works like this:

Suppose that we have an array of INTEGER values.

18	35	22	97	84	55	61	10	47

selection sort

Search through the array, find the largest value (97), and exchange it with the value stored in the first array location (18).

97	35	22	18	84	55	61	10	47

Next, find the second largest value in the array, and exchange it with the value stored in the second array location. This is identical to the first trip through the array, except that we don't look at the first value—we already know it's the largest.

97	84	22	18	35	55	61	10	47

Now, repeat the 'select and exchange' process, each time beginning the search one value further along the array. As we go along, we'll be building an ordered array of values. Eventually, we'll get all the way to the end of the array—which has to be the smallest stored value—and the array will be ordered.

97 84 61 18 35 55 22 10 47

97 84 61 55 35 18 22 10 47

97 84 61 55 47 18 22 10 35

97 84 61 55 47 35 22 10 18

97 84 61 55 47 35 22 10 18 22 *happens to be OK.*

97 84 61 55 47 35 22 18 10

A pseudocode expansion of the selection sort algorithm is:

selection sort pseudocode

FOR *every 'first' component in the array*
find the largest component in the array;
exchange it with the 'first' component;

Insertion Sort

Algorithm two, *insertion sort*, also concentrates on keeping the left side of the array sorted. However, instead of sorting the very largest values, it sorts the values seen so far, even if this means more rearranging in the long run. The algorithm is:

Start with the same set of values as last time. The leftmost value, 18, can be said to be sorted *in relation to itself.* At the same time, we might anticipate having to move the next value, 35:

18 35 22 97 84 55 61 10 47

insertion sort

Now, remove the second value, 35. To insert it in its proper place, to the left of 18, we slide the 18 over to make room. Once more, we can anticipate the move we'll have to make to properly locate the next value:

35 18 22 97 84 55 61 10 47

Now the leftmost two components are relatively sorted. We can remove the third component, 22, then slide the 18 over again to make room. I've anticipated the proper 97 move:

35 22 18 97 84 55 61 10 47

Continue the process. Notice that the shaded portion is always in correct order. However, inserting a new value may—or may not, in the case of 10—require moving all or part of the shaded portion down to make room:

97	35	22	18	84	55	61	10	47
97	84	35	22	18	55	61	10	47
97	84	55	35	22	18	61	10	47
97	84	61	55	35	22	18	10	47
97	84	61	55	35	22	18	10	47
97	84	61	55	47	35	22	18	10

10 *stays put for a moment.*

Writing out the algorithm in pseudocode results in:

insertion sort pseudocode

FOR *every 'newest' component remaining in the array*
temporarily remove it;
find its proper place in the sorted part of the array;
slide smaller values one component to the right;
insert the 'newest' component into its new position;

a quick analysis

Let me pause for a quick comparison between selection and insertion. As I've stated the problem, we always begin with an array of mixed-up values (well, maybe it's the array that's mixed up, rather than the values). Algorithm one, selecting and repositioning values in order of their size, is a pretty effective way to reorder the array, while minimizing the number of moves that have to be made.

But suppose that the ground rules are different. If we're given an empty or partially sorted array, and told to fill it *in order*, then the insertion algorithm is better. As each new value arrives, we can insert it in the place it should occupy given the values we've gotten already.

Bubble Sort

Algorithm three is called *bubble sort*. Like selection sort, it's usually applied to arrays that are already filled with values. Bubble sort usually requires fewer comparisons, but many more exchanges than a selection sort does. A bubble sort works like this:

Begin with the same array as before:

18	35	22	97	84	55	61	10	47

bubble sort

Compare the first two values. If the second is larger, exchange them.

35	18	22	97	84	55	61	10	47

Next, compare the second and third values, exchanging them if the third is larger.

35	22	18	97	84	55	61	10	47

Then compare (and possibly exchange) the third and fourth values, the fourth and fifth, etc., until you reach the end of the array. Note that the smallest stored value ends up stored in the last position (shaded). However, despite all the movement, the rest of the array isn't ordered.

35	22	97	84	55	61	18	47	10

Now, go back to the beginning of the array and start all over again. Work your way through the array comparing and exchanging values again. However, since the smallest value is already at the far right, you need not compare the final value.

35	97	84	55	61	22	47	18	10

Repeat the process of comparison and exchange without bothering the final *two* values.

97	84	55	61	35	47	22	18	10

As you can see, an ordered list is forming on the right. Continue the process of comparison and exchange, ignoring the last *three* values this time.

97	84	55	61	47	35	22	18	10

This array is already ordered. If it weren't, it *would* be when we got to the point of only comparing the first two values.

In a sense, the smallest values 'bubble' to the right side of the array. A pseudocode version of the algorithm is:

bubble sort pseudocode

FOR *every 'last' component*
  FOR *every component from the first to the 'last'*
    compare that component to each remaining component;
    exchange them if necessary;

Three Implementations of Sort

It's time to make an important point:

Deciding to sort in the first place, and
➥ *choosing a particular sorting algorithm, and*
➥ *implementing the algorithm, and*
➥ *defining the underlying data structure and primitives*

can each be discussed in relative independence, although not necessarily in total ignorance, of each other.

Obviously, choices at one level can affect our judgment in another. The sorting methods I've described would not be very practical if we were sorting files, rather than arrays. Nevertheless, we can work with an eye toward making effective choices at each and every stage. Each of the sorting methods I described has its own set of advantages:

each method's advantages

- The selection sort doesn't require any wholesale moves—only two records are exchanged on any trip through the array.
- The insertion sort minimizes unnecessary travel through the array. If the records are sorted to begin with, a single trip through the array establishes that fact. In contrast, selection sort requires the same number of trips no matter how well-organized the array is.
- The bubble sort is worse than selection sort for a jumbled array—it will require many more component exchanges—but it's just as good as insertion sort for a pretty well-ordered array. More important, it's usually the easiest one to write correctly.

As you read procedures Selection, Insertion, and Bubble on the next few pages, I'd like you focus on three points of comparison:

comparing the algorithms

1. At best, and at worst, how many comparisons are made between components?
2. Again, at best, and at worst, how many times will two components' positions be changed?
3. Finally, will there be any difference between the starting and ending order of two or more components that have the exact same value? Sorts that leave such components in order are called *stable*, while sorts that may change order are *unstable*.

To help you keep track, I've written the procedures to count the number of switches and comparisons each method requires. All output reflects the test data (18 35 22 97 84 55 61 10 47) of the examples.

(*procedures follow*)

selection sort procedure

```
PROCEDURE Select (VAR Data: ARRAY OF INTEGER);
  (*Uses a selection sort algorithm to order an array of INTEGERs.*)
  VAR Temp, First, Current, Largest, Comparisons, Swaps: INTEGER;
  BEGIN
    Comparisons := 0;
    Swaps := 0;
    FOR First := 0 TO HIGH(Data)-1 DO
      Largest := First;
      FOR Current := First TO HIGH(Data) DO
        Comparisons := Comparisons + 1;
        IF Data[Current] > Data[Largest] THEN
          Largest := Current
        END
      END; (*Current FOR*)
      (*Postcondition: Largest subscripts the largest Count from First
       through HIGH.*)
      IF Largest <> First THEN    (*We have to make a swap.*)
        Swaps := Swaps + 1;
        Temp := Data[Largest];
        Data[Largest] := Data[First];    (*Make the swap.*)
        Data[First] := Temp
      END
      (*Postcondition: 0 through First subscript the array's largest values,
       as ordered by Count.*)
    END; (*First FOR*)
    IO.IntOut (Comparisons,1); IO.StringOut ("  comparisons,  ");
    IO.IntOut (Swaps,1); IO.StringOut ("  swaps.")
  END Select;
```

```
44 comparisons, 8 swaps.
```

A procedure for the insertion sort algorithm is shown below. Note that I've counted each 'slide' as a swap. Since the slide takes just one assignment, and a call of procedure Swap requires three, do you think this is fair?

insertion sort procedure

```
PROCEDURE Insert (VAR Data: ARRAY OF INTEGER);
  (*Uses an insertion sort algorithm to order an array of INTEGERs.*)
  VAR Newest, Current, Comparisons, Swaps: INTEGER;
      NewItem: INTEGER;
  BEGIN
    Comparisons := 0;
    Swaps := 0;
```

(procedure continues)

```
    FOR Newest := 1 TO HIGH(Data) DO
      Current := Newest;
      NewItem := Data[Newest];
      LOOP (*until we've found a home for NewItem.*)
        Comparisons := Comparisons + 1;
        (**)IF (Current = 0) OR (*then if*)
          (Data[Current – 1] >= NewItem) THEN EXIT END;
        Data[Current] := Data[Current – 1]; (*Slide one value to the right.*)
        Swaps := Swaps + 1;
        Current := Current – 1
      END; (*LOOP*)
      (*Postcondition: NewItem belongs in Data[Current].*)
      Data[Current] := NewItem
    END; (*Newest FOR*)
    IO.IntOut (Comparisons,1); IO.StringOut ("  comparisons,  ");
    IO.IntOut (Swaps,1); IO.StringOut ("  swaps.")
  END Insert;
```

```
27 comparisons, 19 swaps.
```

Finally, the procedure for the bubble sort algorithm is:

bubble sort procedure

```
PROCEDURE Bubble (VAR Data: ARRAY OF INTEGER);
  (*Uses the bubble sort algorithm to order an array of records on Count key.*)
  VAR Temp, Last, Current, Comparisons, Swaps: INTEGER;
  BEGIN
    Comparisons := 0;
    Swaps := 0;
    FOR Last := HIGH(Data) TO 1 BY –1 DO
      FOR Current := 0 TO Last–1 DO
        Comparisons := Comparisons + 1;
        IF Data[Current] < Data[Current+1] THEN
          Swaps := Swaps + 1;
          Temp := Data[Current];
          Data[Current] := Data[Current + 1];
          Data[Current + 1] := Temp
        END (*IF*)
      END (*Current FOR*)
      (*Postcondition: Components Last through the end of the array are ordered.*)
      END; (*Last FOR*)
    IO.IntOut (Comparisons, 1); IO.StringOut ("  comparisons,  ");
    IO.IntOut (Swaps, 1); IO.StringOut ("  swaps.")
  END Bubble;
```

```
36 comparisons, 19 swaps.
```

Modification Problems

1. If you haven't already done so, write a test harness for the sorting procedures. You'll need to generate test data, of course. As I pointed out earlier, the test data should try to make the sorts *fail*, not show that they can work.
2. It's always nice to see an algorithm in action. Embed debugging code in any of the three procedures to print the contents of the array after each pass. Make sure that your output is decipherable.
3. All of the sorting code arranges values from largest to smallest. Modify it so that values are ordered from smallest to largest.

Zen Study Guide

Big Ideas

The big ideas in this chapter are:

1. *It's important to study the details of searching and sorting, but it's important to ignore them, too.* For many programs, searching and sorting are the only significant CPU 'costs,' so methods have to be chosen carefully. But from the overall viewpoint of program design, they are no more important than other basic program steps—input, output, error checking, and so on.

2. *Fundamental program improvements come from better algorithms, not from better implementations.* Sometimes it's best to stand back and ask yourself just how well a program will work under the best reasonably expected circumstances. There's no point in grinding out slight improvements in a method that will never be good enough in the final analysis. Or, as Dijkstra puts it, if one blunt axe will not sharpen a pencil, ten blunt axes will do no better.

I think that you can think up a few sorting exercises on your own. Additional problems that involve sorting accompany the next chapter's exercises.

'New types used to be put together haphazardly, as the demands of a particular program dictated.' (p. 413)

Chapter 10 The **RECORD** Type

Chapter 10 introduces records, and a new approach to dealing with data abstractions—layered design. Records come to Modula-2 from COBOL, via Pascal, where they were included, in Wirth's words, to 'help erase the mystical belief in the segregation between scientific and commercial programming methods.'*

Scientific programming tends to revolve around just two data types: REAL and INTEGER. Arrays are fine for science because they provide the kind of structure that's needed—a way to hold lots of anonymous INTEGER or REAL values. Non-science programming, in contrast, is built around collections of values that have distinct names and types.

Records come to the rescue. A record uses individually named *fields*, which can have different types, to package information in a way that models the real world. In practice, programmers often define arrays of records, and get the best of both worlds: large numbers of components, each capable of storing a package of disparate values.

Chapter 10 begins with practical concerns. But our main focus, as you'll see in sections 10-2 and 10-3, continues to be data abstraction, and the object-oriented approach to program design.

---

* Niklaus Wirth, *The Programming Language Pascal*, Acta Informatica, Vol. 1, No. 1, 1971, pp. 35–63. Copyright 1971 Springer-Verlag.

Section 10-1

Strictly Modula-2

A RECORD IS A COLLECTION OF VALUES. OF COURSE an array is a collection of values too, but records have a few unique aspects:

- First, the stored values can have different types. This makes records potentially *heterogeneous*—composed of values of different kinds. Arrays, in contrast, hold values of just one type, so they're said to be *homogeneous*.
- Second, a record's stored values are called *fields*, and are given identifiers. A record's fields are individually named, like ordinary variables, whereas an array's components are selected according to their subscripts.
- Third, a single record type can, in effect, have two or more definitions. These *variant* definitions aren't important to us now, but they're described in Chapter 20.

field lists

Record types are defined by naming the type as a whole, then naming and typing the individual fields. The *field list* follows the general form of a variable declaration. It goes between the words **RECORD** and **END**, which lets the compiler know where a field list ends, and where the next definition or other declaration begins. Fields can have any type—standard or user-defined, simple or structured. For example, consider this definition of StatTYPE:

a very basic record

```
TYPE StatTYPE = RECORD
                  Letter: CHAR;
                  Count: INTEGER
                END;
VAR Most, Least: StatTYPE;
```

The declaration creates two record variables, each with two fields: Most.Letter, Most.Count, Least.Letter, and Least.Count.

dot notation

> Notice the period (usually read 'dot') between the variable and field name. It serves the same purpose for a record that square brackets serve for an array—it helps single out one component of a more complicated structure.

an array of records

I can also use the StatTYPE records as components of an array:

```
TYPE StatTYPE = RECORD
                  Letter: CHAR;
                  Count: INTEGER
                END;
     StatArrayTYPE = ARRAY [1..26] OF StatTYPE;
       (*Holds twenty-six of the StatTYPE records.*)
VAR  Most, Least: StatTYPE;
     Storage: StatArrayTYPE;
```

This declaration is more sophisticated, but it's more typical. The twenty-six components of Storage are records. Storage[1], Storage[2], and the others each have type StatTYPE, just like Most and Least. And, just as Most and Least each have two fields, so do each of the array's components: Storage[1].Letter, Storage[1].Count, and so on.

Records can appear within arrays, or a record can hold arrays. Indeed, a record's fields can have any type, simple or structured, and can even be records themselves. This nesting can go as deeply as your compiler will allow. Fortunately, the identifiers used for fields have limited scope, and don't conflict with identifiers used for variables, or the fields of other records.

limited scope of field names

Self-Check Questions

Q. Define a type that might be used to represent a deck of playing cards. Include types for the suits, one card, and the entire deck.

A. I've made individual card values 1 through 13, below. Can you think of a better way to handle card names?

```
TYPE SuitTYPE = (heart, club, spade, diamond);
     ValueTYPE = INTEGER[1..13];
     CardTYPE = RECORD
                    Suit:  SuitTYPE;
                    Value: ValueTYPE
                END;
     DeckTYPE = ARRAY [1..52] OF CardTYPE;
```

Access to Fields

There are three ways to access the values stored in a record variable.

accessing fields

1. An individual field can be identified using the 'dot' notation.
2. The complete record can be assigned to in a single statement—all the fields of one record variable can be assigned to the corresponding fields of another record with the same type in a record-to-record assignment.
3. The **WITH** statement lets us access individual fields without bothering with the dot notation.

I'll define a somewhat more complicated record to experiment with:

```
TYPE ConditionsTYPE = (Clear, Cloudy, Raining);
     WeatherTYPE = RECORD
                     Temperature: [-25..125];
                     Barometer: REAL;
                     Present, Outlook: ConditionsTYPE
                   END;
VAR Morning, Noon: WeatherTYPE;
```

The series of assignments to Morning, below, uses the dot notation. The record variable's identifier is followed by a period, and the name of a field. Formally, the trio of variable, period, and field name is known as a *qualified identifier*—the variable name and period *qualify*, or specifically identify, a particular field.

qualified identifiers

```
Morning.Temperature := 73;
Morning.Barometer := 30.16;
Morning.Present := Cloudy;
Morning.Outlook := Raining;
```

If weather conditions (and predictions) are identical at midday, we can take a shortcut and assign all the Weather fields in one fell swoop, like this:

record-to-record assignment

```
Noon := Morning;
```

This single assignment is equivalent to a sequence of field by field assignments. Every field of Noon gets the value of its counterpart in Morning. Naturally, all of Morning's fields should be defined before making such an assignment.

> Remember that record-to-record assignments can only be made between records of an *identical* type, e.g. declared with the same type name.

Although record-to-record *assignments* can be made, record-to-record *comparisons* cannot. You can't compare two complete records to see if they're equal or not. Instead, you have to compare them field by field.

The **WITH** Statement

Now, suppose that we want to gain access to individual fields, but don't want to have to go to the trouble of qualifying—preceding with a record name and period—each field name. For convenience, Modula-2 includes a 'statement' that lets a record be named once, then accessed directly. **WITH** is really just a syntactic scope-establishing device (like the 'dot'), but everybody calls it a statement because that's what it looks like. The series of assignments below is equivalent to the last example assignment:

using **WITH**

```
WITH Noon DO
   Temperature := 73;
   Barometer := 30.16;
   Present := Cloudy;
   Outlook := Raining
END
```

In a sense, the fields inside the **WITH** above are 'attached' to Noon. If we wanted to make assignments to another WeatherTYPE variable, we'd have to use the dot notation:

```
WITH Noon DO
    Temperature := 73;  (*refers to Noon.Temperature*)
    Morning.Temperature := 69;
END
```

Frankly, this last example is something you're only likely to create by accident. I would skip using the **WITH** if I were making the assignments above.

Self-Check Questions

Q. Show two different ways (i.e. using dot and **WITH** notation) of reading a value for the Temperature field and printing the value of the Barometer field of Morning.

A. As long as it's properly named, a field can be inspected or passed as a parameter just like any other variable. After all, it *is* a variable.

```
IO.IntOut (Morning.Temperature, 1);
IO.RealIn (Morning.Barometer);
  (*is the same as . . . *)
WITH Morning DO
    IO.IntOut (Temperature, 1);
    IO.RealIn (Barometer)
END
```

Q. Below are a few examples that have been exaggerated to make a point. Which ones are legal? What are the names of the legal example's fields?

```
a) TYPE HAVE = RECORD
                  HAVE: INTEGER
               END;
   VAR Save: HAVE;
b) TYPE HAVE = RECORD
                  Save: INTEGER
               END;
   VAR Save: HAVE;
c) TYPE HAVE = RECORD
                  HAVE: INTEGER
               END;
   VAR HAVE: HAVE;
```

A. The inside of a record creates a new scope for names, just as the inside of a subprogram does. That makes definition *a* legal—the field name is in a different scope than the record's type name. It's referred to as Save.HAVE.

Definition *b* is also all right. As in *a*, the field Save and the variable Save are in different scopes. Within the record, the field name takes precedence; outside the record, the other usage of a name holds sway. Here, the field is called Save.Save.

Finally, definition *c* is illegal because of the variable declaration. It errs because a type and a variable have the same identifier, but are in the same scope. The compiler has no rule for giving one use precedence over the other, so it announces an error.

Constructed Types

WeatherTYPE was a simple definition; it was good for examples, but not very realistic. In real life, records are usually the main building block of *databases*—collections of information that have been organized to make finding things easy. In such applications, records usually show up as array components. Since an array of records is built from separately defined structures, I usually think of it as a *constructed* type.

Let's construct a more complicated type to practice with: an address book. What information should it store? Names, obviously, along with telephone numbers and addresses. We could treat all the values of one entry as though they were part of a long sequence of CHAR values stored in an array:

An address book *could* be an array of these arrays, but it wouldn't be very convenient to program with. A record type, with distinct fields for each entry, is much more appropriate:

a basic telephone book

```
TYPE StringTYPE = ARRAY [0..50] OF CHAR;
     EntryTYPE = RECORD
                   Name, Street, Town: StringTYPE;
                   Zipcode: INTEGER;
                     (*telephone number fields*)
                   Area, Prefix, Number: INTEGER
                 END;
     BookTYPE = ARRAY [1..100] OF EntryTYPE;
VAR OneNumber: EntryTYPE;
    Book: BookTYPE;
```

Making assignments to a stand-alone record like OneNumber is pretty straightforward:

records as variables

```
OneNumber.Name := "Doug Cooper";
OneNumber.Street := "33 Washington Avenue";
OneNumber.Town := "Northampton, Massachusetts";
OneNumber.Zipcode := 01060;
OneNumber.Area := 413;
OneNumber.Prefix := 584;        and so on.
```

The assignments to the Name, Street, and Town fields all take advantage of the special rule for string assignments: that 'short' literal values can be assigned to string variables.

Since the components of Book have the same type as OneNumber, the values I assigned to OneNumber fields can be copied in a record-to-record assignment:

```
Book[1] := OneNumber;
```

A series of direct assignments to the second record of the Book array is:

records as components

```
Book[2].Name := "Prince Fan Club";
Book[2].Street := "City Hall";
Book[2].Town := "Minneapolis, Minnesota";
Book[2].Zipcode := 55415;
Book[2].Area := 612;
Book[2].Prefix := 348;
Book[2].Number := 3000;
```

I might also use a **WITH** statement, perhaps with the ordinary IO input procedures:

```
WITH Book[3] DO
    IO.StringOut ("Please enter the name:    ");
    IO.StringIn (Name);
    IO.StringOut ("Please enter the zip code:    ");
    IO.IntIn (Zipcode);
      . . .
END
```

Now let's try the tough ones. How would I assign "X" to the third character of the Name field of the fourth record of Book? Like this:

```
Book[4].Name[2] := "X";    (*The Name subscripts start at zero.*)
```

How about making every zip code in Book equal Bruce Springsteen's old zip code in Asbury Park?

```
FOR i := 1 TO 100 DO
    Book[1].Zipcode := 07712
END
```

Let's make every record in Book equal the first one:

initializing Book

```
FOR i := 2 TO 100 DO
    Book[i] := Book[1]
END
```

problem: lookup

Putting information into our phone book is just half of it, of course. We can also get data out. Suppose I want to know who was at the other end of that $87 telephone call; I have the number, but need the name. Finding it requires a typical linear array search. In pseudocode:

get the number from the program user;
look for it;
print the name or a failure message

The loop specification of *look for it* is:

loop specification

Goal: Find the stored name that goes with a stored number.
Intentional bound: We've found a matching number.
Necessary bound: We've run out of numbers to check.
Plan: Increment a counter variable.

And, in Modula-2 we have:

looking up a name

```
PROCEDURE LookUp (Book: BookTYPE);
  (*Asks for and tries to look up a particular phone number.*)
  VAR Area, Prefix, Number, i: INTEGER;
  BEGIN
    IO.StringOut ("Enter the number in three parts: XXX XXX XXXX");
    IO.IntIn (Area);
    IO.IntIn (Prefix);
    IO.IntIn (Number);
    i := 1;
    LOOP
      (**)IF ( (Area = Book[i].Area) AND (Prefix = Book[i].Prefix)
              AND (Number = Book[i].Number) )
                OR (i = 100) THEN EXIT END;
      i := i + 1
    END;
    (*Postcondition: If the name was found, it's in the ith component.*)
    IO.StringOut ("The mystery name was  ");
    IF (Area = Book[i].Area) AND (Prefix = Book[i].Prefix)
                             AND (Number = Book[i].Number)
      THEN IO.StringOut (Book[i].Name)
      ELSE IO.StringOut ("not in your book.")
    END
  END LookUp;
```

Modification Problems

1. "Enter a number in three parts"? Get serious—LookUp has a pretty terrible user interface. Write a procedure that can read a phone number entered in any reasonable format, and return it separated into the area code, prefix, and number fields.

Self-Check Questions

Q. BookTYPE holds one hundred of the EntryTYPE records. This is fine for a bookish type like me, but a true party animal would have to be a little more creative. Use the definition of BookTYPE to define a structure that holds hundreds and hundreds of numbers, and which makes it easy to store entries in alphabetical order.

A. Piece of cake—we can define an array that is subscripted by the alphabet. There's a separate array of one hundred names for each letter.

```
TYPE PartyListTYPE = ARRAY ["A".."Z"] OF BookTYPE;
```

Q. Which of these code segments actually initializes the Name field of each record to ????

```
Current := 1;  (*Segment A*)
LOOP
   WITH Book[Current] DO
      Name := "???";
      Current := Current + 1
   END;  (*WITH*)
   (**)IF Current > 100 THEN EXIT END
END  (*LOOP*)

Current := 1;  (*Segment B*)
WITH Book[Current] DO
   LOOP
      Name := "???";
      Current := Current + 1;
      (**)IF Current > 100 THEN EXIT END
   END  (*LOOP*)
END  (*WITH*)
```

A. Segment *A* works, but segment *B* doesn't. Why not? Because the reference to Book[Current] is established when Counter equals 1. Even though Current changes *within* the **WITH** statement, the record assignment continues to be made to Book[1].Name.

A Few Definition Exercises

I'll wrap up the introduction to record syntax by looking at some sample definitions. We won't usually have to define complicated record types out of the blue, but you need to be relaxed enough with definitions to knock out a few alternatives if a problem demands it.

Data structures that depend on records can be pretty interesting. Most games will be programmed with records because of their duplicated parts:

games require duplication

A baseball game consists of nine innings.
A poker game usually has a half-dozen or so players, each holding a handful of playing cards.
A chessboard has 64 squares whose color and contents vary.
A Monopoly board has squares that represent properties, and usually include schedules of rents and buildings.
Computer games like Adventure, Rogue, and Zork contain rooms that are filled with unknown objects and are connected in various ways.
Animated video games use records to hold the details and positions of objects that appear on the screen.

Programming games poses problems of keeping score, remembering positions, preventing cheating, and the like. Do they require real algorithms? Well, although winning play might need some kind of algorithm, the programmer's basic job is to manage data by tracking scores and board positions. A suitable abstraction—that is, an underlying data type and collection of primitive procedures that 'understand the moves'—make the programmer's life easy.

What data type might underlie a program that plays *Wheel of Fortune*? As noted in Chapter 8, players are shown a phrase whose letters have been hidden. Each player has the chance to name individual letters, which are shown if they appear in the phrase. We can hold the phrase with:

holding the phrase

```
CONST PHRASE = 100;  (*Maximum phrase size.*)
TYPE SquareTYPE = RECORD
                    Letter: CHAR;
                    Showing: BOOLEAN
                  END;
     GameTYPE = ARRAY [1..PHRASE] OF SquareTYPE;
```

Let's turn to a game that's a little more sophisticated—chess. Each square of a chessboard can be empty, or occupied by one of a half-dozen different pieces. The pieces themselves belong to either of two owners. It isn't hard to decide that it makes sense to store a chessboard as a two-dimensional array. What's tough is designing the contents of each square:

chessboard type definition

```
TYPE ColorTYPE = ( Black, White);
     PieceTYPE = (Empty, Pawn, Knight, Bishop, Rook, Queen, King);
     SquareTYPE = RECORD
                    Occupied: BOOLEAN;
                    Piece: PieceTYPE;
                    Owner: ColorTYPE
                  END;
     BoardTYPE = ARRAY [1..8], [1..8] OF SquareTYPE;
VAR Board: BoardTYPE;
```

How about an example that's nearer and dearer to a student's heart?

grade records

```
TYPE GradeTYPE = "A".."F";
     QuarterTYPE = (Fall, Winter, Spring);
     StatusTYPE = (Passed, Failed, Incomplete, Withdrawn);
     NameTYPE = ARRAY [0..25] OF CHAR;
     WhenTakenTYPE = RECORD
                       Term: QuarterTYPE;
                       Year: INTEGER[1990..2000]
                     END;
     ClassTYPE = RECORD
                   Name: NameTYPE;
                   Hours, Units: INTEGER[1..5];
                   Grade: GradeTYPE;
                   Results: StatusTYPE;
                   Taken: WhenTakenTYPE
                 END;
```

Or a definition that might underly a data structure used for graphics:

graphics

```
TYPE HueTYPE = (red, blue, green, black, white);
     DotTYPE = RECORD
                    xPos, yPos: INTEGER;
                    Color: HueTYPE
               END;
```

Or the data type that a personnel record structure uses:

personnel records

```
TYPE NameTYPE = ARRAY [0..50] OF CHAR;
     NumberTYPE = ARRAY [0..10] OF CHAR;
     PersonTYPE = RECORD
                     First, Last: NameTYPE;
                     Born, Hired, SSN: NumberTYPE;
                     Salary: REAL
                  END;
     EmployeesTYPE = ARRAY [1..1000] OF PersonTYPE;
```

lateral thinking

Because the design of data types and structures is largely a mental exercise, a problem solving technique called *lateral thinking* can be put to good use. Lateral thinking is a name Edward DeBono invented to describe the process of repeatedly exploring and reconsidering possible solutions before committing ourselves to one particular method. A lateral approach to digging for buried treasure would entail digging many shallow holes, instead of a very deep one.

A lateral programmer might propose several potential data types before basing a data structure on one particular version. Experienced programmers can do this in their heads (or at least they think they can) but novices should sketch out some proposals on paper.

For example, consider these two possible data type definitions for a program that scores a baseball game.

baseball data type 1

```
TYPE TeamTYPE = (Pirates, Mets, Astros, Giants, Yankees, Angels);
     InningTYPE = (Top, Bottom);
     AtBatTYPE = RECORD
                        TeamUp: TeamTYPE;
                        Runs, Hits, Errors: INTEGER
                 END;
     GameTYPE = ARRAY [1..9], [Top..Bottom] OF AtBatTYPE;
```

baseball data type 2

```
TYPE NameTYPE = ARRAY [0..14] OF CHAR;
     InningTYPE = (Top, Bottom);
     StatisticsTYPE = ARRAY [1..9] OF INTEGER;
     TeamStatTYPE = RECORD
                           Name: NameTYPE;
                           Runs, Hits, Errors: StatisticsTYPE
                    END;
     GameTYPE = ARRAY [Top..Bottom] OF TeamStatTYPE;
```

The definitions create two different ways of viewing and storing the exact same information. A preference for one over the other will be strongly influenced by choices we make about primitive procedures. Nevertheless, if need be, either can be made to work.

Section 10-2 An Extended Example: Counting Letter Pairs

problem: letter pair frequency

LET'S WRITE A FAIRLY LARGE PROGRAM THAT'S SIMILAR to one from Chapter 8, but which can't be written without records. There, I posed the problem of counting appearances of single letters. Here, we'll deal with the problem of letter *pairs*.

> Write a program (call it Doubles) that reads a sample of text, and counts the number of times each pair of letters appears. Disregard differences between lower-case and upper-case letters. Summarize the program's findings by printing the 50 most frequent pairs, along with each pair's percentage of the total count.

For example, the text sample **"VANNA"** contains four letter pairs: **VA**, **AN**, **NN**, and **NA**. Note that order matters; **AN** and **NA** are different pairs. Since there are twenty-six possibilities for each letter, there are 26*26, or 676, possible pairs in any text sample. This makes a summary of program results a necessity. Possible applications for our program's output include:

some applications

- Breaking codes, especially if the encrypted sample is short—letter pair frequencies will be much more accurate than single letter frequencies.
- Compressing text by allowing substitution of unused single characters for frequently appearing pairs.
- Most obviously, developing a winning game plan for *Wheel of Fortune*.

A goal-oriented approach to solving the letter pair problem would have us begin by splitting up the problem in procedural terms, and starting to pseudocode actions. The object-oriented approach, in contrast, focuses on program data—where it comes from, and where it's going. The 'data outline,' below, shows the flow. It suggests that any solution to the letter pair problem has to make a step-by-step transformation of data:

outlining the flow of data

Uncounted letters from the raw text sample
(*which will become*)
Counts for each pair of letters
(*which will become*)
Ordered pairs of letters and their counts
(*which will become*)
Final frequency statistics

In its first stage, as *uncounted letters*, data is the raw text awaiting processing. To be transformed into *counted letter pairs*, we have to chop the input into pairs, then count each pair. At this point, it makes sense to hold the pairs in some way that will make it easy to look up any particular pair so that we can add one to its count.

Turning the pair counts into *ordered letter pairs* requires another transformation—out of some kind of alphabetical order, and into order of frequency of appearance. That will make it easy to move to the last step: picking out the most frequently appearing pairs to come up with *frequency statistics*.

What is remarkable is how much of the work can be hidden by the primitive operators of the data structures that ultimately store all the information. For example, when we save a pair's count of appearances, we might know nothing more than the two letters and their number. But if we save it properly, we might be able to retrieve it with one additional piece of information—the relative position (e.g. most or least frequent) of the pair among all the counts.

In contrast, the assignment operator of an ordinary variable (that is, ':=') just specifies *what* is going to be saved. When we inspect the variable, we'll just get the same value back. But the assignment primitive of a data abstraction also explains *how* and *where* the value should be saved, just as its inspection primitive defines how and where the value should be looked for. We can add information to the stored value by the rules we follow in saving it or looking it up again.

Two Data Abstractions

Let's turn to outlines of our data abstractions. I think that we need two of them because we want to store information in two basically different ways:

handling raw data

- *Counted letter pairs* associate a number with each pair of letters. However, the numbers are not in any particular order—we just want to be able to count each appearance of a new pair, and to get each pair's count back.

handling counted data

- *Ordered letter pairs* associate a pair of letters with the number of times it has appeared. Here, the numbers *are* in order—we want to find the count *and* the letters in first, second, or *n*th place.

Each of the abstractions will need its own special primitives for initialization, assignment, and inspection.

separating specification from implementation

> Recall that a data abstraction is defined in terms of its primitive operations—what they will do, not how they will do it. Although primitives can be described in English or pseudocode, they can't be written without knowing details of a type's underlying implementation.

First, let's have rough descriptions of the primitive operations we'll want to carry out on the raw data. The first draft of the primitives is purely descriptive. It should give us an idea of what we want the data abstraction to do, but it can't commit us to any particular way of using the primitives or getting it done:

raw data primitives, first draft

RawTYPE *will hold a count of the number of letter pairs;*
InitializeRaw *will initialize the table that holds the counts;*
Count *will increase the count for a given letter pair;*
GetRaw *will return the count of a given letter pair;*

The second abstraction will handle the ordered letter pairs and counts. Note that we don't need a Sort primitive. Instead, we can think of the SortTYPE abstraction as being an ordered list because of the way Insert is defined. If a pair and count are going to be 'assigned' to the sorted list, they're automatically put into the right place:

sorted count primitives, first draft

SortTYPE *will be a list of the 50 most frequent pairs;*
InitializeSort *will initialize the table that holds pairs and counts;*
Insert *will put a pair, and its frequency count, into the right position among the top 50;*
GetSort *will get the pair and count in a given position;*

Sooner or later, I'm going to have to figure out how Insert is going to work. But right now I'm satisfied to be breaking a big problem (finding letter pair frequencies) into relatively small problems like Insert.

It's perfectly reasonable to stop at this point and ask a question: are we getting any closer to solving the problem of counting letter pairs? Well, let's go back to the more traditional goal-oriented design approach for a minute, and try to develop a pseudocode outline of the final program Doubles. A first draft would be:

first pseudocode

count the pairs of letters;
select and order the fifty most common pairs;
print the results;

It doesn't say much. But remember why we started with the object-oriented approach:

> Rather than asking *what will our program do?*, we posed the question *what information will it use?* Then we began to define the primitives we'd need to solve our problem, almost as though we were extending Modula-2 to meet our specific needs.

Let's try a second pseudocode refinement that makes the most of the primitives outlined above:

```
(*'Count the pairs of letters' becomes ...*)
call InitializeRaw to initialize the pair counts;
LOOP
   get a pair of characters;
   (**)IF we ran out of data then EXIT;
   IF they're letters
      call Count to add 1 to the pair's count;
      add 1 to a Total count
END of the loop;
(*Postcondition: We've counted each appearance of every character pair.*)

(*'Select and order the fifty most common pairs' becomes ...*)
call InitializeSort to initialize the table of ordered figures;
FOR every possible pair of characters
   call GetRaw to get the pair's raw count from the raw count table;
   IF the raw count is among the fifty highest THEN
      Insert the pair and its count in the table of ordered figures
(*Postcondition: We have the fifty most frequent pairs, in order.*)

(*'Print the results' becomes ...*)
FOR each of the top 50 positions
   call GetSort to get the pair and count in that position;
   print the results
(*Postcondition: We've printed the most common pairs and frequencies.*)
```

second pseudocode

We hardly know anything, yet we've practically finished the program. Ain't abstraction wonderful?

Layered Design

Now, we've managed to use data abstraction pretty informally. But as our programs start to get more complicated, I'd like to give the approach we've been using a more formal name—*layered design*.

> In layered design, a data abstraction is developed in two stages—*define*, them *implement*. First, define the abstraction by writing its *name layer*—a specification of the abstraction's requirements and performance. Second, implement the abstraction by writing its *code layer*—the Modula-2 code that makes the abstraction concrete.

The name layer has to be a little more formal than the rough outlines of primitives we've been using so far. After all, it serves two different masters:

the name layer

- It is a *performance specification* for using the abstraction. It explains how the primitives will work—what kind of performance and results they promise.
- It is a *requirements specification* for implementing the abstraction. It

explains what the primitives must do—what kind of performance and results they have to deliver.

the implementation layer

The code layer of a data abstraction contains the declarations of any primitives it provides. It also has the definitions of any new data types. Even though new types may be named in the name layer, we don't commit to any particular implementation until we come to the implementation layer.

calling forms of subprograms

What does the name layer look like? Since operations in Modula-2 are supplied as procedures and functions, it's convenient to provide the name layer as a series of mock subprogram headings. They provide the *calling forms* of the subprograms—their names, their results, and the kinds of arguments they require. If you look back over Chapters 8 and 9, you'll see that we've already done this many times.

The Name Layer

Layered design helps split up the work of programming, although maybe in an unexpected way. Imagine that you're half of a two-person programming team. Layered design *doesn't* let you split the jobs of writing the name layer and implementation layer. Instead, if you share the work of designing the name layer, you can split what remains: writing the code layer and main program.

For example, let's put together a name layer for the RawTYPE and SortTYPE abstractions. We can outline the calling forms of their primitives even though we don't know any details of the underlying type definitions. In effect, we're insisting that what we *want* to do will determine the definition, and not settling for letting the definition decide what we *can* do.

The name layer of RawTYPE is:

RawTYPE name layer

```
TYPE RawTYPE doesn't have to be defined just yet;
PROCEDURE InitializeRaw (VAR Raw: RawTYPE);
    (*Initialize Raw to zeros.*)
PROCEDURE Count (VAR Raw: RawTYPE; C1, C2: CHAR);
    (*Increment the count for pair C1, C2.*)
PROCEDURE GetRaw (Raw: RawTYPE; C1, C2: CHAR) : INTEGER;
    (*Return the number of appearances of pair C1, C2.*)
```

The equivalent name layer for SortTYPE is:

SortTYPE name layer

```
TYPE SortTYPE doesn't have to be defined just yet;
PROCEDURE InitializeSort (VAR Sort: SortTYPE);
    (*Initialize Sort to some non-value.*)
PROCEDURE Insert (VAR Sort: SortTYPE;
                          C1, C2: CHAR; Count: INTEGER);
    (*Insert the C1, C2 pair and Count ranking, if among the top fifty.*)
PROCEDURE GetSort (Sort: SortTYPE; Position: INTEGER;
                       VAR C1, C2: CHAR; VAR Count: INTEGER);
    (*Return the C1, C2 pair and Count in Positionth place.*)
```

> The name layer acts as a layer of insulation. As long as the operations perform as advertised, the user of a data structure doesn't have to concern herself with details of implementation. By the same token, the implementor can write and test her operations without worrying about—or waiting for details of—their ultimate application.

If we wanted, we could begin either of the two jobs now: writing the main program, or writing the code layer of the abstractions. A third refinement of the main program pseudocode shows just how close to completion we've come. Assume that RawTable is the RawTYPE variable, and that SortTable has type SortTYPE.

third pseudocode

```
InitializeRaw (RawTable) to initialize the pair counts;
LOOP
    get a pair of characters, C1 and C2;
    (**)IF we run out of data EXIT;
    IF they're letters
        Count (RawTable, C1, C2) to increment the pair's count;
        add 1 to a Total count
END of the loop;
(*Postcondition: RawTable holds counts for all Total characters.*)
InitializeSort (SortTable) to initialize the ordered figures;
FOR every pair of characters C1, C2 in the raw count table
    Count := GetRaw (RawTable, C1, C2) to get the pair's count;
    Insert (SortTable, C1, C2, Count) if they're among the top fifty;
END of the FOR;
(*Postcondition: SortTable holds the fifty most frequent pairs, in order.*)
FOR each Position from 1 to 50
    GetSort (SortTable, Position, C1, C2, Count);
    print the results
END of the FOR;
(*Postcondition: We've printed the most common pairs and frequencies.*)
```

Notice that as detailed as this pseudocode is, it doesn't depend on knowing anything about the definitions of RawTYPE or SortTYPE, or the declarations of their primitives.

RawTYPE's Code Layer

The name layer sets up some pretty explicit expectations for the RawTYPE and SortTYPE definitions. Let's begin by defining RawTYPE, and then pseudocoding its primitive procedures.

clusters

> A Modula-2 program can have as many definition and declaration parts (i.e. **TYPE** and **VAR** parts) as it needs. To make programs easier to read, I'll group the definitions and primitives of data abstraction into separate *clusters* of definitions and declarations.

RawTYPE has to hold the overall counts: how many **AA** pairs, how many **AB** pairs, and so on. When we counted single characters, we used a one-dimensional array subscripted by CHAR. Here, a two-dimensional array can serve the same purpose:

RawTYPE definition

```
TYPE RawTYPE = ARRAY ["A".."Z"],["A".."Z"] OF INTEGER;
```

RawTYPE can hold one INTEGER for each letter pair. The component subscripted by ["A","A"] can hold the number of times the pair **AA** has appeared, ["A","B"] can hold the **AB** count, and so on.

RawTYPE's first primitive, Initialize, prepares a RawTYPE variable for later use. Since each of the components has to be initialized to 0, we have to traverse the two-dimensional array. I'll pseudocode it as:

InitializeRaw pseudocode

```
PROCEDURE InitializeRaw (VAR Raw: RawTYPE);
    FOR each row in the array
        FOR each column in the array
            Raw[Row, Col] := 0
```

Procedure Count, in turn, takes just one assignment. Nevertheless, it has to be written as a primitive. Why? Because making the assignment requires knowledge about the exact definition of RawTYPE.

Count pseudocode

```
PROCEDURE Count (VAR Raw: RawTYPE; C1, C2: CHAR);
    Raw [C1, C2] gets increased by 1
```

Finally, function GetRaw returns the value stored in one component. Like Count, it isn't very complicated. But, like Count, it has to be a primitive because it relies on details of the RawTYPE definition:

GetRaw pseudocode

```
PROCEDURE GetRaw (Raw: RawTYPE; C1, C2: CHAR) : INTEGER;
    RETURN the value of Raw[C1, C2]
```

SortTYPE´s Code Layer

Let's turn to the SortTYPE cluster. Thanks to our hard work with RawTYPE, we can assume that we've stored the counts for each letter pair. Unfortunately, part of what we need to know—the exact letters that each count is associated with—is actually built into the underlying data type.

If we want to rearrange the counts *and* remember what pair each count belongs to, we'll need something new. We can hold two letters (the pair) and one number (the count) in a record. Then, we can hold fifty records with an array:

SortTYPE definition

```
CONST LIMIT = 50;          (*we're saving the fifty most common pairs*)
TYPE ShareTYPE = RECORD
                   C1, C2: CHAR;        (*the letters*)
                   Share: INTEGER        (*the count*)
                 END;
     SortTYPE = ARRAY [1..LIMIT] OF ShareTYPE;
```

Like most abstractions, SortTYPE has primitives for initialization, assignment, and inspection. The initialization primitive stores empty records in SortTYPE. What does 'empty' mean? Well, we can initialize the Share fields to 0. I'll initialize the C1 and C2 fields to the ASCII NUL character, 0C. Recall from the file reversal program (in Chapter 8) that although it's a two-character sequence, 0C is a CHAR-type value. It never appears in ordinary input, and it doesn't print anything when it's output.

InitializeSort pseudocode

```
PROCEDURE InitializeSort (VAR Sort: SortTYPE);
  FOR each of the LIMIT records in Sort
    initialize the CHAR fields to 0C;
    initialize the Share field to 0
```

The pseudocode of the assignment primitive, Insert, may strike you as being a bit vague. Well, it is. I'll return to it in a few moments:

Insert pseudocode

```
PROCEDURE Insert (VAR Sort: SortTYPE;
                  C1, C2: CHAR; Count: INTEGER);
  IF Count is bigger than the smallest count already in Sort THEN
    find the proper position for the C1,C2 pair and count
```

Finally, SortTYPE's inspection procedure has to retrieve values from one of the stored records. GetSort is written as a procedure (and not a function) because it returns more than one value.

GetSort pseudocode

```
PROCEDURE GetSort (Sort: SortTYPE; Position: INTEGER;
                   VAR C1, C2: CHAR; VAR Count: INTEGER);

  make the appropriate assignments to C1, C2, and
  Count from the record in Sort[Position]
```

A Coding Detail: Insertion

The second procedure, Insert, is the only one that requires an algorithm of any substance. When we sort the raw data to single out and order the fifty most frequent pairs, we will, in effect, be given each of the 676 possible pairs (and its count) in turn. Our job is to insert each of the large counts in its proper place.

The algorithm that turns the trick turns out to be a variation of *insertion sort* from Chapter 9. Each value is inserted into its proper location, taking into consideration the values that have already been sorted. For all practical purposes, this is what happens when you sort a hand of playing cards. As you pick up each new card, you insert it into the correct position amongst the cards already in your hand.

It's not quite so simple with arrays, of course. Let me illustrate with a simple array of INTEGER values. In order to insert a new value, say 8, we'll probably have to slide some previously stored values over.

inserting an 8

For example, if 8 is to be inserted properly, a hole has to be opened up between 7 and 9. We must patiently work from the right, and slide each component over—first 2, then 4, then 5, and so on—until there's room for the 8. We may even find that the smallest component vanishes, knocked off the end of the array.

Now, before you can open a hole you have to decide where it belongs. There are two alternative strategies that can be followed:

- slide values to the right one at a time while you look for the proper position;
- find the proper position for the new value, then do all the sliding at once.

Suppose that we take the first approach. In effect, we create a moving hole that heads toward the largest stored value. Naturally, we won't even consider the prospect of insertion unless the new Count value has occurred more frequently than the Share stored in the LIMIT*th* component. The loop specification must be stated rather carefully—it's easier to achieve the first half of the goal than to get the first *and* the second!

insertion loop specification

Goal: The assignment 'Sort[i] := *the new* C1, C2, *and* Count' *puts the new values in their proper place, without losing any values.*
Intentional bound: The Share *field in* Sort[i–1] *is* >= *the new* Count.
Necessary bound: There can't be a larger or equal stored value, because i = 1.
Approach: Copy Sort[i–1] *to* Sort[i] *to open a hole.*
Subtract 1 *from* i *to move left.*

Note that the necessary bound must be checked first, to avoid the possibility of inspecting Sort[0], which doesn't exist. In pseudocode, the loop is:

insertion pseudocode

```
(*Assuming that the new Count is greater than the value Sort[LIMIT].Share... *)
Initialize i to LIMIT;  (*i always points to the 'hole'*)
LOOP
    (**)IF (i equals 1) OR (*then if*)
        Sort[i-1].Share isn't smaller than Count EXIT;
    Slide the Sort[i-1]th component to the right, into Sort[i];
    Subtract 1 from i
END of the loop
(*Postcondition:  Count, C1, and C2 belong in Sort[i].*)
```

The Final Doubles Program

The complete program Doubles occupies the next few pages. I've supplied the first thirty thousand characters from this chapter as input. The definitions and declarations that underly each abstraction are clustered separately, and the main program's variable declarations can be found just before the statement part.

The only feature of Doubles that wasn't explained in detail earlier comes from the pseudocode statement:

get a pair of characters, C1 *and* C2;

There is a potential bug lurking here because each incoming character actually appears in two letter pairs: first as a pair's first letter, and then as its second. If we implemented the pseudocode literally, as:

```
IO.CharIn (C1);
IO.CharIn (C2)
```

we'd only use each letter once. The problem isn't hard to solve, though, if we use the *moving window* technique. In its bare bones form (let's assume that at least one character is present) it is:

the moving window

```
IO.CharIn (C1);
LOOP
    IO.CharIn (C2);
    IF it wasn't done then EXIT;
    Do what you want with C1 and C2;
    Let C1 := C2 to advance the window
END of the loop
```

In Doubles, I've converted the characters to capital letters, using the standard function CAP, before counting the pairs.

(program follows)

letter-pair counting program

```
MODULE Doubles;
  (*Reads an input text sample, and prints frequency counts for the LIMIT
   most common letter pairs.*)
IMPORT IO;
(*Defining the Raw Data cluster \/\/\/\/\/\/\/\/\/\/\/\/\/\/\/\/\/\/\/\/\/\/    *)
TYPE RawTYPE = ARRAY ["A".."Z"],["A".."Z"] OF INTEGER;
PROCEDURE InitializeRaw (VAR Raw: RawTYPE);
  (*Initialize Raw to zeros.*)
  VAR Row, Column: CHAR;
  BEGIN
    FOR Row := "A" TO "Z" DO
      FOR Column := "A" TO "Z" DO
        Raw [Row, Column] := 0
      END
    END
  END InitializeRaw;
PROCEDURE Count (VAR Raw: RawTYPE; C1, C2: CHAR);
  (*Increment the count for pair C1, C2.*)
  BEGIN
    Raw [C1, C2] := Raw [C1, C2] + 1
  END Count;
PROCEDURE GetRaw (Raw: RawTYPE; C1, C2: CHAR) : INTEGER;
  (*Return the number of appearances of pair C1, C2.*)
  BEGIN
    RETURN Raw [C1, C2]
  END GetRaw;
(*Defining the Sorted Data cluster \/\/\/\/\/\/\/\/\/\/\/\/\/\/\/\/\/\/\/\/\/\/    *)
CONST LIMIT = 50;        (*since we're saving the fifty most common pairs*)
TYPE ShareTYPE = RECORD
                   C1, C2: CHAR;        (*the letter pair*)
                   Share: INTEGER        (*its count*)
                 END;
     SortTYPE = ARRAY [1..LIMIT] OF ShareTYPE;
PROCEDURE InitializeSort (VAR Sort: SortTYPE);
  (*Initialize Sort to some non-value.*)
  VAR i: INTEGER;
  BEGIN
    FOR i := 1 TO LIMIT DO
      Sort[i].C1 := 0C;
      Sort[i].C2 := 0C;
      Sort[i].Share := 0
    END
  END InitializeSort;
```

```
PROCEDURE Insert (VAR Sort: SortTYPE;
                      C1, C2: CHAR; Count: INTEGER);
  (*Insert the C1, C2 pair and Count ranking, but only if among the top LIMIT.*)
  VAR i: INTEGER;
  BEGIN
    IF Count > Sort[LIMIT].Share THEN   (*make the insertion*)
      i := LIMIT;
      LOOP
        (**)IF (i = 1) OR (*then if*) (Sort[i-1].Share >= Count)
         THEN EXIT END;
        Sort[i] := Sort[i-1];
        i := i - 1
      END; (*LOOP*)
      (*Postcondition:  Count, C1, and C2 belong in Sort[i].*)
      Sort[i].C1 := C1;
      Sort[i].C2 := C2;
      Sort[i].Share := Count
    END  (*IF we made an insertion.*)
  END Insert;
PROCEDURE GetSort (Sort: SortTYPE; Position: INTEGER;
                      VAR C1, C2: CHAR; VAR Count: INTEGER);
  (*Return the pair and count in Positionth place.*)
  BEGIN
    C1 := Sort[Position].C1;
    C2 := Sort[Position].C2;
    Count := Sort[Position].Share
  END GetSort;

(*Main program variables \/\/\/\/\/\/\/\/\/\/\/\/\/\/\/\/\/\/\/\/\/\/\/\/  *)
VAR C1, C2: CHAR;            (*the two-character input window*)
    Total,                   (*the number of pairs counted*)
    PairCount,               (*the number of appearances of one pair*)
    Position: INTEGER;       (*used to traverse the final SortTable*)
    Percent: REAL;           (*the pair's share of the total*)
    RawTable: RawTYPE;
    SortTable: SortTYPE;
BEGIN  (*main program*)
  IO.StringOut ("Analysing letter pair frequency in file Chapter.10.");
  IO.LineOut;
  IO.OpenInput ("Chapter.10");   (*My source file for this chapter*)
  IF NOT IO.Done() THEN
    IO.StringOut ("Could not open input file.  Exiting.");
    HALT
  END;
```

(program continues)

```
    InitializeRaw (RawTable);
    Total := 0;
    IO.CharIn (C1);
    C1 := CAP(C1);
    (*CAP has no effect unless its argument is lower-case. Assume there was a character.*)
    (*Postcondition: Input is redirected, and the input window is initialized.*)
    LOOP
        IO.CharIn (C2);
        (**)IF NOT IO.Done() THEN EXIT END;
        C2 := CAP(C2);
        IF (C1>="A") AND (C1<="Z") AND
                (C2>="A") AND (C2<="Z") THEN
            Count (RawTable, C1, C2);
            Total := Total + 1
        END;
        C1 := C2  (*Advance the window.*)
    END; (*LOOP*)
    (*Postcondition: All Total letter pairs have been counted.*)
    InitializeSort (SortTable);
    FOR C1 := "A" TO "Z" DO
        FOR C2 := "A" TO "Z" DO
            PairCount := GetRaw (RawTable, C1, C2);
            Insert (SortTable, C1, C2, PairCount)
        END
    END;
    (*Postcondition: SortTable holds the fifty most frequent letter pairs.*)
    FOR Position := 1 TO 50 DO
        GetSort (SortTable, Position, C1, C2, PairCount);
        Percent := 100.0 * (FLOAT(PairCount) / FLOAT(Total));
        IO.CharOut (C1);
        IO.CharOut (C2);
        IO.StringOut (":  ");
        IO.FixOut (Percent, 4, 1);
        IO.StringOut ("%    ");
        IF (Position MOD 5) = 0 THEN IO.LineOut END
    END
END Doubles.
```

```
Analysing letter pair frequency in file Chapter.10.
TH:  2.6%  IN:  2.4%  RE:  2.3%  ER:  2.0%  HE:  1.9%
OR:  1.7%  CO:  1.6%  EN:  1.5%  NT:  1.5%  ON:  1.4%
AR:  1.4%  AT:  1.4%  TE:  1.3%  PE:  1.3%  TI:  1.2%
AN:  1.2%  ST:  1.2%  ES:  1.2%  IT:  1.2%  LE:  1.1%
RA:  1.1%  TO:  1.1%  AL:  1.0%  ND:  1.0%  EC:  1.0%
NG:  1.0%  TY:  1.0%  ME:  0.9%  OF:  0.9%  DE:  0.9%
YP:  0.9%  OU:  0.9%  ED:  0.9%  HA:  0.8%  FI:  0.8%
RD:  0.8%  AM:  0.8%  BE:  0.8%  ET:  0.8%  IO:  0.8%
SE:  0.8%  AS:  0.8%  VE:  0.7%  PR:  0.7%  EA:  0.7%
RI:  0.7%  NE:  0.7%  TA:  0.7%  CA:  0.6%  IE:  0.6%
```

Modification Problems

1. In designing Doubles I intentionally avoided goal-oriented *procedural* abstractions so that I could concentrate on object-oriented *data* abstractions. The penalty I paid is that the main program's statement part is longer and more detailed than it might be.

Use the layered design approach we discussed to develop procedural abstractions for Doubles. Try to define a name layer for main-program procedure calls, then 'implement' the procedures by moving code out of the main program. You can continue to use the primitives we created for the two data abstractions.

2. The method we used to implement procedure Insert was just one of two possibilities (reread the A Coding Detail section). Rewrite Insert so that it locates the proper position for the new values *before* any records slide down, instead of sliding them down while we look for the right position.

3. Suppose that instead of having two CHAR-type fields, the ShareTYPE record held an array of two CHAR components:

```
LettersTYPE = ARRAY [0..1] OF CHAR;
ShareTYPE = RECORD
               Letters: LettersTYPE;      (*the letter pair*)
               Share: INTEGER           (*its count*)
            END;
```

What modifications will Doubles require? Is the main program affected?

Section 10-3 Standard Practices: Layered Type Design

RECORDS ARE INCREDIBLY USEFUL FOR PUTTING together new *type abstractions*. Calendar dates, oversized characters, complex numbers, polar coordinates—each of these values has simpler parts, sometimes of different types. They're all ideal candidates for programming with records.

Now, what's a type abstraction? Well, although we've worked with data abstractions for a few chapters I haven't bothered to distinguish between type and structure. Loosely, a data abstraction has been any type or structure that was defined in terms of primitive operations. As it happens, we haven't been too concerned with making up new kinds of values, so the abstractions we've talked about have usually been abstract structures. Our abstractions have focused on the relationship between stored values, rather than on the values themselves.

A *type abstraction* does describe new values. As always, the description is indirect: the values are defined in terms of primitive operations with the values, and not by their Modula-2 definitions.

> The distinction between abstract types and structures isn't an important fact per se. It's useful because it can encourage more careful development of abstractions, so that they'll be easier to reuse in future programs.

New types used to be put together haphazardly, as the demands of a particular program dictated. However, in recent years careful type design has become a basic issue of programming practice, especially as the object-oriented, data first design methods we've been using have become popular.

In this section, we'll look at the development of new types in isolation, independently of solving any particular programming problem. I'll begin by nailing down the difference between a new type as an abstraction and as an implementation:

abstraction vs. implementation

- A type *abstraction* consists of *a*) plain English definitions of its values, and *b*) plain English explanations of the names or symbols that invoke operations with those values.
- A type's *implementation* includes *a*) its Modula-2 type definition, and *b*) the Modula-2 code required to carry out primitive operations with values of the type.

good abstraction hides implementation

> A successful type abstraction lets us work with values without knowing anything about their underlying implementation.

This idea is obviously the basis of the layered approach to design:

name layer vs. code layer

- First, abstract the type by coming up with the type's *name layer*—a requirements and performance specification for the new type.
- Second, implement the type by developing the type's *code layer*—the Modula-2 type definitions and primitive operations that let the new type be used.

Now, in case this is starting to sound rarified, let me point out that you've been working with type abstractions for as long as you've been programming. Think about the built-in data type REAL. What is it, anyway? To us, the type consists of our understanding of what real values are, along with the word REAL, the symbols '+', '–', '*', '/', and the primitive operations they imply. But this is just an abstract definition that ignores the bits and bytes that implement REAL at the machine level, and keep track of the whole and fractional parts, scale factor and signs of every number.

Is REAL a successful abstraction? Marvelously so! We can use REAL values for every conceivable purpose without having the faintest idea of how the type is implemented. Is the compiler's definition of type REAL based on an array, a record, or separate INTEGER and BOOLEAN variables? I don't know, and I don't care. Imagine how much work actually goes on when two REAL values are added or divided! Fortunately, you and I are oblivious to this toil—we deal with the real numbers in everyday abstract terms.

Self-Check Questions

Q. What primitives does Modula-2 supply for INTEGER? Are all of them necessary? Could you do away with any of the built in primitives without limiting your ability to use INTEGERs?

A. The obvious primitive operators are +, –, *, **DIV**, and **MOD**. Although we could use addition and subtraction to fake the others, it would be so inefficient that multiplication and division are supplied as primitives too.

There are also less obvious primitives. MAX and MIN help define the legal range of INTEGER; other built-in functions include ABS and ODD. The relational operators >, =, and so on also define the meaning of type INTEGER.

Some Warmup Abstraction

Later in this section we'll abstract and fully implement a new data type for working with fractions. First, though, let's get a little bit of practice with the first (and more confusing) step—defining a type in terms of its legal values and primitive operations.

abstracting DateTYPE

I'll start with *dates*. What are the values of an imaginary DateTYPE? Well, the year runs from January 1 through December 31. Not every combination of month and date is legal, though. January 32 isn't a date, for instance. Nor are February 30 or April 31. February 29 might or might not be a date, depending on what year it is.

Let me propose a plain English definition:

DateTYPE *represents any legal calendar date.*

This definition is most interesting for what it *doesn't* include. It doesn't say what a date looks like. It doesn't explain how to type in or print out DateTYPE values. It certainly doesn't say how dates are going to be stored. All it does is establish the range of values we'll expect a DateTYPE variable to store correctly.

What operations might be invoked on DateTYPE values? It's not easy to start thinking about how to use values that we can't 'see,' is it? I'll suggest these:

DateTYPE primitives

AssignDate *will store a date supplied as a month, day, and year.*
PrintDate *will print the date in month/day/year format.*
Tomorrow *increases the date of a* DateTYPE *argument by one day.*
NextWeek *increases the argument's date by one week.*
NextMonth *increases the argument's date by one month.*

These are all primitives, because they obviously can't be written without knowing some of the details of DateTYPE. Together, they establish the meaning of DateTYPE by what they allow as well as what they don't provide for. Add two dates? Can't be done, because the primitives don't allow for it.

abstracting TimeTYPE

Let's try a second example—*times*. What are plausible TimeTYPE values? According to my wristwatch, hours can be numbered one through twelve, while minutes and seconds go from zero through fifty-nine. Of course, that only takes care of half the day, so I'll propose this as my plain English definition of TimeTYPE:

TimeTYPE *represents any legal time in hours, minutes, seconds, and AM/PM.*

As before, this definition doesn't say anything about how times are going to be stored, what they look like, how they'll be input or output, and so on. It does establish that times are limited to reasonable values: 13:67 PM isn't in the book.

TimeTYPE's primitives are a little bit different from DateTYPE's. Assignment and inspection primitives are there, of course, but it's also reasonable to talk about doing 'time' arithmetic.

TimeTYPE **primitives**

AssignTime *will store a time supplied as hours, minutes, seconds, and AM/PM.*
PrintTime *will print the time in hours:minutes:seconds:AM/PM format.*
AddTime *will add one* TimeTYPE *value to another.*
SubtractTime *will subtract one* TimeTYPE *value from another.*

The informal descriptions of DateTYPE and TimeTYPE have to be taken a little bit further to establish a name layer. After all . . .

> The name layer should be a requirements and performance specification that's tight enough to let one programmer *use* the new type while her partner *codes* it.

I'll stick with the same method I've used before, and let mock subprogram headings flesh out the specifications:

calling forms of DateTYPE **primitives**

```
TYPE DateTYPE will represent any legal date;
PROCEDURE AssignDate (VAR Date: DateTYPE;
                          Month, Day, Year: INTEGER);
   (*Stores the new date if it's legal, stores 1,1,1 otherwise.*)
PROCEDURE PrintDate (Date: DateTYPE);
   (*Prints the date in month/day/year format.*)
PROCEDURE Tomorrow (VAR Date: DateTYPE);
PROCEDURE NextWeek (VAR Date: DateTYPE);
PROCEDURE NextMonth (VAR Date: DateTYPE);
   (*Advances the stored date by one day, week, or month.*)
```

The equivalent collection for TimeTYPE is:

calling forms of TimeTYPE **primitives**

```
TYPE TimeTYPE is a legal combination of hours, minutes, seconds, and AM/PM;
PROCEDURE AssignTime (VAR Time: TimeTYPE;
                          Hours, Minutes, Seconds: INTEGER;
                           AMorPM: CHAR);
   (*Stores the time if it's legal, stores 0,0,0,A otherwise.*)
PROCEDURE PrintTime (Time: TimeTYPE);
   (*Prints the time in HH:MM:SS:?M format.*)
PROCEDURE AddTime (VAR Time: TimeTYPE; Plus: TimeTYPE);
   (*Adds Plus to Time.*)
PROCEDURE SubtractTime (VAR Time: TimeTYPE; Less: TimeTYPE);
   (*Subtracts Less from Time.*)
```

Modification Problems

1. Make up a type definition for DateTYPE or TimeTYPE. Put the definition and procedure headings (with empty bodies) into a program shell, and see if it compiles.
2. Suppose that TimeTYPE turned out to be INTEGER, rather than some record type. Can you still manage to write its primitives? Will they be more or less complicated?
3. What additional primitives might you require for the above types? Can you devise one or more basic 'increment' primitives that would take a *how many?* argument, and might supersede Tomorrow, NextWeek, and NextMonth? Does it make sense to define relational operators for either DateTYPE or TimeTYPE?

A Larger Example: FractionTYPE

Let's try a larger example, and carry it through to implementation. Our problem will be to create a type for representing fractions. Coming up with an abstract definition of the type—the first step in the process of design—requires answers to two questions:

1. What are the legal values of the new type?
2. What basic operations can be applied to those values?

the range of fractions

The question of values can be dealt with easily. Each fraction, we will agree, consists of a pair of values: a whole numerator, and a whole denominator. Either of them might be negative, but for now, I'll say that both fall within the range of INTEGER. However, we can return to this detail later, when we work on the concrete implementation of a fraction type.

What operations are carried out on fractions? They might include:

identifying the primitives

input/output	*read*	*write*		
inspection	*get numerator*		*get denominator*	
assignment	*store numerator*		*store denominator*	
boolean	*equal*	*less than*	*negative*	
transform	*reduce*	*reciprocal*	*negate*	
transfer	*decimal*	*fractional*		
arithmetic	*add*	*subtract*	*multiply*	*divide*

Now, the table above gives a very broad answer to the question *what operations can be applied to fractions?* From the viewpoint of type design, though, it is too all-inclusive. Will we really have to supply procedures or functions for each and every one of those operations? No. We can narrow the field by remembering the definition of a 'basic' operation—it is *primitive*.

primitive operations

> Writing a primitive operation requires knowledge of a type's underlying implementation. In other words, you can't write it without knowing details of the type definition.

Operations like *store numerator* and *get denominator* are primitives, because we can't possibly write them without knowing how fractions are stored. In contrast, a BOOLEAN function like *negative* isn't really a primitive. If we're able to find out what the actual numerator and denominator values are, then negative *can* be written without knowing how the fraction is stored.

In practice, a type's abstract definition won't always be limited to the bare minimum of primitive operations. Frequently it will also include operations that *could* be put together from simpler primitives, but perhaps not easily or efficiently.

Suppose that we get to work on a name layer for FractionTYPE.

a name layer for FractionTYPE

```
TYPE FractionTYPE doesn't have to be defined yet;
(*Mock headings for manipulating FractionTYPE values.*)
Equal (Frac1, Frac2: FractionTYPE) : BOOLEAN;
LessThan (Frac1, Frac2: FractionTYPE) : BOOLEAN;
GetNumerator (Frac1: FractionTYPE) : INTEGER;
GetDenominator (Frac1: FractionTYPE) : INTEGER;
PutNumerator (Numerator: INTEGER; VAR Frac1: FractionTYPE);
PutDenominator (Denominator: INTEGER; VAR Frac1: FractionTYPE);
Reduce (VAR Frac1: FractionTYPE);
FractionalEquiv (Decimal: REAL; VAR Frac1: FractionTYPE);
Add (VAR Result: FractionTYPE; Frac1, Frac2: FractionTYPE);
Subtract (VAR Result: FractionTYPE; Frac1, Frac2: FractionTYPE);
Multiply (VAR Result: FractionTYPE; Frac1, Frac2: FractionTYPE);
Divide (VAR Result: FractionTYPE; Frac1, Frac2: FractionTYPE);
```

FractionTYPE's Code Layer: The Type Definition

Let's now turn to the actual implementation of the new fraction type—the code layer. We have two jobs:

- choosing the type definition that will underly FractionTYPE, and
- writing code for the primitive operations on FractionTYPE variables.

The name layer outlined above will serve as the requirements specification. How shall we complete the definition of FractionTYPE? The simplest definition would be:

a basic definition

```
TYPE FractionTYPE = RECORD
                      Numerator, Denominator: INTEGER
                    END;
```

Is this the only possible definition of a type suited to representing fractions? No, of course not. This definition is an equally plausible alternative:

```
TYPE PartsTYPE = (numerator, denominator);
     ArrayFractionTYPE = ARRAY PartsTYPE OF INTEGER
```

an array-based alternative

Or, we could return to using a record, but with CARDINAL fields, in order to double the range of the numerator and denominator. Since CARDINALs are nonnegative, we have to include additional fields for the signs:

yet another record alternative

```
TYPE SignTYPE = (positive, negative);
     LongFractionTYPE = RECORD
                           NumSign, DenomSign: SignTYPE;
                           Numerator, Denominator: CARDINAL
                        END;
```

> As long as we meet all abstract requirements specified for FractionTYPE, our particular choice of implementation is not crucial—if we supply all the primitive operations that might require knowing implementation details.

It might be more or less convenient to write the primitives using any given type definition. However, from the point of view of the type's abstract definition, the concrete type definition is a hidden detail. As long as Fraction-TYPE variables behave properly, it doesn't matter how they are put together.

The choice of implementation is not entirely unimportant, of course. For instance, when we discussed the abstract definition of FractionTYPE I limited the numerator and denominator to the range of INTEGER. Had I insisted on a wider range as part of the abstract definition, we'd have to fulfill that promise here.

FractionTYPE's Code Layer: Primitive Declarations

Let's assume that the simplest definition of FractionTYPE will suffice; it can be a record with just two fields. Writing the code of the primitive operations is our next step. I'll just code a few of the subprograms since, aside from the fact that we're working with records, there's nothing radical about them.

partially implementing the primitives

```
PROCEDURE Equal (Frac1, Frac2: FractionTYPE) : BOOLEAN;
  (*TRUE if the numerators and denominators are equal.*)
  BEGIN
    RETURN (Frac1.Numerator*Frac2.Denominator)
              = (Frac2.Numerator*Frac1.Denominator)
  END Numerator;
PROCEDURE GetNumerator (Frac1: FractionTYPE) : INTEGER;
  (*Represents the argument's numerator as an INTEGER.*)
  BEGIN
    RETURN Frac1.Numerator
  END GetNumerator;
```

(code continues)

```
PROCEDURE PutNumerator (Numerator: INTEGER;
                        VAR Frac1: FractionTYPE);
  (*Assigns a numerator value to Frac1.*)
  BEGIN
    Frac1.Numerator := Numerator
  END PutNumerator;
```

Some of the procedures involve more detailed programs. For example Reduce, which is called below in Divide, relies on Euclid's algorithm (described in Chapter 6) for finding greatest common denominators.

```
PROCEDURE Divide (VAR Result: FractionTYPE;
                      Frac1, Frac2: FractionTYPE);
  (*Divides, then reduces, two fractions.*)
  BEGIN
    Result.Numerator := Frac1.Numerator * Frac2.Denominator;
    Result.Denominator := Frac1.Denominator * Frac2.Numerator;
    Reduce (Result)
    (*Do you think I would deprive you of the pleasure of writing Reduce?*)
  END Divide;
```

limit knowledge

> As a general rule, as few primitives as possible should require knowledge of the actual type definition. This makes it easier to modify the implementation details later, if necessary.

Suppose that our goal were to write as few primitives as possible. As a brief mental exercise, try to make an outline of the primitives that are absolutely necessary. For example, we might have written Divide using calls of GetNumerator, and PutNumerator, and *never* referred to the Numerator and Denominator fields directly. Reciprocal, below, is another procedures that's deliberately written in terms of primitives.

using a few primitives

```
PROCEDURE Reciprocal (VAR Frac1: FractionTYPE);
  (*Returns the reciprocal of its argument.*)
  VAR Temp: FractionTYPE;
  BEGIN
    PutNumerator ( GetNumerator(Frac1), Temp);
    PutDenominator ( GetDenominator(Frac1), Temp);
    Frac1 := Temp
  END Reciprocal;
```

A procedure like Reciprocal is a proof of the value of abstraction. Even though it can change the value of a stored fraction, it doesn't depend in any way on the actual definition of FractionTYPE. The underlying definition could be changed to one of the alternatives I suggested earlier without affecting the operation of Reciprocal in any way.

Section 10-4 Program Engineering Notes

WHEN WE DISCUSSED PROGRAM ENGINEERING WITH ARRAYS, our chief concern was sight—seeing data that was hidden within arrays. Our tools were Load and Dump, the basic procedures for initializing and displaying stored data.

When arrays were our only structured type, Load and Dump were primarily used as testing and debugging aids. However, as you begin to work with records you will find that:

- the amount of data programs use tends to increase overall; and
- it is very often necessary to save data between program runs.

For instance, even though one of the simple filters we've written in the past could be used to process reams of data, it might not have to actually hold more than a few characters' worth at a time.

In contrast, programs that use more elaborate data types usually have to manage a great deal of information. Indeed, that's why we're using the fancier types in the first place. If there's enough data involved, it may turn out to be convenient to try to save data between program runs. A game-playing program might save current board positions, while an address-book program would certainly save stored names and numbers.

As a result, Load and Dump become more than convenient procedures for tracking down errors. Instead, they can become a program's main input and output routes. This means that we'll have to think a little more carefully about how they're put together—about what Load expects to read, and what Dump is expected to put out.

transfer data as text

> Programs move and transform data quickly, but not always conveniently. When it is possible, assume that data will start and finish as plain text—ordinary characters in an ordinary file. In other words, use plain text as the medium of exchange *between* programs, as well as between your program and you.

As you become more adept at summoning up the tricks of a particular programming language, you have to remember that writing new programs is not always the answer. Most programs, after all, are tools, rather than all-inclusive systems. A program is usually invoked to do some relatively specialized job, and not to control every possible interaction with stored data.

write tools, not systems

Consider a basic application like maintaining an address book. You'd want to write a program to look up a particular number or address, or to sort data by name or zip code. But some ordinary, everyday operations, like adding a new entry, or correcting an old one—might be more easily handled with a very ordinary, everyday tool—a text editor.

Load and Dump procedures won't necessarily show up in a problem statement. However, they are useful debugging procedures that can end up making your programs work better, too.

Self-Describing Data

Now, what does it mean to use plain text as a medium of exchange? Well, you have to recognize that a file of stored data has two potential users: people, and programs. It has to hold information in a manner that

- makes it easy for a human reader to understand and interpret the meaning of the stored values; and
- makes it easy for a program to load its internal data structures from the file, and to dump data in the same format.

Thus, data should be labeled (for the benefit of humans), but the labels can't interfere with input and output (for the benefit of programs). The phrase *self-describing data* has been used to describe this goal.

Let's look at some examples. Suppose that a record holds a name, an age, and various other assorted data items.

a lot of stored data

```
TYPE InfoTYPE = RECORD
                    Name: StringTYPE;
                    Age: INTEGER;
                    . . .        (*other fields*)
                END;
     EntireTYPE = ARRAY [1..100] OF InfoTYPE;
```

What kind of standardized formats can we establish for Load and Dump? I've tried to attach names that highlight the differences between these examples:

different data formats

```
Doug                                  line-oriented
36
   . . . other fields of one record

User's name is Doug                   embedded
He is 36 years old.
   . . . other fields of one record

User's Name: Doug                     labeled and delimited
Age: 36
   . . . other fields of one record

Doug    #the user's name              positional
36      #the user's age
   . . . other fields of one record
```

The first format is line-oriented—each line contains a single item of data. The second embeds data within a prose description of its meaning. The third format uses a colon (:) as a delimiter to mark the end of a descriptive comment.

Finally, the fourth format has a delimiter (the **#**) to catch your eye, but it's really positional—the first item on the line is the data value.

What are the relative advantages of each approach?

- The basic line-oriented format is easiest from the viewpoint of computer input and output, but it's hard for a human reader to understand what the data means—especially if it's mistaken.

advantages and disadvantages

- Embedding the data in a description errs in the other extreme. Humans can read it easily, but it's difficult to write a Load routine that can extract the actual data.
- Labeled output is typical of interactive programs. A Load procedure can skip to the delimiter that marks the end of the descriptive comment. This is a little restrictive because the comment can't contain the delimiter, and because the space that follows the delimiter might be a meaningful CHAR.
- Positional notation is probably the best method. Actual data values can lead the line, while the delimiter *follows*.

Note that these guides apply to the contents of a single record. How would you separate one record from the next? Most people use one or two completely blank lines, or have yet another delimiter that separates one record from the next, e.g.:

delimiters between entries

```
!
one record's data
!
the next record's data
!
the next record's data
$    end of input
```

The ultimate goal, as noted above, is to embed descriptions of data within the data file itself. Although *programs* that use the data may ignore the descriptions, *people* who read it will be grateful.

Record Bugs

Like arrays, records pose the immediate problem of notation. We have barely mastered square brackets when, out of the blue, we have to deal with dots as well. Is this:

```
This[That].TheOther.Something.Else[3]
```

a variable identifier? Well, it might be. On the other hand, it could be bug city.

An old game that I just made up is custom-tailored to give you practice with notation. If you've been to camp, you probably know how to tell *continuation stories*. Each camper gets a minute or two to add a new chapter to a continuing, usually ever-more-grisly campfire story. Well, let's do some continuation naming. Start with an identifier:

```
Age
```

Make it one field of a record:

```
Facts.Age
```

practice with constructed types

Turn that record into one component of an array:

```
Camper[32].Facts.Age
```

Turn the array into a field:

```
Boys.Camper[32].Facts.Age
```

I think that you get the idea. For more practice, make up an arbitrarily complicated variable name, like:

```
Room[273].NextWeek[Monday].Schedule[3, PM] := Busy;
```

Then, figure out what's going on. Here, it seems that Room is an array of records. One field of each record—the NextWeek field—is also an array of records. One field of the Monday record—the Schedule field—is a two-dimensional array. By reading the assignment one step at a time (and breathing very slowly) we can conclude that on next Monday, Room 273 will not be available at 3 P.M.

Enough notation! Two other common errors in working with records are:

1. forgetting to use the **END** that marks the conclusion of a record type definition, and
2. accidentally using a type identifier in place of a variable identifier.

These are syntax bugs, so the compiler can catch them. Nevertheless, they're annoying. Leaving the **END** off a record is confusing because it may lead to a misdiagnosis of the actual error. For example:

a missing **END**

```
TYPE PropertyTYPE = RECORD
                        Price: REAL;
                        Zoning: CHAR;
VAR ApartmentNumber: INTEGER;
    House: PropertyTYPE;
```

What will the compiler read? Well, it doesn't know that **VAR** isn't intended to be another field for PropertyTYPE. You'll get an error message that points out the futility of using reserved words as field identifiers, along with a host of other presumed transgressions. Compilers freak out because they're built that way. You, on the other hand, should look for the missing **END** before you panic.

Using a type name as part of an expression is almost always a mistake. It's obvious here because of the capitalization of type names:

```
IO.RealOut (PrcpertyTYPE.Price);
  (*This should refer to House.Price.*)
```

> As a rule, type identifiers almost *never* appear in statements. If you see a type identifier lurking about a main program, it's probably up to no good.

The only exception to the rule involves calls to the standard procedures MIN, MAX, and SIZE.

WITH Bugs

The **WITH** statement also provokes a few perplexing situations. **WITH** simplifies naming for just one record. As a result, using **WITH** to simplify the name of a record that's an array component can be tricky. This sequence is wrong:

a buggy sequence

```
WITH AnArray[i] DO
  FOR i := 1 TO Last DO
    references to the fields of the current AnArray component
  END
END
```

Even though i changes inside the **FOR** statement, the particular record that **WITH** is attached to doesn't. The order of the statements has to be reversed:

the correct version

```
FOR i := 1 TO Last DO
  WITH AnArray[i] DO
    references to the fields of the current AnArray component
  END
END
```

Modula-2 Details

Syntax Summary

• record: a structured type whose components—the record's *fields*—can have different types:

```
TYPE ExampleTYPE = RECORD
                     Field1: ItsTYPE;          (*the field lists*)
                     Field2, Field3: TheirTYPE
                   END
VAR Test, Sample: ExampleTYPE;
```

• The fields can have any type, simple or structured.

• dot notation: used to *qualify* a field, i.e. to construct its full identifier:

Test.Field1 *is different from* Sample.Field1

• **WITH** statement: establishes a reference to a particular record variable so that its fields don't have to be individually qualified:

```
WITH Test DO
    Field1 := whatever;   (*this refers to Test.Field1*)
    Sample.Field1 := whatever   (*Sample still requires dot notation*)
END
```

Modula-2 Details

Remember . . .

• Assignment can be made between any two record variables as long as they have the same type. However, none of the relational operators, not even =, are defined for record types. Records must be compared one field at a time.

• A record definition creates a new scope for identifiers, just as a subprogram does. An identifier used as a field name doesn't conflict with an identifier used outside the record definition, e.g. in other record definitions.

• Within a given **WITH** statement, you can't switch from one record to another. In particular, when looking at record-components of arrays, traverse the array outside the **WITH** statement, not inside it.

• A type *abstraction* consists of plain English definitions of its values, and plain English explanations of the names or symbols that invoke operations on those values. Another way of phrasing this is to say that a type is defined, in abstract terms, by its *values* and its *operations*, and not by its implementation.

• A *name layer* is a design tool for formalizing a type's abstract definition. It consists of the type's identifier, and the *calling forms*—names and headings—of subprograms that will carry out its primitive operations. The name layer acts as a *performance specification* for using the type abstraction, and as a *requirements specification* for implementing it.

• A type's *implementation*, or code layer, includes its underlying definition in Modula-2, and the code required to carry out operations on values of the type. It is invariably possible to suggest alternative implementations for any given abstraction.

• By definition, a *primitive* operation on a value can't be implemented effectively without knowing the value's underlying type definition. Since primitives do rely on a particular underlying definition, having large numbers of primitives makes it difficult to change an implementation—all the primitives may have to be rewritten.

Zen Study Guide

Big Ideas

The Big Ideas in this chapter are:

1. *Type design methodologies are meant to separate what a type does from how it's done in Modula-2.* Layered design, with its clear division

between name and code layers, uses a small, well-defined set of primitive operations as a performance specification for using the type, and a requirements specification for implementing it. Using the abstraction means using the primitives; implementing the abstraction means implementing the primitives.

2. *Variables of structured types will usually have constructed names.* Identifiers frequently have to be constructed using arbitrary symbols like '.', '[', and ']'. The identifier FirstName may seem more 'name-like' than List[7].First or Store.Last[3,5], but they are all variable identifiers.

3. *Most record programming is actually array-of-record programming.* The record is a syntactic mechanism that lets arrays store packages of related values that have different types. As a result, record programming is primarily concerned with type design (unlike array programming, which focuses on algorithms).

Zen Study Guide Questions

10-1 Scope rules are intended to let identifiers be reused within programs, while at the same time forbidding any reuse that could lead to ambiguity. The prototype test question for exploring scope confusion is *Do any of these identifiers have to be renamed:*

```
TYPE ATYPE = RECORD
                A: INTEGER;
                B: REAL;
             END;
     BTYPE = RECORD
                A: INTEGER;
                B: REAL;
             END;
VAR A: ATYPE;
    B: BTYPE;
```

This particular example is perfectly legal. Make up three variations that are not so acceptable.

10-2 Once a structured type has been defined, access to its values can be made arbitrarily difficult. The most confusing variables involve records within arrays within records, ad infinitum Define the most hideously complex type you can think of, then ask *Which of these are legal assignments?* to it.

10-3 Initializing a single record isn't too hard. Nor is initializing an array of simple values. But suppose that you have an array of records? You might be given an array of twenty-six records, each with a single INTEGER and CHAR field, and be asked to initialize the records to *a*) 0 and "A", *b*) increasing numbers and letters, *c*) increasing numbers and decreasing letters, etc.

Make up a few mechanical questions of this sort. The two variations to make are *a*) making relatively complicated initializations of simple records, and *b*) keeping the initialization simple, but giving the records a more complex internal structure.

10-4 The **WITH** statement issues two invitations for appearance on a quiz. First, suppose that two record variables called Left and Right have the same type. Each includes a field called Up. If you're given this code segment:

```
WITH Left DO
    WITH Right DO
        Up := 0
    END
END
```

questions you might pose include *a) Is this segment legal?*, *b) Which record's* Up *field is being assigned to?*, and *c) How would you assign to the other record's* Up *field within the inner statement?* Make up another question or two, then create two more potentially confusing code segments of this sort.

10-5 A word might be best suited for storage in an array, whereas a word and its frequency might work best within a record with an array field—and a hundred words and frequencies would be ideal in an array of records with array fields. Because I know that you probably watch *Jeopardy*, please make up the ten questions whose answers are:

a) an array;

b) a record;

c) an array of records;

d) a record with an array field;

e) an array of records with array fields.

10-6 If I were writing a test, I'd love to supply a few primitive operations on a type as a specification, then ask my students to propose two alternative implementations of the underlying type definition. For example, in 10–2, we named primitives before looking at alternative implementations of an underlying definition for FractionTYPE.

Come up with three new data types (say, ColorTYPE or TimeTYPE), and name a few of the primitive operations each one requires. Then, supply two or three alternative definitions for each type. Why two or three alternatives? So that you can ask the question *Which of these definitions would* not *be suitable for the primitive operations listed above?*

10-7 Choose any three board games, and ask the question *Define a type suitable for representing this board.* You can make this a little harder by asking *What are the primitive operations of this game?* as well.

10-8 This address book record is a very common piece of boilerplate:

```
TYPE StringTYPE = ARRAY [0..50] OF CHAR;
     EntryTYPE = RECORD
                    FirstName, LastName: StringTYPE;
                    Street, City, State: StringTYPE;
                    ZipCode: INTEGER;
                    AreaCode, Phone: INTEGER
                 END
```

Many applications—employment records, school records, medical records—require some variation on the basic definition. Pick three, then ask the question *How would you change the boilerplate definition?*

10-9 Developing test data is, by now, as critical a programming skill as coding. Suppose that you are in possession of a Sort procedure, and can run three tests on it. Write six possible sets of test data (an array of ten values each should suffice) for the question *Which three of these six tests should you run on* Sort?

10-10 A *stable* sort leaves the relative order of two equal values unchanged, while an *unstable* sort can exchange them. This isn't important when you're just sorting numbers, but it can be important if the sort key (the number you sort on) is one field of a larger record.

Suppose that you plan to sort this array:

```
TYPE ItemTYPE = RECORD
                    Count: INTEGER:
                    Value: CHAR
                  END;
     SortedListTYPE = ARRAY [1..26] OF ItemTYPE;
```

Write two procedures that initialize a SortedListTYPE array for testing purposes. Then, ask the question *Which of these procedures will let us know if* Sort *uses a stable or unstable sorting algorithm?*

10-11 Suppose that you were given a Sort procedure, but weren't told what algorithm it used. What kinds of statistics (e.g. number of swaps) would you have to have in order to decide? Using what test data?

Exercises

Warmups and Test Programs

10-12 What error messages accompany:

a) a record that's missing its **END**?

b) a record that uses the same field name twice?

c) a **RECRD**?

10-13 Suppose that I make these definitions:

```
TYPE ATYPE = RECORD
                 A: INTEGER;
                 B: REAL;
               END;
     BTYPE = RECORD
                 A: INTEGER;
                 B: REAL;
               END;
VAR First, Second: ATYPE;
    Third: BTYFE;
```

Initialize First, then find out the effects of these assignments:

a) Second := First;

a) Third := First;

10-14 How deeply can record definitions be nested? Does your system impose any limit?

10-15 Is a complete record assignment any faster than assigning the fields one at a time? Try it on an array of 10,000 records.

10-16 Define an array of fifty-two records, where each record holds a number and a character. Initialize the array so that:

a) each number field is 0 and each character field is the letter "B";

b) the numbers ascend from 1 through 52, while the letters go from "a".."z", "A".."Z";

c) the numbers go from 52 down to 1, while the letters go from "a" to "z" twice.

10-17 Suppose that a file contains as many as fifty lines, each of which contains a name and an age, in that order. These definitions create a structure suited for storing the names and ages:

```
TYPE StringTYPE = ARRAY [0..35] OF CHAR;
     ItemTYPE = RECORD
                  Name: StringTYPE;
                  Age: INTEGER
                END;
     ListTYPE = ARRAY [1..50] OF ItemTYPE;
```

Write two primitive procedures: one should read from the file to initialize the array of records, while the other should print the same information in reverse (age followed by age).

10-18 Define primitives, then a data type, suitable for representing the game of Monopoly.

10-19 Define primitives, then a data type, that could be used for finding the distance between geographic locations.

10-20 Define a data type that might be used to represent 100-digit REAL numbers. Don't forget about signs and exponents.

Exercises

Programming Problems

10-21 A sorting method is said to be *stable* if two equal keys maintain the same order in both unordered input and ordered output. The sort is *unstable* if it can change equal values. For example, if our starting data consists of counts of letters:

A(5) B(7) C(7) D(6)

stable and unstable sorts are:

A(5) D(6) B(7) C(7)	A(5) D(6) C(7) B(7)
a stable sort	*an unstable sort*

Consider the procedures for selection sort and bubble sort that I wrote in 9-2. Are they stable or unstable? Can a stable implementation be made unstable, and vice versa?

10-22 Blood pressure is given by a pair of numbers—the systolic and diastolic pressures—in millimeters of mercury. Pulse pressure, in turn, is defined as the difference between the two.

Write a program that is able to:

a) order blood pressure records by systolic, diastolic, or pulse pressures;

b) print the *n* records with the highest pressures of a given type;

c) print any records that appear on all three 'high pressure' lists.

Hint: use a single sorting routine by defining a SortKey field for each record. Don't forget to number (or somehow identify) individual records.

10-23 A fraction consists of a numerator and denominator. Write a program that reads an expression that contains plus or minus signs and numerator/denominator pairs, then:

a) orders them by denominator;

a) adds or subtracts, as appropriate, numerators of like denominators;

a) prints the resulting expression.

For example, the input $+\frac{2}{7}+\frac{3}{5}-\frac{1}{5}+\frac{3}{7}-\frac{1}{9}$ should yield $+\frac{2}{5}+\frac{5}{7}-\frac{1}{9}$. Try your program by adding a *Farey Series*, devised by John Farey in 1816, of reduced fractions between 0 and 1, with denominator $\leq n$. The 7 series is:

$$\frac{1}{7}+\frac{1}{6}+\frac{1}{5}+\frac{1}{4}+\frac{2}{7}+\frac{1}{3}+\frac{2}{5}+\frac{3}{7}+\frac{1}{2}+\frac{4}{7}+\frac{3}{5}+\frac{2}{3}+\frac{5}{7}+\frac{3}{4}+\frac{4}{5}+\frac{5}{6}+\frac{6}{7}$$

(An interestingly difficult problem that I will raise, but not pose, incidentally, is to write a program that produces a Farey series.)

10-24 Use the binary search method described in Chapter 9 to implement a form of reverse telephone directory—one that is ordered by telephone number, rather than by the subscriber's name. Allow these operations:

a) Add a name, number, and address.

b) Look up a number, and print the name and address.

c) Delete a subscriber.

10-25 Write a program that reads twenty *x*, *y* coordinate pairs, and sorts them:

a) In order of increasing *x* values.

b) In order of decreasing *y* values.

c) In order of increasing distance from 0,0.

Why might a mechanical pen plotter find this sorting useful? How difficult is it to sort the points in order of distance from an arbitrary point?

10-26 A problem that often arises in the design of VLSI (very large-scale integration) circuits is to determine whether or not two circuit wires inadvertently intersect. For all practical purposes, the lines are either vertical or horizontal. Each line can be described by a pair of endpoints—(4,7..4,12) is a vertical line, while (2,9..5,9) is horizontal.

A simple check of *x* and *y* coordinates can reveal if two lines intersect. Write a program that reads twenty pairs of coordinates, sorts them into horizontal and vertical lines, then determines which lines intersect.

10-27 A standard problem of computer graphics systems is the suppression of hidden surfaces—surfaces that are behind other surfaces, and should not be shown.

Write a program that lets a user specify the four corner locations of up to five rectangles (you can assume that two corners will never occupy the same location), along with their relative depths—which is closest, and which are further away. Print the visible portions of each rectangle by 'painting' each onto a two-dimensional array (using a different number for each rectangle), then printing the entire array. How difficult is it to change the relative depth of the rectangles? You may find that an auxiliary, one-dimensional 'depth chart' array comes in handy.

10-28 Yet another problem that involves computer graphics deals with the difficulty of using a mouse or joystick to draw lines that are perfectly aligned. This is often a requirement in design work, particularly in the design of integrated circuits, where space is at a premium.

A common technique is called *Manhattanizing*. Suppose we represent a drawing area as a two-dimensional array. A mouse is used to indicate the approximate starting and ending points of a line. A program then makes sure that all approximately vertical lines are in a single column, perhaps by slightly moving a start or end point.

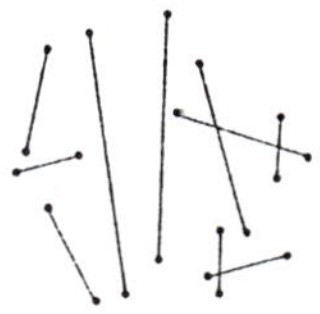

Before Manhattanizing

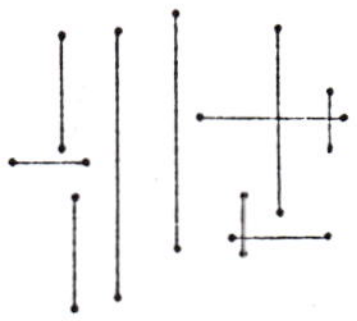

After Manhattanizing

Write a program that lets a user supply (in lieu of a mouse) the starting and ending points of more than one horizontal or vertical line. Allow these Manhattanizing features:

a) *Fixed posting*. Hold the upper and left-hand points fixed, relocate only lower and right-hand points.

b) *Two-point posting.* Move both endpoints, if necessary, to an 'average' vertical or horizontal axis.

c) *Equidistant posting.* Spread vertical and horizontal lines equally over the entire field.

10-29 Let's consider a diffusion problem that is bounded by a maximum level of absorption—for instance, the spread of oil from several different sources onto a standing body of water.

Suppose that a conveniently rectangular lake is 100 meters long and 50 meters across. If three different types of oil are dumped into the lake (at three different points), each will begin to spread evenly in all directions. However, when the oil slicks contact each other, they do not mix. Thus, the thickness of the oil layer within any particular slick at any particular point equals the average of the surrounding points *except that* a boundary with a different oil acts as an impenetrable edge.

Write a program that models the flow of oil across the surface of the lake. Assume that diffusion rates are equal, and the relative quantities of oil spilt are irrelevant. Where will the final boundaries between the slicks be? (Can you write a routine that represents a duck's attempt to remain in an unpolluted area?)

10-30 We usually think of computerized scheduling as an attempt to absolutely minimize usage of time or allocation of resources. In real life, though, such precision is impossible. Consider the problem of defining routes for garbage pickup. Two factors limit the length of each run: the garbage truck may get filled up, or it may be time to take a break. Suppose that we are given data of the following form:

```
block 1 weight 0.3 time 14
block 2 weight 0.1 time 8
  ...
```

Each block has two figures associated with it—the weight of its garbage, and the time it takes to pick it up. Naturally, blocks must be visited in order.

Write a program that can be used to schedule pickups, subject to these constraints:

Garbage-constrained routes are limited by the capacity of the truck.

Time-constrained routes are limited by the length of a shift.

A truck is assumed to be filled (or a shift ended) on the block prior to overflow (or overtime).

Your program should read data for an arbitrary number of blocks. Then, subject to maximum garbage and time constants supplied by the user, print the blocks on each garbage-constrained or time-constrained route. If you really want to help the dispatcher, order the list of time-constrained routes according to how full the truck is, and sort the list of garbage-constrained routes according to how long they take.

10-31 Anybody who has ever gone on vacation is familiar with the problem of *bin packing*—storing objects of varying sizes into a number of containers, or bins. There are actually many ways of picturing situations that involve bin packing—filling freight cars, stocking standard lengths of pipe, scheduling television commercials, buying assorted stamps from a vending machine, and so on—but the underlying problem is always the same: we want to store fixed-size values in the smallest possible number of fixed-size bins.

Packing and unpacking bins will eventually lead to the most effective solution, but the amount of work involved rapidly makes this approach infeasible. Two rote approaches that rely on simple rules are:

a) *First fit increasing* loads the bins starting with the smallest items first.

b) *First fit decreasing* loads bins starting with the largest items first.

Taking for granted that loads are ordered to begin with, ments and tests each method. Assume that we have twenty or each, and are given the following loads:

51	51	51	51	51	51
27	27	27	27	27	27
26	26	26	26	26	26
23	23	23	23	23	23
23	23	23	23	23	23

What are the contents of each bin after it is packed? Which strategy is better. optimal? Can you devise a strategy for loads that arrive in random order? (It ha shown, incidentally, that first fit increasing requires no more than (17/10), plus 2, of the optimum packing, while first fit decreasing is no worse than (11/9), plus 4.)

10-32 Geographic positions are given in degrees, minutes, and seconds of latitude and longitude. Since they are full circles of the Earth, latitude and longitude lines are broken into 360 degrees, where each degree has 60 minutes, and each minute has 60 seconds. Latitude lines are numbered from 0°, at the equator, through 90°, at each pole, which results in Northern and Southern (for their respective hemispheres) latitudes. Similarly, longitude lines begin with 0° at Greenwich, England, then increase toward both East and West until they reach 180° near Napier, New Zealand.

Now, although some variations may apply in the farm states, each fifteen-degree change in longitude usually implies a one-hour difference in standard time. Suppose that it is a clear spring equinox. Write a program that reads three positions of latitude (North or South) and longitude (East or West), along with the time at Greenwich. What is the time at each location? Which location is likely to be coldest?

10-33 We can define the *information content* of a letter as its predictive potential for the letter that follows. Since 'q' is invariably followed by 'u,' for instance, we can say that 'q' has a high information content.

Write a program that reads the contents of a textfile, and creates a table of letter-pair frequencies. You should convert all letters to lower-case. Then, either:

a) print the 100 most common letter pairs, in order (you'll find that keeping an array of records, where each record holds a letter pair and its count, is convenient), or,

b) print, in declining order of information content, a list of letters. For this exercise, say that a letter that can be followed by many letters has a low information content, while a letter that can be followed by few of its fellows has a high information content. (You'll find it's convenient to order 26 records, where each record holds a letter, and the number of different letters that follow.) Can you suggest a method of ordering that doesn't automatically allocate a high information content to letters that appear infrequently?

'At their most basic, library modules provide a way to split up the text of a program. In and of itself, this partitioning has no particular value.' (p. 440)

Chapter 11 Introduction to Library Modules

Any architect will tell you that scale—the relation of parts to each other, and to their environment—is an essential part of design. In building, scale is often based on the size of people's bodies; sometimes in obvious choices (like the size of doorways or ceilings) and sometimes in more subtle ways. Bricks are brick-sized, for example, because they fit into one hand easily and leave the other free for trowel and mortar. Clay roof tiles are shaped like short rain gutters because, for most of history, they were formed by draping a rectangle of wet clay over the builder's thigh.

Choosing a scale for program components has been more troublesome. Programs built to machine scale are too large—computers can hold programs that are far bigger than we can hope to understand. But programs that are limited to the ability of just one person are too small. We need a scale that will let people work together on projects that would overwhelm an individual.

The cure has been to attempt to find the programming equivalent of bricks and tile—proportions that are reasonable for our brains, if not our bodies, to handle. The *library module* seems to be one cure—a way to make things (as Einstein requested) 'as simple as possible—but no simpler.'

Chapter 11 introduces libraries. 11-1 is about ideas—what purpose do modules serve? 11-2 focuses on technique—what are the syntax and system rules? Hopefully, much of its discussion won't be relevant, because most Modula-2 systems have some facility for taking care of the vexing parts automatically. Still (sigh . . .) it's always a good idea to know where to look when bit rot strikes.

Section 11-1

Why Modules? The Basic Design Issues

LET'S START WITH AN EXAMPLE. PROGRAM Importer, below, performs a few of the least astonishing calculations I could think of.

importing a new module

```
MODULE Importer;
   (*Demonstrates imports from a new module named CIRCLE.*)
IMPORT CIRCLE;
IMPORT IO;
BEGIN
   IO.StringOut ("Pi equals approximately:  ");
   IO.FixOut ( CIRCLE.PI , 7, 5);                IO.LineOut;
   IO.StringOut ("The area of a one-inch diameter circle is:  ");
   IO.FixOut ( CIRCLE.Area(0.5) , 7, 5);         IO.LineOut;
   IO.StringOut ("The diameter of a one-inch area circle is:  ");
   IO.FixOut ( CIRCLE.Diameter(1.0) , 7, 5)
END Importer.
```

```
Pi equals approximately: 3.14159
The area of a one-inch diameter circle is: 0.78540
The diameter of a one-inch area circle is: 1.12838
```

Importer imports a library module called CIRCLE. Where did CIRCLE come from? I wrote it, and put it in a file called **`CIRCLE.def`**. (I use the **`funny`** typeface to show the names of actual computer files.) Here's what **`CIRCLE.def`** holds:

the contents of **`CIRCLE.def`**

```
DEFINITION MODULE CIRCLE;
   (*Lists the things that can be imported from the CIRCLE library.*)
   CONST PI = 3.141592;
   PROCEDURE Area (Radius:  REAL) : REAL;
   PROCEDURE Circumference (Radius:  REAL) : REAL;
   PROCEDURE Diameter (Area:  REAL) : REAL;
   (*Definition modules only contain definitions and headings—never statements.*)
END CIRCLE.
```

`CIRCLE.def` holds the *definition module* of a new library. Like the definition modules of other libraries we've used (such as IO or MathLib0), it makes constants and subprograms available. If a program **IMPORT**s the library module, it can use the identifiers.

Naturally, there's no free lunch. I had to write a second file, called **`CIRCLE.mod`**, to hold the *implementation module* of the CIRCLE library. (Your system might conceivably use **`.imp`** as the suffix instead of **`.mod`**.) Here's what **`CIRCLE.mod`** contains:

the contents of **CIRCLE.mod**

```
IMPLEMENTATION MODULE CIRCLE;
(*All identifiers created in the definition module are known here. Moreover ... *)
(*Implementation modules can import other libraries.*)
IMPORT MathLib0;
(*They can include the creation of 'local' constants, types, and variables. Any
  new variables keep their values permanently—until the main program ends.*)
VAR TwicePi: REAL;
(*If 'local' subprograms are needed, they can be declared too. In any case,
  we have to finish the declaration of subprograms from the definition module.
  Note that the headings are identical to those in the definition module, above.*)
PROCEDURE Area (Radius: REAL) : REAL;
  BEGIN
    RETURN Radius * Radius * PI
  END Area;
PROCEDURE Circumference (Radius: REAL) : REAL;
  BEGIN
    RETURN Radius * TwicePi
  END Circumference;
PROCEDURE Diameter (Area: REAL) : REAL;
  BEGIN
    RETURN MathLib0.sqrt(Area / PI) * 2.0
  END Diameter;
(*Finally, there can be, but don't have to be, statements that initialize the
  implementation part. They're run just once at the first IMPORT.*)
BEGIN
  TwicePi := 2.0 * PI
  (*TwicePi will keep this value until the main program ends.*)
END CIRCLE.
```

CIRCLE.mod is the brains of the CIRCLE library. It contains the code of the procedures whose headings we saw in **CIRCLE.def**. Unlike the definition module, it has a statement part, which was automatically (and invisibly) executed when the main program, Importer, imported CIRCLE. And, unlike an ordinary subprogram, its 'local' variables are permanent.

> Every library module (sometimes called an *external module*) has two parts: a definition module and an implementation module. Each belongs in its own file.

Although **CIRCLE.def** and **CIRCLE.mod** are in two separate files, they join forces to form a single CIRCLE library. We haven't had cause to look at them, but if you did hunt them down, you'd find that the IO and MathLib0 libraries had each been created in the exact same way. Each library is a pair of files stored somewhere on your computer system. And, whenever you **IMPORT** IO or **IMPORT** MathLib0, you:

what **IMPORT** does

a) arrange to use the names supplied in a definition module, and

b) create, for the duration of your program, any variables that are declared in the implementation module, and

c) run any initialization code the implementation module contains.

Any Modula-2 program can draw on the contents of libraries that are supplied with the system, or have been written by local programmers. Most high-level languages nowadays have some mechanism for providing libraries, although one language's LIBRARY might be another's UNIT or PACKAGE. But whatever they're called, all these approaches have goals in common with Modula-2's **MODULE**:

- Reducing programs to manageable size by breaking them into small, specialized parts.
- Letting code be shared and reused by different programs and programmers.
- Hiding coding details by making procedure and functions available as though they were built into the language.
- Making programs more portable by letting machine-dependent details be confined to specific libraries.
- Allowing separate compilation of different parts of a program, so that localized changes don't require recompiling the entire code.
- Encouraging new approaches to program design that are better suited for team work on large programs.

why break a program into parts?

Why Use Two Files?

Programming languages have different ways of breaking programs into smaller pieces. Why does Modula-2 use a two-file, definition module/implementation module approach?

> The two-file approach builds layered design into Modula-2. The definition module serves as a performance and requirements specification. It describes what a calling program *can* do, and what the implementation module *must* do.

In Modula-2, layered design—separating specification from implementation during planning—isn't just a good idea. Instead, it's a method of programming that's backed up by the syntax of the language. A definition module is a formalized name layer, while the implementation module is obviously the code layer.

1. The definition module serves as a target for programming. The names a library will create can be agreed on *before they are either used or implemented.*

specifying the interface

The definition module is a handy design tool because it's a fixed reference point. When program development is spread out over time, or when many people are involved, it provides an essential framework for organization. Without the well-defined specification a definition module provides, small differences or misunderstandings can lead to bugs that are serious, but cannot be spotted until the final program is put together.

2. The implementation that supports a definition module can change *without affecting use of the names.* In other words, a familiar name layer can continue to be used with an entirely new—and perhaps better—underlying implementation.

changing implementations

Suppose that a program relies on a sorting routine (call it Sort) that has been imported from a library. If, at a later date, the sorting method is no longer efficient enough for the final application, the underlying algorithm (found in the implementation module) can be improved. However, programs that rely on Sort are still syntactically correct. To them, the library works just as it always has, but faster.

3. The names that are used to achieve certain effects can be changed *without altering the underlying implementation.* In other words, an underlying implementation can be used with an entirely new, and perhaps more convenient, set of access names.

changing access

I wrote the IO library we use in this book for precisely this reason. Most Modula-2 systems have the quasi-standard input/output libraries that Wirth wrote, but they can be awkward to use. The implementation module of IO uses these calls, but hides their idiosyncrasies.

Awkwardness isn't always avoidable or even undesirable, incidentally. It's nearly inevitable that making libraries very general-purpose will require details, often in the form of extra parameters to procedures, that make them less convenient to use. Having to add a BOOLEAN-valued Success argument to every call is inconvenient, but it's common in commercially available libraries, and is essential to the few programs that might need it.

Encapsulation and Abstraction

We're going to have to learn a few syntax rules before we start to write new libraries. But to tell the truth, it's just as important to learn the software engineering *why?* behind library modules as their Modula-2 *how?* Before we get into details, let's look at the ideas behind the whole module system.

At their most basic, library modules provide a way to split up the text of a program. In and of itself, this partitioning has no particular virtue. After all, programs are like buildings—they can be recklessly subdivided in a manner that makes them harder, rather than less difficult, to understand and use.

Nevertheless, if we break things up in the right way, we can make our lives easier. But how are the lines drawn? Do libraries divide a program into separate *clusters* of code, where each cluster happens to consist of two files:

modules as clusters

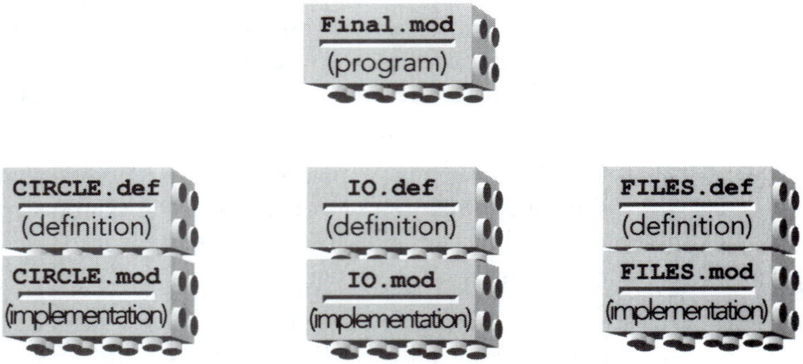

Or do library modules split the program horizontally, so that a program is just the highest level in a series of increasingly abstract layers of code? Here, the top layer is an *application*—the main program. It sits atop a *name layer* of definition modules that rest, in turn, on a *code layer* of implementation modules:

modules as layers

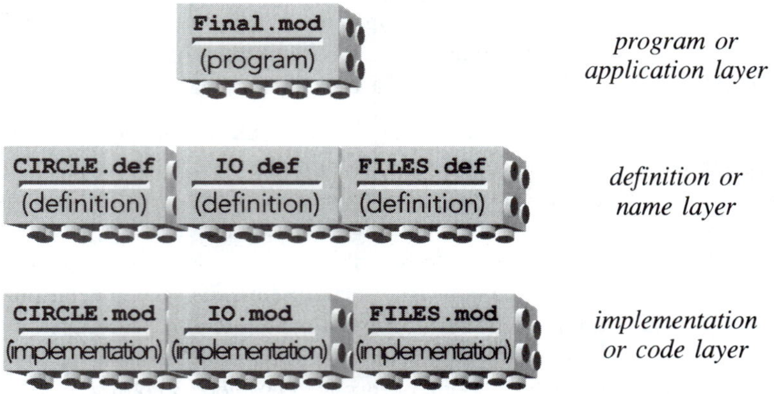

The answer to both questions is *Yes!* Library modules do both. Taken as a single unit, a definition module and implementation module form a package. When viewed separately, though, *many* definition modules make a layer, as do *many* implementation modules.

As clusters, library modules encourage *encapsulation*, and as layers they encourage *abstraction*. Let me explain both of these terms. As clusters, library modules are good for grouping things—actions, identifiers, and so on—that have some common purpose.

encapsulation implies localization

> The goal of clustering is to *encapsulate* program segments—to divide them into distinct parts. A well-encapsulated segment is said to be highly *localized.* It has few connections to the outside world, and can be changed without causing problems elsewhere.

Breaking a program into encapsulated clusters makes modules attractive because:

why cluster?

- It eases the development process. Module boundaries are limited and clearly defined, which lets them be written and tested separately. They can't have unseen side-effects on main-program data.
- It simplifies program maintenance. Clustering related subsystems makes subsequent modification easier, because related subsystems are grouped together, and changes can be localized in effect.
- It reduces the size of programs. The encapsulation mechanism can provide for separate compilation of different modules.
- It encourages software reuse. The operating system can make the contents of stand-alone modules available to other programs.

abstraction implies information hiding

> The goal of layering is to encourage *abstraction*, or the ability to work in high-level terms. Effective abstraction is meant to *hide information*—to separate names (which must be known in order to be used) from implementations (which we don't really have to know anything about).

What kind of information might be kept hidden?

why layer?

- We might release the name of a constant (so that it can be used in an expression) but hide its Modula-2 definition.
- We might release the name of a type (so that variables of that type can be declared) but hide the details of its definition.
- We might release the name and parameter list of a function or procedure (so that it can be called in a program), but hide its code.
- We might release subprograms that access stored data, but hide both the definitions used to store the data *and* any code required to initialize or generate it.

A Taxonomy of Modules: Interface, Package, Data

As Abraham Lincoln once pointed out, you can call a dog a mule, but that won't help him to pull your plow.* Well, we can put code into modules, in a language called Modula-2, but that doesn't make our programs modular.

* All right, maybe he didn't say it. But he should have.

Before we look at modules in action—in fact, before we even look at their syntax—let's establish some guidelines for their content.

A Taxonomy of Library Modules

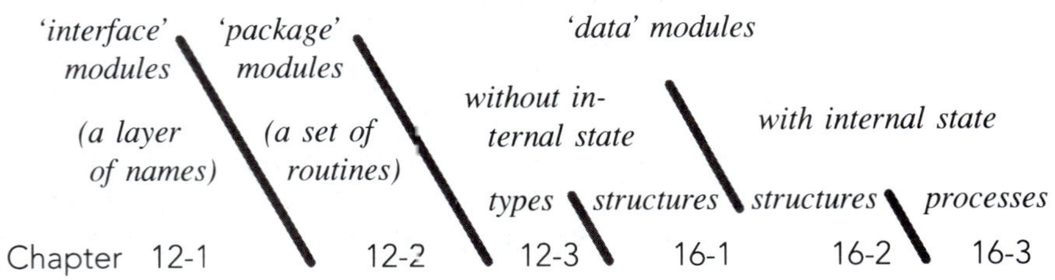

> A library module is most likely to be used and reused if it does just one job, but does it well. The categories I designate are meant to support this goal. They are guidelines, rather than hard and fast rules for module construction.

Libraries won't always be so strictly divided, any more than programs will always be adequately commented, or identifiers mnemonically named. However, following the guidelines will help you write modules that fit together, even when they're put to totally unexpected applications.

For our discussion, I'll divide library modules into three basic categories: *interface* modules, *package* modules, and *data* modules.

name abstraction modules: `interfaces´ or `layers´

- *Interface modules* abstract names; they make identifiers available.

Typically, interface modules rename the contents of existing general-purpose libraries. This means that they'll seldom contain much new code. Instead, an interface module adds a customizing layer to code that already exists, and could even be imported directly. IO, the input/output library we've been using, is a typical interface module. Since it provides a simpler, more consistent use of identifiers, the new layer makes existing libraries easier to use.

algorithm abstraction modules: `packages´

- *Package modules* abstract algorithms; they supply a collection of procedures and functions.

Although any library module might be described as a 'package,' I use the term in a very specific way. A package module delivers a set of procedure calls, but hides all the details of their implementation. It *doesn't* try to hide the details of type definitions. Although package modules may occasionally require a new type definition, they generally collect algorithms that work with values of the basic data types. MathLib0 is a typical package module.

data abstraction modules: types, structures, objects

- *Data modules* abstract data; they supply methods of storing data, and of acting upon stored data.

Data abstraction covers an awful lot of ground, so I'll distinguish between different uses of modules. First, do they provide values or storage? In other words, do they provide new sets of specific values, or do they supply ways to store unspecified values? This is the *type vs. structure* distinction. Second, are the values or structures held in the module, or in the program that uses the module? This is the *internal state* distinction.

type libraries

1. *Type abstraction libraries* make new data types available by providing a type name, an implicit set of values, and a collection of primitive operations.

As you know, a type's primitives define the type's values by limiting what can or can't be done with them. We abstracted a type without using modules when we created FractionTYPE. Other values that are good candidates for type abstraction libraries include strings, complex numbers, times, and dates.

structure libraries

2. *Structure abstraction libraries* supply new data structures. They provide a type name and a set of primitive operations.

A structure's primitives define the relationship between the structure's internal components. But although the structure stores values, it doesn't really have a value. We abstracted a structure without using modules when we discussed sorting—the sorted list was our structure. Other kinds of abstracted data structures that you may have heard of include *stacks* and *queues*, which store and retrieve data in terms of the arrival times of individual items.

> Drawing a distinction between type and structure abstractions is a convenient way to make libraries more reusable. The split makes it relatively easy to create new libraries that combine existing abstractions; in particular, to create new ways to structure existing types.

Now, none of the applications I've described so far have really had to take advantage of a library module's ability to create its own 'hidden' variables. However . . .

object libraries

3. *Data object libraries* maintain data for the benefit of a main program. They *don't* provide a type name. They *do* provide primitive operations that manipulate variables hidden within or by the implementation module.

A window manager (a program that organizes a terminal screen) is a typical data object, as are stream generators (programs that produce a sequence of information on demand) and database managers (that control access to a large amount of stored data). These are relatively advanced applications of libraries that we won't get into in depth.

Let's take library modules one step at a time. I'll devote the rest of this chapter to discussing the basic facts of library creation. Chapter 12, which follows, presents three major applications: interfaces, packages, and type abstractions. More advanced uses of libraries, including structure abstractions and various ways to take advantage of internal state, are discussed in Chapter 16.

Section 11-2 Strictly Modula-2

IT'S TIME TO TANGLE WITH THE SYNTAX OF library modules. Actually, their syntax is pretty easy. But to use libraries, we have to learn some details of program compilation that we've managed to avoid so far.

looking at an **IMPORT**

Let's approach the subject by looking at what happens when an **IMPORT** statement is encountered. Suppose that a program named Test begins like this:

```
MODULE Test;
IMPORT IO;
   ...
```

We'll make Test run in three distinct steps—compilation, linking, and execution:

- *Compilation* translates the English-language version of Test—its Modula-2 *source*—into a form that a computer can understand.
- *Linking* attaches the compiled version of Test to the compiled versions of any libraries that it imports.
- *Execution* makes the program run.

the steps are sometimes joined

Not every computer system requires that you take these steps separately. Some systems join the second two steps, so that after a program is compiled, the 'executor' takes care of linking automatically. Some systems even join all three steps. Nevertheless, you still have to know what goes on at each stage so that you can make changes, and understand error messages that might come up. Let's start with compilation.

First Step: Compilation

First, Test has to be compiled. At this point, the compiler's main interest in library modules is to make sure that their names are being used correctly. Although the exact sequence of steps followed varies slightly from system to system, here is what basically happens when the compiler encounters the statement **IMPORT** IO in program Test:

- It finds the file that contains the IO definition module.
- It makes sure that any IO identifiers used in Test (like IO.CharIn or IO.IntOut) actually appear in the definition module, and that they are called with correctly typed and ordered arguments.
- It goes ahead and compiles the rest of program Test.

Note that the implementation module need not even have been written at this point! That's because the compiled version of Test doesn't include any actual code from IO. The IO code turns up in the next step—linking.

Second Step: Linking

Linking is the second step. The linker is a program, just like the compiler, but its job is simpler. It takes *object files* of compiled code that might be in different places, and merges these separate pieces into a single *executable file*. In our present example, the linker would merge the compiled version of Test with the compiled version of the IO implementation module.

where? Now, where does the linker find the compiled version of IO? Well, every computer has one or more *library directories*—central repositories for the definition modules and compiled implementation modules of libraries. The linker looks there first. If it can't find the file it wants, it might look in your own directory next. Most Modula-2 systems have some mechanism for specifying a *search path* to other places as well.

how? How is an implementation module compiled in the first place? Just like a program. Suppose that you write a new library, which is to say a new definition/implementation module pair. If you compile the implementation part, and keep it among your own files, only you can use it. However, if you put it (and the definition module) in one of the central library directories, everybody can use it.

Can you tell yet why splitting library modules into two parts is such a clever idea?

why? The reason is that as long as the definition module doesn't change, the implementation *can* be changed, and recompiled, without requiring that the main program be recompiled. At worst, it will have to be relinked—an operation that the system often takes care of automatically.

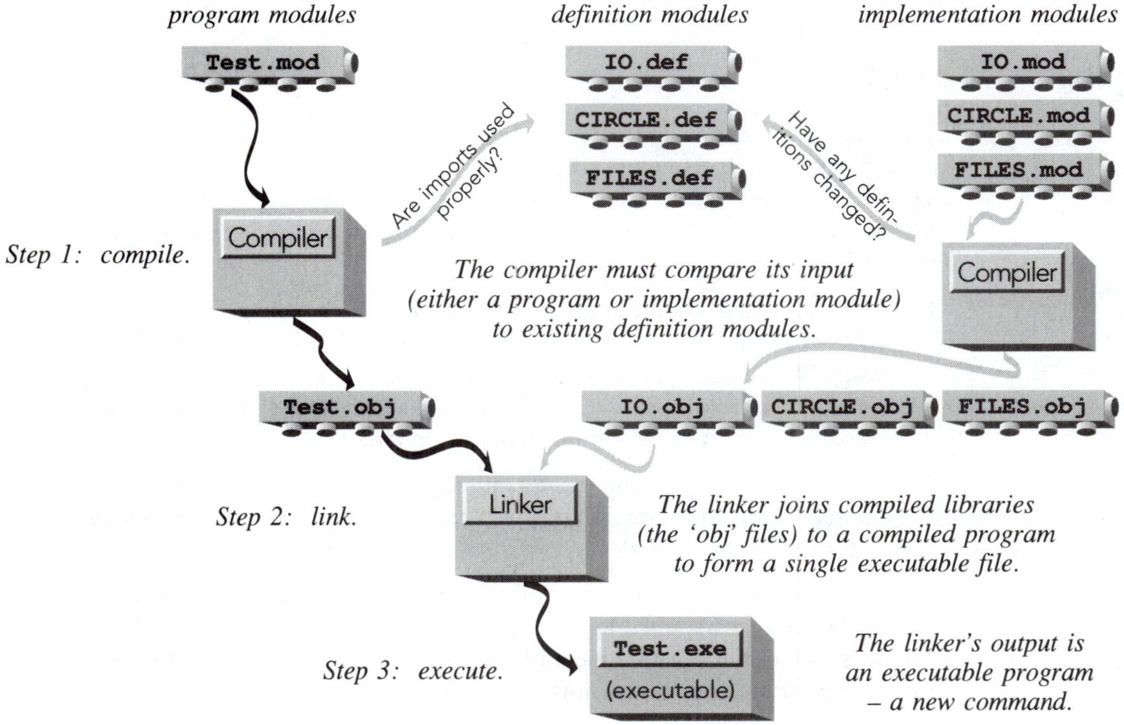

Third Step: Execution

Execution is the third and final step. In the past, we've only worried about the execution of statements we've written ourselves. However, an implementation module can contain statements too. If an implementation module does have a statement part, it's run first—before the rest of our program. In effect, an implementation module is executed right at the moment of **IMPORT** in the main program. Afterward, the main program is executed as usual.

Is there any way to tell whether or not an implementation module contains statements that are executed before your program? Not really—in fact, you can only tell that it's running if it produces some kind of output. Most implementation modules don't.

Besides having statements run, the implementation module shares another main-program feature:

variables in modules are permanent

> Any variables that are created and used within the implementation module will exist for the life of the main program. Unlike the local variables of a procedure, which are disposed of between calls, an implementation module's variables hold their values permanently.

Look back at the implementation module of CIRCLE, on page 437. The implementation module allocates a variable (TwicePi), and assigns it a value. If and when function Circumference is called, it can rely on the value that was assigned to TwicePi when the module's initialization code ran. Once the implementation module has been initialized, program Importer executes normally.

What a Module Can Contain

Library modules follow a number of rules that regulate their contents, usage, and order of compilation. Although why we want to use libraries (and how we can use them best) will occupy us for the next few chapters, I'll summarize the technical details here and now for easy reference later. Let's consider module format and overall content first:

module format rules

1. Each definition module and implementation module belongs in a separate file. The heading begins with the reserved word **DEFINITION** or **IMPLEMENTATION**, as appropriate. Rules vary from system to system, but the filenames are usually required to have a suffix like `.def` (for the definition module) or `.mod` (for the implementation module).

2. Both definition and compilation modules can have **IMPORT** statements, constant and type definitions, and variable declarations. Their scopes (where they can be used) vary; we'll look at details in a moment.

3 Subprogram headings are given in the definition modules, while headings *and* bodies are supplied in the implementation module. The headings of all subprograms must be identical in both modules. All identifiers, types, and order within headings must be precisely the same.

4. A definition module does not contain a statement part. An implementation module does—it is executed when the first **IMPORT** from the module is made—but the statement part can be empty.

```
DEFINITION MODULE Test;
  IMPORT IO;
  CONST LAST = IO.EOL;
  TYPE AnyTYPE = ARRAY . . . ;
  PROCEDURE Call (A: CHAR);
  PROCEDURE Undo ( ) : CHAR;
END Main.
```

```
IMPLEMENTATION MODULE Test;
IMPORT MathLib0;
VAR First: CHAR;
PROCEDURE Call (A: CHAR);
  VAR Time: CHAR;
BEGIN
  procedure statements
END Call;
PROCEDURE Undo ( ) : CHAR;
BEGIN
  function statements
END Undo;
BEGIN
  First := "A";
  other initializing statements
END Main.
```

Procedures and functions are completed in the implementation module. The headings must be identical in each module. The implementation module can also contain strictly local subprograms (although none are shown here).

Module Scope Rules

The second set of library module rules apply to scope, which is the range of a program in which identifiers can be used.

module scope rules

1. All definitions and declarations from a definition module are available, through the **IMPORT** statement, to other modules. Because they can be imported, they are said to be *public*.

2. A definition module's new constant, type, and variable identifiers—all the public identifiers—are automatically known in the implementation module.

3. New definitions or declarations that appear in the implementation module alone are *not* available elsewhere—not even in the definition module. Because they cannot be imported, they are said to be *private*.

4. Any **IMPORT**s to the definition module are only known locally, within the definition module itself. Similarly, all **IMPORT**s to the implementation module are restricted to the implementation module alone. In effect, imported identifiers are private.

Available within the definition module:

identifiers from IO,
LAST
AnyTYPE

```
DEFINITION MODULE Test;
  IMPORT IO;
  CONST LAST = IO.EOL;
  TYPE AnyTYPE = ARRAY . . . ;
  PROCEDURE Call (A: CHAR);
  PROCEDURE Undo ( ) : CHAR;
END Main.
```

Available within the implementation module:

identifiers from MathLib0,
LAST, Call, Undo,
AnyTYPE
First, Local

Available to a module that imports Test:

LAST, Call, Undo
AnyTYPE

```
IMPLEMENTATION MODULE Test;
IMPORT MathLib0;
VAR First: CHAR;
PROCEDURE Call (A: CHAR);
    contents of the procedure
END Call;
PROCEDURE Undo ( ) : CHAR;
    contents of the function
END Undo;
PROCEDURE Local;
    contents of the procedure
END Local
BEGIN
  First := "A";
  other initializing statements
END Main.
```

Module Compilation Rules

The third set of rules describes the order of compilation of program, definition, and implementation modules. This order must be defined precisely so that modules can be compiled separately. The advantage of separate compilation is that library modules can be compiled just once, but used many times in main program modules.

1. The definition module must be written (and compiled, if the system requires it) first. If a definition module **IMPORT**s from another, the imported definition module has to be written (and possibly compiled) first. This means, incidentally, that two definition modules may not import from each other.

compilation order rules

2. The implementation module or modules are compiled next.
3. Finally, the application program that calls on the library module or modules is compiled.

4. If a definition module is changed, then it, its implementation module, and the calling program module must all be recompiled. But if the implementation module alone is changed, without affecting the definition module, only the implementation module has to be recompiled.

The first two steps have already been taken care of for the standard libraries. It should be obvious that these libraries are left alone if a main program module changes. After all, we've been making main program changes for ten chapters without ever worrying about updating IO or MathLib0.

On the other hand, if an implementation module has been changed and recompiled, the main program may have to be relinked, or reconnected to the compiled implementation part. That makes sense—if the code in an implementation module has been improved, you want your program to be able to use it.

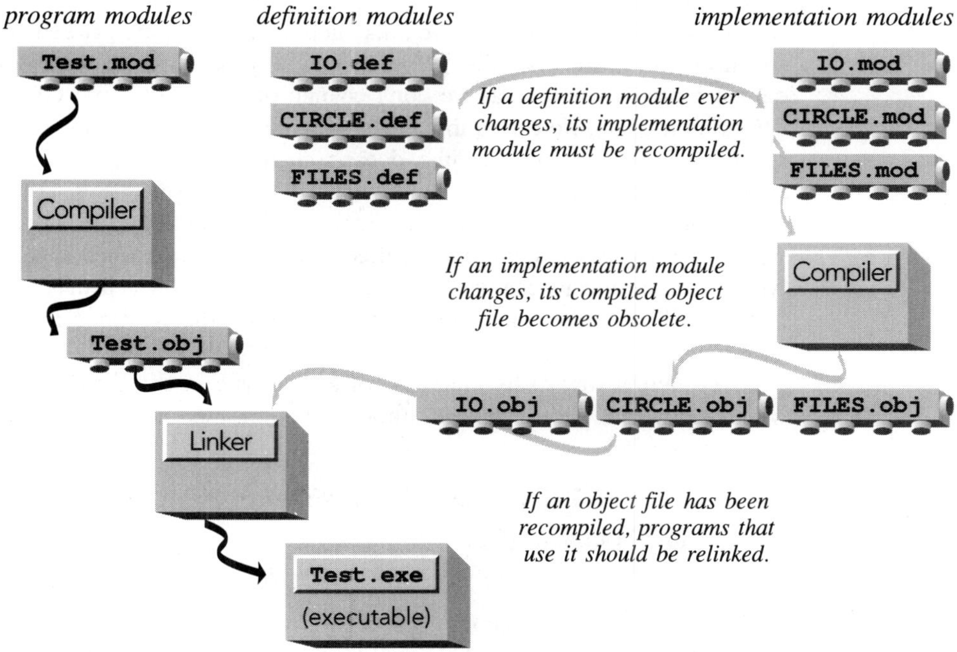

As a technical note, Modula's separate compilation differs from the *independent* compilation of some other languages (e.g. FORTRAN) by its strong type checking. Even if a library module was compiled years earlier, a main program module still has to obey the type rules laid out in the definition module.

The exact method of installing library modules will vary. Rules for naming and compiling a definition module (if necessary) and its implementation module will be very much like those for ordinary programs. However, you will have to find out what your own system's regulations are. Here's some room for your notes:

installing library modules

____

____

____

____

____

____

____

Opaque Type Definitions*

If you take a moment to reflect, you may notice that the split between definition and implementation modules grants a right to procedures that is, seemingly, denied to types. As long as we don't change the name of a procedure in the definition module, we can re-implement it as we choose in the implementation module. As long as the heading doesn't change, a client program uses the revamped procedure as before.

But can we change the implementation of a type? Not yet. When a new type is defined in the definition module, it is fixed for all time. Changing the type definition means changing the definition module—which means that the implementation module and any client that imports the definition module both have to be recompiled.

Is there a way to fix the name of a new type in the definition module, while hiding its details in the implementation module? Yes, through the mechanism of an *opaque* type definition:

an opaque type defined

```
DEFINITION MODULE OpaqueExample;
TYPE HiddenTYPE;
    ...         (*other definitions continue*)
```

HiddenTYPE called an opaque type because we can't 'see through' its name to the details of its definition. They will be found in the OpaqueExample implementation module.

> There is an important restriction on opaque types: they *must* be pointer types.

* Opaque type definitions require some understanding of pointers (Chapter 13). This section is here to keep all the module definition details in one place.

```
IMPLEMENTATION MODULE OpaqueExample;
TYPE HiddenTYPE = POINTER TO SECRET;
     SecretTYPE = RECORD
                     Field1: ItsTYPE;
                     Field2: ItsTYPE;
                  END;
  ∴ etc.
```

the opaque type implemented

If at a later date we wish to change the implementation of SecretTYPE, client programs that use type OpaqueExample.HiddenTYPE will not be affected. It may not seem so important now, but the ability to rewrite implementation modules turns out to be enormously important for software development.

A second advantage of opaque type definitions stems from the fact, mentioned as scope rule 1, that *all* definitions made in a definition module are public. When a new type is defined in the definition module, the details of its definition are just as public as the type name. This is often an undesirable property—we would like to discourage client programs from relying on the 'internal' details of library modules.

```
DEFINITION MODULE TypicalExample;
TYPE NotHiddenTYPE = RECORD
                        Field1: ItsTYPE;
                        Field2: ItsTYPE;
                     (*In theory, these field names could be used by
                       a program that imports TypicalExample.*)
                     END;
  ∴ etc.
```

I won't bother to go into the details here, but a module that imports TypicalExample can actually refer to Field1 and Field2 by name; they are not hidden. In practice, we won't ever do so, because the definition module will also contain procedures and functions to manipulate the values of NotHiddenTYPE for us. In summary, defining the record as an opaque type guarantees the privacy of its contents.

Zen Study Guide

Big Ideas

The Big Ideas in this chapter are:

1. *Abstraction and encapsulation are the* result *of good design, rather than* being *good design.* Putting parts of a program into modules doesn't make a program modular—layering and clustering may even be at cross-purposes. Modules don't make a program modular; how we use them does.

2. *Rules of library module syntax and compilation order permit a good deal more than good practice allows.* In a sense, the rules permit complexity so that the libraries themselves can be written and used simply.

Brief summary comments follow Chapter 12.

'Package modules are intended to be extremely portable.' (p. 462)
Christo, master of packaging, wraps up the Whitney Museum.

Chapter 12 Interfaces, Packages, Data Abstractions

'When will they come up with *the* programming language?' is a question asked despairingly by non-programmers. The answer, unfortunately, is that the wrong question is being asked. A language that's fixed is dead—something that's as true for programming languages as for spoken ones.

There are two approaches to keeping programming languages alive. One way is to occasionally revitalize the language by redefining it. We might call this the Menudo approach, after the pop group that regularly replaces its over-the-hill (which is to say, over 18) singers. But while changing singers may improve a band, adding features seldom improves programming languages. The mistakes of the earlier incarnations are almost invariably preserved in the latter ones, while the language gets bigger and bigger.

An alternative approach is to start with a small language, but provide ways to extend it. That's the approach Modula-2 takes. 'Raw' Modula-2 is useless; programming really requires a set of Modula-2 libraries. However, you're not stuck with the same libraries forever. If improved versions come along, the superseded libraries can be jettisoned.

The hard part is creating libraries that *can* be tossed out. In a sense, the examples in Chapter 12 are more notable for what they don't allow than for what they do permit. As you read, try to get a feel for the difference between writing a library that does as much as it can, and one that does as much as it *should*.

Section 12-1

Adding Layers: Interface Modules

AN INTERFACE MODULE, LIKE A THREE-PRONG ADAPTOR, might not always be essential, but it's nice to have one around. Interfaces customize. They make programming easier by making existing tools more comfortable to use.

The most basic custom order we can place is a constant definition. Constant identifiers don't invent values, but they do make them a little bit easier to work with by giving them easily remembered names. For those of you who, like me, have trouble (especially after a salmarazd or two) recalling that a rehoboam of wine plus a methuselah is not quite a belshazzar, I present WINE:

contents of **wine.def**

```
DEFINITION MODULE WINE;
  (*An interface module that lets oenophiles program in their own terms.*)
  CONST BOTTLE = 1;
        MAGNUM = 2 * BOTTLE;
        JEROBOAM = 2 * MAGNUM;
        REHOBOAM = MAGNUM + JEROBOAM;
        METHUSELAH = REHOBOAM + MAGNUM + BOTTLE;
        SALMARAZD = 2 * REHOBOAM;
        BELSHAZZAR = 4 * JEROBOAM;
        NEBUCHADNEZZAR =  5 * JEROBOAM;
END WINE.
```

The companion implementation module doesn't do anything, but it still has to be written. This is a minor pain for us, but keeping a consistent format for modules makes life much easier for the compiler.

an empty implementation part, **wine.mod**

```
IMPLEMENTATION MODULE WINE;
BEGIN
  (*No initializing actions are required.*)
END WINE.
```

exporting to clients

> For convenience in discussion, definition modules are said to *export* identifiers to *client* modules.

In other words, definition module WINE can be described as the exporter of BOTTLE, MAGNUM, and the other constants. Any program modules that **IMPORT** WINE can be described as WINE's clients, because they make use of a service that WINE provides.

WINE, I grant, is a bit facetious. However, there are useful modules that contain only constant definitions. One that's often encountered exports the mnemonics (brief names) of the ASCII control characters. Some of the more frequently employed characters are:

the ASCII module

```
DEFINITION MODULE ASCII;
   (*Exports mnemonics for the ASCII character set.*)
CONST NUL = 0C;   (*null character*)
      BEL = 7C;   (*the beep*)
      BS = 10C;   (*backspace*)
      HT = 11C;   (*tab character*)
      ESC = 33C;  (*escape character*)
        . . .
END ASCII.
```

Like WINE, ASCII requires an implementation module even though it takes no action:

```
IMPLEMENTATION MODULE ASCII;
BEGIN
END ASCII.
```

Other interface modules that export only constants might include:

other constant-only interface modules

- LIMITS, a description of your Modula-2 system's built-in bounds: the smallest and largest values it can deal with, the number of files that can be in use at one time, and so on.
- LETTERS, a collection of the appropriate sequences needed to print foreign letters or symbols: ¿, ç, ¥, and so on.
- PHYSICAL, a collection of real-world constants like pi, the speed of light, and the ever-popular Avogadro's number.

Modification Problems

1. As you know, the implementation module can have initializing code that will be executed by any program that imports WINE. Although it's an uncommon practice, you might have the implementation module print the values of all the constants that WINE defines. Do it, then write (and run) a program that imports WINE, but does nothing else.
2. Change WINE so that only BOTTLE is defined as a constant. Declare MAGNUM, JEROBOAM, and the others as variables (where should this declaration be?), then initialize them in the implementation module. Write a short program that imports WINE, then prints the values of the imported variables.

Self-Check Questions

Q. The program below imports the WINE module, and also defines a constant named BOTTLE. Is it legal? Can we refer to both BOTTLEs, or will one name take precedence?

```
MODULE Tipsy;
   (*Reuses an identifier that has appeared in WINE.*)
   IMPORT WINE;
   CONST BOTTLE = 2;
   . . .
```

A. Not only is the definition legal, but we can use both BOTTLE definitions in the main program. The imported identifier is *qualified* by the library's name:

```
BEGIN
    IO.IntOut ( WINE.BOTTLE, 1 );       (*The WINE identifier.*)
    IO.IntOut ( BOTTLE, 1 )             (*The Tipsy identifier.*)
END Tipsy.
```

A Software Interface: TRIG

One of the most useful jobs of an interface module is making existing software more convenient to use. Just because software works doesn't mean that it's easy to work with.

> A *software interface* doesn't change the functional ability of a lower-level module. It just presents an existing capability in a form that's clearer or more convenient.

For example, consider the MathLibØ module, which supplies a limited number of trigonometric functions: MathLibØ.sin, MathLibØ.cos, and MathLibØ.arctan. Although these three can do as much trigonometry as you please, MathLibØ is inconvenient in two ways:

why MathLibØ might need an interface

1. Even though the remaining trigonometric functions can be expressed in terms of the three primitives that MathLibØ supplies, the programmer of a $90,000 computer has to look up the exact conversions in her $3.95 book of mathematical tables.
2. Even though most day-to-day usage of angles is conducted in degrees, the MathLibØ functions do their business in radians. Once again, the programmer is forced to translate her needs into terms the computer can understand.

I'll be the first to confess that my voluntary recall of trigonometry ends with the vague memory that tangent equals sine over cosine. As for radians—well, I know that a circle has 2π of them, but I'd be in trouble if I had to plan a bank shot in radians rather than degrees.

Nevertheless, I've used my limited recollection to begin a TRIG module, below. Although you're certainly free to view TRIG as a new package of functions, I view TRIG as a software interface because it doesn't really do anything new. It doesn't implement any algorithms, or even make any calculations beyond simple conversion-factor arithmetic. It's just a customizing layer that makes up for my fading memory.

TRIG does differ from WINE and ASCII because it exports subprograms, rather than constants. As a result, TRIG's implementation module has to include the full declarations of the exported functions.

defining TRIG

```
DEFINITION MODULE TRIG;
   (*This interface module provides trigonometric functions that
    take degree-valued arguments.*)
   PROCEDURE Sin (Angle: REAL) : REAL;
   PROCEDURE Cos (Angle: REAL) : REAL;
   PROCEDURE Tangent (Angle: REAL) : REAL;
      ...   (*Other headings might follow.*)
END TRIG.
```

TRIG's implementation module, below, needs a little extra help to actually implement the functions.

implementation module scope

> Imports, definitions, and declarations that are made in an implementation module, like TRIG's, are local to the module. They couldn't be used in TRIG's definition module, or in any program that imports TRIG.

The implementation module below has three new local identifiers: MathLibØ, PI, and DegToRad. PI and DegToRad can't be used in either the main program or the MathLibØ definition module. MathLibØ could be used, but only if it were imported again.

implementing TRIG

```
IMPLEMENTATION MODULE TRIG;
IMPORT MathLibØ;
CONST PI = 3.1415926;
PROCEDURE DegToRad (Degrees: REAL) : REAL;
   (*This function is strictly local to this module.*)
   BEGIN
      RETURN Degrees * (PI / 180.0)
   END DegToRad;
PROCEDURE Sin (Angle: REAL) : REAL;
   BEGIN
      RETURN MathLibØ.sin (DegToRad(Angle))
   END Sin;
PROCEDURE Cos (Angle: REAL) : REAL;
   BEGIN
      RETURN MathLibØ.cos (DegToRad(Angle))
   END Cos;
PROCEDURE Tangent (Angle: REAL) : REAL;
   BEGIN
      RETURN MathLibØ.sin(DegToRad(Angle))
                  / MathLibØ.cos(DegToRad(Angle))
   END Tangent;
BEGIN
END TRIG.
```

Modification Problems

1. Modify TRIG so that it exports the constant PI and the function DegToRad. In addition, add a new private constant called TWICEpi to TRIG's implementation module. Naturally, it should use PI to get its value.
2. A complaint about libraries (as well as languages) designed by Niklaus Wirth is their lack of an exponentiation operator or function. Modify TRIG so that it exports a function, called Power, that raises a number to an exponent.

A Hardware Interface: ANSI

As you know by now, computer systems have two different kinds of resources: software and hardware. Although we usually think of programming as being a way to control software, in the final analysis our programs control hardware—input and output devices—as well.

> A *hardware interface* module controls a physical device that's attached to the computer. It lets us treat hardware as though it were programmable software.

In practice, the interface we write is usually removed from hardware by at least one layer of software already. We're really writing an interface to an interface.

For example, the American National Standards Institute, or ANSI, is an organization that is devoted to making the world a more standard place to live in. One of its more noteworthy achievements was to adopt a set of *escape sequences* that control printing on a terminal screen.

escape sequences

> An *escape sequence* is a special character or two that directs a terminal to obey commands contained in the letters that follow, instead of simply printing them. In other words, the letters 'escape' normal processing so that they can be interpreted as commands.

The ANSI set was originally developed by the Digital Equipment Corporation for their VT100 terminal. It's simple, easily extended, and widely used. You'll find a summary of the full set inside the back cover.

Why are escape sequences necessary? Well, under normal circumstances a terminal listens for characters from the computer, and prints them as they arrive. Sending a character to the screen (as part of output) prints the character, and moves the cursor one position to the right. As you may know, a handful of control characters do other things. For example, we saw earlier (in the ASCII module) that 7C makes your terminal beep, 10C backspaces, and 11C tabs.

But suppose that you wanted to blank the entire screen, or make the cursor jump to some particular row and column? There aren't standard control characters for these. However, almost all terminals have such capabilities built into them. The ANSI escape sequences provide a uniform set of names for invoking them, so that terminals made by different manufacturers can be controlled in the same way.

The complete set of escape sequences defined by ANSI is rather detailed. It controls:

what the escape sequences control

- moving the cursor to arbitrary screen positions,
- erasing various parts of the screen's contents,
- defining the effects of the function keys,
- the use of color on the screen, and
- the size, color, and intensity of printed characters.

Now, let me ask an easy question. Do you really want to know what the ANSI escape sequences are? The answer I'm looking for, of course, is *No way*. No matter how easy it is, an escape sequence is a hardware detail that no mortal should have to learn.

Instead, what we really want is to be able to use the escape sequences without having to know what they are. Obviously, this will require a hardware interface module. Like a software interface, it won't change the functional ability of a piece of hardware. It will, however, make the hardware easier to use.

Since I want to focus on method, I'll just implement a few of the ANSI terminal features here.

definition of ANSI

```
DEFINITION MODULE ANSI;
  (*Provides a few of the ANSI escape sequences.*)
  PROCEDURE Blank;
    (*Erases the screen, and leaves the cursor in the upper left corner.*)
  PROCEDURE GotoXY (Line, Column: INTEGER);
    (*Moves the cursor to Line, Column.*)
  PROCEDURE Blink;
    (*Makes text blink on and off on the screen.*)
  PROCEDURE EraseLine;
    (*Erases from the cursor position to the end of the line.*)
  PROCEDURE CursorUp (Rows: INTEGER);
    (*Moves the cursor up by Rows.*)
  PROCEDURE CursorLeft (Columns: INTEGER);
    (*Moves the cursor backward by Columns.*)
END ANSI.
```

The actual escape sequences are easy to learn, but they're arbitrary and hard to remember. The escape command is the two-character sequence esc[—the control character named 'escape,' then a left square bracket—followed by arguments (if necessary), and a closing character. As it happens, the escape character's value as a Modula-2 CHAR is 33C. I've shown a few of the ANSI escape sequences below:

some ANSI escapes

Escape sequence	*Effect on the screen*
33C[X;YH	*move the cursor to screen coordinates* X, Y
33C[NA	*move the cursor up by* N *lines (*B *moves down)*
33C[NC	*move the cursor right by* N *spaces (*D *moves left)*
33C[2J	*erase the screen, then go to the upper left corner*
33C[K	*erase from the current position to the end of the line*
33C[5m	*blinking text*
33C[0m	*set video back to normal*

For example, to move the cursor to line 10, column 15, you'd send the sequence:

moving the cursor

```
IO.CharOut (33C);      (*sends the escape character*)
IO.CharOut ("[");      (*sends the left square bracket*)
IO.IntOut (10,1);      (*sends the row number*)
IO.CharOut (";");      (*sends the semicolon*)
IO.IntOut (15,1);      (*sends the column number*)
IO.CharOut ("H");      (*sends the 'H'*)
```

Note that I say *send*, rather than *print*. Since it is an escape sequence, the terminal interprets it as a command, rather than as an arbitrary sequence of characters.

In any case, I think that you can begin to appreciate just why we'd like to have an interface module. The implementation module of the ANSI definitions I made above is shown below. Notice that for the first time, I've included initialization code in its statement part.

implementation of ANSI

```
IMPLEMENTATION MODULE ANSI;
  (*Provides a few of the ANSI escape sequences.*)
IMPORT IO;
CONST ESC = 33C;
PROCEDURE Blank;
  (*Erases the screen, and leaves the cursor in the upper left corner.*)
  BEGIN
    IO.CharOut (ESC);
    IO.StringOut ("[2J")
  END Blank;
PROCEDURE GotoXY (Line, Column: INTEGER);
  (*Moves the cursor to Line, Column.*)
  BEGIN
    IO.CharOut (ESC);
    IO.CharOut ("[");
    IO.IntOut (Line,1);
    IO.CharOut (";");
    IO.IntOut (Column,1);
    IO.CharOut ("H")
  END GotoXY;
```

```
PROCEDURE Blink;
   (*Makes text on the screen blink on and off.*)
   BEGIN
      IO.CharOut (ESC);
      IO.StringOut ("[5m")
   END Blink;
PROCEDURE EraseLine;
   (*Erases from the cursor position to the end of the line.*)
   BEGIN
      IO.CharOut (ESC);
      IO.StringOut ("[K")
   END EraseLine;
PROCEDURE CursorJump (Direction: CHAR; Distance: INTEGER);
   (*Moves the cursor up, down, right, or left by Distance.*)
   BEGIN
      IO.CharOut (ESC);
      IO.CharOut ("[");
      IO.IntOut (Distance, 1);
      IO.CharOut (Direction)
   END CursorJump;
PROCEDURE CursorUp (Rows: INTEGER);
   (*Moves the cursor up by Rows.*)
   BEGIN
      CursorJump ("A", Rows)
   END CursorUp;
PROCEDURE CursorLeft (Columns: INTEGER);
   (*Moves the cursor backward by Columns.*)
   BEGIN
      CursorJump ("D", Columns)
   END CursorLeft;
BEGIN
   IO.CharOut (ESC);
   IO.StringOut ("[0m");     (*Turns off any graphics that might be on.*)
   Blank
END ANSI.
```

Modification Problems

1. As you can see, I've left a few easy procedures unwritten. Add CursorDown and CursorRight procedures to ANSI. You should take advantage of procedure CursorJump.
2. Modify ANSI so that the terminal beeps whenever the cursor jumps. Hint: import ASCII.
3. Implement the other ANSI escape sequences (found inside the back cover). Testing is essential; just because escapes have been standardized doesn't mean that every terminal will support the full set.

Self-Check Questions

Q. Exactly when is the initialization code in the ANSI implementation module carried out?

A. When the module is imported into a main program.

Q. The ANSI module uses IO. In the example below, is the import of IO necessary? Must it precede the import of ANSI?

```
MODULE FancyGraphics;
IMPORT IO;
IMPORT ANSI;
```

A. FancyGraphics needs to import IO if, and only if, it's going to use it directly. If you reread the ANSI implementation module, you'll see that it includes its own import of IO, as it must. The order of imports in FancyGraphics doesn't matter.

Section 12-2 Collecting Algorithms: Package Modules

THE EXAMPLES WE'VE JUST SEEN TOOK ADVANTAGE OF library modules for layering. Aside from the very simplest cases (collecting constant definitions), we used modules to insert an interface, or new layer of names, between our programs and an underlying set of primitive operations. Let's go down one layer and look at *package* modules that collect and abstract algorithms.

Package modules are intended to be extremely portable. This goal is achieved partly by minimizing dependence on existing libraries and partly by having efficient implementations. In summary, package modules:

characteristics of package modules

- implement procedures and functions, rather than simply renaming the contents of existing libraries;
- don't abstract any new types whose implementation is intended to be hidden from the user;
- aren't used to hold data for the user.

In effect, package modules abstract algorithms. They separate the names that describe and invoke algorithms—the contents of the definition module—from the code that gives them meaning—the contents of the implementation module.

The best example of this application is, I think, the standard MathLibØ package. It operates on 'known' data types, particularly REAL, that don't have to be specially provided to the program user. It's self-contained, and doesn't rely on calls to other high-powered math libraries. Finally, it performs its calculations pretty efficiently. Other typical package applications include:

- Supplying collections of statistical routines, or more complicated mathematical functions.

typical package applications

- Providing routines that sort files or arrays, or search them for particular values.
- Providing procedures for conversion between units.
- Implementing packages of financial, engineering, or scientific calculations.

Let's look at two examples. The first, AREA, is a fairly simple package of routines for geometric calculations. The second, EZstats, is more typical of real-life applications.

A Geometry Package: AREA

For our first example, let's put together a package that makes functions for calculating the areas of plane figures available. We'd like to relieve the programmer of the burden of remembering formulas like:

an informal outline of AREA

Figure	*Area*
triangle	(*base* ∗ *height*) / 2
square	$side^2$
rectangle	*base* ∗ *height*
trapezoid	(*sum of parallel sides* ∗ *altitude*) / 2
circle	πr^2
. . . *and so on*	

The definition module of AREA serves to document the function calls. It's important that any potential ambiguity be dealt with here. For instance, the area formula of a trapezoid requires the lengths of the two parallel sides—not just whatever sides happen to be on top and bottom. The definition module of AREA is:

definition of AREA

```
DEFINITION MODULE AREA;
   (*A package that supplies a set of functions that calculate areas.*)
   PROCEDURE Triangle (Base, Height: REAL) : REAL;
   PROCEDURE Square (Side: REAL) : REAL;
   PROCEDURE Rectangle (Base, Height: REAL) : REAL;
   PROCEDURE Trapezoid (Top, Bottom, Altitude: REAL) : REAL;
      (*Top and Bottom represent the two parallel sides.*)
   PROCEDURE Circle (Radius: REAL) : REAL;
END AREA.
```

Like an earlier example (module TRIG), the implementation module of AREA requires a local constant definition. Technically, the definition of PI could have been placed in the definition module, since it would have been usable in the implementation module. I didn't put it there because PI doesn't need to be public—it's only used in the implementation module.

implementation of AREA

```
IMPLEMENTATION MODULE AREA;
   (*Implements the set of area-finding function calls.*)
CONST PI = 3.1415926;
PROCEDURE Triangle (Base, Height: REAL) : REAL;
   BEGIN
      RETURN (Base * Height) / 2.0
   END Triangle;
PROCEDURE Square (Side: REAL) : REAL;
   BEGIN
      RETURN Side * Side
   END Square;
PROCEDURE Rectangle (Base, Height: REAL) : REAL;
   BEGIN
      RETURN (Base * Height)
   END Rectangle;
PROCEDURE Trapezoid (Top, Bottom, Altitude: REAL) : REAL;
   (*Top and Bottom represent the two parallel sides.*)
   BEGIN
      RETURN (Altitude * (Bottom + Top)) / 2.0
   END Trapezoid;
PROCEDURE Circle (Radius: REAL) : REAL;
   BEGIN
      RETURN PI * Radius * Radius
   END Circle;
BEGIN
END AREA.
```

Modification Problems

1. Change AREA into a package of PERIMETER functions.
2. How robust is AREA? Not very. Garbage in (like negative measurements) will produce garbage out. Modify AREA so that it calls procedure HALT if it detects a negative argument. If you're not so vindictive, add a function named Absolute to AREA's implementation module instead. Have each of the calculating functions call it to correct arguments as necessary.

Self-Check Questions

Q. Recall that the implementation module of TRIG began like this:

```
IMPLEMENTATION MODULE TRIG;
IMPORT MathLib0;
CONST PI = 3.1415926;          etc.
```

Instead of defining its own PI, could AREA have imported TRIG and used TRIG.PI? If not, is there any way to change TRIG's definition so that it would be legal?

A. As written, PI can't be imported by AREA because any definitions or declarations made in an implementation module are private. However, if PI had been defined in TRIG's definition module instead, it could be imported along with TRIG's other procedures.

A Statistics Package: EZStats

Package modules usually work with 'known' data types, and don't contain the definition and abstraction of new data types. That's the job of the data modules we'll encounter in a few pages. But are packages limited to working with simple types like CHAR and REAL? No. Thanks to open array parameters, they can provide an astonishing variety of general-purpose routines.

For example, we can write a small package of general-purpose statistical routines. No matter what identifier a client program gives to its own data types, a one-dimensional array of INTEGER-valued components can always be passed to an open array parameter—an **ARRAY OF** INTEGER.

An informal outline of the contents of an EZStats module might call for:

an informal outline of EZStats

functions that represent the highest and lowest values
a function that computes and represents the average value
a function that represents the values' standard deviation

Although I hate to invoke this fearsome formula for calculating the standard deviation, you'll soon see that the code it represents isn't especially complicated. You can probably steal it for an assignment later on.

$$\textit{standard deviation} = \sqrt{\frac{\sum_{i=1}^{n}(X_i - \bar{X})^2}{n-1}}, \quad \text{where } \bar{X} = \textit{mean} = \frac{\sum_{i=1}^{n} X_i}{n}$$

For our purposes now, though, I'm much more concerned with an issue of design, rather than mere coding. Look at the division by $n-1$ inside the large square-root sign. Can you anticipate a problem? Well, suppose that an incautious programmer passes an array that's just one component long to a function provided by an EZStats module. This won't cause any problems when it comes to determining the array's largest value, or its smallest, or even its average.

It would, however, cause a divide-by-zero crash in computing the array's standard deviation. Obviously, when we get to the implementation module we'll have to figure out a way of spotting, and avoiding the error. Right now, though, the definition module must document the way in which the situation will be handled.

the definition of EZStats

```
DEFINITION MODULE EZStats;
   (*A package that supplies a few simple statistical functions.*)
   PROCEDURE Most (Data: ARRAY OF INTEGER) : INTEGER;
   PROCEDURE Least (Data: ARRAY OF INTEGER) : INTEGER;
      (*Returns the greatest or smallest value in Data.*)
   PROCEDURE Mean (Data: ARRAY OF INTEGER) : REAL;
      (*Returns the average of the values in Data.*)
   PROCEDURE StdDev (Data: ARRAY OF INTEGER) : REAL;
      (*Returns 0.0 if given a one-component array argument.*)
END EZStats.
```

Using the definition module as a specification of the requirements of EZStats, we can proceed to the implementation module, below. Be sure to look for the way that potential division by zero is handled in procedure StdDev.

the implementation of EZStats

```
IMPLEMENTATION MODULE EZStats;
   (*A package that implements some statistical functions.*)
IMPORT MathLib0; (*We'll need MathLib0.sqrt below.*)
PROCEDURE Most (Data: ARRAY OF INTEGER) : INTEGER;
   (*Returns Data's largest stored value.*)
   VAR i, Biggest: INTEGER;
   BEGIN
      Biggest := MIN(INTEGER); (*Take a wild (and bad) guess.*)
      FOR i := 0 TO HIGH(Data) DO
         IF Data[i] > Biggest THEN Biggest := Data[i] END
      END; (*FOR*)
      (*Postcondition: Biggest represents the largest value stored in Data.*)
      RETURN Biggest
   END Most;
PROCEDURE Least (Data: ARRAY OF INTEGER) : INTEGER;
   (*Returns Data's smallest stored value.*)
   VAR i, Smallest: INTEGER;
   BEGIN
      Smallest := MAX(INTEGER); (*Take another terrible guess.*)
      FOR i := 0 TO HIGH(Data) DO
         IF Data[i] < Smallest THEN Smallest := Data[i] END
      END; (*FOR*)
      (*Postcondition: Smallest represents the least value stored in Data.*)
      RETURN Smallest
   END Least;
```

```
PROCEDURE Mean (Data: ARRAY OF INTEGER) : REAL;
   (*Returns the average of the values in Data.*)
   VAR i, Sum: INTEGER;
   BEGIN
      Sum := 0;
      FOR i := 0 TO HIGH(Data) DO
         Sum := Sum + Data[i]
      END; (*FOR*)
      (*Postcondition: Sum is the sum of the values stored in Data.*)
      RETURN FLOAT(Sum) / FLOAT(HIGH(Data) + 1)
   END Mean;
PROCEDURE StdDev (Data: ARRAY OF INTEGER) : REAL;
   (*Returns standard deviation of Data*)
   VAR i: INTEGER; (*The array index variable.*)
      Sum, Average: REAL;
   BEGIN
      Average := Mean(Data);
      Sum := 0.0;
      FOR i := 0 TO HIGH(Data) DO
         Sum := Sum + (( FLOAT(Data[i]) - Average )
                          * ( FLOAT(Data[i]) - Average ))
      END; (*FOR*)
      (*Note that the formula calls for n-1, which equals HIGH(Data).*)
      IF HIGH(Data) < 1 (*Avoid division by zero.*)
         THEN RETURN 0.0
         ELSE RETURN MathLib0.sqrt( Sum / FLOAT(HIGH(Data)))
      END
   END StdDev;
BEGIN
END EZStats.
```

Modification Problems

1. Build a test harness for EZStats. What kind of test data should you try on it? What other tools (like Load and Dump procedures) might you build into EZStats?
2. EZStats assumes that it will be given INTEGER data. Create a package called REALStats that manipulates REAL data. How *little* work can you get away with doing?

Section 12-3 Encapsulating Types: Data Abstractions

LET'S TAKE A MOMENT TO REVIEW THE advantages of library modules in general. Among other things, they:

- simplify the process of developing and testing new program components,
- let program components be shared and reused, and
- ease long-term problems of program maintenance and improvement.

These benefits are just as applicable to data abstraction as they were to procedural abstraction. We've already seen the benefits of layering, which was typical of interface modules:

layering, as in interfaces

Layering *of program segments encourages*
Abstraction *or the definition and use of high-level working terms, which promotes*
Information Hiding *or the separation of definition and implementation.*

Clustering, in turn, was illustrated by package modules:

clustering, as in packages

Clustering *program segments implies*
Encapsulation *into distinct packages, which leads to*
Localization *of effects and changes.*

Data abstraction will draw on both aspects:

data abstraction uses both

Data Abstraction

Clustering — **Layering**

type names and primitive operations are collected in one place

Encapsulation — **Abstraction**

but underlying implementation can be changed without affecting usage

Localization — **Information Hiding**

Abstract Type Modules

An abstract type module holds the definitions and primitives that create an abstract type. It's important to understand that encapsulation alone is not what brings a new data type into the league of the abstract data types. Nor is the separation of definition and implementation sufficient. Instead, the type is abstracted by being defined in terms that are distinct from any particular implementation, and, insofar as possible, from any particular application.

abstraction separates operations from implementation

> What values will it include? What operations will be permitted on those values? A data type is only fully abstracted when we've separated the names and operations used to manipulate values from any underlying representations used to hold the values, or algorithms used to carry out the operations.

The fact that Modula-2 splits modules into two parts is clearly to our advantage. The type name and the names of operations on values of the type can go into the definition module. Implementation details, in turn, can be hidden in the implementation module.

Here are some data types that are commonly abstracted. As you read the list, try to imagine a set of operations that might apply to each type.

common data type abstractions

- *Fractions*, as we saw in Chapter 10.
- *Complex numbers*, which, like fractions, are two-piece values.
- *Cartesian coordinates*—again, two-piece values.
- *Extended ranges* of more prosaic types like REAL or INTEGER.
- *Strings*, or sequences of characters.
- *Graphics* and *typesetting* characters whose meaning may depend on the availability of input or output devices.
- *Colors* and other enumerable types that may have meaningful operations on their values.
- *Vectors* and other values that have some characteristics of structures, but are commonly understood and used as values.

For each of these, it's easy to imagine declaring variables *of* the appropriate type (e.g. FRACTION or STRING), and to anticipate naming a collection of primitive operations *on* values of the type.

> Data abstraction focuses on primitive operations, rather than type definitions. It answers the question 'How do we use it?' In doing so, the abstraction is able to hide the 'what's it made of?' details.

Goal-oriented design, by its nature, has a relatively narrow focus on getting results. Object-oriented design, in contrast, will be more general:

full abstraction

- An abstraction's primitives should *fully abstract* a type or structure. They should make its operation independent of any underlying implementation.

full realization

- The primitives must *fully realize* the abstraction. A data abstraction has to consider the meaning of a new type or structure above and beyond the demands of a particular application.

A data abstraction should strive for both characteristics: full abstraction, and full realization. Merely putting a type definition in a module does not a data abstraction make. Primitives, as well as values, define the abstract type. Operations that cannot be carried out with the primitives, or some combination of primitives, simply do not exist for the type abstraction.

Designing Data Abstractions

The design of data abstraction modules essentially follows the discussion of layered type design in Chapter 10. The three basic steps are *draft, formalize,* and *implement:*

draft

1. Define the new type or structure informally, by describing its legal values (for types) or relations between stored values (for structures), and its basic operations.

2. Make an *abstract definition* of the new type or structure by:

formalize

   a) drafting a *name layer* that contains the calling forms of the primitive operations; then

   b) formalizing the name layer by writing the module's definition part.

implement

3. Using the name layer (the definition module) as a requirements specification, implement the data abstraction. In practice, we will often put off the final definition of any types until now.

Thus, the definition module makes all the primitive operations, which might require knowledge of the underlying type definition, available. It serves as the performance specification of the data abstraction.

The implementation module, in turn, acts as the hidden code layer. It contains the code of the primitive subprograms listed in the definition module. It can also include additional routines that might be useful for carrying out the primitive operations, but which aren't meant to be made available publicly.

avoiding name pollution

Once we start programming for the ages, naming becomes quite important; 'name pollution' is one of the side effects of abstraction. As a rule, the external module's name should describe the data abstraction itself: Color, perhaps, or Fraction, or List. Primitive names, in contrast, should be consistent and self-descriptive: In, Out, or Assign. They should describe their own operation, rather than the name of the abstraction. Finally, the type name itself can be a simple T.

Paradoxically, the goal of the primitive naming rule is to avoid confusion between modules, even though it means that the same names may be used repeatedly. Consistency in usage is the goal:

A module that contains:

IMPORT Color	**IMPORT** Fraction	**IMPORT** LongFraction

should refer to like operations consistently:

use consistent primitive names

Color.T	Fraction.T	LongFraction.T
Color.In	Fraction.In	LongFraction.In
Color.Out	Fraction.Out	LongFraction.Out
Color.Add	Fraction.Add	LongFraction.Add

Additional primitives may differ, of course:

Color.Complement	Fraction.Reduce	LongFraction.Digits

This rather mechanical approach to naming makes it much easier to keep names straight. Moreover, it simplifies substitutions (LongFraction for Fraction) that may occur late in the game.

An Ordinal Type Abstraction: Color

Let's begin by abstracting a relatively simple type. The primary colors of light are red, blue, and green. Taken in combination, they can form any of the other shades; in equal parts they produce white light. Take a very close look at your television screen if you want to confirm this.

If all we need is a type definition, it's easy enough to enumerate a few colors:

```
TYPE PrimaryTYPE = (red, blue, green);
```

But instead of being satisfied with the ability to declare variables that can have colors as values, let's really try to abstract a Color type. What does a full realization of Color require? At the very least, we need primitives to:

operations on Color values

read and write the legal colors
mix two colors together
find the complement (opposite) of a given color

These operations are the very first draft of a name layer for Color. A second draft begins to lead us into an abstract definition of the type:

drafting the name layer

the type name: T
an input primitive: In
an output primitive: Out
a primitive for mixing colors: Add
a primitive for finding a complementary color: Complement

The abstract definition is formalized in the Color definition module, below.

abstract definition of Color

```
DEFINITION MODULE Color;
  TYPE T = (undefined, red, yellow, green, cyan, blue, magenta);
    (*red, green, and blue are the light (additive) primaries.
     yellow, cyan, and magenta are the pigment (subtractive) primaries.
     All procedure results that are unknown in this implementation are
     returned as undefined.*)
  PROCEDURE In (VAR Value: T);
    (*Reads the string equivalent of a color value.*)
  PROCEDURE Out (Value: T);
    (*Prints the string equivalent of a color value in exact space.*)
  PROCEDURE Complement (Value: T) : T;
    (*Returns the 'opposite' of the color Value.*)
  PROCEDURE Add (First, Second: T) : T;
    (*Returns the result of combining colors First and Second.*)
END Color.
```

Pay special attention to the comment that accompanies the definition of T.

document the abstract definition

> Since an importing program's only visible point of contact with Color is through its definition module, all results must be well-documented.

As noted, the definition module has to serve as a performance specification for use of the abstraction, as well as a requirements specification for its implementation.

Self-Check Questions

Q. The individual values of Color.T are automatically known in the Color definition and implementation modules. Can they be used in a program that imports Color as well?

A. Yes. If a program imports Color:

```
IMPORT Color;
```

it can refer to Color.undefined, Color.red, Color.yellow, and the others. Naturally, references to procedures, like Color.In, are legitimate too.

Q. As matters stand, we've written the definition part of Color, but haven't gotten to its implementation part yet. Could a program import Color before the implementation part is written?

A. Yes—a program that imports Color can be written, and compiled (it's a good way to check its syntax). All the compiler uses Color for at this point is to check on the legality of calls. However, a program that imports Color can't be linked and executed until the implementation module has been written and compiled as well.

The Implementation of Color

The final step in developing the type abstraction is to implement it. As long as the definition module's contents don't change, writing, compiling, and linking the implementation module can be delayed until the last possible moment.

The implementation is pretty straightforward. Aside from the Add function, it relies on techniques that you're already familiar with. Add is different because it uses the **SET** data type (Chapter 15's topic). Normally, I wouldn't make an out-of-order introduction like this, but the alternative nested **CASE** was too horrible to contemplate. Briefly, the definition and declaration:

a set definition . . .

```
TYPE HueSET = SET OF T;
VAR Mix: HueSET;
```

creates a variable, called Mix, that can represent any combination of the values of Color.T:

and assignment . . .

```
Mix := HueSET{First, Second};    (*Assigns Mix two Color.T values.*)
```

Once Mix has been initialized, it can be compared to specific subsets of the Color.T colors. This expression:

and expression

```
Mix = HueSET{red, blue}
```

is TRUE if Mix represents only red and blue, and is FALSE otherwise.

the Color implementation module

```
IMPLEMENTATION MODULE Color;
IMPORT IO;
IMPORT STRINGS; (*or your own system's equivalent*)
PROCEDURE In (VAR Value: T);
    (*Reads the string equivalent of a color value.*)
  TYPE STRING = ARRAY [0..7] OF CHAR;
  VAR Temp: STRING;
  BEGIN
    IO.StringIn (Temp);
    IF STRINGS.Equal(Temp, "red") THEN Value := red
      ELSIF STRINGS.Equal(Temp, "yellow") THEN Value := yellow
      ELSIF STRINGS.Equal(Temp, "green") THEN Value := green
      ELSIF STRINGS.Equal(Temp, "cyan") THEN Value := cyan
      ELSIF STRINGS.Equal(Temp, "blue") THEN Value := blue
      ELSIF STRINGS.Equal(Temp, "magenta") THEN Value := magenta
      ELSE Value := undefined
    END (*IF*)
  END In;
```

(module continues)

```
PROCEDURE Out (Value: T);
  (*Prints the string equivalent of a color value.*)
  BEGIN
    CASE Value OF
        red: IO.StringOut ("red")
      | yellow: IO.StringOut ("yellow")
      | green: IO.StringOut ("green")
      | cyan: IO.StringOut ("cyan")
      | blue: IO.StringOut ("blue")
      | magenta: IO.StringOut ("magenta")
      | undefined: IO.StringOut ("undefined")
    END (*CASE*)
  END Out;
PROCEDURE Complement (Value: T) : T;
  (*Returns the 'opposite' of the color Value.*)
  BEGIN
    CASE Value OF
        red: RETURN cyan
      | yellow: RETURN blue
      | green: RETURN magenta
      | cyan: RETURN red
      | blue: RETURN yellow
      | magenta: RETURN green
      ELSE RETURN undefined
    END (*CASE*)
  END Complement;
PROCEDURE Add (First, Second: T) : T;
  (*Returns the result of combining colors First and Second.*)
  TYPE HueSET = SET OF T;
  VAR Mix: HueSET;
  BEGIN
    Mix := HueSET {First, Second};
    IF Mix = HueSET{red, blue} THEN RETURN magenta
      ELSIF Mix = HueSET{red, green} THEN RETURN yellow
      ELSIF Mix = HueSET{red, red} THEN RETURN red
      ELSIF Mix = HueSET{blue, green} THEN RETURN cyan
      ELSIF Mix = HueSET{blue, blue} THEN RETURN blue
      ELSIF Mix = HueSET{green, green} THEN RETURN green
      ELSE RETURN undefined
    END (*IF*)
  END Add;
BEGIN
END Color.
```

Modification Problems

1. Write a program that imports Color. Have it do something simple, like reading two colors, then finding and printing their sum. Compile it (you'll have to compile the implementation module of Color, too) and run it.

Now, add one of the new Klingon colors, blurk, to Color.T. Modify the implementations of In, Out, Complement, and Add to deal with the new color. You'll have to decide what the appropriate results of mixing and complementing blurk are, of course.

2. Oops—it turns out that blurk doesn't mix very well. Adding it to any color yields blurk, and the most complimentary description of its complement is undefined. Modify the Color implementation module to reflect these findings. Does your test program have to be recompiled to be run again? What must you do?

A Structured Type Example: XYPoints

Let's turn to a type that can represent two-dimensional Cartesian coordinate values, i.e. *x*, *y* coordinate pairs. Assume that the range of REAL will be sufficiently, but not overly, large and precise for our purposes. What operations might we want to carry out on coordinates?

operations on Cartesian coordinates

read and print coordinate values
separately assign the x and y parts of a coordinate value
find the midpoint between two coordinates, or their distance
determine what quadrant a coordinate is located in
calculate a coordinate's distance to, and angle from, the origin

and so on.

Informally listing operations on coordinates gives us an idea of where we're heading. In a second draft of the name layer, we can start to choose names for the type and its primitives.

drafting the name layer

the type name: T
the input and output primitives: In, Out
assignment primitives: AssignX, AssignY
a midpoint-calculation primitive: Midpoint
a distance-to-origin primitive: ToOrigin
calculating the angle to the x-axis: Angle
a point-to-point distance finder: Distance
a quadrant-locating primitive: Quadrant

Formalizing the Abstract Definition

Our next goal is to stabilize the abstraction, by fixing the form of the primitive calls. Note that even at this late point, there is no need to fill in the details of the definition of T. The abstract definition of the type should not, and does not, depend on its representation in Modula-2.

It is, however, essential to make sure that the primitives are clearly defined. Procedures like In and Out are particularly important when it comes to documentation.

the XYPoints abstract definition

```
DEFINITION MODULE XYPoints;
  (*A library module that abstracts a two-dimensional coordinate type.*)
  CONST WIDTH = 5; (*Printing field and decimal accuracy. See Out.*)
        PLACES = 2;
  TYPE T = ? (*We'll complete this definition below.*)
  PROCEDURE In (VAR Point: T);
    (*Reads blank-separated REAL values for the x and y coordinates.*)
  PROCEDURE Out (Point: T);
    (*Prints a T value. Format: "(x.xx, y.yy)" *)
  PROCEDURE AssignX (XValue: REAL; VAR Point: T);
  PROCEDURE AssignY (YValue: REAL; VAR Point: T);
    (*Assigns x and y coordinates to a Point.*)
  PROCEDURE Midpoint (Point1, Point2: T; VAR Point3: T);
    (*Returns the point that's in the middle.*)
  PROCEDURE ToOrigin (Point: T) : REAL;
    (*Returns the distance between Point and 0.0, 0.0.*)
  PROCEDURE Angle (Point: T) : REAL;
    (*Returns the angle formed by Point, 0.0, 0.0, and the x-axis.*)
  PROCEDURE Distance (Point1, Point2: T) : REAL;
    (*Returns the distance between two points.*)
  PROCEDURE Quadrant (Point: T) : INTEGER;
    (*Returns the quadrant (1..4) that Point lies in.*)
  ... other primitives
END XYPoints.
```

Pay special attention to AssignX and AssignY. Why are they included as primitives? So that assignments to XYPoints.T variables won't require knowledge of the details of the type definition. Such primitives are exceptionally important for information hiding.

Now, you can see that I did something a little peculiar in the definition module of XYPoints—I didn't bother to define the underlying type T. Not surprisingly, I did this to make a point:

> The representation chosen for an abstract type should be picked for the benefit of its primitives.

As it happens, Cartesian coordinates don't really have very much in the way of alternatives. In this case, the obvious representation is the best:

the definition of XYPoints.T

```
DEFINITION MODULE XYPoints;
   (*A library module that abstracts a two-dimensional coordinate type.*)
   CONST WIDTH = 5;
         PLACES = 2;
   TYPE T = RECORD
              x, y: REAL
            END;
        . . .   etc.
```

Self-Check Questions

Q. Suppose that a program contained this definition and declaration:

```
IMPORT XYPoints;
VAR Location: XYPoints.T;
```

Would these assignments be legal?

```
Location.x := 1.0;
Location.y := 2.0;
```

A. Yes, but we'll never make them. In technical terms, a definition module *transparently exports* the details of any full type definitions it contains. This applies to both structured and enumerated types. The assignments could be made.

However, we won't, because such assignments make the calling program dependent on a particular implementation of what is supposed to be an abstract type. Programmers shouldn't take advantage of 'hidden' knowledge, and definition modules should export a set of primitive operations sufficient to remove temptation.

The Implementation Layer

With the definition module as a requirements specification, we can draft the implementation module of XYPoints. Note that:

- The type definition of XYPoints.T, complete with internal details, is automatically known within the implementation module, as are the constant definitions FIELD and PLACES.
- Additional imports (here, from MathLib0 and IO) must be made within the implementation module if they are needed there.
- As always, the headings of the subprograms supplied in the definition module must be copied exactly when they are implemented.

(code follows)

the XYPoints implementation

```
IMPLEMENTATION MODULE XYPoints;
IMPORT MathLibØ;
IMPORT IO;
PROCEDURE In (VAR Point: T);
   BEGIN
      IO.RealIn (Point.x);
      IO.RealIn (Point.y)
   END In;
PROCEDURE Out (Point: T);
   BEGIN
      IO.CharOut ("(");
      IO.FixOut (Point.x, WIDTH, PLACES);
      IO.CharOut (",");
      IO.FixOut (Point.y, WIDTH, PLACES);
      IO.CharOut (")");
   END Out;
PROCEDURE AssignX (XValue: REAL; VAR Point: T);
   BEGIN
      Point.x := XValue
   END AssignX;
PROCEDURE AssignY (YValue: REAL; VAR Point: T);
   BEGIN
      Point.y := YValue
   END AssignY;
PROCEDURE Midpoint (Point1, Point2: T; VAR Point3: T);
   BEGIN
      Point3.x := (Point1.x + Point2.x) / 2.0;
      Point3.y := (Point1.y + Point2.y) / 2.0
   END Midpoint;
PROCEDURE ToOrigin (Point: T) : REAL;
   BEGIN
      RETURN MathLibØ.sqrt( (Point.x*Point.x) + (Point.y*Point.y) )
   END ToOrigin;
  . . .
BEGIN
   (*This particular implementation doesn't require any initialization code.*)
END XYPoints.
```

A program that operates on Cartesian coordinates (perhaps it can do your analytic geometry homework) imports the XYPoints module. Variables of type XYPoints.T can be used as conveniently as variables of type REAL or INTEGER.

using the type abstraction XYPoints

```
MODULE Geometry;
    (*Take advantage of the type abstraction defined in XYPoints.*)
IMPORT IO;
IMPORT XYPoints;
VAR A, B, C: XYPoints.T;
BEGIN
    IO.StringOut ("Give me a REAL x,y coordinate pair (x  y):  ");
    XYPoints.In (A);
    XYPoints.Out (A);   (*Echo the entry.*)
    IO.LineOut;
    IO.StringOut ("How about another pair?  ");
    XYPoints.In (B);
    XYPoints.Out (B);   (*Echo the entry.*)
    IO.LineOut;
    IO.StringOut ("The midpoint of the two coordinates is:  ");
    XYPoints.Midpoint (A, B, C);
    XYPoints.Out (C);
    IO.LineOut;
    IO.StringOut ("Its distance to the origin is:  ");
    IO.FixOut ( XYPoints.ToOrigin(C), 5, 3 );
END Geometry.
```

```
Give me a REAL x,y coordinate pair (x  y): 1.0  3.0
( 1.00, 3.00)
How about another pair? 3.0  1.0
( 3.00, 1.00)
The midpoint of the two coordinates is: ( 2.00, 2.00)
Its distance to the origin is: 2.828
```

Modification Problems

1. Procedure XYPoints.In is a weak link in the design of XYPoints, because it relies on particular field identifiers (x and y) in variables of type T. Modify it to use calls of AssignX and AssignY instead.

Section 12-4 Program Engineering: Design Strategies

KISS, OR KEEP IT SIMPLE, STUPID, ISN'T A VERY elegant name, but it's probably the best piece of programming advice there is. Don't take it personally, by the way—it isn't aimed at you! Instead, it's a tip for *us*—for everybody who has to write programs. As is usually the case, software manages to become complicated all by itself, but keeping things simple takes a fair amount of work.

Overall, our goal is to limit the purpose of a module to one thing, but to do that one thing well. There is no special virtue inherent in making a distinction between using libraries as interfaces or as packages. However, when a program deals with ten or twenty libraries, life is much simpler if we can readily identify a problem as 'We need a different sort of interface to this library.'

In the long run, you'll find that clearly identifying the role a module will play—as distinct from the specific things its procedures will do—helps make better modules. You may sit down to write a few procedures because of the demands of a particular project, but coding them in a shareable library makes them an addition to your Modula-2 programming environment.

the moral

> Program code may be quick and dirty, but you should write module code for the ages.

Let's look at some of the other program engineering considerations that underly the use of libraries.

Interfaces and Layered Implementation

We take the trouble of looking at interfaces as a distinct category of modules because they help support a style of programming called *layered implementation*. As you start to develop the skills that let you add to your own Modula-2 system's stock of libraries, there's a useful lesson to be learned from layers.

In a layered implementation, the programmer accepts the fact that the perfect interface can't be planned in advance. Instead, she makes a tradeoff: if she can be provided with libraries modules that are powerful, but perhaps inconvenient, she'll be willing to write specialized interfaces when she needs them.

A non-layered implementation, in contrast, might be 'user-friendly,' but it wouldn't necessarily be programmer-friendly. For example, WYSIWYG—what-you-see-is-what-you-get—text processors are, indeed, quite friendly to their users. However, most programmers deride them as what-you-see-is-ALL-you-get systems! There's no way to get at the system's low-level primitives; we can't make it do things that didn't occur to the text processor's designers.

> Knowing that later programmers can write interface modules to add customizing layers to their systems makes life considerably easier for those who are engaged in providing and managing programming systems. They do not have to anticipate precise user requirements, nor do users have to adapt to an unyielding standard interface.

Experience has shown that it is nearly impossible to write library modules that will please everybody. But why bother? Why trade power for ease of usage in a futile attempt to write the perfect module? Instead, why not try to provide it all, in the knowledge that an interface module can do whatever customizing is needed?

Interface modules use three basic customizing strategies, sometimes simultaneously:

- *Simplifying calls:* suppressing parameters that might be useful for error-checking or general-purpose control, but which might not be needed in a particular application.
- *Merging libraries:* making it easier to call procedures that might have originally been put in separate modules. As it happens, this is what IO does.
- *Splitting libraries:* selecting a particular part of a large and detailed library, so that its relevant procedures will be easier to use and understand.

layered implementation styles

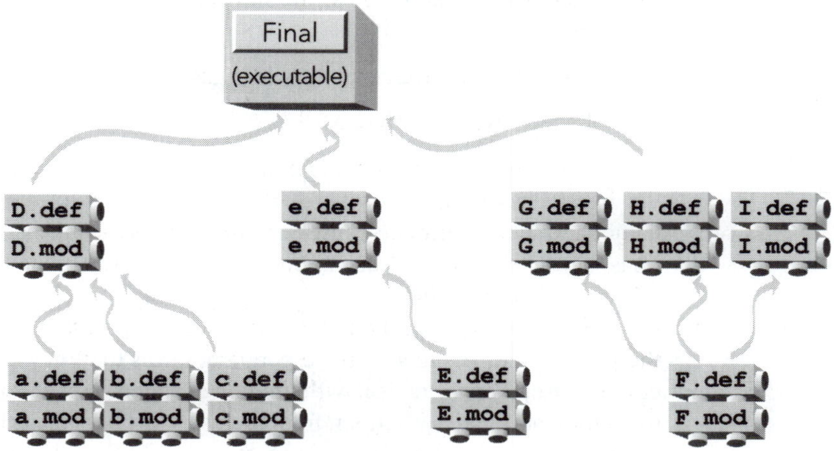

Merging separate libraries. *Simplifying complex libraries.* *Splitting a complex library.*

For example, the International Standards Organization (ISO) is like ANSI, but more so. It has defined a standard for graphics operations known as the *graphical kernel system*, or GKS. A library that provides GKS will be enormous—it's required to include more than two *hundred* procedures. GKS is great for system programmers because it touches all the bases. However, it's unwieldy to work with because it's so detailed. In practice, most systems with graphics capabilities come with one or more libraries that simplify GKS by providing a select subset of its procedures. The main library is split, and its procedure calls are simplified.

In contrast, most input/output libraries merge procedures from several modules. For example, the IO library that we use joins REAL I/O procedures from a module called RealInOut with other I/O procedures from a module named InOut. These libraries, in turn, are written atop even lower-level libraries that control access to the terminal and file system, but often require extremely detailed procedure calls. Thus, IO merges and simplifies.

keep interfaces shallow

> Whichever approach you use, try to keep interface layers relatively *shallow*. Special-case calls to libraries at different levels may make a library easier to write, but they make it much more difficult to understand and maintain.

Library Quality Issues

But effective interfaces depend on libraries that are good enough to encourage software reuse. And 'goodness' is not an easy quality to obtain! Libraries should be:

- Correct—they have to perform as advertised.
- Robust enough to deal with any reasonably anticipated problems.
- Portable, so that they can be moved along with applications that rely on them.
- Efficient, since they may be widely used.
- Easy to use correctly!

release, don't unleash

Correctness is the fundamental issue. Once it is put out into the world, the library has to stand on its own. The promises implied by its procedure names and their comments must be upheld. This means that the library's limitations must be as well-documented as its features. Libraries should be released—not unleashed!

Robustness is closely tied to correctness. As we saw in developing EZStats, the programmer has to anticipate potential errors that may arise in the course of ordinary use, then try to deal with them in advance. Never forget that the implementation of libraries is hidden from the user—who may, in consequence, have a difficult time in finding bugs. Where appropriate, include an error-indicating parameter as part of the procedure call. If necessary, the client can write an interface module that suppresses the parameter.

Portability is the third requirement of library programming. System managers are seldom interested in installing library modules that are not self-contained—a library should not be the tip of an iceberg of extensive **IMPORT**s. Nor should a library consciously exploit the peculiarities of a particular system. To the extent possible, library modules should be based on commonly available 'standard' libraries. When possible, use standard types in preference to defining new types. As we've seen, the open array parameter was included in Modula-2 for precisely this application.

Efficiency is a fourth issue. In providing a library, the Modula-2 system creates a bond of trust with the client programmer. In return for not peeking at a library's implementation, the client deserves to know that she will be given access to reasonably efficient code. It is the library programmer's responsibility not to abuse that trust.

Ease of use is the final essential requirement. Obviously, the standard rules that apply to the creation of any set of procedures—meaningful naming and clear ordering of parameters, in particular—apply. However, I have to make the point, once more, that the definition module is all that an importing programmer will ever see. It must be as well-documented as possible.

Data abstractions, in particular, take on a life of their own that's quite independent of the program they were originally intended for. This means that careful planning will be essential. Usually, the programmer tries to step back from immediate program concerns when she designs a module for data abstraction. Instead, she looks at the needs of the abstraction itself. Why?

- Building reusable software helps free programmers from having to reinvent the wheel each time a particular type or structure is required.

When a module has been written in a limited manner, just to meet the needs of one program, it will not just be inconvenient to use elsewhere. The sad fact is that it will not even be considered for use, because the programmer has never even considered advertising its availability. Why should she?

- Taking into account future demands that may be placed on the abstraction helps create *open designs*.

Open designs are easily extended and adapted. They're suited to accommodate new requirements that, like as not, are suggested by the success of the current implementation.

- Defining types in plain English, in terms of the primitive operation's names, encourages proper implementation and usage.

Misunderstanding of specialized requirements is the most pernicious of the flaws that arise from limiting the implementation of data abstraction modules. Although the need for documenting specifications and software *and reading other people's documentation* is as accepted as the need for highway speed limits, they are strictly observed about as often.

In practice, most people assume that data-oriented modules will fulfill their own common-sense expectations, and rely on them accordingly. But if modules have been written in a limited or job-specific manner, mismatches between modules and the programs that import them can and do occur. Unfortunately, such mismatches are often not discovered until the worst possible time—when the final program is at its largest and most unwieldy.

Background: Why Worry About the Next User?

Hearing a disturbance, the master programmer went into a novice's cubicle.

'Curse these personal computers!' cried the novice in anger. 'To make them do anything I must use three or four editing programs. Sometimes I get so confused that I erase entire files. This is truly intolerable!'

The master programmer stared at the novice. 'And what would you do to remedy this state of affairs?' he asked.

The novice thought for a moment. 'I will design a new editing program,' he said, 'a program that will replace all these others.'

Suddenly the master struck the novice on the side of his head. It was not a heavy blow, but the novice was nonetheless surprised. 'What did you do that for?' exclaimed the novice.

'I have no wish to learn another editing program,' said the master.

And suddenly the novice was enlightened.

From *The Zen of Programming*, by Geofrey James, © 1988, InfoBooks, Santa Monica, Ca. 90406

Modula-2 Details

Syntax Summary

• library module: a pair of files (definition and implementation modules) whose contents can be imported by any Modula-2 program. Sometimes called *external module*.

• definition module: supplies the identifiers that a library wishes to make available. These *public* identifiers can be imported by other programs or libraries, and are automatically known in the implementation module.

> **DEFINITION MODULE** *its name;*
>
> **IMPORT** *as necessary (usable in definition module only);*
>
> **CONST** *definitions as desired;*
>
> **TYPE** *definitions as desired;*
>
> **VAR** *declarations as desired;*
>
> **PROCEDURE** *headings as desired;*
>
> **END** *its name* **.**

• implementation module: contains the *bodies* or code of any subprograms from the definition module, along with any *private* definitions or declarations that are used solely within the implementation module.

> **IMPLEMENTATION MODULE** *its name;*
>
> (**These are all local to the implementation module:* *)
> **IMPORT** *as necessary;*
> **CONST** *definitions as desired;*
> **TYPE** *definitions as desired;*
> **VAR** *declarations as desired;*
> **PROCEDURE** *declarations as desired;*
>
> (**Subprograms from the definition part must be completed as well.**)
> **PROCEDURE** *headings are repeated, then completed;*
> **BEGIN**
> *initializing statements, if desired*
> **END** *its name* **.**

• opaque type definition: (not discussed in detail) Separates the availability of a type name from the details of its definition. In the definition module, the type is simply presented:

> **DEFINITION MODULE** Opaque;
>
> **TYPE** NameTheTYPE; *etc.*

In the implementation module, the opaque type must be defined as a pointer type:

> **IMPLEMENTATION MODULE** Opaque;
>
> **TYPE** NameTheTYPE = **POINTER TO** *a privately defined type;* *etc.*

Modula-2 Details

Remember . . .

- Any code in an implementation module will be run when the library is first imported. This code is usually strictly limited to statements needed to initialize the library.
- Variables declared in an implementation module have the same lifetime as main program variables, but have an entirely separate scope. They keep their values for the entire program, but they can only be changed indirectly by calling library procedures.
- All the details of a type definition made in the definition module are public, including the constants of new ordinal types, and the names of fields in records. However, good programming practice is to refrain from using these details in a program that imports the library. Naturally, the library must supply subprograms that make direct references to the 'details' unnecessary.
- Splitting a Modula-2 system's supporting code into separate libraries—*encapsulation*—helps localize system- or machine-specific details.
- Splitting libraries into distinct definition and implementation parts—*abstraction*—separates *what?* from *how?* By encouraging information hiding (keeping implementation details out of application programs), it eases changes in implementation.
- Libraries aren't inherently separated into classifications. But distinguishing between different library goals (for instance, by asking if the library will focus on names, actions, or data types) is useful as the means to an end—writing libraries that will work well together.
- *Interface* modules typically provide a new way to access existing identifiers. Their importance is indirect; the fact that interfaces can be written encourages full and detailed implementations of packages and data modules.
- *Package* modules provide new functions and procedures, but don't hide the details of new type definitions or abstract types. They encourage the collection and use of general-purpose routines.
- *Data* modules provide methods of storing information, and of acting on stored information. I've described three major groups: *type abstractions* (defined as a set of primitive operations and values), *structure abstractions* (defined by primitive operations and the relations between values, rather than the values themselves, and discussed in 16-1), and *hidden objects* (which maintain just one instance of a data abstraction, and are discussed in 16-2).
- Whether data abstractions involve new types or structures, they depend on primitives in two ways: the primitives should fully abstract the type or structure (to make it independent of underlying implementation), and the primitives should fully realize it (to make it useful beyond any current application).
- Object-oriented design is the 'data first' approach to problem solving and programming. It differs from more traditional goal-oriented design by focusing on actions that *may* be taken with program data, rather than on steps that *must* be taken in order to solve a problem. We build data-oriented tools before working on strictly goal-oriented ones.

• Layered design is a methodology that's used in support of object-oriented design, and which happens to fit well with Modula-2's split-module approach to libraries. After informally drafting the new type's primitives and properties, we make a more formal specification (i.e. in the definition module), then implement it (in the implementation module).

• Consistent naming of types and primitives (type name T, input/output primitives IN and OUT, etc.) helps minimize name pollution. It's one way to obey the Law of Least Astonishment.

Zen Study Guide

Big Ideas

The Big Ideas of chapters 11 and 12 are:

1. *Rules of library module syntax and compilation order permit a good deal more than good practice allows.* In a sense, the rules permit complexity so that the libraries themselves can be written and used simply. Choose the clearest alternative, rather than relying on the reference manual to save the day.

2. *Abstraction and encapsulation are the* result *of good design, rather than* being *good design.* Putting parts of a program into Modula-2 modules doesn't make a program modular—layering and clustering may even work at cross purposes. Modules don't make a program modular; how we use them does.

3. *A good module, like a good waiter, maintains its integrity by serving its 'profession' rather than its customer.* A little bit of distance between the problem (as supplied by a particular program) and the solution (as provided by a library module) is a good thing. Perhaps the best way to ensure this is to imagine that you are writing modules to be used with and by other modules, rather than by specific programs.

Exercises

Warmups and Test Programs

12-1 Write and compile a library module (it doesn't have to really do anything). What suffix must its name have? What suffix does the compiled output have? Can you write a test program that imports it? What happens if you then add a blank space to the implementation module? To the definition module?

12-2 If the constant DEBUGGING exists, and if IO has been imported, then the following statement can legally appear within a implementation module:

```
IF DEBUGGING THEN
    IO.StringOut ("Importing and initializing the TEST library.");
END
```

Where should DEBUGGING be defined? What is the difference between defining it in the definition module and in the implementation part? What must be done if the value of DEBUGGING ever changes?

12-3 Where does your system keep its library modules? Are the definition and implementation modules in the same place? Does the library source code (i.e. the definition and implementation modules) always have to be available?

12-4 Find a computer lab (or go to your nearby computer store) and see how many different input and output devices you can find. Next, imagine that you are writing input or output libraries based on some of these devices. Which ones might allow for identical definition modules, as long as there were appropriate (but different) implementation modules? Which ones are, in effect, simplified versions of others? For example, a touch screen and light pen might allow essentially the same kinds of input—an *x,y* coordinate—whereas a keyboard and a bar code reader are quite different.

12-5 What happens when two libraries import each other? Yes, I know it's illegal, but what happens?

12-6 The notion of printing field widths for numerical output is confusing at first. How much work would be involved in writing an EasyIO module that would have the same powers as IO, but which wouldn't require field width arguments for its output procedures? Outline the new library's definition and implementation parts.

Exercises

Programming Problems

12-7 The universe is filled with fundamental constants that should always be at your fingertips. In addition to mathematical constants (like π and *e*), chemical constants (like Avogadro's number), and physical constants (like the speed of light), there are also the social constants (two's company, three's a crowd), the artistic constants (the golden ratio, or ratio between successive Fibonacci numbers) the philosophical constants (one, of course), and such ephemeral constants as the time lapse between the moment you enter the shower and the doorbell rings, or the moment you begin a journey and the first chorus of 'Are we almost there yet?' is heard from the back seat.

Write a library of useful constants.

12-8 Write a CONVERSIONS library that supplies functions that represent English measures as their metric equivalents, and vice versa.

12-9 Complete and test the ANSI library. Then, use it to extend or enhance some of the output procedures supplied by the IO library. For instance, your new SuperIO library might include HiLiteStringOut and BlinkingIntOut procedures.

12-10 Almost every output device worth its salt obeys some set of escape sequences. Like the ANSI sequences, they control character printing. Most hardcopy devices allow various combinations of overstriking, some font control, line spacing, choice of character set, use of graphics characters, and so on. The most common sets of escape sequences I know of are the Epson FX and IBM ProPrinter sets, which largely overlap.

Write a library that is similar to ANSI, but is used for printer control instead. You will have to find out which set of escape sequences your own system uses, of course. Hint 1: Be sure that you write the 'reset' command first, otherwise you will lose your mind during debugging. Hint 2: Do *not* assume that you can inspect your library's output by printing it on your terminal screen—try redirecting it to a file, then editing (or printing) the file, instead.

12-11 Write a library that supports arithmetic for very long whole numbers (how long should be documented, but it's up to you). At a minimum, you should provide primitives for input, output, and addition. Moreover, you must decide and document how your library will deal with errors (such as input or results that are too long for your implementation).

12-12 (To be read by engineering majors only!) The basic operations that are performed on vectors and matrices are the calculation of vector magnitude, dot product, cross product, and matrix multiplication. Write a library that supports these operations. Naturally, you should use open array parameters if possible. In particular, you should see if and how your Modula-2 system supports multi-dimensional open arrays.

'A chain of pointers isn't like a ball of string. If the end gets lost, it's really gone.' (p. 540)

Chapter 13 The Pointer Types

In his fascinating book *The Medium is the Massage*, Marshall McLuhan argued that information is changed by the medium that conveys it. Our understanding, said McLuhan, is affected by the way that information is massaged into our brains. As a result, a medium—radio, print, television—becomes part of any message it delivers.

Does the medium matter in computer science? Certainly, because the method used for storage can add to the information that's being stored. Indeed, most of our work with arrays and records has revolved around taking as much advantage of this as we can.

In this chapter, we'll see how to build structures whose components can be put into *any* order that's convenient and meaningful. Our basic tool will be a *node*—a record that has some fields for information, and one or more *pointer* fields that locate the node's neighbors. I won't kid you—pointers can be confusing at first. But they are the basis of all advanced programming, so don't give up.

Section 13-1
Strictly Modula-2

I HATE ROUNDABOUT INTRODUCTIONS, BUT I'M willing to make an exception for pointers. Life will be easier later if you have a feel for why pointers are necessary, and why certain details of usage are an inevitable consequence of what pointers are used for.

Let's begin with an example that seems as though it can be solved with our old tools arrays and records, but winds up giving us the motivation we'll need to 'invent' pointers on our own. Suppose that we're working with this array of records:

```
TYPE ItemTYPE = RECORD
                    Data: StringTYPE;  (*Holds a short word.*)
                    Next: INTEGER
                END;
     ListTYPE = ARRAY [1..20] OF ItemTYPE;
VAR List: ListTYPE;
    First, Last, CurrentPtr: INTEGER;
```

We might use the array to hold a sequence of words in alphabetical order:

the words are alphabetized

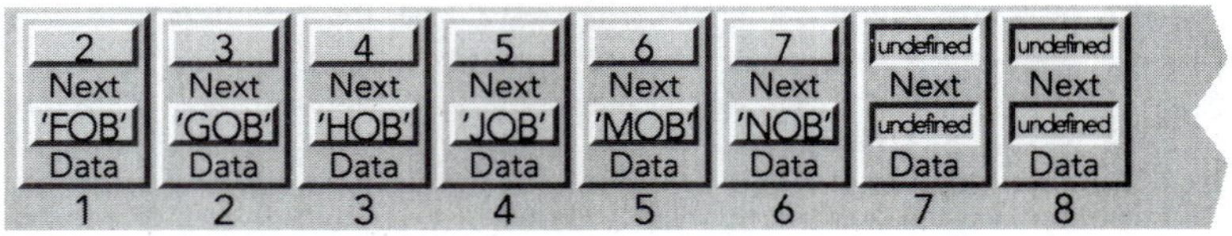

First = 1, Last = 6, CurrentPtr = 7

physical and logical order

As you can see, the Next field of each component holds a subscript. If you follow the trail of Next fields, you'll read the words in alphabetical order. However, since the physical (subscript) order and logical (alphabetical) order of the components are the same, the Next field doesn't seem to serve any purpose.

But Next is useful if we add new words to the end of our carefully ordered array. Depending on what arrives, physical and logical order may break down. Not every word causes problems, mind you. If a lucky word like "Sob" arrives, it can be stored in the seventh record without messing up the array's order.

But what if "Lob" shows up instead? Since it follows "Job", components "Mob" and "Nob" have to move to the right to make room. A word like "Bob" is much worse, since every word has to slide down.

problems with insertion

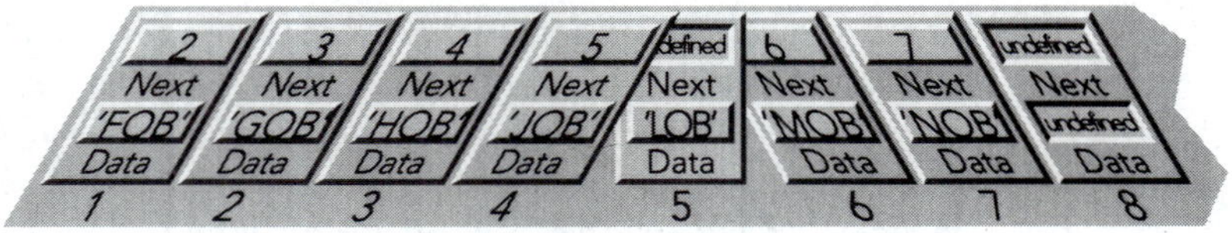

It's time to get smart and take advantage of the Next field. Even if we jumble the contents of the array, as I have below, we can still use Next's value to keep track of alphabetical order:

the array is jumbled, but ordered

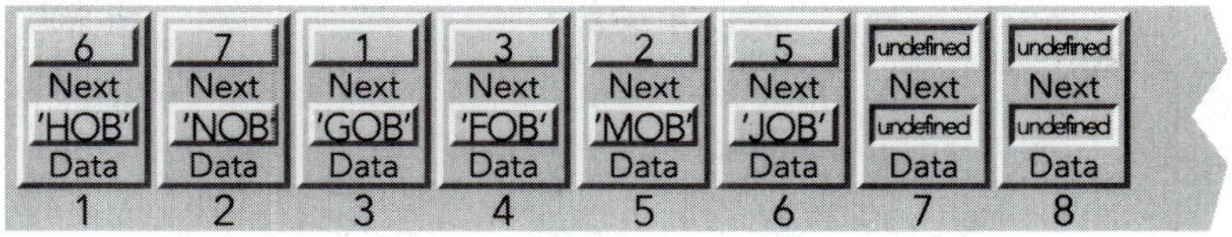

First = 4, Last = 2, CurrentPtr = 7

I'll show you how easy it is to store a new word at the end of the array (in List[CurrentPtr], which happens to be List[7] now) without messing up the ordered sequence of Next fields.

New word	*Changes*	*Inserts . . .*
"Sob"	List[7].Data := "Sob"; List[7].Next := 8; CurrentPtr := CurrentPtr + 1; Last := 7;	*a new 'highest' word*
"Bob"	List[7].Data := "Bob"; List[7].Next := 4; CurrentPtr := CurrentPtr + 1; First := 7;	*a new 'lowest' word*
"Lob"	List[7].Data := "Lob"; List[7].Next := 5; List[5].Next := 8; CurrentPtr := CurrentPtr + 1;	*a new word in the middle*

subscripts as links

> Physical and logical order are independent now. Even though the physical ordering of the components is jumbled, we use the Next fields to keep the sequence alphabetized. The records are connected by their subscripts in a 'linked' list of components.

the catch

Although it makes a very pretty picture, there is a major flaw in the method I've used to store words. It depends on an array declaration, and an array's size is fixed at compile time. If words keep arriving, we'll eventually run out of room to store them in the array. Although we could try to avoid problems by the simple expedient of declaring enormous arrays, real-life limits on computer memory rule this dodge out—we can't declare an array that's big enough for every eventuality.

We Invent Pointers!

What we would like, then, is to have records that can be tied together without being locked into an array. In some way, we'd like each record's Next field to hold the next record's name, instead of holding its subscript in an array.

But at the same time, we don't want to have to declare each record as a separate variable. How many records will we need? We can't tell, because we don't know how many words there will eventually be. As you might expect, pointers come to the rescue.

pointers represent addresses

> Pointers let us *dynamically allocate* record variables—create them one at a time, as we need them, while a program is running. A pointer variable represents the 'subscript,' in computer memory, of a dynamically allocated variable. This 'subscript' is called an *address*.

Take something on faith: computer memory is nothing more than a big array, whose subscripts are called addresses. When we declare program variables, we're just grabbing little pieces of memory. Record and array variables are just slightly larger hunks of the big memory array.

A pointer's actual value (which we can't look at) is an address—a subscript in computer memory. When we dynamically allocate, say, a record variable with a pointer, two things happen:

- First, the computer sets aside enough storage in its memory to hold the record. This is a variable, but it's anonymous, since it doesn't have a fixed name. I'll usually refer to it as a *location*.
- Then, the address that marks this location, or particular piece of memory, is given to the pointer variable.

An ordinary variable represents the value stored in a location in memory.

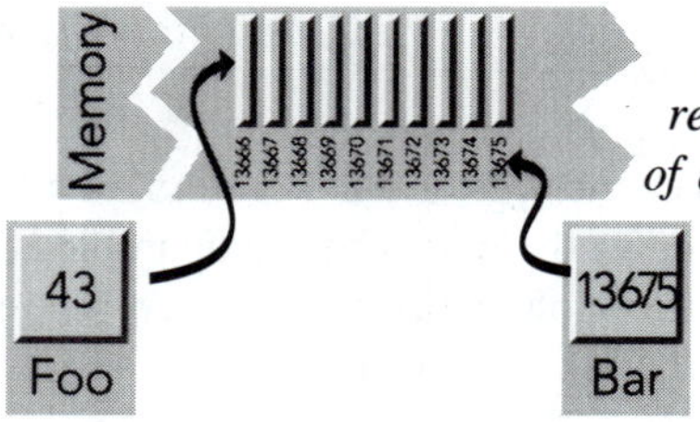

A pointer variable represents the address of a location in memory.

Now, an ordinary variable *names* its piece of memory. An assignment to an ordinary variable changes the value stored there. For all intents and purposes, the name *is* the variable, and we can only make assignments *to* it.

A pointer variable, in contrast, *addresses* its piece of memory. There are two ways to make an assignment with a pointer—either *to* it, or *through* it.

assigning to pointers

- An assignment *to* a pointer changes the address it represents, and not the value stored *at* the address.

assigning through pointers

- An assignment *through* a pointer changes the value stored at the location the pointer addresses. To go through a pointer, follow its name with an up-arrow (↑) or circumflex (^).

Although a pointer's value acts as a subscript it isn't one literally, of course. Even though it's a number, we can't add to it, or subtract from it, or even find out its value. However, there are a few interesting things that we *can* do:

- We can inspect the location that the pointer addresses (or its fields, if it's a record).
- We can change that location's value (or its field's values, if it's a record).
- We can see if two pointers have the same value—if they address the same location or stored variable.
- We can assign a pointer's value to another pointer variable (or to a pointer-type *field* in a record), so that they both point to the same address.

The last point is the most useful for us now. If a record's Next field has a pointer type we can use it to store a pointer's value. Each record can have a Next field that holds the address of another record.

Most programs that employ pointers use them with records. The combination is powerful, but notation (the combination of arrows and dots) makes it a little confusing at first. Let's start with an program that doesn't do anything practical, but provides a very clear example of pointer declaration and allocation.

pointers to CHAR variables

```
MODULE Pointers;
   (*Demonstrates pointer allocation and assignment.*)
IMPORT IO;
IMPORT Storage;  (*This is one of Wirth's semi-standard libraries.*)
TYPE CharPOINT = POINTER TO CHAR;                                  (*1*)
VAR FirstPtr, SecondPtr, ThirdPtr: CharPOINT;                      (*2*)
BEGIN
   (*ALLOCATE a CHAR-sized location for each pointer to address.*)
   Storage.ALLOCATE (FirstPtr, SIZE(CHAR));                        (*3*)
   Storage.ALLOCATE (SecondPtr, SIZE(CHAR));
   FirstPtr↑ := "A";   (*Give the location a value.*)              (*4*)
   SecondPtr↑ := "B";   (*Store a different value here.*)
   ThirdPtr := FirstPtr;   (*Give the first location a second name.*)  (*5*)
   IO.CharOut (FirstPtr↑);
   IO.CharOut (SecondPtr↑);
   IO.CharOut (ThirdPtr↑);
   IO.LineOut
END Pointers.
```

```
ABA
```

Here's what to look for in Pointers:

1. *Definition* establishes the Modula-2 type of the location the pointer addresses.

2. *Declaration* creates the pointer variables.
3. *Allocation* gives the pointer the address of the location (i.e. the anonymous variable) it will name.
4. *Assignment to a location*, e.g. FirstPtr↑ := "A", stores a value in the variable the pointer addresses.
5. *Assignment to a pointer*, e.g. ThirdPtr := FirstPtr, gives the pointer a different address value.

Modification Problems

1. Ready already? Add whatever is needed to make all three pointers address the same location—say, the one that holds a "B" now.
2. Change Pointers so that is uses pointers to INTEGER values instead of CHAR. Don't forget to change the argument to SIZE.

Definition, Declaration, Allocation

As you know, ordinary variables can only hold values of a specific type. In consequence, they can only be assigned values of that one type. Pointers, in addition, can only *address* variables of a particular type. A pointer's type definition says what that type is. A pointer can address either a simple variable (like a CHAR) or one that's structured (like a record type). For example:

some typical definitions

```
TYPE CharPOINT = POINTER TO CHAR;  (*from the example above*)
     NodePOINT = POINTER TO NodeTYPE;  (*a more practical definition*)
     NodeTYPE = RECORD
                   Letter: CHAR;
                   Next: NodePOINT
                END;
VAR FirstCharPtr, SecondCharPtr: CharPOINT;
    FirstNodePtr, LastNode: NodePOINT;
```

There are two differences between CharPOINT and NodePOINT:

- They *have* different types (even though they're both pointers).
- They *address* different types.

Although CharPOINT and NodePOINT are both pointer types, they're not the *same* type, any more than two different arrays would have the same type. Nor do they address the same type of variable—one addresses a CHAR variable, while the other addresses a record.

allocation

Our next step is to allocate locations in memory—the 'anonymous' variables our pointers will address. Memory allocation is very system-specific, so Modula-2 relies on a special module called Storage:

```
IMPORT Storage; (*Must appear in every program that uses pointers.*)
```

Storage provides at least two procedures to deal with memory: one allocates a location, and the other returns unneeded memory back to the system.

Storage.ALLOCATE and SIZE

> Storage.ALLOCATE takes two arguments: the name of a pointer variable, and the size of the location it's going to address. SIZE(*type name*) is a standard function whose result is the amount of memory a variable of type *type name* will require.

For example, these calls allocate storage of the appropriate sizes for two of our pointers:

```
BEGIN
   Storage.ALLOCATE ( FirstCharPtr, SIZE(CHAR) );
   Storage.ALLOCATE ( FirstNodePtr, SIZE(NodeTYPE) );
```

The call Storage.ALLOCATE(FirstCharPtr, SIZE(CHAR)) tells the computer to set aside a space big enough to hold a CHAR variable, and to have FirstCharPtr address that space. The second call allocates a space big enough for a NodeTYPE record, and gives FirstNodePtr its address.

Procedure Storage.DEALLOCATE frees space for reallocation. Most programs won't require deallocation (the applications we'll look at don't use all that much memory), but it's important to know how it's done:

deallocating pointers

```
(*Return locations to the computer.*)
Storage.DEALLOCATE (FirstCharPtr, SIZE(CHAR) );
Storage.DEALLOCATE (FirstNodePtr, SIZE(NodeTYPE) );
```

To summarize, we prepared to use pointers by:

1. Importing a module named Storage.
2. Defining a new type as a **POINTER TO** variables of some type.
3. Declaring variables with the pointer type.
4. Allocating locations for the pointer variables to address.

Self-Check Questions

Q. Quick: what's wrong with this call?

```
(*an incorrect call*)
Storage.ALLOCATE ( FirstNodePtr, SIZE(NodePOINT) );
```

A. It makes a common mistake—it allocates a location big enough for a NodePOINT. It should have requested enough room for a NodeTYPE record, instead:

```
Storage.ALLOCATE ( FirstNodePtr, SIZE(NodeTYPE) );
```

Always allocate the variable that's pointed to, not the one that does the pointing.

Q. Is this sequence of calls legal? What does it do?

```
Storage.ALLOCATE ( FirstCharPtr, SIZE(CHAR) );
Storage.ALLOCATE ( FirstCharPtr, SIZE(CHAR) );
```

A. The sequence is legal, but I can't think of any reason you'd want to make it. The first call allocates a location, and gives FirstCharPtr its address. The second call allocates a second location, and reassigns FirstCharPtr. Thus, the address of the first location is lost—no pointer addresses it, and it can't be used for anything.

Assignment To and Through Pointers

A pointer variable can get a value in three different ways:

1. A call of Storage.ALLOCATE will give it the address of a new location in memory.
2. It can be assigned the value of another pointer variable, so that they both address the same location.
3. It can be assigned a 'non-value' called NIL. NIL can be assigned to *any* pointer-type variable.

For example, let's allocate a location for FirstCharPtr, then give SecondCharPtr the same address, and finally make FirstCharPtr a NIL-valued pointer.

assignments to pointers

```
Storage.ALLOCATE (FirstCharPtr, SIZE(CHAR) );
SecondCharPtr := FirstCharPtr;
FirstCharPtr := NIL;
```

Although FirstCharPtr and SecondCharPtr are both defined—they both have values—the location that SecondCharPtr addresses is still undefined. We haven't assigned anything to it yet.

The actual address values that pointers hold can only be compared for equality, or to NIL. They cannot be printed or used in any other kind of expression. Only pointers of the same type can be compared.

A pointer that equals NIL doesn't address a location:

```
(*Determine if FirstCharPtr addresses a storage location.*)
IF FirstCharPtr = NIL
    THEN IO.StringOut ("Warning!  Nil pointer.  No location to access.")
    ELSE IO.StringOut ("A location has been allocated for this pointer.")
END
```

comparing pointers

If two pointers equal each other, it means that they address the same location *or* they're both NIL.

```
(*What does it mean that FirstCharPtr and SecondCharPtr have the same value?*)
IF FirstCharPtr = SecondCharPtr
   THEN IO.StringOut ("They have the same address or are both NIL.")
   ELSE IO.StringOut ("The pointer values differ.")
END
```

So much for assignment to pointers. Assignment *through* a pointer stores a value in the location that the pointer addresses.

how pointers name variables

> A pointer identifier that is followed by an up-arrow (↑) or circumflex (^) names a dynamically allocated variable. Without the arrow, the pointer identifier names an address in memory. With the arrow, the pointer identifier names the variable location at that address.

I'll always use the up-arrow because it's more readable in this typeface, but your keyboard probably uses a circumflex.

Allocating and assigning a value to a CHAR location is pretty easy.

allocating simple variables

```
Storage.ALLOCATE ( FirstCharPtr, SIZE(CHAR) );
FirstCharPtr↑ := "T";
SecondCharPtr := FirstCharPtr;
```

Now, both FirstCharPtr↑ and SecondCharPtr↑ equal "T".

Working with a pointer to a record is more interesting. When a pointer addresses a record, the pointer identifier followed by an up-arrow is the record's name. Dot notation will be needed to make an assignment to a single field of that record:

allocating structured variables

```
Storage.ALLOCATE (FirstNodePtr, SIZE(NodeTYPE) );
FirstNodePtr↑.Letter := "W";
```

Names are going to get a little more confusing:

FirstNodePtr *is a pointer that can represent an address or be* NIL
FirstNodePtr↑ *is the record at that address*
FirstNodePtr↑.Letter *is one field of that record*

Now let's try the test question that will see if you understand pointers. Suppose that we've allocated and initialized FirstNodePtr. What's the difference between this assignment:

```
SecondNodePtr := FirstNodePtr;
```

a test question

and this call and assignment?

```
Storage.ALLOCATE (SecondNodePtr, SIZE(NodeTYPE) );
SecondNodePtr↑ := FirstNodePtr↑;
```

The first assignment makes SecondNodePtr address the same location (which holds a record) as FirstNodePtr. In effect, there's only one variable, but it has two names. In contrast, the second set of statements allocates a new location for SecondNodePtr to address. Then, we make a record-to-record assignment between the two locations (since they're both records). The two records have the same values, but they're different variables.

Finally, let's get a little bit ahead of ourselves. Recall that the record FirstNodePtr addresses has two fields. We've been working with the Letter field. Now let's try something with the Next field of FirstNodePtr↑—the one that has type NodePOINT:

```
(*We can initialize it . . .*)
FirstNodePtr↑.Next := NIL;
(*. . . or allocate it.*)
Storage.ALLOCATE ( FirstNodePtr↑.Next, SIZE(NodeTYPE) );
```

FirstNodePtr is a pointer that addresses a record. One field of that record, Next, is also a pointer that addresses a record. By allocating a new record for that field to address, we've just invented the *linked list*! We'll spend most of the next section looking at links.

Self-Check Questions

Q. Write a program segment that prints the value in the location that FirstCharPtr addresses—but make it robust, just in case FirstCharPtr is NIL.

A. Since a NIL pointer doesn't address a location, it can't be followed by an up-arrow. In other words, this would be illegal:

```
(*An illegal example.*)
FirstCharPtr := NIL;
IO.CharOut (FirstCharPtr↑);
```

To avoid the mistake of trying to look at a location that hasn't been allocated, we have to see if FirstCharPtr is NIL first:

```
IF FirstCharPtr <> NIL
   THEN IO.CharOut (FirstCharPtr↑)
   ELSE IO.StringOut ("Nil pointer.")
END;
```

Some Practice with Names

It's absolutely essential to understand the distinction between pointers, and the locations they address. Without introducing anything new, let's just look at a few examples of notation. Note that even though using arrows is unusual, we have had to use extra characters to construct variable names before:

constructing variable names

Kind of variable	*Extra notation it requires*
unstructured	*none*
array	*square brackets:* [ ]
record	*period:* .
pointer	*up-arrow:* ↑

If the variable a pointer addresses is structured, things get more complicated. But with patience, you can always put together every variable's name:

pointers to variables

Kind of pointer	*Example*
to an INTEGER	NumberPtr↑
to an **ARRAY**	RowPtr↑[3]
to a **RECORD**	FractionPtr↑.Numerator

Now let's take another look at pointer assignments. I'll use these pointers:

```
TYPE NumberPOINT = POINTER TO INTEGER;
VAR OnePtr, TwoPtr, ThreePtr: NumberPOINT;
```

Suppose that I allocate two new variables:

```
Storage.ALLOCATE ( OnePtr, SIZE(INTEGER) );
Storage.ALLOCATE ( TwoPtr, SIZE(INTEGER) );
```

Although OnePtr and TwoPtr are pointers, OnePtr↑ and TwoPtr↑ are INTEGER-type variables. The distinction is a little confusing at first, but you've got to learn it. This assignment changes OnePtr↑ the INTEGER variable:

changing the location

```
OnePtr↑ := 17;
```

This assignment, in contrast, changes OnePtr the pointer:

changing the pointer

```
OnePtr := TwoPtr;
```

OnePtr and TwoPtr now have the same value—the same address. This means that OnePtr↑ and TwoPtr↑ both refer to the exact same anonymous INTEGER variable in memory. Incidentally, what happened to the location that OnePtr used to address, the one that we stored 17 in? It's in limbo. We can't use it because we don't know where it is.

Sometimes, it's easier to understand what's right by looking at what's wrong. Using the same pointers, consider these incorrect statements, and their corrected versions:

some incorrect statements

	Wrong	*Right*
a)	OnePtr := 5;	OnePtr↑ := 5;
b)	TwoPtr↑ := NIL;	TwoPtr := NIL;
c)	IO.IntOut (ThreePtr, 5);	IO.IntOut (ThreePtr↑, 5);
d)	OnePtr↑ := TwoPtr + ThreePtr;	OnePtr↑ := TwoPtr↑ + ThreePtr↑;

a) Tries to assign 5 to OnePtr instead of assigning it to the location OnePtr addresses.

b) Errs because NIL may only be assigned to a pointer—not to the location the pointer addresses (unless it too is a pointer).

c) Attempts to print the value represented by Third—which is the unprintable address of a location within the computer—instead of printing the value stored in that location.

d) Tries to add two addresses instead of adding the values stored at each address. Addresses may only be compared for equality, and aren't used in arithmetic expressions.

Self-Check Questions

Q. Try some harder examples. Consider this definition:

```
TYPE RowPOINT = POINTER TO RowTYPE;
     RowTYPE = ARRAY [1..10] OF CHAR;
VAR RowPtr: RowPOINT;
```

What is the effect of each of the statements, below? Consider them independently, as though each one follows this allocation call:

```
Storage.ALLOCATE ( RowPtr, SIZE(RowTYPE) );  then, in turn ...
a) RowPtr := NIL;
b) Storage.ALLOCATE ( RowPtr, SIZE(RowTYPE) );
c) RowPtr↑ := NIL;
d) RowPtr := "X";
e) RowPtr↑ := "Y";
f) RowPtr↑[1] := "Z";
g) RowPOINT[2] := "A";
```

A. And, the answers are:

a, b) Both are legal. However, the effect of the original call to Storage.ALLOCATE is undone. Any memory that was previously allocated is now unreachable.

c) Illegal. Although RowPtr is a pointer variable, the location it addresses, RowPtr↑, is an array. An array can't be assigned the value NIL.

d) Also illegal. Since RowPtr is a pointer variable, it can only be assigned NIL, or the value of another pointer.

e) Alas, illegal as well. As in statement *c*, RowPtr↑ is an entire array. It can't be assigned a single CHAR value.

f) Legal. It assigns "Z" to the first component of the array that RowPtr addresses.

g) Illegal, because RowPOINT is a array type identifier, not the name of a variable (like RowPtr).

Section 13-2

Standard Practices: Basic Link Operations

POINTERS TO SIMPLE VALUES LIKE CHAR ARE EASY to understand, and convenient for practice. In real programs, though, pointers just about always address records. The type definition we'll see most frequently is of a record that has some data fields, and at least one pointer field. Why have the pointer field? So that it can address another record of the same type.

nodes and links

> A record that includes a pointer to another record of the same type is usually called a *node*. The pointer field that ties one record to the next is called a *link*.

The true beauty of pointers is expressed by the idea of a node. If each record can point to another record through one of its fields, then we can put together chains of records that are as long as we like. In this section we'll look at linked nodes, and write code for most of the basic operations we'll work with them.

For example, suppose that FirstNodePtr is a pointer to a record that has one or more Data fields (their type doesn't really matter) and a Next field that's also a pointer with the same type as FirstNodePtr. Now, FirstNodePtr might be NIL:

FirstNodePtr → NIL

Or, if we allocate FirstNodePtr, it will point to a record. Most people draw pictures to help them follow what pointers are doing, so I will too.

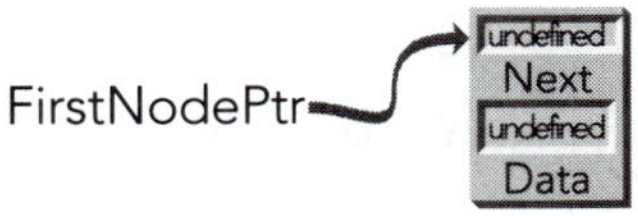

Since the record's Next field is a pointer too, it gets an arrow as well. If we allocate FirstNodePtr↑.Next, then allocate the Next field of *that* record, we'll wind up with a sequence like this:

I think you get the idea. A definition that could underly the basic node is:

a basic node

```
TYPE NodePOINT = POINTER TO NodeTYPE;
     NodeTYPE = RECORD
                   Next: NodePOINT;
                   Data: DataTYPE (*This could even be another record.*)
                END;
VAR FirstNodePtr: NodePOINT;
```

We can use FirstNodePtr and the Next field to allocate away to our hearts' content:

```
Storage.ALLOCATE ( FirstNodePtr, SIZE(NodeTYPE) );
Storage.ALLOCATE ( FirstNodePtr↑.Next, SIZE(NodeTYPE) );
Storage.ALLOCATE ( FirstNodePtr↑.Next↑.Next, SIZE(NodeTYPE) );
Storage.ALLOCATE ( FirstNodePtr↑.Next↑.Next↑.Next, SIZE(NodeTYPE) );
```

The series of allocations gives us this structure:

where:

checking the names

```
FirstNodePtr                  (*Represents an address.*)
FirstNodePtr↑                 (*The record at that address.*)
FirstNodePtr↑.Data            (*One field of that record—a stored value.*)
FirstNodePtr↑.Next            (*Represents an address.*)
FirstNodePtr↑.Next↑           (*The record at that address.*)
FirstNodePtr↑.Next↑.Data      (*One field of that record—a stored value.*)
```

The illustrated list above shows one of the peculiarities of linked structures. Although the computer has allocated four different locations in memory for nodes, only a single identifier—FirstNodePtr—is associated with them. This is a source of convenience and confusion. For instance, we can access the entire list through FirstNodePtr to make the last record's Next field NIL instead of merely undefined.

```
FirstNodePtr↑.Next↑.Next↑.Next↑.Next := NIL;
```

At the same time, a misstep might cause us to lose contact with part of the list. The assignment below advances the pointer variable, so that it addresses the second record in the list.

```
FirstNodePtr := FirstNodePtr↑.Next;
```

Unfortunately, this leaves us with no way of accessing the very first list node. We'll see how to use auxiliary pointers to avoid this problem in a minute.

Self-Check Questions

Q. Given the situation in the illustration above, what are the effects of these assignments?

a) FirstNodePtr↑ := FirstNodePtr↑.Next↑;
b) FirstNodePtr := FirstNodePtr↑.Next;
c) FirstNodePtr↑.Next := FirstNodePtr;

A. I'm assuming that the assignments are entirely independent of each other; i.e. we revert to the starting situation after each assignment.

a) Makes the first record's contents equal the second record's contents, *and* cuts the second record out of the list (because the first record's Next field now points to the third record).

b) Makes FirstNodePtr point to the second record *and* cuts the first record out of the list (because nothing points to it).

c) Makes the Next field of the first record point to the first record *and* loses the rest of the list (because the first record's Next field points to the first record).

Q. The basic node definition on the last page includes the declaration of FirstNodePtr as a variable of type NodePOINT. Could you also declare a variable of type NodeTYPE?

A. Certainly, as I've done below. Note that since HeadOfList is a regular variable, a NodeTYPE record exists immediately. In contrast, CurrentNodePtr has to be dynamically allocated.

```
TYPE definitions of NodePOINT and NodeTYPE from above;
VAR CurrentNodePtr: NodePOINT;    (*must be allocated*)
    HeadOfList: NodeTYPE;         (*is allocated automatically*)
```

Q. Suppose that I actually do declare a HeadOfList record along with a pointer to a record, as above. Are either of these assignments legal? Why or why not?

```
a) CurrentNodePtr := HeadOfList;
b) CurrentNodePtr↑ := HeadOfList;
```

A. Assignment *a* is illegal. HeadOfList is a record, not a pointer like CurrentNodePtr, so the assignment violates the basic assignment compatibility rule. There's no way to use ALLOCATE to make CurrentNodePtr address the HeadOfList record.

We can, however, copy HeadOfList. Assignment *b* is perfectly all right as long as CurrentNodePtr points to a record (that is, as long as it has been allocated). The type of CurrentNodePtr↑ is the record NodeTYPE—the same as the type of HeadOfList.

Abstractions That Use Linked Lists

The individual nodes of a linked structure may be attached to each other in a variety of patterns, as though they were Lego blocks or Tinkertoys. A *linked list* is a series of nodes that are connected by pointers into a more-or-less straight line. We created a linked list in the examples above.

The linked list is the chief building block of a variety of data abstractions. Recall that an abstract structure is defined in terms of its primitive operations—*what is it?* is answered by *what can you do with it?* We can come up with several different abstract structures simply by varying the rules that govern how a node is added or taken away:

Queues have new records added to one end, and taken from the other end. Since all travel is in one direction, queues are FIFO—*first in, first out*—structures.

a few abstractions

Stacks have new records added to ('pushed on') or taken from ('popped off') just one end of the list (the 'top'). Stacks are LIFO—*last in, first out*—structures.

Deques can have records added to, or taken from, either end. Deques are often employed when we're likely to bend the rules and make insertions in the middle, as well (strictly forbidden for queues and stacks).

Each of these abstract structures can be defined in terms of primitives for initialization, assignment, inspection, and so on. We'll do that for stacks and queues in the next section. However, abstraction without the ability to implement is useless. Lest we take a false pride in our talents, we'll have to master a few nitty-gritty operations:

basic link manipulations

- maintaining auxiliary pointers to various parts of a list, especially its first and last nodes;
- adding to the end of a linked list;
- traveling through an entire list;
- searching for a particular node;
- inserting into the middle of a linked list;
- merging two lists together, or inserting one list into another;
- deleting individual nodes or sublists;
- disposing of lists.

Maintaining Pointers and Extending Lists

Our first concern will be the auxiliary pointers that allow list manipulation. Linked structures usually have several auxiliary pointers that maintain contact with the beginning of a list, its end, our current position, and so on.

> Auxiliary pointers are your only means of finding a list, so use them early and often.

For example, suppose that FirstNodePtr points to the beginning of a list of our generic NodeTYPE records. If we had an auxiliary pointer named CurrentNodePtr (with the same type as FirstNodePtr), the assignment:

```
CurrentNodePtr := FirstNodePtr↑.Next↑.Next↑.Next
```

doesn't change the list, but it does *advance* a pointer through the list:

advancing the auxiliary pointer

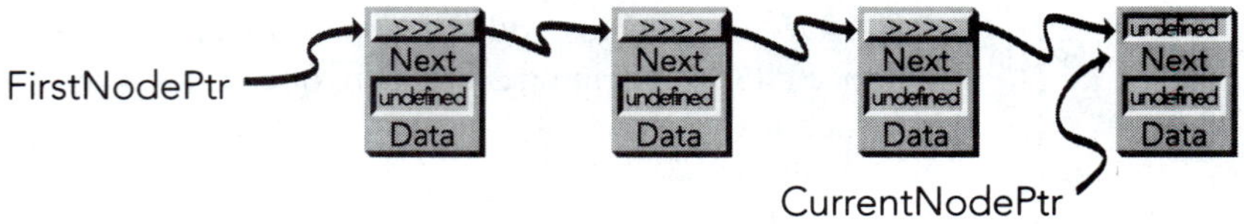

The auxiliary pointer CurrentNodePtr makes assignments far from FirstNodePtr much easier to follow:

```
CurrentNodePtr↑.Data := Value;
CurrentNodePtr↑.Next := NIL;
```

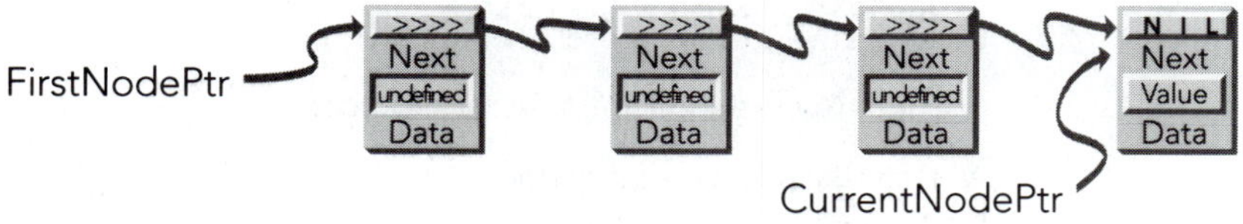

Extending a list is our second concern. What is the purpose and effect of these statements? Assume the situation of the illustration above.

list extension

```
Storage.ALLOCATE ( CurrentNodePtr↑.Next, SIZE(NodeTYPE) );
CurrentNodePtr := CurrentNodePtr↑.Next;
CurrentNodePtr↑.Next := NIL;
```

These statements extend the chain, but keep CurrentNodePtr pointing to the very last link. The result looks like this:

Let's consider the three statements in turn. The first statement:

```
Storage.ALLOCATE ( CurrentNodePtr↑.Next, SIZE(NodeTYPE) );
```

allocates a new location. CurrentNodePtr↑.Next now addresses an undefined record, and the NIL value is lost. The second statement:

```
CurrentNodePtr := CurrentNodePtr↑.Next;
```

is potentially the most confusing—it advances the current position pointer, moving it to the end of the list. The illustration below shows how the pointer is reconnected.

The final statement of the list extension makes the new list end NIL.

```
CurrentNodePtr↑.Next := NIL;
```

using NIL

> In general, NIL should always be used to mark the end of linked structures. Every pointer, or pointer field, should either have an address or be NIL.

Self-Check Questions

Q. Given the situation above, what would be the effect of making the assignment:

```
CurrentNodePtr↑.Next := FirstNodePtr;
```

A. Draw the pointer—it turns the sequence of nodes into a *circular* list that doesn't really have an end.

Q. How would you do the opposite of extending the list? In other words, instead of adding a new node to the end, get rid of the node at the beginning.

A. Doing the job requires an additional temporary pointer:

```
TempNode := FirstNodePtr; (*Save a pointer to the head.*)
FirstNodePtr := FirstNodePtr↑.Next; (*Advance the head by one node.*)
Storage.DEALLOCATE ( TempNode, SIZE(NodeTYPE) ); (*Get rid of the node.*)
TempNode := NIL; (*Keep the pointer defined.*)
```

Searching for a Node

The precaution of making the last pointer in a linked list NIL simplifies our third concern—traveling through a list. It's important to state the goals and bounds of list-traversal loops very carefully. Are you looking for an item, or for the node before an item? Will you be bounded by the loop's last node, or by the NIL value that ends the list?

Suppose that we want to find a list's end, which is to say that we want a pointer to the last node in the list. A loop specification is:

a list search specification

Goal: Have a pointer to the list's last node.
Bound: The current node's Next *field is* NIL.
Plan: Advance a pointer to the next node in the list.

In the code segment, below, assume that CurrentNodePtr is pointing to a node in a linked list. We can advance it to the list's end by searching for the NIL-valued Next pointer—the last node. Note that we check before advancing, since we might be at the last node already.

travel through a linked list

```
LOOP
   (**)IF CurrentNodePtr↑.Next = NIL THEN EXIT END;
   CurrentNodePtr := CurrentNodePtr↑.Next
END
(*Postcondition: CurrentNodePtr addresses the last node, or is
 NIL if the list was empty.*)
```

finding a particular node

Our fourth concern is to find a particular node. We want to traverse the linked list, but we stay prepared to leave the loop early. As in an ordinary array search, a list-searching loop's basic action is to move to the next component. Its **EXIT** condition, if it is to be correct, must contain a double test—either we find the sentinel we want, or we run out of data.

loop specification

Goal: Have a pointer to the node that has a particular Data *field.*
Intentional bound: The current node's Data *field is the one we want.*
Necessary bound: The current node's Next *field is* NIL.
Plan: Advance a pointer to the next node in the list.

Since there are two reasons for leaving the loop, an additional check has to be made afterward. The appropriate code segment is:

searching for SoughtValue

```
CurrentNodePtr := FirstNodePtr;  (*Start at the beginning.*)
LOOP
   (**)IF (CurrentNodePtr↑.Data = SoughtValue) OR
                    (CurrentNodePtr↑.Next = NIL) THEN EXIT END;
   CurrentNodePtr := CurrentNodePtr↑.Next
END;
(*Postcondition: If SoughtValue is there, it's in the CurrentNodePtr.*)
IF CurrentNodePtr↑.Data = SoughtValue
   THEN IO.StringOut ("Value found.")
   ELSE IO.StringOut ("Value not found.")
END
```

Self-Check Questions

Q. Suppose that you wanted to put the searching code above into a procedure. What parameters would you need?

A. Well, if we were searching an array, we'd pass the array, and the value we want to find. To search a linked list, we need a pointer to the head of the list, along with the value we're looking for. For the example above, we'd like to make a call along the lines of:

```
Find (CurrentNodePtr, SoughtValue);
```

This would correspond to a heading like:

```
PROCEDURE Find (CurrentPtr: NodeTYPE; Value: DataTYPE);
```

We'll look at more detailed examples of pointers as parameters later in this chapter.

An Example: Creating and Printing a List

Let's write a program that uses what we've learned so far—creating a list, adding to it, maintaining auxiliary pointers, and finally inspecting and printing the list. Our problem will be:

problem: making a linked list

Read and save a sequence of positive numbers that ends with a zero or negative sentinel. Print the sentinel that ends the sequence, then echo the sequence in order.

We could, if we wanted, decide to describe our solution as a data abstraction. As usual, the abstraction will be defined in terms of its primitives for initialization, assignment, and inspection:

the list primitives

TYPE NodeTYPE, NodePOINT *will underly the list of numbers.*
PROCEDURE Initialize *creates the list.*
PROCEDURE Store *extends the list and adds a new value to it.*
PROCEDURE Print *prints the list's contents, in order.*

Drafting a more formalized name layer is useful mainly because it makes us consider how auxiliary pointers will be used:

a more formal name layer

TYPE NodeTYPE, NodePOINT *don't have to be detailed yet.*
VAR FirstNodePtr, CurrentNodePtr: NodePOINT;
PROCEDURE Initialize (FirstNodePtr);
(**Returns a pointer to an empty, but allocated, node.**)
PROCEDURE Store (CurrentNodePtr, TheNumber);
(**Extends the list by one node, and stores a new value there.**)
PROCEDURE Print (FirstNodePtr);
(**Prints the numbers stored in the linked list.**)

However, I'm going to write this program without using primitive procedures because we're going to take a special look at pointers as parameters later on.

For now, we can assume that the linked list will consist of nodes. Each node will have a Data field that holds the number, and a Next field that either points to the next node or is NIL. The heart of a program that uses the linked list will have three parts—a few statements that initialize the list, then one loop to read and store values, and a second loop that prints the list's contents.

Since input ends with a zero or negative number, the read-and-store loop has a sentinel bound:

read-and-store loop specification

Goal: Hold a series of non-negative numbers in the order they arrive.
Bound: Reading a zero or negative number.
Plan: Read a number.
Add a new node to the end of the list.
Advance CurrentNodePtr *to the new node.*
Make its Next *field* NIL.
Make its Data *field equal the number.*

Making the new node's Next field NIL is quite important, because it marks the end of the list.

Let's start to pseudocode the list program. I can outline the first two steps, which initialize and create the list, as:

initialize the list by allocating its first node, via FirstNodePtr;
point an auxiliary pointer CurrentNodePtr *to the first node;*
LOOP
    read TheNumber;
    (**)**IF** *it's the sentinel then* **EXIT**;
    add a new node to the end of the list;
    advance the CurrentNodePtr *pointer to it;*
    make its Next *field* NIL;
    save TheNumber *in it*
END *of the loop;*
    (**Postcondition: The first stored value is in the node that follows* FirstNodePtr.
    The last stored value is in the node whose Next *field is* NIL.*)
echo the sentinel value;

reading the values

Note the specific mention, in the loop's postcondition, of the first and last values' positions.

the postcondition

> Because there are no fixed recipes for creating linked lists, postconditions assume unusual importance. It isn't enough to say why the loop ended. Loops that generate linked lists should clearly document the structures they create.

Once we spot (and echo) the sentinel it's time to print the list's contents.

printing loop specification

Goal: Print the numbers held in the list headed by FirstNodePtr.
Bound: The current node's Next *field is* NIL.
Plan: Advance a CurrentNodePtr *pointer, if possible.*
    Print the current node's Data *field.*

In the pseudocode version of the specification, I've reused the CurrentNodePtr that helped create the list:

printing the list

(**Remember that the first node's* Data *field is undefined.**)
point CurrentNodePtr *to the first node again;*
LOOP
    (**)**IF** *the current node's* Next *field is* NIL *then* **EXIT**;
    advance the CurrentNodePtr *pointer to the* Next *node;*
    print the current node's Data *field*
END *of the loop*
(**Postcondition: We've printed every defined* Data *field in the list.**)

Note that we could have traversed the list with the FirstNodePtr pointer. However, that would leave us without a pointer to the head of the list—a colossal mistake in most applications.

The final Modula-2 code of LinkEcho is shown below. It illustrates some of the basic linked list methods, so you should be sure to understand it.

linked list program

```
MODULE LinkEcho;
    (*Stores positive numbers in a linked list, then echoes the sentinel and list.*)
IMPORT IO;
IMPORT Storage;
TYPE NodePOINT = POINTER TO NodeTYPE;
     NodeTYPE = RECORD
                  Next: NodePOINT;
                  Data: INTEGER
                END;
VAR FirstNodePtr, CurrentNodePtr: NodePOINT;
    TheNumber: INTEGER;
BEGIN
  Storage.ALLOCATE ( FirstNodePtr, SIZE(NodeTYPE) );
  FirstNodePtr↑.Next := NIL;
  CurrentNodePtr := FirstNodePtr;
  (*We've initialized the list by creating the first node.*)
  IO.StringOut ("Please enter numbers to echo (negative sentinel).");
  IO.LineOut;
  LOOP
    IO.IntIn (TheNumber);
    (**)IF TheNumber <= 0 THEN EXIT END;
      (*Since the number wasn't a sentinel, we have to add a node ...*)
    Storage.ALLOCATE ( CurrentNodePtr↑.Next, SIZE(NodeTYPE) );
      (* ... then advance the CurrentNodePtr pointer ...*)
    CurrentNodePtr := CurrentNodePtr↑.Next;
      (* ... then initialize its Next field ...*)
    CurrentNodePtr↑.Next := NIL;
      (* ... and finally save the new number.*)
    CurrentNodePtr↑.Data := TheNumber
  END; (*LOOP*)
  (*Postcondition: FirstNodePtr↑.Next and CurrentNodePtr hold the first and last
   stored values. CurrentNodePtr↑.Next is NIL, and we've read the sentinel.*)
  IO.IntOut (TheNumber, 4);          (*Print the sentinel.*)
  CurrentNodePtr := FirstNodePtr;  (*Its Data field is empty.*)
  LOOP
    (**)IF CurrentNodePtr↑.Next = NIL THEN EXIT END;
    CurrentNodePtr := CurrentNodePtr↑.Next;
    IO.IntOut (CurrentNodePtr↑.Data, 4)
  END (*LOOP*)
  (*Postcondition: We've printed every defined Data field in the list.*)
END LinkEcho.
```

```
Please enter numbers to echo (negative sentinel).
12  59  33  18  3  65  -82  111
 -82  12  59  33  18   3  65
```

Modification Problems

1. Modify program LinkEcho so that it saves a sequence of negative numbers, then prints only the sequence without the sentinel.
2. Write an EndOfList function that could provide the exit condition of the list-printing loop of LinkEcho.
3. Change the construction of LinkEcho by putting its code into Initialize, Store, and Code procedures.

Inserting Nodes

We've seen how to maintain, search, travel, and add to the end of a list. Let's move on to inserting or deleting one or more nodes. For example, suppose that we wanted to save a list of values (like the list above), but keep them in numerical order. After reading in a new value, we'd search through the list for its proper position, then insert it between two nodes. Assume that CurrentPtr addresses the next lower number. A single new node, addressed by the pointer TempPtr, can be appended after CurrentPtr with:

inserting after the current pointer

```
Storage.ALLOCATE ( TempPtr, SIZE(NodeTYPE) );
TempPtr↑.Next := CurrentPtr↑.Next;  (*This order is crucial.*)
CurrentPtr↑.Next := TempPtr;
```

Note that the order of the assignments is critical—we have to attach the new node before disconnecting CurrentPtr from its predecessor.

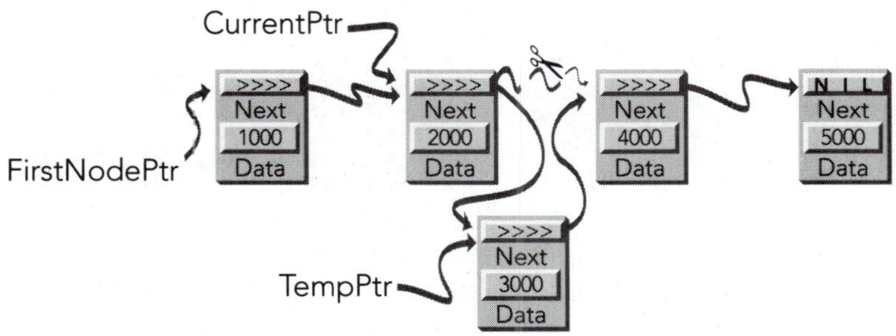

Although it seems unlikely that it's possible (think about it before you read the code), an existing node addressed by NewNode can be inserted *before* the current pointer position node with:

inserting before the current pointer

```
Storage.ALLOCATE ( TempPtr, SIZE(NodeTYPE) );
TempPtr↑ := CurrentPtr↑;
CurrentPtr↑.Next := TempPtr;
CurrentPtr↑.Data := NewNode↑.Data;
Storage.DEALLOCATE ( NewNode, SIZE(NodeTYPE) );
NewNode := CurrentPtr;
CurrentPtr := CurrentPtr↑.Next;
```

As you can see, I engaged in sleight-of-hand and didn't really insert the node addressed by NewNode into the list. Instead, I created a new, blank node:

```
Storage.ALLOCATE ( TempPtr, SIZE(NodeTYPE) );
```

gave it the same Data and Next fields as CurrentPtr:

```
TempPtr↑ := CurrentPtr↑;   (*a record-to-record assignment*)
```

inserted it after CurrentPtr:

```
CurrentPtr↑.Next := TempPtr;
```

the most complicated pointer maneuver we'll see

and copied the new node's data in CurrentPtr:

```
CurrentPtr↑.Data := NewNode↑.Data;
```

Since the new node's data has been stored in CurrentPtr↑.Data, the node itself is superfluous. We can deallocate it:

```
Storage.DEALLOCATE ( NewNode, SIZE(NodeTYPE) );
```

arrange for the NewNode pointer to address the duplicate copy:

```
NewNode := CurrentPtr;
```

and finally, advance CurrentPtr to the node that holds the data that CurrentPtr originally addressed:

```
CurrentPtr := CurrentPtr↑.Next;
```

It will take a little effort to follow that sequence, but once you do, you shouldn't have any further trouble with pointers. Here's an outline of the nodes involved in the insertion, and a pair of scissors—try filling in the pointers.

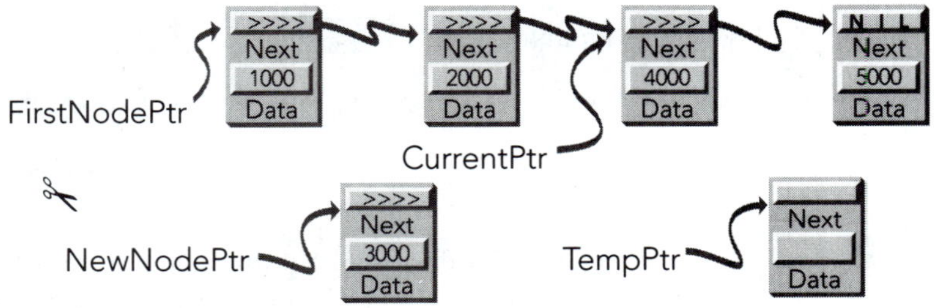

I'll leave it to you to figure out how to insert a node before the *first* node in a list.

Self-Check Questions

Q. Ok, how *do* you make an insertion before the first node in a list?

A. It takes essentially the same code as above. But in the sequence above, we made sure to advance CurrentPtr so that it ended up pointing to the same stored values. If we were inserting a node before the FirstNodePtr, we wouldn't want to advance the pointer—its name says that it should always point to the first node in the list.

Deleting Nodes

Let's look at the final group of basic list manipulations. They involve inserting or deleting *sublists*, which may be one or more nodes long. As always, having plenty of auxiliary pointers makes the job easier.

sublist insertion

Suppose that we want to insert a new list between two nodes of a currently existing list. Two reconnections do the trick. The new list goes between the nodes addressed by CurrentNodePtr and CurrentNodePtr↑.Next.

```
NewEndPtr↑.Next := CurrentNodePtr↑.Next;
CurrentNodePtr↑.Next := NewHeadPtr;
```

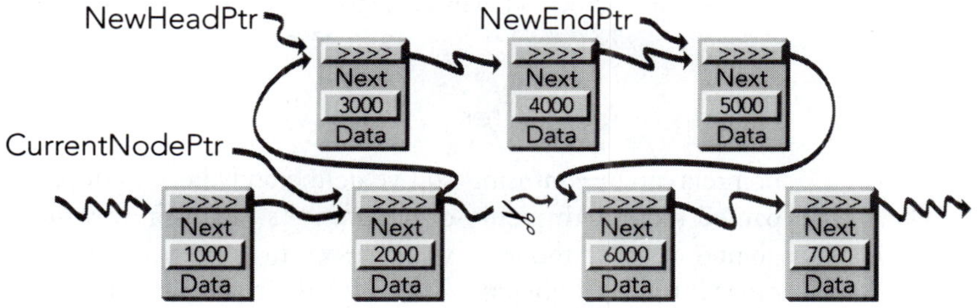

Auxiliary pointers are also useful for deleting one or more nodes from a list. However:

don't let go!

> Don't let go of the list! Don't mistake a firm grip on the deleted segment for a pointer to the list itself. If the list gets away, it is impossible to retrieve.

When done correctly, list deletions take only a reconnection or two. Suppose this is the situation.

sublist deletion

We can delete all nodes from (not including) CurrentNodePtr↑ through (and including) LastBadNodePtr↑ with:

```
CurrentNodePtr↑.Next := LastBadNodePtr↑.Next;
```

LastBadNodePtr↑ would be retained in the new list were we to make this assignment instead:

```
CurrentNodePtr↑.Next := LastBadNodePtr;
```

Note that I haven't bothered to DISPOSE of the nodes I cut out. In a program that does a great deal of pointer manipulation, though, a *list disposal* scheme is essential to ensure that we don't inadvertently run out of memory.

Self-Check Questions

Q. Suppose that two pointers—CurrentNodePtr and PreviousNode—address the first node in a list. If the last node of the list is NIL, what is the effect of this code?

```
LOOP
    (**)IF CurrentNodePtr = NIL THEN EXIT END;
    CurrentNodePtr := PreviousNode↑.Next;
    Storage.DEALLOCATE ( PreviousNode, SIZE(NodeTYPE) );
    PreviousNode := CurrentNodePtr
END;
```

A. The code segment carries out a simple scheme for list disposal. It requires two pointers; one lags behind so that the record it addresses can be deallocated.

Pointers as Parameters

Pointers can be confusing, so I've deliberately kept examples simple and accompanied them with plenty of pictures. As we begin to write programs based on pointer types, though, we'll have to package the basic operations in procedures—it's obviously easier to draft programs using primitive procedure calls than with detailed pointer assignments.

It's important to really understand how pointers are used as parameters. The problem is that we usually think of variable parameters as being changeable, while value parameters are 'safe,' in the sense that incorrectly changing a value parameter won't screw up a main-program variable.

location contents can always be changed

But pointers play by a much looser set of rules. When a pointer is passed as a variable parameter, its address can be changed, as well as the contents of the location at that address. This is expected. When a pointer is passed as a value parameter, changing the address it contains is just a local assignment. This is expected too. But suppose I define a pointer type and make some declarations:

some definitions to work with

```
TYPE NodePOINT = POINTER TO NodeTYPE;
     NodeTYPE = RECORD
                     Next: NodePOINT;
                     Data: CHAR
                END;
VAR CurrentPtr: NodePOINT;
PROCEDURE VarExample (VAR VarPointer: NodePOINT);
  . . .  (*body of the procedure*)
PROCEDURE ValueExample (ValuePointer: NodePOINT);
  . . .  (*body of the procedure*)
BEGIN   (*main program*)
    . . .  (*ALLOCATE and do things with CurrentPtr, then call . . .*)
  VarExample (CurrentPtr);
  ValueExample (CurrentPtr);
```

> Even if a pointer has been passed as a value parameter, the contents of the location it addresses can be changed permanently.

After CurrentPtr has been allocated, assignments to or through CurrentPtr change CurrentPtr, or the location it addresses, as expected. Within procedure VarExample, assignments made either to, or through, VarPointer are just like assignments to or through CurrentPtr. Variable parameters are useful for traditional reasons—they let us write general-purpose procedures without having to know the names of main-program variables.

It is procedure ValueExample that is of special interest. Within ValueExample, changes made to ValuePointer are local, no matter what argument initializes ValuePointer. Suppose, for example, that ValueExample is called with CurrentPtr as its argument. These assignments:

. . . to value parameters

```
(*Inside procedure ValueExample.*)
ValuePointer := NIL;
Storage.ALLOCATE ( ValuePointer, SIZE (NodeTYPE) );
(*These assignments are strictly local.*)
```

make it impossible to access any of the locations addressed by CurrentPtr from within the procedure. However, on return to the calling program, CurrentPtr (and any list it heads) is in its original condition.

On the other hand, assignments *through* ValuePointer are permanent. For instance:

. . . through value parameters

```
(*Inside ValueExample.*)
ValuePointer↑.Data := "X";
Storage.ALLOCATE ( ValuePointer↑.Next, SIZE (NodeTYPE) );
(*These assignments change globally accessible locations.*)
```

On return to the main program, CurrentPtr↑.Data will hold a new value, while CurrentPtr↑.Next points to a new location.

> If you really don't want to change the variable a pointer addresses, pass the variable—not the pointer to it—as a value parameter.

Even though we'll usually employ pointers to NodeTYPE variables, once in a while we'll create a value parameter of type NodeTYPE—the record—as well.

Example: List Disposal

Let's look at a concrete example. Procedure ListDispose, below, disposes of locations that aren't needed anymore. Its argument points to the head of such a list; by convention, the list's last pointer will be NIL.

As you read ListDispose, ask yourself this important question: Why is it necessary to declare Head as a variable parameter? After all, calls of Storage.DEALLOCATE *through* Head would have the desired effect even if Head were a value parameter. But what might go wrong if I used a value parameter?

list disposal procedure

```
PROCEDURE ListDispose (VAR Head: NodePOINT);
  VAR TempPtr: NodePOINT;
  BEGIN
    LOOP
      TempPtr := Head;          (*Initialize TempPtr.*)
      (**)IF Head = NIL THEN EXIT END;
      Head := Head↑.Next;          (*Advance the list head.*)
      Storage.DEALLOCATE (TempPtr, SIZE(NodeTYPE) )
        (*Dispose of the previous node.*)
    END
  END ListDispose;
```

In ListDispose, the local pointer variable TempPtr identifies locations that are slated for removal. The condition of TempPtr at the end of the procedure doesn't really matter. Head, on the other hand, must wind up with the value NIL—not only because that provides the exit condition of the loop, but because it should be defined back in the main program. If Head had been declared as a value parameter, its argument would, technically, be undefined on return to the main program.

Self-Check Questions

Q. What is the output of program Passage? What conclusions can you draw about passing pointers as value parameters?

value parameter pointer program

```
MODULE Passage;
  (*Demonstrates some effects of passing a pointer as a value parameter.*)
IMPORT IO;
IMPORT Storage;
TYPE NodePOINT = POINTER TO NodeTYPE;
     NodeTYPE = RECORD
                  Next: NodePOINT;
                  Data: CHAR
                END;
VAR CurrentPtr: NodePOINT;
PROCEDURE Change (TempPtr: NodePOINT);
  BEGIN
    TempPtr↑.Data := "C";          (*Which of these are*)
    TempPtr := TempPtr↑.Next;   (*local assignments?*)
    TempPtr↑.Data := "D"
  END Change;
```

```
BEGIN
    Storage.ALLOCATE ( CurrentPtr, SIZE(NodeTYPE) );
    CurrentPtr↑.Data := "A";
    Storage.ALLOCATE ( CurrentPtr↑.Next, SIZE(NodeTYPE) );
    CurrentPtr↑.Next↑.Data := "B";
    IO.CharOut (CurrentPtr↑.Data);
    IO.CharOut (CurrentPtr↑.Next↑.Data);
    Change (CurrentPtr);
    IO.CharOut (CurrentPtr↑.Data);
    IO.CharOut (CurrentPtr↑.Next↑.Data)
END Passage.
```

A. Passage demonstrates the hazards of passing pointers as parameters. Its output is:

```
ABCD
```

rather than the expected `ABAB`. Since TempPtr is a pointer, it addresses the same location in memory as its argument, CurrentPtr. Changes to TempPtr don't affect CurrentPtr, but any change to the location TempPtr addresses do. After all, TempPtr↑ and CurrentPtr↑ are the exact same record. We could even have used TempPtr to extend—or remove—the entire list!

Section 13-3 Extended Examples

LET'S TURN TO A FEW LONGER PROGRAMS THAT rely on pointers. As will often be the case in advanced programming, we'll focus on tools as much as programs. In other words, as soon as we write the tools, we'll take credit for all the problems that could be solved using them.

The tools we'll create are *stacks* and *queues*, which I've described as *structure* abstractions. This term serves as a reminder that the abstraction is defined in terms of the relationship between stored values, rather than by the kinds of values that it stores. The structure's primitive operations serve to order and rearrange the structure, rather than to calculate values.

Stacks

What does it mean to specify relations between values, while ignoring the values themselves? Well, consider the abstract structure known as a *stack*. A stack obeys two rules:

1. A new value can be added to the stack.
2. The newest of the remaining values can be retrieved from the stack.

These operations have common names—we save a value with a *push*, and remove a value with a *pop*. Stacks get their name from a kind of stack we've all pushed and popped—a spring-loaded stack of trays in a cafeteria.

What's interesting is that the stack operations are totally independent of the sort of values that are being saved. Without knowing anything about the 'value' of stacked data, a stack can be defined in terms of the sequential relation of values—they may be saved in order, and retrieved in reverse order.

underflow

In practice, push and pop aren't quite sufficient to define a stack. A stack might be *empty*, which would cause a pop to fail. Emptiness (which might lead to *stack underflow*) is a possibility for any stack, regardless of its implementation. Three operations, then, define an abstract stack no matter how it is programmed:

primitive stack operations

1. *Push* means 'save a value.'
2. *Pop* means 'get the remaining value that was saved most recently.'
3. *Empty* means 'there are no values to pop.'

We'll generally see two or three more operations added to the definition of computer stacks. These additional terms are more closely related to implementation. For instance, an array-based stack is more likely to run the risk of fullness (or *stack overflow*) than one based on pointers. On the other hand, 'remove' might only be relevant for a pointer-based stack, since only pointers can actually free unneeded memory. I'll just list them all:

three additional stack primitives

4. *Create* means 'give a name to a new, empty stack.'
5. *Remove* means 'remove the stack's contents.'
6. *Full* means 'there is no more room to push values onto the stack.'

primitives require implementation knowledge

> All six of these operations are primitive: we can't write them unless we know something about how the stack is actually implemented, e.g. as an array or linked list. It is possible to define other stack-related operations, of course, but they can be written using the primitives.

Let's settle on a name layer for a stack. I'll make a partial implementation commitment—we'll write it using pointers—but I'll still leave out as many details as I can. Although the list of headings below might seem suspiciously well-formed, it really didn't take very much insight to design it. All we need to know is what kind of information each primitive will require: at least, a pointer to the stack; at most, a pointer and an item to be put on or taken off:

the stack name layer

TYPE StackPOINT *doesn't have to be defined yet.*
TYPE DataTYPE, *of stored items, doesn't have to be defined yet.*
PROCEDURE Create (**VAR** TopPtr: StackPOINT);
  (**Initialize a new stack.**)
PROCEDURE Empty (TopPtr: StackPOINT) : BOOLEAN;
  (*TRUE *if there are no pushed values.**)
PROCEDURE Push (**VAR** TopPtr: StackPOINT; Item: DataTYPE);
  (**Add* Item *to the stack.**)
PROCEDURE Pop (**VAR** TopPtr: StackPOINT; **VAR** Item: DataTYPE);
  (**Remove the top item from the stack and return it as* Item.*)
PROCEDURE Remove (**VAR** TopPtr: StackPOINT);
  (**DEALLOCATE any current stack contents.**)

Self-Check Questions

Q. Procedure Pop actually removes the top value from the stack. Using the primitives described above, show how you might program a new operation called TopPtr—'get a copy of the value on top of the stack, but *don't* remove it.'

A. Let's assume that Stack has type StackPOINT, and that Data and TopCopy have type ItemTYPE. We pop the top value, copy it, then push it back on the stack.

```
(*The heart of a non-primitive procedure that copies a stack's top value.*)
IF NOT Empty(Stack) THEN
   Pop (Stack, Data);        (*Pop the top value . . .*)
   TopCopy := Data;          (*. . . copy it . . .*)
   Push (Stack, Data)        (*. . . then push it back on the stack.*)
END
```

The Implementation Layer

As an abstraction, the stack is done. But as a practical matter (and it's the practical matters that will be causing problems for the next few months), we still have to implement the stack in Modula-2.

Stacks are often built with pointers because a linked list can get longer and longer, while an array is limited to its defined size. A node that's suited for stack work differs from the nodes we've used for linked lists previously in the subtlest way possible. Can you spot it?

a stack's underlying data type

```
TYPE DataTYPE = whatever we want to stack;
     StackPOINT = POINTER TO NodeTYPE;
     NodeTYPE = RECORD
                   Previous: StackPOINT;
                   Data: DataTYPE
                END;
VAR Item: DataTYPE;
    StackPtr: StackPOINT;
```

The difference between this node and the others is that the node's internal pointer is called Previous, rather than Next. Why? Because it always points to an existing past node, rather than to some potential node we might add in the future. A picture of a stack might, when compared to our earlier lists, show its pointers in reverse:

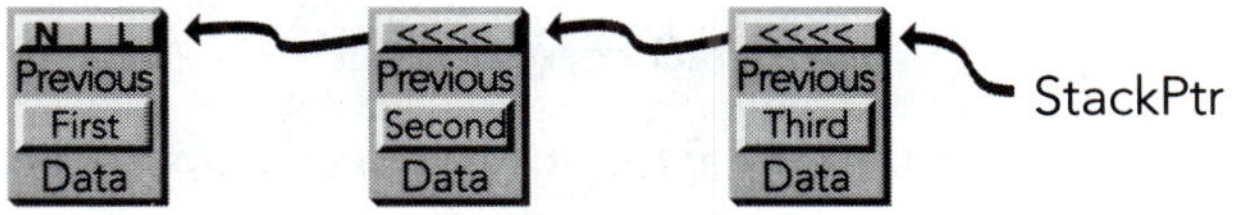

bottom of the stack *top of the stack*

Let's start implementing the primitives. The first question to ask is *what is an empty stack?* Suppose that I say it's a stack whose TopPtr, or top node, is NIL-valued. That means that procedure Create will be:

Create **procedure**

```
PROCEDURE Create (VAR TopPtr: StackPOINT);
  (*Initialize a 'top-of-stack' pointer.*)
  BEGIN
    TopPtr := NIL
  END Create;
```

Sticking to the convention *empty means that* TopPtr *equals* NIL makes function Empty easy to write:

Empty **function**

```
PROCEDURE Empty (TopPtr: StackPOINT) : BOOLEAN;
  (*TopPtr is NIL if the stack is empty.*)
  BEGIN
    RETURN TopPtr = NIL
  END Empty;
```

The definitions of Push and Pop, in turn, are:

Push, Pop **procedures**

```
PROCEDURE Push (VAR TopPtr: StackPOINT;
                    Item: DataTYPE);
  (*Add Item to the stack.*)
  VAR TempPtr: StackPOINT;
  BEGIN
    Storage.ALLOCATE ( TempPtr, SIZE(NodeTYPE) );
    TempPtr↑.Data := Item;     (*Create and fill a new node . . . *)
    TempPtr↑.Last := TopPtr;   (* . . . point it to the current top . . . *)
    TopPtr := TempPtr          (* . . . then 'advance' the top pointer.*)
  END Push;
PROCEDURE Pop (VAR TopPtr: StackPOINT;
               VAR Item: DataTYPE);
  (*Return the stack's top item, then remove its node.*)
  VAR TempPtr: StackPOINT;
  BEGIN
    Item := TopPtr↑.Data;
    TempPtr := TopPtr;         (*Save a pointer to the old top node . . . *)
    TopPtr := TopPtr↑.Last;
    Storage.DEALLOCATE ( TempPtr, SIZE(NodeTYPE) )
          (* . . . so that it can be disposed of later.*)
  END Pop;
```

The fifth and final procedure is Remove. It's a list-removal routine much like the earlier example ListDispose:

Remove procedure

```
PROCEDURE Remove (VAR TopPtr: StackPOINT);
   (*Remove the stack, leaving the TopPtr node NIL.*)
   VAR TempPtr: StackPOINT;
   BEGIN
      LOOP
         (**)IF TopPtr = NIL THEN EXIT END;
         TempPtr := TopPtr;          (*Save a pointer to the old top node.*)
         TopPtr := TopPtr↑.Last;     (*Move the top pointer down.*)
         Storage.DEALLOCATE ( TempPtr, SIZE(NodeTYPE) )
                  (*Dispose of the former top node.*)
      END (*LOOP*)
   END Remove;
```

Stack Applications

What's a stack good for? A basic application is reversing input.

Push *all incoming* Item *values onto stack*;
Pop Item *values from the stack until it's empty*;

If values are pushed onto a stack in order of receipt, they will be popped in reverse. The exact type definition of Item is important for a Modula-2 program, but it's really irrelevant to the design of our algorithm. The pseudocode below applies equally well to reversing digits in a number, characters in a string, words in a line, or lines in a file:

reversing input

```
Create (Stack);
LOOP          (*create the stack*)
   try to read Item;
   (**)IF NOT IO.Done() or we get to the end of the input EXIT;
   Push (Stack, Item)
END the stacking loop;
LOOP          (*print the stack*)
   (**)IF Empty(Stack) EXIT;
   Pop (Stack, Item);
   print Item
END the unstacking loop;
```

Another standard stack application comes in evaluating expressions. You might already be familiar with a stack notation (called *postfix* or *reverse Polish notation*) commonly used for stating expressions. In postfix, operators follow their operands, rather than coming in between them as in ordinary *infix* notation.

infix vs. postfix

Infix	*Postfix*
(A+B) * C	AB+C*
(A-B) / (C*D)	AB-CD*/
((A* (B/C)) - (D*E))	ABC/*DE*-

Postfix gets rid of parentheses; operands are pushed onto a stack as they appear, then popped off, as needed, as operators arrive. Although it's tempting to write a program that translates infix into postfix, you'll be able to do it in your head with a few minute's practice.

Suppose that we have a ready supply of postfix expressions. How can we evaluate such an expression in Modula-2? Well, assuming we can read the input sequence of operands and binary (two-operand) operators one value at a time, there are just two basic rules:

evaluating postfix expressions

1. If we read an operand value, push it onto the stack.
2. If we read an operator, pop the top two values from the stack, perform the operation, then push the result back onto the stack.

In Modula-2 pseudocode, the evaluation rules become:

stack-oriented expression evaluation

```
LOOP
    try to read the next Item;
    (**)IF there wasn't anything EXIT;
    IF Item was an operand
        THEN Push (Stack, Item)
        ELSE it's an operator, so
                Pop (Stack, Term1);
                Pop (Stack, Term2);
                carry out the appropriate operation;
                Push (Stack, TheResult);
    END the IF statement
END of the loop;
Pop (Stack, RemainingTerm);
print RemainingTerm
```

Note that the pseudocode doesn't contain any checks for incorrect input. Were input to arrive in infix form (say, as **A * B**), a premature Pop would cause a program crash. You should figure out how checks for an empty stack could deal with this kind of problem.

In addition to simple applications like the reversals and arithmetic described above, the stack is an extremely important abstraction for many facets of computer and program operation, including:

other stack applications

- *Language translation.* When programming languages are translated into a form the computer understands, meaning may be unclear, particularly when statements or definitions are nested. Stacks are used to delay processing of 'outer' sequences until 'inner' sequences have been translated.
- *Subprogram calls.* Calling a procedure or function implies that another ongoing process (e.g. the main program) will have to be temporarily suspended. Stacks are used to preserve the current state of uncompleted processes, and to let them be reinvoked in the proper order.

- *Delayed evaluation*. Suppose that evaluating a function requires a call of the function itself. This special instance of stacked subprogram calls, called *recursion*, is a particularly exciting application of the computer's stack. We'll explore recursion in detail in Chapter 17.
- *Delimiter matching*. Checking that paired delimiters, like parentheses or comment brackets, are supplied in sufficient quantity and proper order requires stacking the delimiters. Whenever a closing delimiter is encountered, the top of the stack should contain its matching opener; if not, there's an error.

Because of their conceptual simplicity, stacks are also employed when order is unimportant. For example, a stack is often used to maintain a list of free storage space (e.g. the subscripts of unused array components). The order in which free space is retrieved is irrelevant. However, procedures we might name Save and Retrieve and function NoneAvailable can easily be seen as revisions of Push, Pop, and Empty.

Self-Check Questions

Q. How would you use the primitives we've written to reverse the order of a stack's contents?

A. Easily—just use the original stack's pops as the reversed stack's pushes:

```
Create (CopyStack);
LOOP
    (**)IF Empty (OriginalStack) THEN EXIT END;
    Pop (OriginalStack, Item);
    Push (CopyStack, Item)
END;
(*Postcondition: OriginalStack is empty.*)
```

Queues

The queue is the second of the classic abstract structures. It differs from the stack by a slight variation in its underlying rules: we can remove the oldest stored value, rather than the newest.

queue abstraction rules

1. A new value can be added to the queue.
2. The oldest remaining value can be removed from the queue.

We have all waited in queues at one time or another. Queues are lines; we enter at the rear, and wait with more or less patience as earlier arrivals remove themselves from the front.

Queue primitives parallel those provided for stacks. Three basic operations are required:

basic queue operations

1. *Enqueue* means 'save a value.'
2. *Retrieve* means 'get the oldest value that remains.'
3. *EmptyQ* means 'there are no values in the queue.'

Additional functions and procedures may be defined as well, as long as they don't contravene (break) the first-in, first-out basis of the queue:

4. *CreateQ* means 'give a name to a new, empty queue.'
5. *RemoveQ* means 'remove the queue's contents.'
6. *FullQ* means 'the queue can't hold any more values.'

As was the case with the stack abstraction, we can decide on a name layer for the queue without making too many choices about implementation. One important difference to bear in mind, though, is that a queue has two ends—the front and the rear—whereas stacks only have a top.

the queue name layer

```
TYPE QueuePOINT doesn't have to be defined yet.
TYPE DataTYPE, of stored items, doesn't have to be defined yet.
PROCEDURE CreateQ (VAR FrontPtr, RearPtr: QueuePOINT);
  (*Initialize a new queue.*)
PROCEDURE EmptyQ (FrontPtr, RearPtr: QueuePOINT) : BOOLEAN;
  (*TRUE if the queue is empty.*)
PROCEDURE Enqueue (VAR RearPtr: QueuePOINT; Item: DataTYPE);
  (*Add Item to end of the queue.*)
PROCEDURE Retrieve (VAR FrontPtr: QueuePOINT; VAR Item: DataTYPE);
  (*Remove the oldest item from the queue and return it as Item.*)
PROCEDURE Remove (VAR FrontPtr: QueuePOINT);
  (*DEALLOCATE any current queue contents.*)
```

As before, I haven't bothered with a 'structure is full' procedure in this implementation.

Self-Check Questions

Q. Is it possible to use the primitives described above to write a NextItem procedure—a procedure that lets us inspect the item at the front of the queue without removing it?

A. Not easily. Once an item has been removed, it can only be returned to the end of the queue. Inspecting the front item would require copying the entire queue. If it's needed, NextItem is a good candidate for implementation as a primitive.

The Implementation Layer

The implementation of queues is aided immensely by the fact that we can steal most of our code from earlier examples. The basic node will be:

a queue's underlying data type

```
TYPE DataTYPE = whatever we want to queue;
     QueuePOINT = POINTER TO NodeTYPE;
     NodeTYPE = RECORD
                   Next: QueuePOINT;
                  Data: DataTYPE
                END;
VAR Item: DataTYPE;
    FrontPtr, RearPtr: QueuePOINT;
```

Let me ask a question, though. Which of these two possibilities—both based on NodeTYPE—will best serve our purposes? Which way should the pointer go—this way:

first alternative

or is this arrangement easier to deal with? In both cases, we add values to the rear, and remove them from the front.

second alternative

Take a piece of scrap paper and draft the code you'll need to Enqueue and Retrieve in each case. The first alternative, which is a lot like the simple linked list we practiced with earlier, turns out to be deceptive. Although adding values to the rear is a piece of cake, keeping the FrontPtr pointer correctly positioned as items are retrieved would be quite tedious. The pointers only go in one direction—how can FrontPtr back up?

However, the simple change introduced in the second alternative makes life much easier. Why not have each component point to the component behind, rather than the one in front? It's easy to add a new node to RearPtr, and to advance FrontPtr down the queue as nodes are removed.

In the code of the CreateQ procedure below, notice that the queue contains a component even when it's empty. This design decision is invisible to the user, but it simplifies some aspects of implementing the data abstraction. Indeed, I could have gone even further, and sandwiched all new components between permanently empty FrontPtr and RearPtr nodes.

CreateQ procedure

```
PROCEDURE CreateQ (VAR FrontPtr, RearPtr: QueuePOINT);
  (*Do the initialization required to set up an empty queue.*)
  BEGIN
    Storage.ALLOCATE ( RearPtr, SIZE(NodeTYPE) );
    RearPtr↑.Next := NIL;
    FrontPtr := RearPtr
  END CreateQ;
```

In the queue I've initialized, FrontPtr and RearPtr both address the same empty node. If we stick with that rule, we can detect an empty queue with:

EmptyQ function

```
PROCEDURE EmptyQ (FrontPtr, RearPtr: QueuePOINT) : BOOLEAN;
  (*TRUE if the queue is empty.*)
  BEGIN
    RETURN FrontPtr = RearPtr
  END EmptyQ;
```

Note that the node's contents, if it has any, are irrelevant. We're checking to see *where* the pointers go, not *what* they point to.

Having committed ourselves to this method of representing and detecting the empty queue, we are obligated to keep one end of the queue occupied by an empty node. I've kept it at the rear in procedure Enqueue, below:

Enqueue procedure

```
PROCEDURE Enqueue (VAR RearPtr: QueuePOINT; Item: DataTYPE);
  (*Add an Item value to the queue.*)
  BEGIN
    RearPtr↑.Data := Item;          (*'Fill' the currently empty rear node.*)
    Storage.ALLOCATE ( RearPtr↑.Next, SIZE(NodeTYPE) );
                                    (*Create and add a new empty node.*)
    RearPtr := RearPtr↑.Next        (*'Advance' the rear pointer.*)
    RearPtr↑.Next := NIL            (*The empty last node is NIL.*)
  END Enqueue;
```

The definition of RetrieveQ, in turn, is:

RetrieveQ procedure

```
PROCEDURE RetrieveQ (VAR FrontPtr: QueuePOINT;
                     VAR Item: DataTYPE);
  (*Retrieve the front (i.e. oldest) value from the queue.*)
  VAR TempPtr: QueuePOINT;
  BEGIN
    TempPtr := FrontPtr;            (*Save a pointer to the old front node.*)
    Item := FrontPtr↑.Data;         (*Get and save the oldest value.*)
    FrontPtr := FrontPtr↑.Next;     (*Advance the front pointer.*)
    Storage.DEALLOCATE ( TempPtr, SIZE(NodeTYPE) )
                                    (*Remove the old front node.*)
  END RetrieveQ;
```

Finally, as with the stack, a pointer-based implementation makes a FullQ function superfluous for our purposes. The fifth and last procedure is RemoveQ.

RemoveQ procedure

```
PROCEDURE RemoveQ (VAR FrontPtr: QueuePOINT);
  (*Return each node in the queue to computer memory.*)
  VAR TempPtr: QueuePOINT;
  BEGIN
    LOOP
      (**)IF FrontPtr = NIL THEN EXIT END;
      TempPtr := FrontPtr;          (*Save a pointer to the old leading node*)
      FrontPtr := FrontPtr↑.Next;   (*so that it can be disposed of later.*)
      Storage.DEALLOCATE ( TempPtr, SIZE(NodeTYPE) )
    END; (*LOOP*)
  END RemoveQ;
```

Self-Check Questions

Q. Procedure RemoveQ only gets one argument—the FrontPtr pointer. What is the value of RearPtr after a RemoveQ call?

A. As the code now stands, RearPtr is undefined after a call of RemoveQ because the node it used to address has been deallocated. It would be smart to set RearPtr to NIL.

Q. EmptyQ is TRUE if FrontPtr and RearPtr address the same node. Will EmptyQ compare the same addresses each time it's called?

A. No. Every time an item is removed from the queue FrontPtr points to a different node. Eventually, when all the items are removed, FrontPtr has worked its way over to the same node that RearPtr addresses.

Queue Applications

In programs, as in real life, queues are found wherever supply and demand (in jargon, *servers* and *users*) cannot be assured to stay in lockstep. Why does a queue form at the bank? Because demand for services, in the form of customers, sometimes exceeds the available supply of tellers. The queue serves as a *buffer*; it keeps the supply and demand processes at arm's length.

why do queues form?

Programs need queues because statement-by-statement program execution isn't always suited to problem solving. Although we usually think of program operation as a three-step waltz of input, process, and output, this model breaks down when input or processing are unpredictable. Data may come too quickly, or in irregular spurts, or might have to be held for additional processing later. A technical term that can be applied to such uncoordinated activity is *asynchronous processing*—the input, processing and output steps are not synchronized, and may even be running at the same time.

queue applications

For example, think about the operation of a time-shared computer system. Jobs that need the undivided attention of an unshareable resource (like a printer) can be queued until their turns arrive. Tasks (like users' programs) that can be interrupted without harm are more interesting. Here, the necessary resource of CPU power can be shared. The operating system maintains a *run queue* of unfinished jobs. Each job is given a small amount of central processor time (its *time slice*), then returned to the run queue if it's still unfinished. If time slices are appropriately chosen, and the queue is not too long, each user has the illusion of having the computer's undivided attention.

A few word games will help illustrate the use of queues as buffers, or temporary holding places. Notice anything peculiar about these sentences?

> *Just a white shark.*
> *Space, time, and relativity with a ridiculous script.*
> *A nauseating nitwit inneffably embalmed.*

Suppose that we supply each sentence, in turn, as input to the following code segment. Naturally, I'll assume that the queue-oriented procedures have been written to hold CHAR-type data items. What will the output be?

mystery code segment

```
NewWord := TRUE;
LOOP  (*until the end of a line or of all input*)
   IO.CharIn (Ch);
   (**)IF NOT IO.Done() OR (Ch = IO.EOL) THEN EXIT END;
   IO.CharOut (Ch);
   IF NewWord          (*TRUE when the first letter of a new word is read.*)
      THEN Enqueue (Letters, Ch);
           NewWord := FALSE
      ELSE NewWord := Ch = " "          (*Assume one space between words.*)
   END (*IF*)
END; (*LOOP*)
(*Postcondition: There's no more input . . . but what does the queue hold?*)
IO.StringOut ("--");
LOOP  (*until the queue is empty*)
   (**)IF EmptyQ(Letters) THEN EXIT END;
   Retrieve (Letters, Ch);
   IO.CharOut (Ch)
END; (*LOOP*)
```

Within the first loop, every character is echoed but certain characters are queued. The second loop prints the queue's contents. Under what circumstances will a character be added to the queue? If it's the first letter of a new word. The segment's output is:

These self-describing phrases are called *mynorcas*. An *automynorcagram* is a more complex variation on the same theme. It's a phrase that is generated by its own letters; e.g. *This happy instance shows how a pleasing phrase yields its new sentence* and so on. Have fun in the exercises!

detecting palindromes

A second such example uses a stack as well as a queue to detect palindromes—sequences that are the same both backward and forward. The ground rules of text palindrome formation generally allow spaces and punctuation to be ignored. Letter-oriented examples are such a routine exercise that I'll try a word-oriented variation: palindromic sentences. How might we decide if these are palindromes?

So patient a doctor to doctor a patient so.
Girl, bathing on Bikini, eyeing boy, finds boy eyeing bikini on bathing girl.

We'd really like to be able to read each sentence forward and backward simultaneously. Well, since a stack of words effects a reversal, we can.

spotting a palindromic sentence

```
LOOP          (*stack and queue the words*)
   read a word, ignoring capitalization and punctuation;
   (**)IF there was no word to read EXIT;
   Push the word onto a stack;
   Enqueue the word in a queue
END;
LOOP          (*compare the stored values—assume non-empty lists*)
   Pop a word from the stack;
   RetrieveQ a word from the queue;
   (**)IF they don't match OR Empty stack or queue EXIT
END;
decide why we left the loop, and report the results;
```

Each word is both stacked and queued. Then, since the stack is a LIFO (last-in, first-out) structure, while the queue is FIFO (first-in, first-out), we can compare the input sequence backward and forward simultaneously.

A Long Program: Adding Polynomials

Since most of the pointer-based procedures we've written have only been a few lines long, you may feel that you're not getting your money's worth of coding practice. Let's write one last long program, then. Our problem will be to merge two linked lists into one, connecting nodes together while we combine nodes that hold the same kind of information.

One way to pose the problem is to say that two graduate students have been taking a census of the insect population of a small plot of (not a lot of) land in the Amazon. Their laptop computers will ultimately spit out alphabetized lists of the number and kinds of insects they've seen. Our problem is to merge the two lists.

However, computer ecosystems are a bit more hospitable to numbers than they are to insect names. I'll present a list merging question as a problem in mathematics, instead:

problem: polynomial addition

> A *polynomial* is expressed as a series of *terms*:
>
> $$a_n x^n + a_{n-1} x^{n-1} + \ldots + a_1 x^1 + a_0 x^0$$
>
> The important features of each term are its REAL *coefficient* (the *a*'s) and its INTEGER *degree* (the exponents). Write a program that adds two polynomials to form a third.

Adding polynomials is easy. The terms are merged together in order of each term's exponent. If two terms happen to have the same exponent, their coefficients are added together before the term joins the list. For instance:

$$5.0x^4 + 3.0x^2 + 6.0x^1 \textit{ plus } 3.0x^3 - 5.0x^2 + 2.0x^1 - 7.0x^0$$
$$\textit{equals } 5.0x^4 + 3.0x^3 - 2.0x^2 + 8.0x^1 - 7.0x^0$$

As you can see, not every exponent has to appear, but they do have to be in order. It's not too different from adding insects after all, is it?

Now, if every term were always present, it would be tempting to use an **ARRAY** [0..n] to hold the coefficients. We could just assume that each subscript represented an exponent. But imagine a polynomial with a very big degree but only two terms; say, $x^{1000} + 1$. Most of the array would go to waste.

In contrast, the linked list almost seems to have been invented for representing polynomials. We can easily store a REAL coefficient and an INTEGER exponent in a record, along with a pointer to the next term—the next record—in the polynomial.

Let's return to object-oriented design to solve the polynomial addition problem. Although we've already decided to use linked lists to store the polynomials, we'd still like to outline a data abstraction before we get into coding details. What primitives will we need to add two polynomials? I can only think of three—pretty much the same three that we always seem to need:

what primitives will we need?

an input primitive to read each polynomial;
a primitive that adds two polynomials together to create a third;
an output primitive that prints the value of the answer polynomial;

In fact, we can come up with a pretty solid name layer. I'll name TermPOINT now, and put off the details of what it points to for a moment.

the name layer

```
TYPE TermPOINT is a pointer to a term or list of terms.
PROCEDURE ListIn (VAR HeadPtr: TermPOINT);
    (*Stores a polynomial as a linked list of individual terms.*)
PROCEDURE AddLists (First, Second: TermPOINT;
                                   VAR Sum: TermPOINT);
    (*Creates a linked list that represents the sum of two polynomials.*)
PROCEDURE ListOut (HeadPtr: TermPOINT);
    (*Prints the stored polynomial' s value.*)
```

As luck would have it, procedures ListIn and ListOut are practically solved problems. We wrote list input and output loops back in program LinkEcho. Then, I used a number equal to or less than zero to end the list. Here, I'll say that each polynomial always ends with a term whose exponent is zero.

Even the type definition is just a minor variation on the node we defined back then. I'll define our underlying terms as:

type definition

```
TYPE TermPOINT = POINTER TO TermTYPE;
     TermTYPE = RECORD
                  Next: TermPOINT;
                  Degree: INTEGER;
                  Coefficient: REAL
                END;
VAR First, Second, Sum: TermPOINT;
```

Adding the polynomials requires some thought. Suppose that we work from the largest exponents down to the smallest. The loop specification is:

AddLists loop specification

Goal: Add First *and* Second *to create* Sum.
Bound: Reaching the 0-exponent terms of First *and* Second.
Plan: Do the right thing for each pair of terms.

Well, maybe the plan is a little vague! Let's try a case analysis: what are the possibilities we'll encounter as we add terms from two polynomials?

a case analysis

- Both terms' exponents will be the same, so we should add their coefficients before attaching a new term to the Sum polynomial.
- The First term's exponent will be bigger than Second's, so we should attach the term to Sum as is, and get another term from First.
- The Second term's exponent will be bigger than First's, so we should attaching the term to Sum as is, and get another term from Second.

This translates into the pseudocode shown below. Note that I'll use an auxiliary pointer to build the linked list—I want to keep a firm grip on Sum, the pointer to its head:

pseudocode

```
allocate Sum, the head of the polynomial linked list;
make SumPos an auxiliary pointer to the same list;
LOOP
   IF Degree fields are equal, THEN
         copy either term to the new list, but add both coefficients
      ELSIF the First exponent is higher, copy the term to the new list
      ELSE the Second exponent is higher, so copy the term to the new list
   END the if statement;
   (**)IF both pointers equal NIL then EXIT;
   add another empty term to the new list
END the loop;
make the last pointer field in the new list NIL
```

The importance of the last step can't be overemphasized. The NIL pointer that ends the new linked list must be included, or else a list-inspection procedure that looks for a NIL end will crash. The completed program is shown over the next couple of pages.

polynomial addition program

```
MODULE PolyAdd;
  (*Uses linked lists to maintain and add two n-degree polynomials.*)
IMPORT IO;
IMPORT Storage;
TYPE TermPOINT = POINTER TO TermTYPE;
     TermTYPE = RECORD
                  Next: TermPOINT;
                  Degree: INTEGER;
                  Coefficient: REAL
                END;
VAR First, Second, Sum: TermPOINT;
PROCEDURE AddPolynomials (First, Second: TermPOINT;
                          VAR Sum: TermPOINT);
  (*Create a linked list that adds the terms of two polynomials.*)
  VAR SumPos: TermPOINT; (*An auxiliary pointer to the Sum list.*)
  BEGIN
    Storage.ALLOCATE ( Sum, SIZE(TermTYPE) );
    (*We've been passed the pointer, not the list.*)
    SumPos := Sum; (*Sum will stay at the new list's head.*)
    LOOP (*until both First and Second are NIL*)
      IF First↑.Degree = Second↑.Degree THEN (*equal exponents*)
          SumPos↑.Degree := First↑.Degree;
          SumPos↑.Coefficient := First↑.Coefficient
                                 + Second↑.Coefficient;
          First := First↑.Next;
          Second := Second↑.Next
        ELSIF (First↑.Degree > Second↑.Degree) THEN
          SumPos↑.Degree := First↑.Degree;
          SumPos↑.Coefficient := First↑.Coefficient;
          First := First↑.Next
        ELSE (*First↑.Degree was less than Second↑.Degree*)
          SumPos↑.Degree := Second↑.Degree;
          SumPos↑.Coefficient := Second↑.Coefficient;
          Second := Second↑.Next
      END; (*IF*)
      (**)IF (First = NIL) AND (Second = NIL) THEN EXIT END;
      Storage.ALLOCATE ( SumPos↑.Next, SIZE(TermTYPE) );
      (*Advance the end of the list.*)
      SumPos := SumPos↑.Next
    END; (*LOOP*)
    SumPos↑.Next := NIL
  END AddPolynomials;
```

```
PROCEDURE MakeList (VAR HeadPtr: TermPOINT);
  (*Read the coefficient and exponent of each term.*)
  VAR TempPtr: TermPOINT;
  BEGIN
    TempPtr := HeadPtr;
    IO.StringOut ("Please enter coefficients and exponents."); IO.LineOut;
    IO.StringOut ("Start with the largest exponent, and end"); IO.LineOut;
    IO.StringOut ("with a zero-valued exponent."); IO.LineOut;
    LOOP (*until a Degree is zero and we end the list by making Next NIL*)
      IO.RealIn (TempPtr↑.Coefficient);
      IO.IntIn (TempPtr↑.Degree);
      IF (TempPtr↑.Degree <> 0)        (*either extend the list . . .*)
        THEN Storage.ALLOCATE ( TempPtr↑.Next, SIZE(TermTYPE) )
        ELSE TempPtr↑.Next := NIL      (* . . . or end it.*)
      END; (*IF*)
      (**)IF TempPtr↑.Next = NIL THEN EXIT END;
      TempPtr := TempPtr↑.Next
    END (*LOOP*)
  END MakeList;
PROCEDURE WriteList (HeadPtr: TermPOINT);
  (*Print the sign, coefficient, and exponent of each term.*)
  BEGIN
    LOOP (*until HeadPtr is NIL*)
      IF HeadPtr↑.Coefficient > 0.0 THEN
        IO.StringOut ("  + ") (*A necessary detail for positive terms.*)
      END;
      IO.FixOut (HeadPtr↑.Coefficient, 3, 1);
      IO.StringOut ("^"); (*This symbol is often used to show powers.*)
      IO.IntOut (HeadPtr↑.Degree, 1);
      HeadPtr := HeadPtr↑.Next;
      (**)IF HeadPtr = NIL THEN EXIT END
    END; (*LOOP*)
  END WriteList;
BEGIN (*PolyAdd*)
  Storage.ALLOCATE ( First, SIZE(TermTYPE) );
  MakeList (First); (*Read each polynomial . . .*)
  WriteList (First); (* . . . then echo it.*)
  IO.LineOut;
  Storage.ALLOCATE ( Second, SIZE(TermTYPE) );
  MakeList (Second);
  WriteList (Second);
  IO.LineOut;
  AddPolynomials (First, Second, Sum);
  IO.StringOut ("The sum of the polynomials is:   ");
  IO.LineOut;
  WriteList (Sum)
END PolyAdd.
```

(output follows)

```
Please enter coefficients and exponents.
Start with the largest exponent, and end
with a zero-valued exponent.
3.0 4    2.0 3    -1.0 1    10.0 0
 +3.0^^4 +2.0^^3   -1.0^^1 +10.0^^0
Please enter coefficients and exponents.
Start with the largest exponent, and end
with a zero-valued exponent.
4.5 1000    -1.0 4    2.0 3    -5.0 2    -2.0 0
 +4.5^^1000   -1.0^^4 +2.0^^3   -5.0^^2   -2.0^^0
The sum of the polynomials is:
 +4.5^^1000 +2.0^^4 +4.0^^3   -5.0^^2   -1.0^^1 +8.0^^0
```

Modification Problems

1. Since a coefficient can be negative, it's possible to wind up printing a term as `0.0` raised to some power. Modify PolyAdd so that it doesn't print such terms.
2. PolyAdd is nice to us if we're nice to it, but it's sure not robust or user-friendly. Improve it by letting the program user change a term after the entire polynomial has been entered. Three operations to consider are *a*) add a missing term, *b*) delete an existing term, *c*) modify an existing term. Hint: what the user doesn't know won't hurt her—you can get the *modify* command for free.
3. Modify PolyAdd so that it can be used to add two arbitrarily long INTEGER values. You can make life easy for yourself by allowing a space between each digit. Even though you'll read the numbers in from left to right, be sure to add them from right to left.

A Few Definition Exercises

When I learned to program pointers I found that understanding something in the book was a lot easier than recreating it when I had to pound terminal keys.

Even though our examples will involve abstract structures based on simple nodes as above, let's look at some more involved definitions. They'll help you see how pointers fit into the pantheon of types, and give you some practice with their definition syntax. Don't worry—you won't have to write programs that solve the problems I'll pose. You *should* feel up to defining an underlying type for any picture you'd care to draw.

problem: editor data type

> Define a data type suited to building a text editor. Each line must hold up to 80 characters, but the total number of lines shouldn't be restricted. Common editing operations, like deleting or moving lines, or printing subsections of the text, should be convenient to perform.

a file?

What are the considerations to weigh in coming up with a definition? Well, an ordinary file of characters meets the 'unlimited overall size' criterion, but changing files by reading and writing them is extremely inconvenient—changing the order of a few lines requires a considerable amount of work.

an array?

How about a two-dimensional array of CHAR, or a one-dimensional array of eighty-character strings? Arrays let us access lines at random, and defining each line as a string makes it relatively easy to delete, move, and print lines. On the other hand, we face a sticky choice in defining the array—either we grab all available memory by declaring the largest array possible, or we limit the text editor's capacity in violation of the problem specification.

As you might expect, pointers allow a neat solution. Why not define a record that holds a full line of eighty characters, along with a field that points to another line? The definition is:

a linked list of lines

```
TYPE StringTYPE = ARRAY [0..79] OF CHAR;
     LinePOINT = POINTER TO LineTYPE;
     LineTYPE = RECORD
                   Next: LinePOINT;
                   Data: StringTYPE
                END;
VAR FirstPtr, CurrentPtr, Last, other auxiliary pointers: LinePOINT;
```

This is almost identical to the NodeTYPE definition from the beginning of the chapter. Instead of holding three-character words, though, we'll store eighty-character lines. The illustration below shows a typical editing situation:

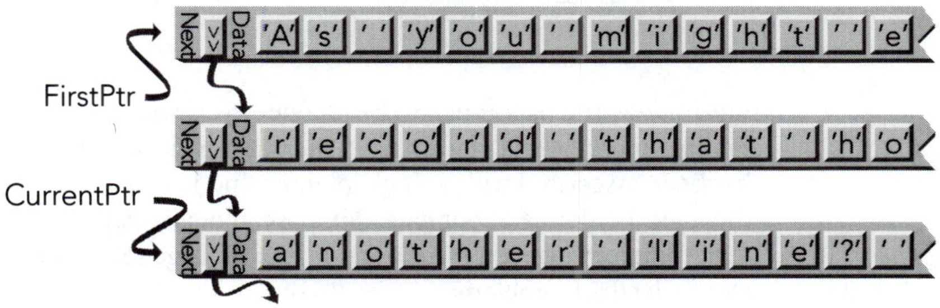

Auxiliary pointers won't always be simple variables, as our next problem shows:

problem: array with buckets

> A manufacturer has labeled her product line with single letters, followed by specific product numbers. In other words, there is an 'A' line, a 'B' line, and so on through the 'Z' line. She wants a data type definition that will allow quick access to any particular line, but doesn't limit the number of products in each line. Assume that product information is stored in a variable of type InfoTYPE.

Once again, we will rely on using records (to hold each product's information) that also contain pointer fields (so that we can link a line of products together).

How can we separate the 'A', 'B', etc. lines, though? Quick access is the clue. An array subscripted by ["A".."Z"] provides random access to each line—if the array's components are pointers.

an array of lists

```
TYPE NodePOINT = POINTER TO NodeTYPE;
     NodeTYPE = RECORD
                    Next: NodePOINT;
                    Data: DataTYPE
                END;
     InfoTYPE = ARRAY ["A".."Z"] OF NodePOINT;
VAR Info: InfoTYPE;
    CurrentPtr: NodePOINT;
```

This data structure is known, informally, as an *array with buckets.* There is an application called *hashing* in which the image of tossing information into empty buckets, which can be added as necessary, is especially vivid. Once again, I can draw a picture of Info's appearance after it's been initialized:

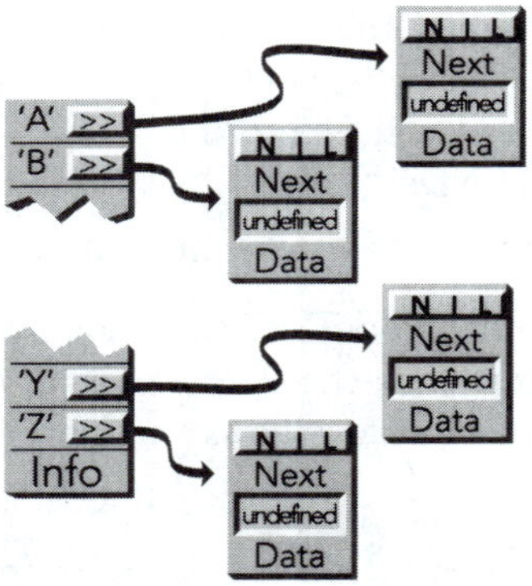

Each of the 26 array components is a pointer to the head of a linked list.

Now, the data types we've defined so far have kept pointers on an equal footing, more or less, with data. Ignoring auxiliary pointers, every item of data has been associated with a single pointer, and just one path of pointers described the logical ordering of our data. Suppose, though, that we want to use pointers to allow different logical descriptions of a single large set of data? We've arrived at a classic database problem:

problem: database type

> A company that keeps voluminously detailed records on its employees has a problem. For various reasons, different divisions within the company want to keep the records in different kinds of order: the Layoff division wants employees ordered by nearness to retirement or tenure, the Health division wants them ordered by frequency of absence, Payroll wants them ordered by social security numbers, and so on. Assuming that records are too large for duplicate sets to be kept, define a data type that will allow each division reasonable access to the data base.*

Employee records are 'big,' in the sense that such records usually take a

* This problem brings to mind former Prime Minister Asquith's crack about the WW I British military command: 'They keep three sets of records—one to mislead the War Office, one to mislead the public, and one to mislead themselves.'

fair amount of space in memory. After all, they hold a lot of information. Pointers, in contrast, are small. What we'd like to do is have just one copy of the master list of employee records, but have differently ordered lists of pointers to the master.

Although the database problem sounds difficult, we can solve it with a very simple definition of a record that holds two pointers. One of the pointers will address another record of the same type—as before, we'll make a linked list. But instead of storing the employee data in our node records, we'll just store pointers *to* the data.

In the definition below, BaseNodeTYPE holds *only* pointers. It provides the skeleton of the database structure. A second record type, which does not include a pointer field, holds each employee's personal history.

```
TYPE DataTYPE = RECORD
                    whatever the employee records contain
                END;
     DataPOINT = POINTER TO DataTYPE;
     NodePOINT = POINTER TO BaseNodeTYPE;
     BaseNodeTYPE = RECORD
                        NextNode: NodePOINT;
                        Employee: DataPOINT
                    END
VAR MainList, WatchList, HealthList,
    PayrollList, CurrentPtr: NodePOINT;
```

The list of records headed by MainList can be set up in any convenient order—say, in alphabetical order of one of the fields in the data record. It is paralleled by other lists that access the same data, but maintain different logical orderings.

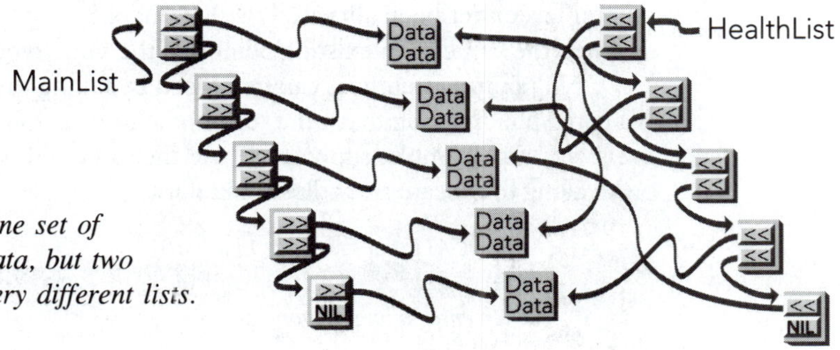

One set of data, but two very different lists.

Background: Are There Limits to Linked Structures?

Thus I have heard.

The greatest mistake made in Computer/Human interfaces is the denial of the computer. Systems that are backfitted to previous conceptions of the universe are always limited by what has gone before. Computers should not simulate reality—they should transcend it.

From *The Zen of Programming*, by Geoffrey James, © 1988, InfoBooks, Santa Monica, Ca. 90406.

Section 13-4 Program Engineering Notes

PROGRAM ENGINEERING WITH POINTERS REQUIRES LESS in the way of new knowledge than it does the conscious application of old knowledge. Take everything you know about the importance of 'seeing' data stored in arrays and records, and start to believe it.

The most important tool there is for working with pointers is a *print the structure* procedure. Unfortunately, any problems you have with pointers may show up sooner than you expected! Try to look at writing and debugging your Dump procedure as a shakedown cruise that takes place before the ship is built.

In time, you'll come to understand pointers via the same road that leads to Carnegie Hall—practice, practice, practice. Don't worry if you start out copying a lot of your code from the book. I did, too, when I first started.

One of the ideas that's hardest to follow at first is that pointer variables can be in many different states. A pointer variable can be:

- *Unallocated, and undefined*—its start-of-program state. The pointer addresses nothing, and it has no value.
- *Defined as* NIL—it can be compared to NIL or to another pointer of the same type, but it still doesn't address anything.
- *ALLOCATEd, but addressing an uninitialized location*—right after allocation, the variable it points to hasn't been initialized yet.
- *ALLOCATEd, and pointing to an initialized variable*—ready to be used.

nil pointer bugs

> A pointer that doesn't address a location can't be followed by an up-arrow or circumflex, even if it has been defined as NIL. Attempting to do so will cause a program crash.

This is sometimes called a *nil pointer reference*. To avoid it, be sure that the pointer has either been allocated (with a call of Storage.ALLOCATE) or has been assigned the value of an existing pointer of the same type.

Nil pointer references can be caused by failing to allocate pointers. There is a corresponding bug that's the result of allocating too many. It's called a *heap overflow*. For example, suppose we use the same old node pointer we've relied on all along to allocate an endless linked list:

overallocation bug

```
(*ALLOCATE the first node in a list.*)
Storage.ALLOCATE ( CurrentPtr, SIZE(NodeTYPE) );
(*Then enter a buggy loop—there's no EXIT.*)
LOOP
   (*ALLOCATE another link in the list.*)
   Storage.ALLOCATE ( CurrentPtr↑.Next, SIZE(NodeTYPE) );
   (*Advance the auxiliary pointer there.*)
   CurrentPtr := CurrentPtr↑.Next
END
```

```
ABNORMAL TERMINATION:  HEAP OVERFLOW
```

As you can see, the loop allocates endlessly; the linked list gets longer and longer. 'Heap' is a technical term that describes the computer's supply of as yet unallocated memory. When a pointer is allocated, the location it uses comes from the heap; when it's deallocated, the location is returned there. In any case, trying to allocate a location once the heap has been used up is called an overflow. Naturally, it causes a crash.

Importing and using the Storage module may be confusing. Any program that uses pointers has to include this statement:

```
IMPORT Storage;
```

The call of Storage.ALLOCATE that provides a location for a pointer to address is, to be perfectly frank, unnecessarily detailed. ALLOCATE is defined in a way that makes life a little easier for the compiler writer. Unfortunately, it's at the programmer's expense, because of a slip that's easy to make. Can you see what the bug is here? The identifiers should be self-explanatory.

```
(*a buggy call*)
Storage.ALLOCATE ( CurrentPtr, SIZE(PointerTYPE) );
```

allocate nodes, not pointers

Look closely at the second argument. Do we want to allocate a location that's the size of the pointer? No. It should be the size of the variable the pointer will address. Wrong-sized allocations cause strange program problems, so be careful:

```
(*the correct version*)
Storage.ALLOCATE ( CurrentPtr, SIZE(NodeTYPE) );
```

Storage.DEALLOCATE can also lead to a hard-to-debug problem. Suppose that two pointers address the same location, then one is deallocated. What has happened to the other pointer?

```
Storage.ALLOCATE ( FirstPointer, SIZE(ItsNodeTYPE) );
SecondPointer := FirstPointer;
Storage.DEALLOCATE ( FirstPointer, SIZE(ItsNodeTYPE) );
FirstPointer := NIL;
(*What has happened to SecondPointer?*)
```

set dangling pointers to NIL

In most compilers, *nothing* has happened to it—it still addresses the deallocated location. SecondPointer is said to be a *dangling* pointer—it's an auxiliary pointer that's still hanging around. It's generally a good programming practice to explicitly set pointers to NIL whenever they're not being used.

Linked Structure Bugs

A general problem associated with linked structures is the inadvertent loss of a pointer.

hold on tight

> It *is* possible to lose a location. And, once a location or chain of locations is gone, there's no way to recapture it.

A chain of pointers isn't like a ball of string. If the end gets lost, it's really gone. The best way to avoid this problem is to draw a picture that duplicates your code, and check the order of program statements carefully.

use auxiliary pointers

Don't worry about using an extra auxiliary pointer or two, either. Programmers often read about overflows, then become needlessly stingy when it comes to declaring pointers. But there's no need to be chintzy—overflows occur when you allocate variables in tens or hundreds of thousands, not when you declare a handful of auxiliary pointers. When one pointer serves two purposes, bugs happen. Extra pointers are cheap, so use them.

Once a linked structure has been created, there are a few typical errors to watch out for. Most common is the attempt to boldly go where no one has gone before—past the end of a linked list. The problem is simply the pointer version of the classic search-loop bug: knowing why we *want* the loop to end doesn't guarantee that it will terminate before disaster strikes.

necessary bounds on searches

- When you search the input stream, check for **NOT** IO.Done().
- When you search an array, watch for the last component.
- When you search a linked list, look out for NIL-valued pointers.

The appearance of this necessary bound varies, but it's purpose is always the same—look before you loop. An incorrect loop hopes for the best:

```
(*A loop with a potential bug—Sought might not be there.*)
CurrentPtr := HeadPtr;  (*Start CurrentPtr at the head of the list.*)
LOOP
   (**)IF (CurrentPtr↑.Value = Sought) THEN EXIT END;
   CurrentPtr := CurrentPtr↑.Next
END
```

One correct version has a double **EXIT** condition, and an extra check that follows the loop. Can you see why this won't always be correct, though? Try to figure out the loop's precondition, below.

one possible alternative

```
(*A possibly correct version of the same loop.*)
CurrentPtr := HeadPtr;  (*Start CurrentPtr at the head of the list.*)
(*What precondition has to exist here, before the loop?*)
LOOP
   (**)IF (CurrentPtr↑.Value = Sought) OR
         (CurrentPtr↑.Next = NIL) THEN EXIT END;
   CurrentPtr := CurrentPtr↑.Next
END;
(*Postcondition: if Sought is there, CurrentPtr addresses it.*)
IF CurrentPtr↑.Value = Sought
   THEN IO.StringOut ("Found the value.")
   ELSE IO.StringOut ("Didn't find the value.")
END
```

The missing precondition is that the list always contains at least one allocated node. If it doesn't, either of the checks in the **EXIT** condition will cause a crash. Now, this isn't an unreasonable precondition, because in many instances, linked structures *will* always have at least one node. We used that approach in the queue abstraction we implemented.

But what if the HeadPtr pointer, and thus, the CurrentPtr pointer, might be NIL? We need this alternative version of the search loop:

a second alternative

```
(*A different version of the same loop.*)
CurrentPtr := HeadPtr;  (*Start CurrentPtr at the head of the list.*)
(*Now, CurrentPtr might be NIL.*)
LOOP
   (**)IF (CurrentPtr = NIL) OR (*then if*)
         (CurrentPtr↑.Value = Sought) THEN EXIT END;
   CurrentPtr := CurrentPtr↑.Next
END;
(*Postcondition:  if CurrentPtr isn't NIL, it addresses Sought.*)
IF CurrentPtr <> NIL
   THEN IO.StringOut ("Found the value.")
   ELSE IO.StringOut ("Didn't find the value.")
END
```

Note the (**then if**) comment that documents our reliance on Modula-2's partial evaluation of BOOLEAN expressions. We'll leave the loop immediately if CurrentPtr is NIL, before we try to inspect the Value field.

> Neither implementation style is right or wrong. It's *your* abstraction—as long as the primitives perform, you can put it together any way you like. What is important is remembering your own implementation decisions!

As you might expect, there are a number of bugs that are associated with structures that may be based on linked lists, rather than with pointers per se. Look out for boundary conditions; ask yourself if your procedure will work:

boundary bugs

- at the beginning of the list?
- at the end of the list?
- if the list is empty?

For example, a common bug is attempting to pop a value from an empty stack. A call of Empty(Stack) avoids the issue.

In closing, let me stress the importance of data abstraction, and the object-oriented design methods we've used. You may feel, at first, that it is your own weakness as a programmer that makes a page of assignments to and through pointers look complicated. It ain't so. In large quantities, pointers are confusing to everybody. Use procedures and functions to cut them down to size.

Modula-2 Details

Syntax Summary

- pointer: a variable that holds the *address* of a record or other variable. Its type definition is:

    TYPE NamePOINT = **POINTER TO** *the type it addresses*;

- A pointer can address any type, but it will usually address a record.
- node: the building block of a linked data structure. Typically, it's a record that has one or more fields that store data, and at least one pointer-typed field that links it to the next node:

```
TYPE NodePOINT = POINTER TO NodeTYPE;
NodeTYPE = RECORD
              Data: INTEGER;
              Link: NodePOINT
           END;
VAR HeadPtr, Tail: NodePOINT;
```

- Storage: is a standard library that must be imported by any program that uses pointers:

```
IMPORT Storage;
```

- dynamic allocation: creates a location for a pointer to address. It takes two arguments: a pointer, and the size of the location (which can be found by giving the built-in procedure SIZE the location's type):

```
Storage.ALLOCATE ( HeadPtr, SIZE(NodeTYPE) );
```

- dynamic deallocation: returns the location a pointer addresses to the computer for reallocation:

```
Storage.DEALLOCATE ( HeadPtr, SIZE(NodeTYPE) );
```

- NIL: a non-value that can be assigned to a pointer of any type. NIL is used to initialize pointers that don't address locations:

```
Tail := NIL;
```

- up-arrow or circumflex: used to construct the name of the location a pointer addresses:

    HeadPtr↑ *is the entire* NodeTYPE *record*;
    HeadPtr↑.Data *is one field of the record*;

Modula-2 Details

Remember . . .

- 'Location,' 'anonymous variable,' and 'value addressed by a pointer' are three different ways of describing the same thing—a dynamically allocated variable.
- An address can be changed by making an assignment *to* the pointer; but the

variable at that address is changed by making the assignment *through* the pointer.

• The value—an address—that a pointer represents can't be inspected. It can only be compared for equality to NIL, or to another pointer of the same type.

• Pointers can only be assigned NIL or the value of another pointer of the same type.

• A linked list is a chain of nodes held together by pointers. Stacks and queues can be built with linked lists, but they're defined as abstractions in terms of their primitive operations: push, pop, and so on.

• A *stack* is a LIFO, or last-in, first-out, structure. The *queue*, in contrast, is FIFO, or first-in, first-out.

• Losing track of the end of a list is one of the most common pointer bugs. Use auxiliary pointers early and often.

• When you deal with linked structures, the first procedure you write should be a Dump procedure that prints the structure's contents.

Zen Study Guide Big Ideas

The Big Ideas in this chapter are:

1. *All the elaborate rules for definition, creation, and access of pointers have just one purpose—to let a programmer work with variables that don't have permanent names.* Pointers separate variables from their identifiers, and vice versa. A single variable can have more than one name, or a single name can refer to different variables. In effect, a pointer is an identifier that isn't tied down to just one variable.

2. *Pointers are prized because they can link records.* Most pointer applications involve lists or chains of records. Pointers make the lists easy to add to, and easy to reconfigure. They let us mimic the structure of data, rather than forcing it into one of the basic types.

3. *Data structures are characterized by the rules they follow and the operations they allow—not by the way they are implemented, or by the values they store.* A linked list is usually built with pointers, but it could be made from an array of records. The primitives define the structure, not its implementation.

Zen Study Guide Questions

13-1 The first idea that has to be nailed down is that a pointer can have *any* address type—it can point to an INTEGER, a record, an array, etc. One way to make this point is to ask for a program that reads and adds two numbers, then prints their sum, but uses only pointer variables. Make up three other questions of this type.

13-2 The syntax of a legal pointer type definition is a detail, true, but it's one you have to know. Here's one that isn't legal:

```
TYPE TestPOINT = POINTER TO RECORD;
     RecordTYPE = RECORD
                    Data: CHAR;
                    Last: TestPOINT
                  END;
```

Why not? The pointer should be to RecordTYPE, not to **RECORD**.

Make up five definitions, of varying degrees of legality, as *Fix the bug* questions. Make them harder by venturing away from the vanilla *record-with-a-pointer-field* model.

13-3 Suppose that the definition above had been done correctly. Then, you might declare a few variables:

```
VAR Test, Trial: TestPOINT;
    HeadPtr, Tail: RecordTYPE;
```

and proceed to have fun with them. There are two ways for you to join in:

a) What can you ALLOCATE and DEALLOCATE?

b) What can be assigned to what?

Write a number of statements in each category, and ask *Which of these are legal?* For example, which of *these* are legal:

```
Storage.ALLOCATE (HeadPtr);
Test := Tail;
```

Neither, of course! For an interesting variation, try putting together potentially illegal sequences of two statements.

13-4 Refer back to the definition examples from the end of section 13-3. In each case, we pose a problem, define one or more data types, and draw a picture of the final structure. Make up three more questions of this sort. For instance:

> A computer science teacher who is seeking tenure has found that the best way to ensure high teaching ratings is to tell students that they are above average. Since she doesn't wish to lie, she organizes her records in several different ways: by homework scores, quiz grades, exam grades, and class attendance. In this way, she's almost always able to find some way in which each student excels. Define a data type that might be used for the job.

13-5 What can you do in the course of traversing a simple linked list? Well, you might just have to locate the end. You might have to print every data value in the list. You might have to print every *other* value. You might have to find a particular value. You might even have to make up three more questions of this sort.

13-6 Usually, the last pointer in a linked list is NIL. But for purposes of examination, it might point anywhere—back to the head of the list, perhaps, or to its third component. Make up two *Reattach the end of the list* questions. *For fun:* Which of these letter shapes can you duplicate with a linked list: D P B R? What about with two lists?

13-7 You know what the basic linked list operations are: traverse, find, insert, and so on. But how good would you be at attaching proper names to the Modula-2 code that does each job? Take appropriate code segments from the text, mix them up in a hat, then see how long it takes you to properly identify the first one you pull out.

13-8 Consider the following code segment. We can assume that FirstPtr points to the head of a linked list, and that CurrentPtr, FirstPtr, and NextPtr are all pointer variables of the same type.

```
CurrentPtr := FirstPtr;
LOOP
    (**)IF CurrentPtr = NIL THEN EXIT END;
    TempPtr := CurrentPtr;
    CurrentPtr := CurrentPtr↑.NextPtr;
    Temp↑.NextPtr := FirstPtr
END
```

What does it do? It causes every link in the list to point back to the very first node.

Make up three *What does it do?* questions by writing three variations on the code segment. You'll find that they don't have to be complicated!

13-9 Draw a straightforward, garden-variety, linked list. This will be your *before* picture. Then, draw three *after* pictures (a picture of the example above will be a good start). Then, ask yourself *Which of these three would be a good test question?*

13-10 Take any of the basic linked list handling code segments that appear in the text, and put it into a procedure. Make it into a test question by presenting the heading in as many variations of value and variable parameters as seems appropriate. The question, of course, is *Which heading is correct?*

Exercises

Warmups and Test Programs

13-11 What's the difference between a NIL pointer and an uninitialized pointer?

13-12 Can a pointer point to a pointer? Can it point to itself? Can you think of any possible reason for doing so?

13-13 Write a program that determines if *a)* two pointers to INTEGERs refer to the same value, and/or *b)* if they refer to the same location.

13-14 Many Modula-2 systems have extended the language slightly by including built-in procedures, called NEW and DISPOSE, that replace ALLOCATE and DEALLOCATE. Does your system do this? What syntax do they use?

13-15 All right—the moment we've all been waiting for. How many items can you allocate before your system runs out of gas? Hint: print the current count every thousand allocations or so, so that you can be sure your program is working.

13-16 What error message results from giving Storage.ALLOCATE a second argument that is the size of a pointer variable, rather than giving it the size of the item (e.g. the record) the pointer locates?

13-17 What error message results from attempting to print the value of a pointer? From trying to add 1 to it? From inspecting a NIL pointer?

13-18 It may seem obvious, but you should be able to do any of these:

a) Add a node to a linked (or doubly-linked, see the problems below) list.

b) Delete a node from a list.

c) Insert a node, or series of nodes, into a list.

d) Find the end of a list.

e) Search through a list for a particular value.

f) Combinations of the above, e.g. join two lists, starting with pointers to their first nodes.

13-19 How would you reverse the contents of a stack? Using a queue? Using another stack?

13-20 Write a procedure that checks two linked lists to see if their contents are identical. Can you have your procedure check to see if they include any of the same locations as well?

13-21 Write a procedure that removes the first *n* nodes from a list. Can you remove the last *n* nodes as well?

Exercises

Programming Problems

13-22 In section 13-3 I described a solution to the problem of spotting palindromic sentences. Implement the solution, using the method (simultaneously stacking and queueing the words) I described.

13-23 In a *doubly-linked list*, each node has pointers to the front and rear. Define, then implement, then test, a set of primitive operators for a doubly-linked list. At the very least, you should be able to add, delete, insert, and locate nodes at arbitrary positions in the list.

13-24 In a *circular* linked list, the first node is attached to the last. Thus, a single pointer can address both the head and tail.

We can use a circular linked list to solve an old problem about a ship caught in a terrible storm. Although there were thirty passengers on board, plus the captain, there was only enough room in the lifeboats for fifteen of them. As the captain was reluctant to leave anybody behind, she resolved to throw half the passengers overboard before loading the boats.

As it happens, half of the passengers had slighted the captain by not dining at her table during the cruise. The captain, in revenge, arranged all the passengers in a circle and began to count. Every *n*th passenger went overboard; naturally, the captain's friends were never chosen. Here's how the passengers were originally arranged (I've shown the friends with **O** and the enemies with **X**):

```
   O  O  O  O  O  X  X  X  X  O  O  X  X  O
X                                            X
   X  O  O  X  X  O  X  X  O  X  O  X  O  O
```

The question is what number was *n*? Start counting at the upper-left **O**.

13-25 Recall that an *automynorcagram* is a sentence whose words are determined by its letters:

Each additional character has a distinctly desirable idiosyncrasy that is . . .

Write a program that represents an automynorcagram as a linked list. In addition to linking the letters that form the sentence, you should link individual letters to the start of the words they generate. In the example above, the *E* of *each* would point to itself, the *a* to *additional*, the *c* to *character*, the *h* to *has*, and so on.

13-26 Some of the more athletic among you may be familiar with the exercise known as *running stadium steps*. Repeatedly running up and down steep flights of stairs has no practical application, of course, but it is believed to build character.

One programming equivalent of running stadium steps is to base a one-line text editing system on a linked list. Write a program that has primitives for adding, deleting, printing, measuring, and locating characters and substrings within a line. *Option:* Use the ANSI escape sequences to display the line on the screen, along with an arrow that points to your current position on the line.

13-27 'The only good stack is an empty stack' is an expression that is often heard in mathematical programming. This thought is especially applicable to parentheses and brackets—a solitary delimiter is invariably up to no good.

Write a program that reads an expression and stacks the delimiters {, [, and (. As their matching counterparts },], and) appear, they should be popped. Your program should detect two error conditions—unpaired delimiters (like {5/[3+7]), and misordered delimiters (like {5/[3+7}–2])—and announce which delimiter was expected. All operands and operators should be ignored.

13-28 Suppose that three linked lists are each internally ordered, from least to greatest. Write a procedure that merges them to form a fourth list that is ordered from greatest to least.

13-29 Most computer systems use various kinds of *buffers*, or temporary holding places, to minimize the amount of work required of the CPU. Buffers are like semi-autonomous worker bees; they can do simple tasks that do not require any appreciable computation.

In one common application, buffers do the grunt work of holding a line of characters until the user hits the return key. Suppose that a line is a maximum of 128 characters long. A buffer should:

a) let additional characters be added to the end of the line, up to the maximum;

b) let the user backspace to delete single characters;

c) let the user 'rubout' or delete the entire line;

d) let the user end the line with a carriage return or equivalent.

Implement a line buffer, using a doubly-linked list as the underlying data structure. Note that this is an input-restricted deque, since values can only be added on the user's side, and not on the system side.

13-30 Assignment of computer memory to running processes is the basic *dynamic storage allocation* problem. It is somewhat similar to the bin packing problem discussed in the exercises accompanying Chapter 10, but with two crucial differences. First, we must allocate memory as jobs arrive, and thus cannot predict the amount of resources required. Second, our work is never finished. Jobs are periodically completed and relinquish their hold on memory that can then be reassigned.

Modeling the performance of memory allocation schemes is a fascinating problem. Although in reality computer memory is very much like a long array, it is convenient to represent it as a linked list of 'free' and 'occupied' areas. Memory is initially a single element labeled 'free' with a very large *Size* field. As each new job arrives, an additional 'taken' element, with its size, is added to the list at the expense of the 'free' element.

The allocation problem becomes interesting when we let jobs release memory. Free space is no longer found only at the end of the list. Instead, the list holds interspersed segments of open memory and used memory. How do we decide where to place a new job? Two simple strategies, similar to those employed in packing bins, are *first fit*—stick the new job in the first area that's big enough, and *best fit*—put the new job in the smallest open area that will do.

Write procedures capable of modeling each strategy. They should be able to:

a) search a linked list for the first 'free' element with *Size* equal to or greater than some required n,

b) search for the 'free' element with *Size* closest to, but not less than, n,

c) recognize, and merge, contiguous 'free' elements.

Hint: you may want to consider keeping an auxiliary list of pointers to 'free' elements.

13-31 An interesting problem proposed by R. Hamming is to print a list, in numerical order and with no duplicates, of numbers that have no prime factors other than 2, 3, or 5. The appropriate sequence begins 2,3,4,5,6,8,9,10,12,15,16,18,20... Each term is a number of the form $2^i 3^j 5^k$, for all non-negative integer i, j, k values.

One approach to solving this problem involves the repeated creation and merger of lists. Suppose we start with a $List_{final}$, empty to begin with, of all the terms of the answer. A $List_{temp}$, initially 2, 3, 5, contains some, but not all, of the terms that go into a $List_{final}$. Finally, we can begin to generate a $List_{2terms}$, a $List_{3terms}$, and a $List_{5terms}$ by multiplying the current contents of $List_{temp}$ by 2, 3, and 5.

Now we have a *final* list, a *temp* list, and three *term* lists. The final list grows by being merged with the current temp list. The temp list, in turn, is replaced by the merger of the three term lists. New term lists are generated by multiplying the new temp list by 2, 3, and 5.

Write a program that carries out all this list generation and merging. How do you know how much of the final list can be safely printed at any point?

'Thus, the file is the elephant of data types. Its potential size makes it capable of prodigious feats, but we have to allow for a certain amount of lumbering and unwieldiness.' (p. 551)

Chapter 14 Files

It may come as a shock to you, but the variables we've used so happily for the past dozen chapters are fictions; they are merely the imaginary creations of our programs. When our programs are complete our variables—arrays and INTEGERs, records and REALs, with their own supposed names and destinies—all go back to being anonymous bits and bytes in computer memory.

Not so with files. Files are real; you can pick them up and put them in your pocket. You can spill coffee on them. Although files can be modified by our programs, they're not dependent on them—they're not variables. Instead, they belong to the computer. Unfortunately, this makes files the problem children of modern programming languages.

File implementation is closely tied to each computer's operating system. When the operating system improves, which happens frequently, the rules for working with files often change slightly as well. This creates tremendous conflicts for programming languages, particularly since they are expected *not* to change, but still, miraculously, to work for all time on all computers.

In this chapter, we'll meet the basic file-manipulation algorithms in the context of textfiles—files of characters. For a much more detailed look at file applications, please continue on to Chapter 18.

Section 14-1
Strictly Modula-2

LET'S BEGIN BY WRITING A VERY BIG BOOK. WELL, maybe just the first letter of the first word of the first page:

```
MODULE Wake;
   (*Gets off to a slow start writing Finnegans Wake.*)
IMPORT IO;
IMPORT FILES;   (*Our file-handling library.*)
VAR Book: FILES.T;   (*Declare a file variable.*)
BEGIN
   IO.StringOut ("For a vivid demonstration of how famous authors get");
   IO.LineOut;
   IO.StringOut ("started, read file WAKE.txt when this program is done.");
   FILES.OpenWrite (Book, "WAKE.txt");   (*Create a file . . .*)
   FILES.CharOut (Book, "r");            (*. . . then put something in it.*)
   FILES.Close (Book)
END Wake.
```

```
For a vivid demonstration of how famous authors get
started, read file WAKE.txt when this program is done.
```

Were we to look inside the new file **`WAKE.txt`**, we'd find the single letter **`r`**—the first letter of the first line (*riverrun, past Eve and Adam's, from swerve of shore to bend of bay* and so on) of James Joyce's *Finnegans Wake*.

When is a variable not a variable? When it's a *file*. In some ways files are like variables: they hold data, they can be changed, and they have names. But in other ways, files aren't variables at all: they don't have any particular type, they're permanent, and they're not even part of Modula-2.

> Files are controlled by the computer's operating system, and exist independently of any individual program. However, by using one or more special-purpose library procedures, a program can temporarily take control of a file in order to read it or change it.

Files have two natures: one concrete, and one abstract. (Yes, I know it sounds Zen.) As physical things, files are subject to specific rules imposed by operating systems. However, there is also a data abstraction called a file, which is defined in terms of can be done with a file-handling library's procedures. Thus, a concrete file is a sequence of values that's stored on some permanent medium, like a hard disk or magnetic tape. However, we can often treat files as though they're abstract variables, which can be manipulated through primitives.

The important characteristics of files are:

they are sequential

1. Files store values sequentially. Files have to be read, component by component, from beginning to end. They can only have new components

added after the last of the current components. (In practice, though, most operating systems bend the rules to simulate a degree of random access.)

they are unbounded

2. File length is not limited. While arrays, say, are usually restricted to some fraction of the computer's memory, a file can occupy as much of a hard disk, floppy disk, or backup tape as the operating system will allow.

they are read- or write-only

3. Files have a *state* that depends on whether they are being inspected (read) or generated (written). The legality of an input or output operation depends on the file's current state.

Thus, the file is the elephant of data structures. Its potential size makes it capable of prodigious feats, but we have to allow for a certain amount of lumbering and unwieldiness. We'll see that some tasks that are relatively straightforward when expressed in English and programmed with arrays—for example, adding a value after a sentinel is spotted—may require an elaborate minuet of file operations.

Unfortunately, the fact that files belong to the operating system and aren't defined as part of the Modula-2 programming language brings to mind another comparison with the pachyderm.

> Where does an elephant sit down? Anywhere it wants to. How does a Modula-2 programmer work with files? Using whatever tools her operating system provides. There is no standard set of file-handling libraries.

Although each Modula-2 system has a library of file-oriented procedures, the contents of that library will vary slightly, depending on the insight, programming ability, and intended market of the Modula-2 system's manufacturer. Every Modula-2 system is sure to have some set of file libraries, but the details of how those libraries are used will vary slightly from system to system.

So how are we going to discuss the basic algorithms of file manipulation? By pretending that there is such a thing as a simple, standard interface to every filesystem. I'll make this fantasy more plausible by limiting it to a bare minimum of features that every system is sure to have. You'll leave this chapter equipped to deal with the basic file problems:

what we'll learn

- inspecting a file character by character;
- comparing two or three files for common contents;
- removing duplicated lines or entries;
- merging two sorted files;
- sorting an unsorted file.

what we'll ignore

However, we won't be concerned with other problems that are more dependent on each computer's operating system, like changing a file's read/write protection, accessing the file at an arbitrary location, or dealing with errors that derive from the computer's filesystem.

The FILES Library

Let's work on the assumption that the FILES library shown below exists on every computer. FILES is very much like IO, but it makes two additional contributions. First, it lets us deal with more than one input or output file at a time. This is accomplished rather simply by requiring a filename argument to each of the input and output procedures. In contrast, IO only lets the default input and output streams be changed.

Second, FILES supplies tools for input and output of values of type SYSTEM.WORD. This creates the potential for working with files of numbers, records, arrays, and arbitrarily structured types. Since this is a rather specialized application, we'll ignore these procedures until Chapter 18.

definition of FILES

```
DEFINITION MODULE FILES;
  (*A summary of the primitives we'll use with files.*)
  IMPORT SYSTEM;
    (*We'll eventually use type SYSTEM.WORD to match non-CHAR values.*)
  IMPORT FileSystem;
    (*The module that underlies the local filesystem.
     This import will certainly vary between computers.*)
  CONST EOL = FileSystem.EOL:
    (*EOL will usually appear at the end of lines of characters.*)
  TYPE T = FileSystem.File;
    (*Rename FileSystem.File as FILES.T. Within this module, we refer to T.*)
  PROCEDURE Done () : BOOLEAN;
    (*A general-purpose 'was the last operation successful?' function.*)
  PROCEDURE eof (F: T) : BOOLEAN;
    (*TRUE if we have have reached the end of file F. More specifically, eof(F)
     is TRUE if the last read from F was unsuccessful.*)
  PROCEDURE OpenRead (VAR F: T; Name: ARRAY OF CHAR);
    (*Prepare to read F, starting with its first component.*)
  PROCEDURE OpenWrite (VAR F: T; Name: ARRAY OF CHAR);
    (*Prepare to add to F. Remove any current contents.*)
  PROCEDURE Close (VAR F: T);
    (*The filesystem may require closing processing on open files.*)
  PROCEDURE CharIn (VAR F: T; VAR Ch: CHAR);
    (*Read one character, if one is available, from F.*)
  PROCEDURE CharOut (VAR F: T; Ch: CHAR);
    (*Append the value of Ch to file F.*)
  PROCEDURE WordIn (VAR F: T; VAR Val: SYSTEM.WORD);
    (*Read a single WORD-size value from F.*)
  PROCEDURE WordOut (VAR F: T; Val: SYSTEM.WORD);
    (*Append a single WORD-size value to F.*)
END FILES.
```

FILES includes three unfamiliar lines:

```
IMPORT SYSTEM;
IMPORT FileSystem;
TYPE T = FileSystem.File;
```

The SYSTEM library is found on every Modula-2 system, because it's one of Wirth's semi-standard libraries. We need the import because it makes a very low-level type called SYSTEM.WORD available. We're not going to use this type until Chapter 18.

what FILES.T is

FileSystem, in contrast, may vary from system to system. Although the job it does is essential—it manages the Modula-2 system's interaction with your computer's filesystem—its name, along with the names it creates, isn't standardized at all. I have to import it because I need to use the identifier it gives to file-type variables. Frequently, but not always, this name will be File. The purpose of the type definition T = FileSystem.File is to rename FileSystem.File (a name I'm not sure of) with a simple name I can use elsewhere within the FILES module (as T), and in any program that imports FILES (as FILES.T).

The Basic File Operations

Let's look at FILES in action. A main program begins by importing FILES:

import the library...

```
MODULE FileExample;
    (*Demonstrates some basic file manipulations that use FILES.*)
IMPORT FILES;
    . . .
```

To use files, we have to declare variables of type FILES.T:

declare the file variables...

```
VAR Data, Results: FILES.T;
    . . .
```

Next, we have to *open* each file. Unlike other variables, files have specific states. There are two possibilities: the file has to be in the 'read' state if we want to look at its contents, and it must be in the 'write' state if we want to change the file:

open them...

```
FILES.OpenRead (Data, "info.in");
FILES.OpenWrite (Results, "answers.out");
    . . .
```

what 'open' means

> Although there is no strict terminology on this point, we'll say that a file is *open* when:
>
> *a*) a Modula-2 variable of type FILES.T is associated with an actual file that's under the control of the operating system, and
>
> *b*) the file is either in the read state or the write state.

After the calls above, Data is in the read state. It is attached, so to speak, to an actual file whose name is **`info.in`**. Similarly, Results is in the write state, and is attached to a file named **`answers.out`**. Incidentally, as long as it's a legal identifier, the file *variable* can have the exact same name as the physical file.

The simplest thing that we can do with our open files is to read a single character from Data, and write it to Results:

. . . use them . . .

```
FILES.CharIn (Data, TempCh);
FILES.CharOut (Results, TempCh);
  . . .
```

Finally, procedure FILES.Close should be called whenever we're through writing to a file.

. . . and close them

```
FILES.Close (Results);
```

This makes sure that all output actually reaches the file, and isn't abandoned somewhere in-between.

Now, file operations won't always be successful. For instance, you can't open a file for reading if it doesn't exist. Nor can you read a value from a file, even if it's open, if there's nothing in it. FILES provides two 'post-test' functions that let you know, after the fact, if a procedure call worked:

two post-tests

- FILES.Done() is a general-purpose 'success or failure?' function. It represents TRUE if the last call of any FILES procedures went smoothly, and FALSE if the call failed for some reason.
- FILES.eof(*filename*) is a special-purpose BOOLEAN function. It is TRUE if the last attempt at input from *filename* was a success, and is FALSE otherwise. eof stands for *end of file*.

Strictly speaking, FILES.eof isn't necessary—we could just use FILES.Done for every kind of test. However, FILES.eof is so convenient for implementing file algorithms that just about every file-handling library includes it.

Let's use what we've seen so far to write a basic process filter. Program FileCopy, below, makes a duplicate of an existing file. It can be outlined in pseudocode as:

copy pseudocode

open the existing file for reading;
open the copy file for writing;
make the copy;

The actual code includes a few additions, including a call of HALT (to end the program) if the existing file can't be opened. Pay particular attention to how a file name is obtained from the user.

reading and writing program

```
MODULE FileCopy;
   (*Copies a file of CHAR values.*)
IMPORT IO;
IMPORT FILES;
TYPE STRING = ARRAY [0..35] OF CHAR;
VAR Old, New: FILES.T;
    Name: STRING;
    Letter: CHAR;
BEGIN
   IO.StringOut ("Enter original file name:   ");
   IO.StringIn (Name);     (*Get the existing file's name.*)
   FILES.OpenRead (Old, Name);
   IF NOT FILES.Done() THEN
      IO.StringOut ("Could not open that file.  Quitting.");
      HALT
   END;
   IO.StringOut ("Enter new file name:   ");
   IO.StringIn (Name);     (*Get the destination file's name.*)
   FILES.OpenWrite (New, Name);
   LOOP
      FILES.CharIn (Old, Letter);
      (**)IF FILES.eof(Old) THEN EXIT END;
      FILES.CharOut (New, Letter)
   END;
   (*Postcondition:  there's nothing more to read from Old.*)
   FILES.Close (New);   (*Make sure that New has everything.*)
   IO.StringOut ("Copy complete.")
END FileCopy.
```

```
Enter original file name:  Kristin.pro
Enter new file name:  Marla.pro
Copy complete.
```

Modification Problems

1. FileCopy assumes that New was successfully opened for writing. However, a program that opened many files might not be able to open one more. Modify FileCopy to make an additional check (using FILES.Done()) before it jumps to any conclusions.

2. Does FileCopy handle carriage returns properly? Assuming that when an end-of-line character is read from Old it equals IO.EOL, modify FileCopy appropriately.

3. Modify FileCopy so that it sends even-numbered characters (the second, the fourth, and so on) to file Lefty, and odd-numbered characters to file Righty.

4. Have FileCopy echo each line of Old to New in reverse.

Self-Check Questions

Q. Program Mistakes, below, is meant to write the letters "A" and "Z" in file **Save**. What are the actual contents of **Save** after program Mistakes has run?

```
MODULE Mistakes;
IMPORT FILES;
VAR Sample: FILES.T
BEGIN
    FILES.OpenWrite (Sample, "Save");
    FILES.CharOut (Sample, "A");
    FILES.OpenWrite (Sample, "Save");
    FILES.CharOut (Sample, "Z");
    FILES.Close (Sample)
END Mistakes.
```

A. The second call of FILES.OpenWrite was in error, and should be removed entirely. Every call of FILES.OpenWrite destroys any contents a file currently has. Thus, **Save** ends up only holding the single letter "Z".

Q. Can we copy a file without FILES? Certainly—the IO library has the procedures we need to redirect input and output. But what advantage do we obtain by using FILES to make the copy? What can we do in FileCopy that IO alone wouldn't allow? Hint: can you talk out of both sides of your mouth at once?

A. As the rather cryptic hint points out, FILES lets our program carry on two conversations at the same time. The IO procedures remain connected to the standard input and output (the keyboard and screen) no matter what other file operations are going on. Had we used IO alone, we would have been limited to a single source of input, and a single repository for output.

Section 14-2 Standard Practices

MANY FILE-HANDLING PROBLEMS ARE REALLY *stream*-handling problems. They can be solved using design tools and techniques we learned about in Chapter 6.

However, the applications we'll investigate now are sometimes more complicated to program. Although we'll still be doing things to and with streams of values, we'll have to take into account the prospect that more than one file may be open at a time. Now, the basic process filters involve:

low-level file processing

- finding a particular component within a file;
- adding components to a file;
- deleting components from a file;
- synchronizing these three operations between two or more files.

We've already used the most common combination of low-level file operations in writing our first process filter—Copy. Three other applications, which we'll explore in this section, are merging sorted files, deleting file segments, and finding common file entries. Our extended example in 14–3 takes advantage of our other stream tool, the *state filter*, for file sorting.

Merging Files

File mergers, like corporate mergers, are intended to combine the contents of two separate entities. A merger can be very simple, of course; we might just add one file to the end of another. But in most cases, we're interested in merging files in a way that keeps them ordered. For instance, two previously alphabetized files of words might be merged with the intention of keeping them in alphabetical order.

Let's try a statement of the problem that's just complicated enough to demonstrate the basic file-merging algorithm:

problem: file merging

> Suppose that files Main and Extra both contain CHAR values in alphabetical order. Write a program that creates a third file, called Merged, that holds all the letters in alphabetical order.

Merging files of single letters is really the kernel of the problem of merging files of words. Merging files of words, in turn, is the heart of the problem of merging files of records. If we manage to merge two files of characters, then, we will mentally take credit for merging the employee files that would result from a corporate union of IBM and AT&T.

Mergers of two previously organized files are common in computer systems because of the practical complications of dealing with large main files. In a paper-based filing system, files are usually updated immediately. New records are entered right away so that work won't pile up. Computer filing systems, in contrast, tend to delay updating of main files for several reasons:

why file merging is common

- The entire file may have to be copied in order to insert a single new record. If several new records (in alphabetical order) are available, though, they can be inserted in a single pass through the main file.
- The main file may be so long that other tapes or disks have to be mounted in order to put the new record in its proper place. Since mounting and unmounting is tedious and expensive, it pays to wait and see if other records will require the same file portion.
- Managers of large databases are usually reluctant to allow changes to be made at the user's discretion. They prefer to collect all *transactions*, or requests for changes, and verify their correctness before modifying the master file.
- Finally, a certain amount of delay in inserting records actually adds to overall efficiency—the new record may have to be inspected again, and may wind up not being inserted at all.

For computerized filing systems, then, letting records pile up—in alphabetical order, of course—is very common, and leads to the need to merge files.

How are ordered files merged? A direct analogy to paper files is appropriate. Imagine that you're merging the contents of two partially-filled file cabinets by hand into a third empty cabinet. You open all the cabinets, and get the first record from each of the full ones. The alphabetically 'lower' of the two records goes into the third cabinet, then you pick up another record to refill the empty hand.

developing the algorithm

The process of alphabetical comparison, moving, and replacing goes on until one of the original cabinets is empty. Then, since all the records in the remaining cabinet belong at the end of the third cabinet, and are in alphabetical order already, you move them to the third cabinet without making comparisons.

A Modula-2 algorithm is much the same. We'll have to prepare Main and Extra for reading, and Merged for writing. After obtaining the first value from each of the existing files, we'll compare them, add the lower value to Merged, then get another value. Naturally, we have to repeat this process until Main or Extra is exhausted.

Now, our first filter's process, the one in program FileCopy, was simply *echo*. The process this filter implements might be called *compare*. Instead of merely getting the next value from a single stream, we want to get the lower value from one of two streams. Then, we have to replace the value we took, so that we'll have a basis for comparison next time around. A loop specification is:

`compare´ loop specification

Goal: Merge two ordered input streams into a single ordered output stream.
Bound: The end of the Main *or* Extra *files.*
Plan: Decide which file's current value is lower, then
Add it to the end of the Merged *file.*
Replace it with the next value from the same input file.

A pseudocode outline puts this central loop into an appropriate context:

file-merge refinement

prepare to read Main *and* Extra;
prepare to write Merged;
get the first values from Main *and* Extra;
LOOP
(**)**IF** FILES.eof(Main) **OR** FILES.eof(Extra) *then* EXIT;
add the lower record to Merged;
get the next record from that file
END *of the loop*;
add the non-empty file's remaining records to Merged

Note the placement of the loop's **EXIT** condition. It's essential to leave the loop when an attempted read fails—no sooner, and certainly no later.

checking for empty files

The FILES module, like those on most computer systems, is not aware that a file is empty until we've attempted to inspect it. Thus, FILES.eof isn't TRUE until after a read has failed. However, file-handling programs must always be aware of the possibility that a file will be empty.

The completed program is shown below. After you convince yourself that it works perfectly, try to find the bug it contains. (Hint: what happens at the end of a line?) I'll leave it to you to find a fix.

file merging program

```
MODULE Merger;
   (*Merges Main and Extra into Merged, preserving ordering.*)
IMPORT FILES;
IMPORT IO;
VAR Main, Extra, Merged: FILES.T;
    ChMain, ChExtra: CHAR;
BEGIN
   IO.StringOut ("Merging 'Main' and 'Extra' into 'Merged'.");
   IO.LineOut;
   FILES.OpenRead (Main, "Main");
   FILES.OpenRead (Extra, "Extra");
   FILES.OpenWrite (Merged, "Merged");
   (*We're ready to go.*)
   FILES.CharIn (Main, ChMain);
   FILES.CharIn (Extra, ChExtra);
   (*We think we have the first values from the two existing files.*)
   LOOP
      (**)IF FILES.eof(Main) OR FILES.eof(Extra) THEN EXIT END;
      IF ChMain < ChExtra
         THEN FILES.CharOut (Merged, ChMain);
                FILES.CharIn (Main, ChMain);
         ELSE FILES.CharOut (Merged, ChExtra);
                FILES.CharIn (Extra, ChExtra);
      END (*IF*)
   END; (*LOOP*)
   (*Postcondition: One source file is empty, so flush the other into Merged.*)
   LOOP
      (**)IF FILES.eof(Main) THEN EXIT END;
      FILES.CharOut (Merged, ChMain);
      FILES.CharIn (Main, ChMain)
   END; (*the Main-flushing loop*)
   LOOP
      (**)IF FILES.eof(Extra) THEN EXIT END;
      FILES.CharOut (Merged, ChExtra);
      FILES.CharIn (Extra, ChExtra)
   END; (*the Extra-flushing loop*)
   (*Make sure all output goes to Merged.*)
   FILES.Close (Merged);
   IO.StringOut ("File merger complete.")
END Merger.
```

Modification Problems

1. Merger isn't very robust. Modify Merger so that it considers the possibility that one of the files can't be opened. In addition, have it report on the number of values that are obtained from each of the source files. Finally, what happens to carriage returns in the program as is? Should it be that way?
2. Modify program Merger so that each pair of *identical* values from Main and Extra are put into a new file called Copies instead of going into Merged.
3. Modify program Merger so that just one copy of a repeated sequence or combination from Main or Extra is put into Merged.

Self-Check Questions

Q. Suppose that the call:

```
Copy (Old, New);
```

adds the contents of file Old to file New. This means it assumes that both files are already open. What is the effect of this code segment?

```
  . . .
Copy (Second, First);
Copy (Third, First);
Copy (Fourth, First);
FILES.Close (First);
FILES.OpenWrite (Fourth, "Fourth");
FILES.OpenRead (First, "First");
Copy (First, Fourth);
FILES.Close (Fourth);
  . . .
```

The segment *concatenates*, or joins end-to-end, files Second, Third, and Fourth, using First as a temporary file. The end result is stored in Fourth.

Deleting From Files

Our second multiple-file problem is *deletion*. It calls for a filter in the literal sense—we want to remove a *target*, or single kind of component, from a stream. Simple deletion code is very much like the code required for file copying:

filtering out a target value

```
LOOP  (*Assume that files Old and New have been opened.*)
   FILES.CharIn (Old, Letter);
   (**)IF FILES.eof(Old) THEN EXIT END;
   IF Letter <> TARGET THEN        (*Only print non-target values.*)
      FILES.CharOut (New, Letter)
   END (*IF*)
END (*LOOP*)
```

However, having an entire file of components slated for deletion makes for a much more interesting problem. Suppose that you've made a master list and, upon checking it twice, have come up with some names that should come off. This is the problem of file deletion:

problem: file deletion

> Assume that you have a sorted *master* file, and a smaller sorted *transaction* file that identifies components to be deleted. Write a program that creates an *update* file that holds the remains of the master file after all targets have been deleted.

In real-world database systems, transaction files are multi-purpose; they also identify components to be added or merely modified in place. Again, you can imagine a human file clerk at work. She'd work through the master file until she found a match to the top file on the transaction pile. Then, she'd toss out the top file from the master file. If the possibility for duplicates existed, she'd check again, and perhaps throw away more master files. As soon as she found a mismatch, though, she'd throw out the top transaction file and begin again.

A rough Modula-2 pseudocode is shown below. Note the strategic positioning of the **EXIT** check—one of the files might be empty.

pseudocode

```
prepare to read the input Master and Transaction files;
prepare to write the output Update file;
get the first input values;
LOOP
    (**)IF we're at the end of the Master or Transaction files EXIT;
    decide whether to copy or delete;
    get more values as needed
END the loop;
take care of any components left in the Master file;
```

As before, a component is deleted by not being copied. But how do we *decide whether to copy or delete*? Suppose that ChMaster and ChTrans represent the current components of the Master and Transaction files. There are only three possibilities: either ChMaster and ChTrans are equal, or one or the other is greater. In pseudocode, we can describe the next decision as:

a case analysis

```
IF ChMaster = ChTrans THEN
        read the next Master component, which might also equal ChTrans
    ELSIF ChMaster < ChTrans THEN
        add ChMaster to the Update file;
        read the next Master component
    ELSE ChMaster is greater than ChTrans, so
        read the next Transaction component
END the if statement;
```

Thus, if the components match, ChMaster is deleted by not being added to the Update file. Note that this pseudocode correctly handles the possibility that two successive components of Master have the same value.

What about the last step, *take care of any remaining components?* We've left the loop because we've reached the end of Master or Transaction. If there are still components in Master they'll have to be copied to Update. Program Delete, below, shows how to handle the cleanup. As with previous examples, I assume for convenience that the files consist only of values of type CHAR.

file deletion program

```
MODULE Delete;
  (*Copies Master.In, less components from Trans.In, into Update.Out.*)
  (*Assumes that the existing files are ordered.*)
IMPORT FILES;
IMPORT IO;
VAR Master, Transaction, Update: FILES.T;
    ChMaster, ChTrans: CHAR;
BEGIN
  (*The first part of this program is identical to Merger.*)
  IO.StringOut ("Copying 'Master.In', less 'Trans.In', to 'Update.Out'.");
  IO.LineOut;
  FILES.OpenRead (Master, "Master.In");
  FILES.OpenRead (Transaction, "Trans.In");
  FILES.OpenWrite (Update, "Update.Out");
  FILES.CharIn (Master, ChMaster);
  FILES.CharIn (Transaction, ChTrans);
  (*Inside the loop, though, our analysis is more sophisticated.*)
  LOOP
    (**)IF FILES.eof(Master) OR FILES.eof(Transaction) THEN EXIT END;
    IF ChMaster = ChTrans THEN
          FILES.CharIn (Master, ChMaster)
      ELSIF ChMaster < ChTrans THEN
          FILES.CharOut (Update, ChMaster);
          FILES.CharIn (Master, ChMaster)
      ELSE   (*ChMaster is less than ChTrans, so*)
            FILES.CharIn (Transaction, ChTrans)
    END  (*IF*)
  END; (*LOOP—we leave when one of the files has been spotted as empty.*)
  (*Note that the next loop is 'backward' in comparison to the standard copy
   routine. ChMaster already has a value, so we write, read—not vice versa.*)
  LOOP
    (**)IF FILES.eof(Master) THEN EXIT END;
    FILES.CharOut (Update, ChMaster);
    FILES.CharIn (Master, ChMaster)
  END; (*the Master-flushing loop*)
  FILES.Close (Update)
  (*A check for, and comment on, a non-empty Transaction file might go here.*)
END Delete.
```

Modification Problems

1. Modify Delete so that instead of working with a Transaction file, it suppresses repeated characters from Master instead: aabccad should turn into bad.

2. What should you do if there are components left in Transaction—or, worse yet, if earlier components from Transaction weren't matched and were skipped without notice? How, when, and where might you report on these likely errors? Modify Delete to do it.

Finding Common Entries: The Wall Street Crook

The third of the basic multiple-file problems involves finding common entries. As with file mergers, most searches for common entries arises from rather dull consequences of the way files are used and updated. However, there is an entertaining form of the problem I'll pose for us:

problem: finding common lines

> The Securities and Exchange Commission has in its possession three files on disk. One lists all employees of firms involved in rendering financial advice to a particular company. The second has the names of all individuals who traded heavily in that company's stock. The third consists of every name found in the personal Rolodex of a recently convicted Wall Street success. Each file is in alphabetical order. Help the SEC search for illegal insider traders by writing a program that finds a name common to all three files.

Thus, we have three streams of names going in, and just one stream—perhaps just one name—coming out. As before, though, I'll simplify this problem by assuming that the three files contain only single-character values. Furthermore, you can take it for granted that there actually *is* a common value to be found.

Let's begin by getting an idea of what our program's central loop is going to do:

loop specification

> *Goal: Find a common value in three ordered input streams.*
> *Bound: The current values in all three streams are equal.*
> *Plan: Get the next value in any 'low' stream.*

The loop's bound condition is unexpectedly simple to state. Suppose that Ch1, Ch2, and Ch3 represent the current value from each stream. Then:

comparing three components

```
IF (Ch1=Ch2) AND (Ch2=Ch3) THEN
    all three characters have the same value;
```

If both terms are TRUE, then Ch1 also equals Ch3.

the loop's goal

Now, if we accept the bound condition as stated above—and there's no reason not to—we agree that Ch1 will have to be advanced if it's less than Ch2. Ch2, in turn, must advance if it's less than Ch3. Finally, Ch3 will also have to move ahead if it's currently less than Ch1. This leads to two interesting points:

1. The steps can be considered in any order—no comparison or advance has to come before any other.
2. Although advancing by just one component in a single file is a tiny step, it brings us closer to the overall goal of finding three matching components.

I can rephrase these ideas in pseudocode with:

common entry pseudocode

```
LOOP
    IF Ch1<Ch2 THEN get the next Ch1;
    IF Ch2<Ch3 THEN get the next Ch2;
    IF Ch3<Ch1 THEN get the next Ch3;
    (**)IF (Ch1=Ch2) AND (Ch2=Ch3) THEN EXIT END
END of the loop
```

This segment seems suspiciously short. Nevertheless, if we have faith in the **EXIT** condition, and believe that the *get the next* . . . statements bring us closer to meeting it, the loop will work.

Before reading the finished program, you should note that the order of the statements—including the **EXIT** statement—makes no difference. Each statement is said to be *guarded*—the **IF** check ensures that it can't be executed in error. The final program is shown below.

guarded statements

common line-finding program

```
MODULE Common;
    (*Finds a common entry in the sorted files First, Second, and Third.*)
IMPORT FILES;
IMPORT IO;
VAR First, Second, Third:  FILES.T;
      Ch1, Ch2, Ch3:  CHAR;
BEGIN
    FILES.OpenRead (First, "First");
    FILES.OpenRead (Second, "Second");
    FILES.OpenRead (Third, "Third");
    FILES.CharIn (First, Ch1);
    FILES.CharIn (Second, Ch2);
    FILES.CharIn (Third, Ch3);
    (*We have the first value from each file.*)
    LOOP
        IF Ch1<Ch2 THEN FILES.CharIn(First, Ch1) END;
        IF Ch2<Ch3 THEN FILES.CharIn(Second, Ch2) END;
        IF Ch3<Ch1 THEN FILES.CharIn(Third, Ch3) END;
        (**)IF (Ch1=Ch2) AND (Ch2=Ch3) THEN EXIT END
    END;
    (*Postcondition:  The current value from each file is identical.*)
    IO.StringOut ("The common value in files First, Second, and Third is:    ");
    IO.CharOut (Ch1)            (*Note that we're writing to the terminal.*)
END Common.
```

Modification Problems

1. Create data files for Common, then test it. What happens if your test files aren't in alphabetical order? What if there isn't a common value?
2. It's more than likely that your Modula-2 system has a library that includes a string-comparison procedure or function. Find out what it is, then modify Common to find the common name in three test files.

Section 14-3 An Extended Example

THE LAST OF OUR FILE EXAMPLES INVOLVES sorting. Suppose that we are provided with an mixed-up file of numbers that need to be put into numerical order. A reasonable first inclination is to use some of the array-sorting code developed in Chapter 9. Unfortunately, it is in the nature of file problems that any file we want to sort will contain more components than can be held in the largest array we can declare. The array solution is out.

problem: sorting a large file

A second natural idea will be to adapt some of the array-sorting *algorithms*, rather than their code. Selection sort—traveling through the file to find the smallest value, then again to find the second smallest, and so on—is quite appealing. But once again, we find that the nature of file-sorting problems is such that reopening a file so that it can be read again is certain to take so long that we wish to avoid it at almost any cost.

the ground rules

These general rules—first, that files will be very large; second, that repeated travel through a file is 'expensive' in terms of computing time; third, that opening and closing files is even more costly—are the basic constraints on file-sorting solutions.

Natural Merge Sorting

Thus, we'll try to manage file sorting with no more than a couple of auxiliary files, while minimizing the number of passes through each file. Fortunately, we can draw on our previous file examples to develop an extremely effective method, called *natural merge sorting*, that meets these restrictions.

runs

Natural merge sorting relies on the prospect that even a hideously jumbled file is likely to contain some *runs*. A run is a stretch of values that happen to be in ascending order: 7 3 5 9 6 includes the run 3 5 9.

Suppose that we read through a file called Source, and evenhandedly distribute any runs we find into two auxiliary files named Left and Right. In the illustration below, I've shaded the alternate runs in Source. When they're distributed, all the shaded runs wind up in Left, while the rest go to Right.

Left: **as | in | ex | l |**

1*st* Source: **asortingexample**

Right: **ort | g | amp | e |**

distributing and merging

Note that I put these markers: '**|**' between each run. They help us in the next step—merging the corresponding runs from Left and Right back into a second incarnation of the Source file.

Left: **as | in | ex | l |**

aorstginaempxel *a 2nd* Source

Right: **ort | g | amp | e |**

Taken together, the distribution phase and merging phase are considered to be a single pass. Let's see what happens in the second pass:

second pass

Left: **aorst | aempx |**

aorstginaempxel **aginorstaeelmpx**

2nd Source Right: **gin | el |** *3rd* Source

Our progress is immense! Just one more pass, starting with the third instance of file Source, will sort the entire file:

third pass

Left: **aginorst |**

aginorstaeelmpx **aaeegilmnoprstx**

3rd Source Right: **aeelmpx |** *final* Source

Let's pseudocode the algorithm. The program's main loop has to take care of each split-and-merge pass. What will the bound condition be?

split-and-merge loop specification

Goal: Order the components of Source.
Bound: There was just one run in Source.
Plan: Split runs between Left *and* Right.
Merge runs from Left *and* Right *back into* Source.

The bound is just a count. We're done when there's only one run found.

Let's agree that the first run of each pass always goes into Left. Then, Left will hold the final sorted list. I can pseudocode the final program as:

first pseudocode

initialize the files;
LOOP
distribute runs from Source *into* Left *and* Right, *starting with* Left;
(**)**IF** *just one run was found then* **EXIT**;
merge Left *and* Right *back into* Source
END *of the loop;*
(**Postcondition: just one run was found, so* Left *is sorted.**)
copy the one-run file, Left, *back into* Source

The Distribution Phase

Consider the distribution phase of each pass. How can we *distribute runs* properly? Suppose that I outline a loop specification:

distributing the runs

> *Goal:* Left *and* Right *contain ordered runs, and*
> *we know how many runs there were.*
> *Bound: There's nothing left in* Source.
> *Plan: Get a character from* Source.
> *Decide what to do with it.*

Decide what to do is easily said, but how is it accomplished? Well, each time we get an item from Source, there is the potential for a change in the state of the input stream. A *state filter* is going to come in handy. (Recall that state filters collect and apply information about the content of the streams they filter. This is in contrast to process filters, which dumbly apply some rule to each value as it passes through.)

Now, the smallest step we take to approach the bound (reading a single character) is also enough to let us recognize a transition, or change, of state. The possible states are:

potential distribution states

- A run has ended, and subsequent values should go to Left.
- A run has ended, and subsequent values should go to Right.

How can we spot each state change? Well, the decision to send a character to Left or Right depends on spotting the end of a run. We can do that by remembering the previously read character:

spotting a state transition

> **IF** *the* Current *character is less than the* Last *one* **THEN**
> *a run has ended*

What do we do when a run ends? We have to count the run, of course. In addition, we should add a marker to the Left or Right file to mark the end of the run. It serves the same purpose as the '**|**' in the illustrations, and is needed later to let us merge runs.

marking the runs

> **IF** *the* Current *character is less than the* Last *character* **THEN**
> *Add a* MARKER *to the current output file;*
> *Add* 1 *to the count of runs*

How do we know which file the current run is going to? Well, if we send the first run to Left, then all the odd-numbered runs will go to Left, while the even-numbered runs are sent to Right.

We're ready to pseudocode a procedure that might take care of a single distribution pass. A rough outline would be:

first refinement

```
(*Precondition: Source holds the entire sequence.*)
initialize the runs counter to 1;
initialize a Last character variable to a value that's too small to end a run;
LOOP
   get the current value from Source
   (**)IF there wasn't one then EXIT;
   decide if there was a state transition;
   send the current value to the proper output file
END of the loop
(*Postcondition: Source has been split up into Left and Right correctly.*)
```

A more detailed version of what will eventually become procedure Distribute is:

second refinement of Distribute

```
initialize the Last variable to 0C, which is NUL, the smallest character;
initialize the Runs counter to 1;
LOOP
   read the Current character from Source;
   (**)IF it wasn't done EXIT;
   IF Current < Last THEN
      IF ODD(Runs) THEN add the MARKER to Left
                 ELSE add the MARKER to Right;
      add 1 to Runs
   END;
   (*We've handled the state transition.*)
   IF ODD(Runs)
      THEN append Current to Left
      ELSE append Current to Right
   assign Current to Last
END of the loop
(*Postcondition: Left and Right hold MARKER-separated runs.
 If Runs equals 1, the sorted values are in Left.*)
```

The Merging Phase

Now, if the distribution phase ends with the Runs count equal to 1, the file is sorted. But it's more likely that there will be more than one run. We'll have to go on to a *merging* phase, and merge together corresponding runs from Left and Right.

Merging runs will be a little more involved than merging files. Left and Right can contain a series of runs, each ending with a sentinel MARKER, that must be merged one pair at a time. However, the same reasoning about state transitions that was so effective in the distribution phase will be equally useful in merging. Let's say that this is the basic loop specification:

run-merging specification

Goal: Paired runs from Left *and* Right *have been merged into* Source.
Bound: Reaching the end of file Left *or* Right.
Plan: Compare items from Left *and* Right.
Decide what to do with them.

How can we *decide what to do* at each step? Well, reading a single value is enough to change the current state. What states, and necessary actions, might prevail along the way?

potential merge states

1. We're out of data in either Left or Right (so quit trying to merge).
2. We've reached the end of run marker in both Left and Right (so get two new values).
3. The Left run ended (so add the Right value to Source and get a new Right value).
4. The current Right value is less than the current Left value (so add the Right value to Source and get a new Right value).
5. The Right run ended (so add the Left value to Source and get a new Left value).
6. The current Left value is less than the current Right value (so add the Left value to Source and get a new Left value).
7. The current Left and Right values are equal (so add both to Source and get two new values).

A veritable thicket of possibilities! However, if you look at the implied actions, in parentheses, there's a great deal of overlap. Frequently, two different states have the same outcome. In pseudocode:

merge procedure pseudocode

```
initialize ChLeft and ChRight from files Left and Right;
LOOP
   (**)IF either Left or Right is empty EXIT;                     (*1*)
   IF ChLeft and ChRight both equal MARKER THEN                   (*2*)
         Replace ChLeft and ChRight
      ELSIF ChLeft is the MARKER                                  (*3*)
         OR ChRight isn't and is less than ChLeft THEN            (*4*)
         Append ChRight to Source and replace ChRight
      ELSIF ChRight is the MARKER                                 (*5*)
         OR ChLeft isn't and is less than ChRight THEN            (*6*)
         Append ChLeft to Source and replace ChLeft
      ELSE ChRight equals ChLeft so                               (*7*)
         Append ChLeft and ChRight to Source and replace them both
      END of the state-checking IF statement
END of the loop
(*Postcondition: matching pairs have been merged. The remainder of the non-
  empty file (if there were an odd number of runs) belong at the end of Source.*)
flush any remaining run into Source;
```

The Natural Program

The completed program Natural is shown below. As you read it, notice the care that must be taken to avoid off-by-one errors—missing values, extra sentinels, extraneous end-of-line characters, and the like. If you can reproduce this program, you can feel awfully confident about your ability to manipulate files.

natural merge sort program

```
MODULE Natural;
   (*Sorts file Source using two auxiliary files.*)
IMPORT FILES;
IMPORT IO;
CONST MARKER = 0C;          (*Marks the end of each run.*)
VAR Source, Left, Right:  FILES.T;
    Ch:  CHAR;
    Runs:  INTEGER;
PROCEDURE Distribute (VAR Source, Left, Right:  FILES.T;
                      VAR Runs:  INTEGER);
   (*Split runs from Source into Left and Right.*)
   VAR NewCh, LastCh:  CHAR;
   BEGIN
      LastCh := 0C;  (*Initialize LastCh to the smallest character.*)
      Runs := 1;
      LOOP
         FILES.CharIn (Source, NewCh);
         (**)IF FILES.eof(Source) OR (NewCh = FILES.EOL) THEN EXIT END;
         IF NewCh < LastCh THEN
            IF ODD(Runs)
               THEN FILES.CharOut (Left, MARKER)
               ELSE FILES.CharOut (Right, MARKER)
            END;  (*We've added the end-of-run marker.*)
            Runs := Runs + 1
         END;  (*We've handled the state transition.*)
         IF ODD(Runs)
            THEN FILES.CharOut (Left, NewCh)
            ELSE FILES.CharOut (Right, NewCh)
         END;  (*We've output the current character.*)
         LastCh := NewCh
      END  (*LOOP*)
      (*Postcondition:  Left and Right hold MARKER-separated runs.
       If Runs equals 1, the sorted values are in Left.*)
   END Distribute;
```

```
PROCEDURE Merge (VAR Source, Left, Right: FILES.T);
  (*Merge runs from Left and Right into Source.*)
  VAR ChLeft, ChRight: CHAR;
  BEGIN
    FILES.CharIn (Left, ChLeft);
    FILES.CharIn (Right, ChRight);
    LOOP
      (**)IF FILES.eof(Left) OR FILES.eof(Right) OR
        (ChLeft = FILES.EOL) OR (ChRight = FILES.EOL) THEN EXIT END;
      IF (ChLeft = MARKER) AND (ChRight = MARKER) THEN
          FILES.CharIn (Left, ChLeft);
          FILES.CharIn (Right, ChRight)
        ELSIF (ChLeft = MARKER) OR
          ((ChRight <> MARKER) AND (ChRight < ChLeft)) THEN
          FILES.CharOut (Source, ChRight);
          FILES.CharIn (Right, ChRight)
        ELSIF (ChRight = MARKER) OR
          ((ChLeft <> MARKER) AND (ChLeft < ChRight)) THEN
          FILES.CharOut (Source, ChLeft);
          FILES.CharIn (Left, ChLeft)
        ELSE (*ChRight equals ChLeft, but they're not markers.*)
          FILES.CharOut (Source, ChLeft);
          FILES.CharOut (Source, ChRight);
          FILES.CharIn (Left, ChLeft);
          FILES.CharIn (Right, ChRight)
      END (*of the state-checking IF statement*)
    END; (*LOOP*)
    (*Postcondition: the non-empty file's contents go to the end of Source.*)
    LOOP
      (**)IF FILES.eof(Left) OR (ChLeft = FILES.EOL) THEN EXIT END;
      FILES.CharOut (Source, ChLeft);
      FILES.CharIn (Left, ChLeft)
    END; (*Left-flushing loop*)
    LOOP
      (**)IF FILES.eof(Right) OR (ChRight = FILES.EOL) THEN EXIT END;
      FILES.CharOut (Source, ChRight);
      FILES.CharIn (Right, ChRight)
    END (*Right-flushing loop*)
  END Merge;
```

(program continues)

```
BEGIN        (*main module*)
  IO.StringOut ("Sorting file 'Source'.");
  IO.LineOut;
  LOOP          (*until there's just one run and we're done*)
    (*Open the files.*)
    FILES.OpenRead (Source, "Source");
    FILES.OpenWrite (Left, "Left");
    FILES.OpenWrite (Right, "Right");
    (*Postcondition: Source is ready to be split.*)
    Distribute (Source, Left, Right, Runs);
    FILES.Close (Left);          (*Close the output files.*)
    FILES.Close (Right);
    (*Postcondition: Left and Right are ready to be merged.*)
    (**)IF Runs = 1 THEN EXIT END;
    FILES.OpenWrite (Source, "Source");
    FILES.OpenRead (Left, "Left");
    FILES.OpenRead (Right, "Right");
    Merge (Source, Left, Right);
    FILES.Close (Source);
    (*Postcondition: Source contains MARKER-separated runs.*)
  END; (*the one-pass loop*)
  (*Postcondition: Left contains the ordered data.*)
  FILES.OpenWrite (Source, "Source");
  FILES.OpenRead (Left, "Left");
  (*Finally, copy Left into Source.*)
  LOOP
    FILES.CharIn (Left, Ch);
    (**)IF FILES.eof(Left) THEN EXIT END;
    FILES.CharOut (Source, Ch)
  END; (*the copying loop*)
  (*Postcondition: Source is ordered.*)
  FILES.Close (Source)
END Natural.
```

Modification Problems

1. When I tested Natural, I found two NUL characters at the end of the sorted file. Why? Can you fix the bug?
2. As written, Natural insists that all of its input appear on a single line. Modify the program so that it skip carriage returns, rather than treating them as the equals of end-of-file.
3. Modify Natural so that it can sort a file of plain English words. You'll have to import a string-handling library, of course. It's up to you to decide if the source file should hold its words one per line, or just separate them by spaces.

Section 14-4
Program Engineering Notes

Just how good is natural merge sorting? Well, in the worst of all possible worlds, a file will be in reverse order to begin with. The first distribution phase will produce as many runs as there are elements in the file. However, the first merging phase will cut the number of runs in half. The merged runs will be very short, mind you—each just two components long—but they set the stage for later success.

analyzing the algorithm

What happens in the second split-and-merge pass? Again, on average, we'll double the length of each run, since all the two-component runs will end up being merged into four-component runs. Since we're increasing the length of the runs, it's inevitable that we'll reduce, by half, the number of runs that can be distributed in a subsequent pass. Eventually, the first run we spot will be as long as the file itself, which means that the file is sorted.

How long is *eventually*? There are two ways to ask the question. First, starting with N runs, how many times can we cut N in half before we get down to 1? Alternatively, how many times can we double a run of length 1 before it equals N?

effectiveness of the natural merge

> In either case, the answer is $\log_2 N$—the power 2 must be raised to before it equals N. In practical terms, this means that the natural merge sort can organize a file of one million jumbled components in twenty passes *or less*.

File Bugs

By now you've seen ample evidence that every type provokes its own set of bugs. They usually turn out to occur in direct proportion to the severity of warnings against them—mild 'Bewares!' are usually heeded, but absolute prohibitions promote a frenzy of crashes. Three common errors that involve files are:

1. Attempting to read from a file that isn't in the 'read' state.
2. Trying to write to a file that isn't in the 'write' state.
3. Trying to work with a file that, as far as the operating system is concerned, doesn't exist.

How your system deals with these errors usually turns out to have been a private matter between its implementor and her conscience. Suppose that you try to OpenWrite a file that doesn't exist. Should the system go ahead and create an empty file? Should the program crash? Should it pretend to create the file, then tell you the bad news after it's all over? Your guess is as good as mine.

One of the most useful program engineering tips to remember is *use the right tool for the job.* Don't forget that most computer systems supply file-handling tools, like editors, that may be much better at file manipulation than the programs we can write. Even when we're set on writing new programs, it makes sense to use existing system programs to check our work.

Modula-2 Details
Syntax Summary

• the FILES module: a typical package of type name, functions, and procedures for controlling interactions with a computer filesystem. Its key contents are:

1. FILES.T — the type of every file variable.
2. FILES.eof(F) — a function that is TRUE if it's no longer possible to read input from file F.
3. FILES.Done() — a function that is TRUE if the last file-related procedure call was successful.
4. FILES.OpenRead (F, Name); FILES.OpenWrite (F, Name) — procedures that let the operating system know if we intend to inspect or modify file F, which is known to the operating system as *Name*.
5. FILES.Close (F) — a procedure that alerts the operating system that we're done with whatever we were doing to F, and makes sure that it has all its contents.

Modula-2 Details
Remember . . .

• Files are abstract data types. They are independent of any program, and of any programming language—they 'belong' to the computer's operating system. A module like FILES acts as an interface between standard Modula-2 and computer systems that are inherently non-standard.

• Files are sequential-access structures, as opposed to random-access structures like arrays and records. Since we step through files one value at a time, components aren't individually named.

• A file is either being inspected, or it is being added to. Each call of FILES.OpenRead on a file will return you to the file's beginning. However, every call of FILES.OpenWrite on a file will remove the file's current contents.

• Because opening a file for writing removes its current contents, adding to a file really means copying a file, then adding to the copy. A value can only be inserted into a file by copying, then adding, then finishing the copy.

• A file variable is a stream of values. Most file-handling programs are either process filters (that modify each value in the stream) or state filters (that decide what to do on the basis of the stream's contents).

Zen Study Guide
Big Ideas

The Big Ideas in this chapter are:

1. *The file is the elephant of data types—prodigious, but unwieldy.* The advantage of working with files is that they're unbounded: a file can hold any number of components. But to get this advantage we must generally give up random access to individual components, and the ability to make changes anywhere besides the file's end.
2. *How we work with files will vary from system to system, but what we want to do to and with them remains the same.* The FILES module that we used

provides typical tools for dealing with stored files. But because files depend so much on particular operating systems, they'll never be defined as strictly as, say, arrays.

Zen Study Guide

Questions

14-1 Several details of file usage are unlike other variables: FILES (or a library like it) must be imported, a name has to be connected to the file, the file's state has to be set for reading or writing, and so on. Take two example programs from the text, and by leaving out a statement, or changing an identifier, create four *Where's the bug?* questions.

14-2 Another way at looking at the question above is to tick off the steps that must be taken before a file can be used. Take a simple file-copying program, leave out a few steps, then ask yourself to *Complete this program.*

14-3 Robustness is an especially important issue for file-handling programs. Although the question *How could you improve this program?* is usually too vague to be a good test question, it's pretty reasonable when the answer is *Make a check for* FILES.Done(). Make up two improvable programs.

14-4 As you've no doubt figured out by now, the IO library is, in a sense, a special case of the more general FILES library. It can be confusing to use both libraries at the same time. For instance, any program that reads user input from the keyboard, then sends it to two or more files, would pose such a problem. Make up three questions that would require both libraries.

14-5 Time for a few essay questions? Well, maybe we'll settle for *Define these terms.* Pick five, starting with *What is a file's state?*

14-6 Interaction between types is the most confusing concept of all. Can you take one hundred INTEGER values stored in a file, put them in an array, then put them back into the file in reverse order? Of course! Make up three questions that involve transferring data between files and other data types.

14-7 In most cases, a file-handling program is meant to process a stream of data. The simplest file program, Copy, does nothing to its data, per se—it just connects an input stream (reading a file) to an output stream (writing a file). The 'process' is just read and echo.

Make up three questions that could be solved with program Copy simply by adding a few statements to its process. For instance, *Write a program that copies every other value* would fall into this category, since it can be solved by adding a single extra FILES.CharIn call.

14-8 There's another way to build from the basic Copy program—make its data stream more complicated. For example, *Three files contains judges' scores (one per line) for a competition. Create a file that holds the average of the first lines, second lines, etc.* Make up three questions that involve merging values from two or three files into a single output file.

14-9 Next, do the reverse. For example, *A file holds letters, digits, and spaces. Send each category to a separate file.* Make up three process questions that cause a single input file to split into three output files.

14-10 What are files good for? Like arrays and records, files are more useful for some tasks than for others. Make up six applications (e.g. sorting twenty numbers), then ask: *Which three of these need files for a reasonable solution?*

Exercises

Warmups and Test Programs

14-11 Secretly, deep down in her heart, every programmer wants to know just how big a file can be. How might you find out?

14-12 Most computer systems won't limit the number of files a user can have, but they will restrict the number of files that can be open during a single program. Find out your system's limits.

14-13 When they're input from a keyboard, output by a program, or stored in a file, numbers are usually sequences of digit characters. This is the most convenient way to do things, but it's not the most parsimonious. Write a program that generates a sequence of INTEGER values, and compare the size of the files created when they're stored as digits (use IO.IntOut for output) and as bits (use FILES.WordOut for output).

14-14 Suppose that a record has three fields: a string, an INTEGER, and a CHAR value. Outline the process you'd follow to:

a) store the data for twenty such records in an ordinary text file;

b) read the data from a text file into an array of records;

c) store the records *as records* in a new file;

d) read the records back into the array from the new file;

e) store the array of records' contents into a text file; and

f) make sure that the first file and the last file were the same.

14-15 What happens on your system if a program that has been writing to a file doesn't call FILES.Close? Are all the file's contents there?

14-16 Is the file of characters you can create with a text editor the same as a file of CHAR values you might create using FILES? How might you test two files to see if they're identical?

Exercises
Programming Problems

14-17 What's an easy way to count the number of different words in a file? Well, the way I do it is to rearrange the file into one-word lines, then sort the file, then get rid of any duplicated lines, and finally count the number of lines that are left. On UNIX or MS-DOS systems the command might look like this (although the exact command names will vary):

```
rearrange data.in | sort | unique | countlines > results.out
```

Write a program that can take care of the third job—finding unique lines. Give your program the following options:

a) Print just one instance of each line that appears one or more times.

b) Print just one instance of each line that is *not* repeated, and ignore lines that are repeated.

c) Print one instance of each line that *is* repeated, and ignore lines that are not repeated.

d) Print an error message and quit if the file is not sorted to begin with.

You can assume that each line is just one word long.

14-18 Restricting access to programs or computer systems by requiring passwords is very common. (After all, if it weren't common, we wouldn't be reading all the time about passwords being cracked, now would we?) Passwords are usually kept in a dedicated password file that holds user's names and passwords. Any program that requires a password check can look up the user's name in the password file, then see if the password she supplies is correct.

Usually, the file is open to public access—anybody can read it—but the passwords are encrypted. Password-checking programs work by encrypting the password the user supplies, then checking it against the encrypted version that's already in the file. In theory (and in practice), such systems can be cracked by trying commonly used passwords, e.g. the user's name, or default passwords like changeme.

Write a set of routines (you can supply it as a library module, of course) that could be used to maintain a password file. It should be able to check passwords, as well as changing and adding them. The method you use for encryption isn't important, but making your routines robust and bullet-proof is. Special question: How can you deal with the prospect that two programs may try to change the password file at the same time?

14-19 The song *Midnight Train to Georgia* is known not only for its moving lyrics by J. Weatherly, and the intensity of the performance by Gladys Knight and the Pips, but also for the beautiful arrangement by T. Camillo. The background vocals are so strong and well-balanced that when the Pips appeared on *Saturday Night Live* without Gladys to sing lead they were, nevertheless, able to perform the song.

If you'd like an assignment that is highly suitable for an all-night project, try solving the problem of printing a lyric sheet for *Midnight Train to Georgia.* Assume that you have two files, one containing the lead vocal parts, and the other holding the backup singers' words. Write a program that merges the two files, line by line, into a single output file. In other words, your output file should look like this:

```
Gladys (lead)                                  Pips (backup)
LA proved too much for the man                 Too much for the man
So he's leaving a life he's come to know       He couldn't make it
    . . .                                          . . .
```

Hint: If you don't know the words, you can get the recording on 45 from Buddah Records.

14-20 *Run-length encoding* is a data compression technique that exploits repetition to shorten overall length. Any series of identical characters:

```
aaaaabbccccdddefg. . .
```

can be replaced by a marker (so that the source text can contain digit characters) and count. Here the marker is a backslash:

```
\5abb\4c\3defg. . .
```

Note that the replacements for '**a**' and '**c**' saved room, while '**d**' had no real effect, and '**b**' and '**e**' through '**g**' were untouched, since compressing them would have actually wasted space. Write programs that use run-length encoding to compress, and uncompress, a file.

14-21 A considerable amount of effort goes into protecting the integrity of digitized sound stored on compact discs. For example, imagine that discrete 'sound values' were simply stored in sequence:

17 15 91 68 52 84 ...

There is essentially no protection against loss of information here—a single lost digit destroys an entire value. Redundancy helps minimize the danger:

17 17 17 15 15 15 91 91 91 68 68 68 52 52 52 84 84 84 ...

Here, two complete numbers, plus part of a third, must be lost before an entire value is destroyed.

However, even the space allocated to storing three consecutive numbers is microscopic, so a tiny nick still causes problems. To further reduce the chance of error, redundancy is combined with a technique called *interleaving*—each triple is spread among its neighbors:

17 .. .. 15 17 .. 91 15 17 68 91 15 52 68 91 84 52 68 .. 84 52 .. .. 84 ...

(I've used each **..** as a space holder for unknown earlier or later values.) This approach is extremely fast to encode and decode. When they're interleaved like this, eight consecutive values, plus part of a ninth, must be lost to destroy a value. Various other techniques, incidentally, are used to help distinguish between a value that's correct, and one that's lost because of a changed bit.

Write a coder, and decoder, to implement redundant interleaving and deinterleaving. For fun, you might want to challenge a friend to destroy random parts of the encoded file (changing numbers to zeros), then see if you can reconstruct the original. Compact discs, incidentally, can withstand the destruction of as many as 14,000 consecutive bits—a hole can be drilled in the disc without losing a note!

14-22 Natural phenomena are often difficult to analyze because a wide range of values may obscure an underlying pattern. For example, consider the measurement of ocean wave heights, which can range from a few centimeters to several meters—all within a single series of measurements. A human observer on the beach can estimate wave height fairly easily and accurately, though, because she ignores smaller waves. In formal terms, she focuses on the *significant wave height*—the height of, say, the tallest third of the waves. A computer, though, needs to have its data selectively filtered.

Write a program that analyzes the top third of the REAL values in a file. You may assume that all values fall between 0 and 5 meters, with 0.1 meter accuracy. However, the data file itself is of unknown length. Answer these questions:

a) How many waves were in the 2.3 — 2.4 meter interval?

b) What was the range of the top-third waves?

c) What was the average height of the top-third waves?

14-23 A *macro* is a brief, uniquely identified sequence of characters that represents a longer text sequence. Macros are usually defined in a *macro file*—a brief macro name is followed by the macro's definition. A *macro preprocessor* spots macro names in input text, looks up their definitions in the macro file, then *expands* them as part of output.

Suppose that our macro file read:

```
\P
Whereas the party of the first part,
\.
\D
the party of the second part,
\.
```

Raw source text might be:

```
\P
an unnamed plaintiff, and
\D
defendant (the University of California, Berkeley) ...
```

Preprocessor output would be:

```
Whereas the party of the first part,
an unnamed plaintiff, and
the party of the second part,
defendant (the University of California, Berkeley) ...
```

Note that a macro call starts with a backslash, and appears on a line by itself. The macro definition begins similarly, and ends with a '\.' on a line by itself.

Write a program that allows the definition and use of one-letter macro names. Be sure to make your program relatively robust—it should at least be able to handle a macro that is called, but isn't defined. You need not allow macros that contain other macros, but your program should be able to recognize attempts to define them.

14-24 Many schools require a term or two of foreign language study. Let's kill two birds with one stone by writing a drill program for verb conjugation.

Suppose that an unnamed foreign language (often used by the author on vacation) has three categories of regular verbs—those that end with *are*, those with *ere*, and those with *ire*. Their regular conjugations are:

infinitive	**compr***are*	**vend***ere*	**part***ire*
I	*-o*	*-o*	*-o*
you	*-i*	*-i*	*-i*
he, she, it	*-a*	*-e*	*-e*
we	*-iamo*	*-iamo*	*-iamo*
you (plural)	*-ate*	*-ete*	*-ite*
they	*-ano*	*-ono*	*-ono*

Create three files (one for each category) of English verbs and their foreign counterparts. Write a program that asks the user to conjugate a verb into a form chosen at random, e.g.:

```
to drink   (you, plural)     Type 'return' for answer.
```

then prints the correct answer on request. Note that the files contain only the English and foreign-language verbs—the conjugation forms should be hardwired into your program.

14-25 Repetitive drill is an ideal problem for computer solution. The computer teacher—singleminded, purposeful, and utterly relentless—thinks nothing of asking questions, again and again, until you are as close to perfection as a mere human can be.

For simplicity, let's consider a drill problem with one-line questions and answers; say, states and their capital cities. A poor drill program simply travels through the entire list, then starts over again. A slightly more effective drill program will repeat only the questions that were answered incorrectly. The most effective drill program will return to incorrectly answered questions more frequently, but not exclusively, and not always in the same order.

One suggested algorithm for repetitive drill is:

a) Take a question from the head of the list.

b) If it's answered incorrectly, reinsert the question among the first *n* questions.

c) If it's answered correctly, put the question on the end of the list.

Write a program that implements this algorithm. How might you propose to set the value of *n*? (Suggestion: when a question is answered correctly, mark it, but reinsert it. When it's been answered correctly twice, put it on the end of the list, and decrement *n*.) For the purposes of this exercise, you can trust the user to confess if her answer was incorrect.

14-26 *Computer-aided instruction* is a classroom assistant whose imminent arrival has been heralded for some time now. Let's try to speed things up with a computer-aided multiple-choice question tester. Write a program that lets an instructor prepare, and a student take, multiple-choice tests. The instructor's options should be:

a) Add a question: she should be prompted for the question, the correct answer, then alternate wrong answers.

b) Print the test: print the test, with questions numbered, and answers labled `a`, `b`, `c`, and so on. `a` is always the correct answer.

c) Delete a question: she should be prompted for the question's number.

The student's options are:

a) Take the test: the possible answers should be presented in random order, of course.

b) Confirm: the correct answer should be shown.

c) Review: as an alternative to immediately confirming, missed questions should be repeated.

The student's name and score (correct answers / questions) should be added to a permanent file of scores.

Chapter 15 The **SET** Type

Type **SET** defines variables with an internal structure, just as types **RECORD** and **ARRAY** do. The structure of a set is quite limited—it can hold a small number of simple values of just one type. However, the primitive operations that are defined for sets make them very powerful—they allow an unexpectedly easy access to some ideas of parallel programming. Section 15-1 describes the set operators and relations, then turns to some basic set applications. We'll pay particular attention to parallel 'set-think' in looking at flag variables, scheduling, and chessboard management.

Modula-2 also includes a special set type, called BITSET. This particular set type is built into the language, just like INTEGER and REAL. Bitsets are used for convenience, as an abstract way of dealing with the individual bits of a single computer word. Chapter 18 shows how bitsets come in handy when maximizing execution speed, or minimizing data size, are important program considerations.

Section 15-1
Strictly Modula-2

LET'S BEGIN BY SEEING AN EXAMPLE OF THE **SET** type in action. Program Groucho, below, is a feeble attempt at duplicating one feature of the Groucho Marx television show *You Bet Your Life*. On the show, accidentally stumbling upon the secret word of the day was worth $100. In our program, typing in any character from a set of 'secret' characters causes some output.

a set example

```
MODULE Groucho;
   (*Spots values in a set of 'secret' letters.*)
IMPORT IO;
TYPE LetterSET = SET OF CHAR;
VAR Secret: LetterSET;
    Ch: CHAR;
BEGIN
   (*Begin by initializing the Secret set variable.*)
   Secret := LetterSET{"g", "r", "o", "u", "c", "h", "o"};
   (*Then, try to prompt some input.*)
   IO.StringOut ("Try to say the secret letter.  Start typing.");
   IO.LineOut;
   LOOP
      IO.CharIn (Ch);
      (**)IF Ch = IO.EOL THEN EXIT END;
      IF Ch IN Secret THEN
         IO.StringOut ("Beep!!  ")
      END
   END
END Groucho.
```

```
Try to say the secret letter.  Start typing.
This is a really dumb program.
Beep!! Beep!! Beep!! Beep!! Beep!! Beep!! Beep!!
```

As a means of data storage, plain and simple, the set's advantage is that it represents and allows access to many values without the need for subscripts or field names. Despite itself, Groucho manages to demonstrate a few set features:

some set features

- They hold values of one type, as Secret holds CHAR values.
- They can represent more than one value at a time, as Secret represents six different letters.
- They hold just one instance of each value, as Secret holds just one "o".
- We can see if a value is included in a set, as we check to see if Ch is **IN** the Secret set.

The maximum *cardinality* of a set type (the maximum number of elements it can have) is implementation-defined—determined by the author of a system's Modula compiler. It may be as small as 16, or as large as the computer's memory. When he defined Modula-2, Wirth did not feel comfortable in requiring any particular minimum cardinality. However, since many common programming tasks require it, most compilers will let sets be at least as large as the cardinality of type CHAR. I'll work on that assumption.

> The definition of a set type contains the set type's identifier, the reserved words **SET OF**, and the type of *elements* the set will contain (the set's *base type*).

Sets can, and do, have a variety of base types. They include CHAR, entirely new enumerations (like VitaminTYPE, below), and subranges of existing types (like INTEGER).

```
TYPE CharSetTYPE = SET OF CHAR;
     VitaminTYPE = (A, B1, B2, B3, B6, B12, C, D, E);   (*an enumeration*)
     VitaSetTYPE = SET OF Vitamins;
     LowNumberTYPE = INTEGER[1..12];                     (*a subrange*)
     LowSetTYPE = SET OF LowNumberTYPE;
VAR InputCharacters, OutputCharacters: CharSetTYPE;
    FruitVitamins, VegetableVitamins: VitaSetTYPE;
    Responses: LowSetTYPE;
```

Constants with set types can also be defined. Two definitions are particularly common, for reasons we'll soon see:

```
CONST EMPTY = LowSetTYPE{ };
      FULL = LowSetTYPE{1..12};
```

Notice that it is necessary to mention the type name of the set constant. Admittedly, it seems silly to say that an empty set, with no values, has a type. Nevertheless, because it helps the compiler run as quickly and correctly as possible, it is an inconvenience we'll endure.

Self-Check Questions

Q. Which of these are illegal definitions. Why?

a) TruthSetTYPE = **SET OF** BOOLEAN;
b) IntegerSetTYPE = **SET OF** INTEGER;
c) VowelSetTYPE = **SET OF** {"a", "e", "i", "o", "u"};

A. Definition *a* is perfectly legal, since it creates a set of the BOOLEAN values TRUE and FALSE. Definition *b*, in contrast, might or might not be legal because its cardinality is so large—the full range of INTEGER. It depends on your implementation of Modula-2. Definition *c* is definitely illegal because the group of vowels listed isn't a type or a subrange. Yes, the vowels are a group of values; no, they're not a set in Modula-2.

Q. A set's *powerset* is the collection of different sets that can be created from a given base type. Suppose that a set's base type contains the numbers 1, 2, and 3. How many sets (including the empty set, or set of no values) are in its powerset? What are they?

A. As a rule, if the cardinality of a set's base type is *n*, its powerset has 2^n different values. This is because each element may, or may not, be represented. The three-element set described above has eight sets in its powerset:

1	1, 2	1, 3	1, 2, 3
2	2, 3	3	*the empty set*

Set Assignment

Set variables with the same type can be assigned to each other by name. However, assignments that involve individual elements are more complicated.

labeling elements

> In set assignments, individually named elements of the set's base type must go between braces, sometimes called curly brackets: { }. To help type checking, a second requirement also applies: a value in braces has to be preceded by the name of its base type.

As usual, two dots (..) mean 'through and including':

set assignments

```
OutputCharacters := CharSetTYPE{"a".."z", "A".."Z", "0".."9"};
FruitVitamins := VitaSetTYPE{A..B3, B12, C, E};
VegetableVitamins := FruitVitamins;
InputCharacters := CharSetTYPE{SomeCharacterValue};
OutputCharacters := CharSetTYPE{VAL(CHAR,74)};
InputCharacters := OutputCharacters;
Responses := LowSetTYPE{ };
```

The final assignment is unusual because it makes Responses an empty set that contains no values at all. Such assignments are as necessary to sets as initializing assignments to 0 might be for INTEGER variables. Note that Responses = LowSetTYPE{ } is a reasonable BOOLEAN expression. It's TRUE if Responses has no members, and FALSE otherwise.

I don't want to dwell on details, but it's important to be able to distinguish between values that have a set type, and values that (coincidentally, perhaps) share a set's base type. Suppose that SomeVitamin is a variable with ordinal type VitaminTYPE. In that case:

set type vs. base type

SomeVitamin	has type VitaminTYPE
B12	has type VitaminTYPE
VitaSetTYPE{B12}	has type VitaSetTYPE
VitaSetTYPE{SomeVitamin}	has type VitaSetTYPE
FruitVitamins	has type VitaSetTYPE

set order

A final point about set values is that they're unordered. Thus, these are equivalent representations of the same set:

CharSetTYPE{"A".."E", "X"}   CharSetTYPE{"X", "B".."D", "E", "A"}

Self-Check Questions

Q. Which of these sets have the same value? Are any of them illegal representations of sets? Assume that LowTYPE consists of the numbers 1 through 10.

a) LowTYPE{1..10}	*b*) LowTYPE{1..5, 6..9, 10}
c) LowTYPE{1..3, "4", 5..10}	*d*) LowTYPE{1..5, 3, 5..10}
e) LowTYPE{10..6, 5..1}	*f*) LowTYPE{10, 9, 8, 7, 6, 5, 4, 3, 2, 1}
g) LowTYPE{0..10}	*h*) LowTYPE{1..5, 1..9}

A. Expressions *a*, *b*, *d*, and *f* are all equivalent, and each represents the entire set. Of the others, only *h* is legal.

BOOLEAN Relations Between Sets

Sets can be inspected or compared in expressions that have BOOLEAN values. The most important of these relations involves the **IN** operator. The expression:

value **IN** *set expression*

is TRUE if a *value* is one of the elements of *set expression*, and FALSE otherwise.

Suppose, for example, that FirstCh, SecondCh, and ThirdCh are all values of type CHAR, while Grades is of the CharSetTYPE (base type CHAR) I defined earlier. These are all legal expressions that have BOOLEAN values TRUE or FALSE:

using **IN**

"X" **IN** CharSetTYPE{FirstCh..SecondCh}
FirstCh **IN** CharSetTYPE{SecondCh, ThirdCh}
SecondCh **IN** CharSetTYPE{"a"..FirstCh}
"A" **IN** Grades

All of the regular relational operators can be given set operands. They're summarized below, along with a TRUE example expression that relies on LowSetTYPE as defined earlier.

set relations

Sign	*Operation*	*A* TRUE *Example*
=	set equality	LowSetTYPE{1..3} = LowSetTYPE{1, 2, 3}
<>	set inequality	LowSetTYPE{2..5} <> LowSetTYPE{ }
>=	set 'contains'	LowSetTYPE{1..10} >= LowSetTYPE{2..9}
<=	set 'is contained by'	LowSetTYPE{5} <= LowSetTYPE{5, 7..9}
IN	set membership	7 **IN** LowSetTYPE{3..11}

As an additional example, suppose that WordLetters has the set type CharSetTYPE defined earlier. The first expression below is TRUE if the WordLetters set includes all the vowels (a vexatious problem without sets), while the second is a barefaced lie unless WordLetters includes the letters "a" through "f".

```
(*Can you think of any words for which these are TRUE?*)
WordLetters >= CharSetTYPE{"a","e","i","o","u"}
CharSetTYPE{"a".."f"} <= WordLetters
```

Self-Check Questions

Q. What words *do* contain all the vowels? How about the letters "a" through "f"?

A. It's obvious that 'vexatious' and 'barefaced' comply. But this is an ultraconservative, hardly authoritative list. More feedback?

Set Operators

New set-valued expressions can be constructed with the *set operators* for union, intersection, difference, and symmetric difference. (Two built-in procedures, INCL and EXCL, are discussed with BITSETs in section 18-2.)

union

> The *union*, +, of two sets (or of two representations of Modula-2 sets) is a set that contains *all* the members of both sets: SetA + SetB.

For example, suppose that the variables below all have CharSetTYPE. Some legitimate set unions are:

```
IncludedLetters := IncludedLetters + OutputCharacters;
IncludedLetters := MissingLetters +
                   OutputCharacters + CharSetTYPE{"D".."T"}
```

intersection

> The *intersection*, *, of two sets is a set that contains all values that belong to both sets: SetA * SetB.

If two sets don't contain any common values their intersection is, of course, the empty set. For example, assume that we've defined the months as an ordinal type, and MonthSetTYPE as a set with a month-valued base type. The value of this set expression:

```
MonthSetTYPE{January..June, August} * MonthSetTYPE{May..September}
```

has three elements—it is MonthSetTYPE{May, June, August}. The intersection of MonthSetTYPE{January..May} and MonthSetTYPE{July..November} is the empty set MonthSetTYPE{ }.

difference

The *difference*, –, of two sets is a set that contains all the members of the first set that are not also members of the second set: SetA – SetB.

difference of empty sets

Can we try to remove a value that's not included in a set variable? Certainly. The second assignment below is fruitless, but legal:

```
MissingLettters := CharSetTYPE{ };  (*MissingLetters is now an empty set.*)
MissingLetters := MissingLetters – CharSetTYPE{"A".."Z"}
```

symmetric difference

Finally, the *symmetric difference*, /, of two sets is the set of values in one set or the other, but not both—their sum, minus their intersection: SetA / SetB.

Symmetric set difference is, more or less, the inverse of set intersection. Two sets' symmetric difference is the set of values held by either set, but not by both. Relying on the MonthSetTYPE examples used above, we have:

MonthSetTYPE{January..June, August} / MonthSetTYPE{May..September}
equals MonthSetTYPE{January..April, July, September}

Since MonthSetTYPE{January..May} and MonthSetTYPE{July..November} have no overlapping membership, their symmetric difference equals their union: MonthSetTYPE{January..May, July..November}. Thus, whereas intersection is useful for finding elements common to two sets (the golden months in which oysters can be eaten with blueberries, perhaps), symmetric set difference finds elements that aren't paired.

Remember that if part of an expression is made of individual set elements, the elements must go between braces, and be preceded by their type names:

elements go between braces

Grades *is a set variable, so it's unadorned.*
CharSetTYPE{"A".."C"} *must include braces.*
Grades * Grades *is an expression made from sets.*
Grades + CharSetTYPE{"E"} *is also an expression made from sets.*

Self-Check Questions

Q. Consider these definitions:

1. A *proper subset* of a set must contain fewer elements than the whole set.
2. Two *disjoint sets* cannot contain any elements in common.

Write expressions that are TRUE if *a*) SetOne is a proper subset of WholeSet, and *b*) SetOne and SetTwo are disjoint. Assume that these sets have type SetTYPE.

A. Are these the only expressions that solve the problem?

a) (SetOne <= WholeSet) **AND** (SetOne <> WholeSet)
b) (SetOne * SetTwo) = SetTYPE{ }

Q. A set's *complement* consists of the values that have the same base type, but aren't in a particular set. Thus, the complement of the empty set is a 'full' set and vice versa.

Suppose that Letters represents part or all of the set of lower-case letters (call it LowerSetTYPE). How could you change the value of Letters to its complement?

A. It's easy—just subtract the current value of Letters from the full set:

```
Letters := LowerSetTYPE{"a".."z"} – Letters;
```

Section 15-2 Standard Practices

Let's look at some examples of the set operators for union, intersection, and so on. Union and difference register the appearance—but not the number of appearances—of a set's elements. For instance:

problem: appearance of letters

> Write a program that reads a text sample, then lets us know which lower-case letters *do* appear, and which upper-case letters *don't* appear.

It's hard to avoid recognizing that the data abstraction a solution is based on bears a suspicious resemblance to a set. Suppose that we begin with two sets—one holds the upper-case letters, and the other is empty.

The set type turns out to be convenient as an underlying type for this abstraction for a simple reason: we don't have to bother checking the character before adding to or deleting from a set. If a lower-case character has already been saved, saving it again won't hurt. Similarly, if an upper-case character has been removed, it can be removed again without error. A specification for the program's main loop might be:

loop specification

Goal: Produce sets of used lower-case and unused upper-case letters.
Bound: All the letters in the sample have been seen.
Plan: Read a letter.
Add it to a set of used lower-case letters.
Delete it from set of used upper-case letters.

Even though our abstraction relies on the properties and operations of a built-in type, we still require primitives:

primitives operations

- *Initialization*: the set variables are given meaningful initial values—IncludedLetters starts out as an empty set, while MissingLetters initially contains every capital letter.
- *Update*: the union and difference operators are used to update the set variables as values are read. As I mentioned, it's no error to remove (or add) a value more than once.
- *Inspection*: a **FOR** loop is needed to inspect the set variable element-by-element. Note that the same loops could be used to establish the final cardinality, or number of elements, of IncludedLetters and MissingLetters.

In pseudocode, we:

initialize the sets;
LOOP
    read, then update the sets;
inspect the sets;

The Modula-2 program is shown below. Note in particular its use of constants for set initialization.

set union/difference program

```
MODULE Tracker;
   (*Uses sets to find which letters do and don't appear in a text sample.*)
IMPORT IO;
TYPE CharSetTYPE = SET OF CHAR;
CONST EMPTY = CharSetTYPE{ };
      CAPITALS = CharSetTYPE{"A".."Z"};
VAR Current: CHAR;
    IncludedLetters, MissingLetters: CharSetTYPE;
BEGIN
   IO.StringOut ("Type in a sample of your wide-ranging vocabulary.");
   IO.LineOut;
   (*Initialization *)
   IncludedLetters := EMPTY;
   MissingLetters := CAPITALS;
   (*Update*)
   LOOP
      IO.CharIn (Current);
      (**)IF Current = IO.EOL THEN EXIT END;
      IncludedLetters := IncludedLetters + CharSetTYPE{Current};
      MissingLetters := MissingLetters - CharSetTYPE{Current}
   END;
   (*Inspection*)
   IO.StringOut ("Lower-case letters included were:");  IO.LineOut;
   FOR Current := "a" TO "z" DO
      IF Current IN IncludedLetters THEN
         IO.CharOut (Current)
      END
   END; (*FOR*)
   IO.LineOut;
   (*Inspection*)
   IO.StringOut ("Upper-case letters not included were:");  IO.LineOut;
   FOR Current := "A" TO "Z" DO
      IF Current IN MissingLetters THEN
         IO.CharOut (Current)
      END
   END; (*FOR*)
END Tracker.
```

(output follows)

```
Type in a sample of your wide-ranging vocabulary.
Pack My Box With Five Dozen Quick Brown Foxes
Lower-case letters included were:
acehiknorstuvwxyz
Upper-case letters not included were:
ACEGHIJKLNORSTUVXYZ
```

Modification Problems

1. Although it isn't necessary, change Tracker so that it checks for each letter's current presence or absence before adding it to, or deleting it from, a set.
2. Suppose that your system doesn't support sets with the cardinality of CHAR—perhaps a set can only contain sixteen values. Modify Tracker to deal with this little annoyance.
3. Create a procedure SetOut that prints the contents of its CharSetTYPE argument. Are any other arguments (like a particular range) useful?

Sets as Flag Variables

One standard application of sets is to use them as flag variables.

> State variables represent program conditions. A *flag variable* is a state variable that can represent more than one state at a time.

For example, consider these definitions:

```
TYPE OptionTYPE = (ErrorRecovery, InputChecks,
                   OutputChecks, Testing, LongMessages);
     OptionSetTYPE = SET OF OptionTYPE;
VAR AllOptions, TestOptions: OptionSetTYPE;
```

A program might begin by initializing AllOptions and TestOptions:

```
AllOptions := OptionSetTYPE{ErrorRecovery..LongMessages};
TestOptions := OptionSetTYPE{InputChecks, Testing};
```

After being initialized, AllOptions contains every value of the ordinal type OptionTYPE—every flag is set. We can reduce its membership by using the set difference operator:

```
AllOptions := AllOptions – TestOptions;
 (*AllOptions is OptionSetTYPE{ErrorRecovery, OutputChecks, LongMessages} *)
AllOptions := AllOptions – OptionSetTYPE{ErrorRecovery..OutputChecks};
 (*AllOptions is OptionSetTYPE{LongMessages} *)
```

Flag variables can be put to some unexpected tasks. For example, most of our programs so far are written expressly to move from state to state. Ordinary state variables suffice for such programs because they only have to represent one state at a time—usually, a clue to the next action we want to take.

But imagine using states in a passive, rather than driving, role. On a coarse scale, flag variables can provide a snapshot of program states as control passes from procedure to procedure. On a finer scale, they can record the states a program has gone through, or describe the legal states a program may attain.

using flags

```
TYPE FeatureTYPE = (WarpSpeep, Lasers, ... );
     FeatureSetTYPE = SET OF FeatureTYPE;
CONST EXPERT = FeatureSetTYPE{ ... };
      AVERAGE = FeatureSetTYPE{ ... };
      NOVICE = FeatureSetTYPE{ ... };
VAR PlayerStatus,        (*e.g. EXPERT or NOVICE, to limit the player*)
    FeaturesUsed: FeatureSetTYPE;
```

During program execution the value of PlayerStatus can limit the commands permitted a player, while FeaturesUsed can help with an end-of-game wrap up.

Section 15-3 Extended Examples

Sets lack the close parallels to real-life data structuring offered by arrays, files, and records. Nevertheless, by focusing first on the set's data holding ability, and then on its operators, we can begin to see the unique contributions sets make to Modula-2 programming.

QWERTY vs. MALTRON Keyboards

Let's use sets to compare the efficiency of two popular typewriter keyboards. The *QWERTY* keyboard, shown below, is the current standard.

QWERTY keyboard

```
Q  W  E  R  T  Y  U  I  O  P
  A  S  D  F  G  H  J  K  L  ;  :  space
Z  X  C  V  B  N  M  .
```

Unfortunately, few of the most frequently used letters (e, t, a, o, n, r, i, and s) appear on the center 'home' row—the row of keys that a typist's fingers normally rest on. As you might expect, continually jumping from one row to another slows and tires even the best typists.

Many new keyboard designs have been proposed. The Linotype machine (patented in 1885, and until recently the standard device for setting type) had as its home row the well-known phrase ETAOIN SHRDLU. The common letters were put up front (despite the inconvenience of using the left little finger for e, the commonest of all), in order to keep the machine's mechanical linkages short and secure.

A more modern keyboard, which some computer terminals can be programmed to use, is the *Maltron* system, shown below. Since the most common characters are in the home row, few jumps are required while typing.

Maltron keyboard

```
  Q  P  Y  C  B  V  M  U  Z  L
A  N  I  S  F  E  D  T  H  O  R  ;  :  .  space
  J  G  W  K  X
```

Just how beneficial is the Maltron keyboard? Suppose we write a program that compares the number of jumps a text sample would require from QWERTY and Maltron typists. The central loop will keep track of which categories each character falls into:

loop specification

Goal: Count the jumps a particular text sample will require.
Bound: The end of the text sample.
Plan: Read a character.
Adjust the counts, as appropriate.

why use sets?

Sets are the data type of choice because we do have to hold a collection of values, but don't have to individually count their exact numbers. If we define CharSetTYPE as a **SET OF** CHAR, we can declare variables that represent the 'home' and 'others' characters of the keyboards above. A variable—call it Valid, of type CharSetTYPE—will help restrict the characters we consider, so that extraneous characters from the input file (like end-of-line) don't affect our count.

What kind of conclusions should the program arrive at? Naturally we want to count the number of jumps that are made. However, this information isn't particularly helpful unless we know the total count of characters considered. The pseudocode below includes these refinements.

pseudocode refinement

get a file of input data ready;
initialize the Valid, *'home' row, and 'other' row set variables;*
LOOP
    read a character;
    IF *there wasn't one* **EXIT**;
    IF *it's in the* Valid *set of characters*
        increment the Total *count;*
        count any jumps from the QWERTY or Maltron home row
    END *the* **IF** *statement*
END *of the loop;*
print the number of jumps for each keyboard, and the size of the sample

The completed program is shown below. As input, I gave it the same data we used to count letter pairs in Chapter 10.

typewriter keyboard program

```
MODULE KeyTypes;
  (*Compares the jumps required by QWERTY and Maltron keyboards.*)
IMPORT IO;
TYPE CharSetTYPE = SET OF CHAR;
VAR QWERTYHome, MaltronHome, QWERTYOthers,
                              MaltronOthers, Valid: CharSetTYPE;
   QWERTYJumps, MaltronJumps, Total: INTEGER;
   Current: CHAR;
BEGIN
  Valid := CharSetTYPE{"a".."z", "A".."Z", ":",";",".", " "};
  QWERTYHome := CharSetTYPE{"a","s","d","f","g","h","j","k","l",";","A",
                  "S","D","F","G","H","J","K","L",":"," "};
  MaltronHome := CharSetTYPE{"a","n","i","s","f","e",".","d","t","h","o","r",";",
                  "A","N","I","S","F","E","D","T","H","O","R",":"," "};
  QWERTYOthers := CharSetTYPE{"a".."z", "A".."Z"} - QWERTYHome;
  MaltronOthers := CharSetTYPE{"a".."z", "A".."Z"} - MaltronHome;
  QWERTYJumps := 0;
  MaltronJumps := 0;
  Total := 0;
  IO.OpenInput ("Chapter.10");
  IF NOT IO.Done() THEN
    IO.StringOut ("Couldn't open file 'Chapter.10'. Quitting.");
    HALT
  END;
  LOOP  (*until the end of data*)
    IO.CharIn (Current);
    (**)IF NOT IO.Done() THEN EXIT END;
    IF Current IN Valid THEN
      Total := Total + 1;
      IF Current IN QWERTYOthers THEN
        QWERTYJumps := QWERTYJumps + 1
      END;
      IF Current IN MaltronOthers THEN
        MaltronJumps := MaltronJumps + 1
      END  (*counting the jumps*)
    END  (*counting the Valid characters*)
  END; (*LOOP*)
  IO.CloseInput; (*tidy up*)
  IO.StringOut ("Total input was:   ");  IO.IntOut (Total, 1);
  IO.LineOut;
  IO.StringOut ("QWERTY jumps:   ");  IO.IntOut (QWERTYJumps, 1);
  IO.LineOut;
  IO.StringOut ("Maltron jumps:   ");  IO.IntOut (MaltronJumps, 1);
  IO.LineOut
END KeyTypes.
```

(output follows)

```
Total input was:  22759
QWERTY jumps:  12777
Maltron jumps:  1386
```

Modification Problems

1. Modify program KeyTypes so that it performs the same of check for the home row (ETAOIN SHRDLU) of a Linotype machine.

Sets as 'Parallel' Variables

So far, we've relied on sets as a convenient alternative to other data types. But if we want, we can take a radically different view by focusing on the set operators. Instead of seeing them as the primitives of the data abstraction we call a set, we can look at them as the primitives of an as yet unnamed abstraction for *parallel processing*, or performing different calculations simultaneously.

Now, thanks to the lingering remnants of New Math, basic set concepts have been drilled into you since the third or fourth grade. A set union or intersection probably doesn't strike you as remarkable. However, these are really abstract names for parallel operations. Since we have a language for working on sets as whole collections, rather than as individual elements, we're implicitly performing operations on elements in parallel.

Suppose we were to use an array of BOOLEAN values as a substitute for a set. Although we could still achieve the same effects—we could, say, determine the complement of such an array, or find the common elements of two arrays—we'd have to program the operations component-by-component. In contrast, the set operators add, subtract, or compare elements wholesale *without additional programming*.

Thus, sets let us think in terms of parallel operations. Not only does this lead to more easily written programs, but some kinds of problems turn out to have elegant solutions that we wouldn't ordinarily think of. Let's look very briefly at two examples: appointment scheduling, and chessboard management.

How are sets useful in scheduling appointments? Consider these:

problem: scheduling appointments

```
CONST FULL = HourSetTYPE{1..16};
      EMPTY = HourSetTYPE{ };
      LIMIT = 25;  (*number of employees*)
TYPE WorkHoursTYPE = [1..16];
     HourSetTYPE = SET OF WorkHoursTYPE;
     EmployeeTYPE = ARRAY [1..LIMIT] OF HourSetTYPE;
VAR CEO, VIP,  (*Chief executive officer, very important person.*)
    Free, Booked, Possible: HourSetTYPE;
    Staff: EmployeeTYPE;
```

As hard-working power tools, CEO and VIP have their hours numbered (1 through 16, that is). We can begin by assigning previously scheduled appointments:

initializing the calendars

```
CEO := HourSetTYPE{1, 3..5};
VIP := HourSetTYPE{3, 6};
```

Recall that the complement of a set is the set of elements it *doesn't* contain. In Modula-2, a set's complement is found by subtracting the set from a variable or constant that holds all the elements. We can use complements to determine CEO's unscheduled time, then arrange a meeting with VIP:

CEO meets VIP

```
(*The complement of CEO yields her free hours.}
Free := FULL - CEO;
(*When can CEO and VIP meet? At the intersection.*)
Possible := Free * (FULL - VIP);
```

All hours are checked at once.

The same technique applies to finding a time when none of the Staff are busy: complement their individual schedules, and see if the intersection that results is EMPTY or not. In pseudocode:

```
put every hour in Possible;
FOR every staff person
    remove her unavailable hours from the Possible pool;
see if any hours are left
```

As you read the Modula-2 version, below, don't forget that each component of the Staff array is a set that holds hours in which meetings are scheduled.

finding free Staff time

```
Possible := FULL;
FOR i := 1 TO LIMIT DO
    Possible := Possible * (FULL - Staff[i])
END; (*Possible holds the intersection of all the complements.*)
IF Possible = EMPTY
    THEN IO.StringOut ("A meeting is impossible.")
    ELSE IO.StringOut ("Potential meeting times are:    ");
        FOR i := 1 TO 16 DO
            IF i IN Possible THEN IO.IntOut(i,3) END
        END (*FOR*)
END (*IF*)
```

Sets in Chess

A second group of examples that take advantage of parallel set operations are found on a chessboard. Indeed, set operations are crucial to the success of some kinds of computer chess programs, which may have to analyze millions of positions in a limited time. Suppose that we were to number the sixty-four squares of a board, and define a set on the basis of this base type. A number of possibilities present themselves:

- Suppose we create a set of 'legal-move' squares for each white piece. The union of the white-piece legal-move sets is a set of squares that are potentially under attack.
- Suppose we create a set of positions occupied by black pieces. The intersection of the black 'occupied' set, and the 'legal-move' union of white sets is the set of currently endangered black pieces.
- Suppose we create a set of 'legal-move' positions for each of the black pieces. A set of safe moves for any black piece is the black legal-move set minus the white legal-move union.
- How might black spot a trap, e.g. white is willing to give up a small piece if it can take a larger piece in return? A potential trap exists if the intersection between one black piece's legal-move set, and the white 'occupied' set, shares one or more elements with the white legal-move union.

basing chess strategy on sets

Note that the actual computer operations that underly the set operators might, or might not, really execute in parallel. In fact, some of the fastest chess playing machines require custom-built hardware for true parallel execution. Nevertheless, we can use the set operators to describe the analysis of the board in terms of single operations. The abstract ideas of set relations, reified by a concrete package of set operators, let us discuss hard ideas in simple terms.

Additional discussion of sets and the predefined type BITSET *are found in Chapter 18.*

Modula-2 Details

Syntax Summary

- set: the underlying type of a variable that can represent more than one value—the set's *elements*—of an ordinal type. The set's *base type* must be a subrange or enumerated type, and its *cardinality* (size) may be limited by your Modula-2 implementation:

```
TYPE BaseTYPE = ["a".."e"];
     SetTYPE = SET OF BaseTYPE;
     CharSetTYPE = SET OF CHAR;
VAR Few: BaseTYPE;
    Lots, Many: CharSetTYPE;
```

- set constants: may be defined as well:

```
CONST EMPTY = BaseTYPE{};
      FULL = BaseTYPE{"a".."e"};
```

- set values: can be constructed from set elements. They must go between curly brackets, and are preceded by the set's type name (to help type checking). Two periods '..' mean 'through and including.'

```
Few := BaseTYPE{"b".."d"};
Lots := CharSetTYPE{"b".."d"};
Many := Lots; (*a set variable doesn't need curly brackets*)
Lots := CharSetTYPE{"a", "c".."e"};
```

- set operators: take set-valued operands and return set values. They are:

+	*union — the merger of two sets*
*	*intersection — the values common to two sets*
–	*difference — the 'subtraction' of set members*
/	*symmetric difference — values in either set, but not both sets*

- set relations: the BOOLEAN operators as applied to set operands. They are:

Sign	*Operation*	*A* TRUE *Example*
=	set equality	LowSetTYPE{1..3} = LowSetTYPE{1, 2, 3}
<>	set inequality	LowSetTYPE{2..5} <> LowSetTYPE{ }
>=	set 'contains'	LowSetTYPE{1..10} >= LowSetTYPE{2..9}
<=	set 'is contained by'	LowSetTYPE{5} <= LowSetTYPE{5, 7..9}
IN	set membership	7 **IN** LowSetTYPE{3..11}

- set procedures: add and remove elements from sets. In each case, Value may be a variable or other expression as long as it has the proper base type. (These procedures weren't used in this chapter; they are are described and employed in Chapter 18.)

INCL (SetVariable, Value);	*add* Value *to* SetVariable
EXCL (SetVariable, Value);	*remove* Value *from* SetVariable

Modula-2 Details

Remember . . .

- A set's *cardinality* is the number of values in its base type—the maximum number of elements a set can have. Maximum set cardinality may be limited by every Modula-2 implementation; in particular, a set of CHAR values might not be allowed.
- A set's *powerset* is the group of different sets that can be formed from a given base type. It always includes the *empty set* of no values.
- A set variable's *complement* is the set of values the variable *doesn't* contain.
- *Flags* are state variables—they let us know something about a program's current condition. Because they can represent more than one value, sets are particularly useful for implementing flags.
- By default, a set expression in curly brackets that isn't preceded by a type name is of type BITSET.

Zen Study Guide
Big Ideas

The Big Idea in this chapter is:

1. *Sets are useful because they take care of bookkeeping automatically.* Anything that can be done with a set of values *could* also be done with an array. However, the set operators manage in one step what might otherwise require a whole procedureful of code.

Zen Study Guide
Questions

15-1 In real-world programming practice, sets usually don't show up until fairly large programs are at hand. As a result, most of the set questions you encounter will be relatively technical. Begin to study by presenting a half-dozen set-type definitions, and asking *Which of these are legal?* As an alternative, think up a half-dozen categories (rap groups, say, or vegetables) for which you might reasonably be asked *Define these as set types*.

15-2 Set-valued expressions are also a good ground for *Which are legal?* or *Which expressions are the same?* questions. Try to confuse the issue by mixing up set-valued expressions with BOOLEAN expressions that use set operands.

15-3 Although Modula-2 provides a variety of operations on sets, output of set contents is not among them. However, if I were a student, I would certainly expect to be asked to write a procedure that takes a set and prints its contents. Take three of the sets defined in this chapter, and propose *Write an output primitive* questions.

15-4 A similar question would ask for a primitive that counts a set's elements. Again, pose the problem with three sets from the chapter.

15-5 Given the details of set syntax, *Where's the bug?* should be an easy problem to develop. But it's important to test understanding of set semantics, too. Make up three *Where's the bug?* test segments that err because they initialize sets improperly (or not at all), or because they use the wrong set operators or relations.

15-6 Certainly you can define the term *powerset*. But can you anticipate a question that asks you to list a set's powerset? By providing simple set-type definitions, pose two such questions. Just how complicated could such a set reasonably be?

15-7 Program Tracker presents a typical set application—tracking the appearance of letters. However, reading random amounts of input looking for upper- or lower-case letters doesn't require much imagination. Instead, why not see how many words must be read before each letter has been seen, or how many characters can be read before a duplicate can be found? Make up three more questions of this ilk.

15-8 Can you define any other kinds of sets (like the set of digit characters, or of punctuation marks) for a search? Ask two questions that ask you to *a*) count appearances of values in two categories, and *b*) print the category of values as they appear.

Exercises
Warmups and Test Programs

15-9 What is the maximum set cardinality your Modula-2 system allows? Is a set of CHAR legal?

15-10 A set of sixteen values can be thought of as the equivalent of an array of sixteen BOOLEAN components. Making every array component TRUE is, more or less, like putting every potential element into a set variable. How would you compare the amount of time the two kinds of assignments take? How does the set assignment compare to a 'complete array' assignment, i.e. initializing the array by assigning it another, already initialized array?

15-11 What error messages are caused by neglecting to specify a set's type in an expression? By using the wrong kind of brackets? By using the wrong type of values as elements (e.g. using single-digit INTEGER values instead of the digit CHAR values)?

15-12 Define an ordinal type that represents the months. Then, define a set that can hold MonthTYPE values. Show how to initialize a set variable to *a*) months that have 31 days, *b*) months that have 30 days, *c*) months that have 28 days, *d*) months that don't have the letter 'r' in them.

15-13 Let us imagine that Movers, Groovers, Quakers, Shakers, Lovers, and Fighters are all set variables that represent various constituencies of the base type PopulationTYPE. Write expressions that represent *a*) the entire group, *b*) the Movers and Groovers, *c*) the Lovers who are not Fighters, *d*) the Shakers who are Quakers but not Groovers, *e*) the Fighters who are not Movers, and Movers who aren't Fighters, and *f*) the Groovers and Lovers who are also Shakers.

15-14 Suppose that you can, in fact, create sets of CHAR. Write procedures that take a set parameter of such a type and let you know:

a) the smallest value in the set;

b) the largest value in the set;

c) the cardinality of the set;

d) the set's contents; and

e) the set's complement.

15-15 Suppose that SingleSetTYPE variables can hold the single-digit numbers 0 through 9. On the assumption that you have two SingleSetTYPE variables, write BOOLEAN functions that let you know if:

a) one set is a proper subset of the other;

b) two sets, between them, have all the elements of the sets' base type; or if

c) two sets, between them, have all the elements of the sets' base type without any duplication.

15-16 Suppose that sets were not part of Modula-2. How would you go about duplicating the set abstraction with arrays of BOOLEAN components? Outline procedures that would allow the basic set operations (union, difference, and so on) on array-type operands.

Exercises

Programming Problems

15-17 Write a program that reads a text sample, and prints every word that contains all the vowels. If you have access to an on-line dictionary, use that as data. If not, try:

> 'If an authoritative argument for the augmentation of coeducational boardinghouses is to be communicated, dialogue to exhaustion is often necessary, since vexatious ultraconservatives take tenacious precautions against the refutation of their reputations. They consider sensible feedback to be simultaneously sacrilegious and revolutionary since, after all, barefaced oneupmanship is an ostentatious form of persuasion.'

Can you modify your program to find two words (included) that contain the letters "a", "b", "c", "d", "e", and "f"?

15-18 Write a program that reads three lines of text and prints:

a) the vowels that appear on all three lines;

b) the consonants that appear on the first two lines;

c) the letters that appear on the first line.

15-19 A freshman programming student invents a simple way to shuffle a deck of cards—she defines a set of the INTEGER values 1..52, then chooses a random number between 1 and 52. Each time she picks a number, she removes it from the set, and adds it to an array. Should she ever pick a number that has already been chosen, she simply picks again.

How long will it take her to deal two ten-card hands using this method? How long do you think it will take her to shuffle 52 cards?

15-20 Modula-2 sets are useful because of the operators defined for set expressions. Sets in general, though, are limited by a basic concept—that a set only contains one instance of a given member. Suppose we correct this deficiency by defining a *bag* as a set that can hold up to five instances of each value.

Write a program that uses an array of sets to implement a bag of the INTEGER subset 1..25. Write procedures that make use of the set operators to do bag addition, subtraction, and intersection.

15-21 For a text-oriented version of the bag problem, above, define a bag of characters. How many sets are required for the bag that contains this input sample (ignoring punctuation and capitalization)? What are the contents of each set?

```
Why jog exquisite bulk, fond crazy vamp,
Daft buxom jonquil, zephyr's gawky vice?
Guy fed by work, quiz Jove's xanthic lamp—
Zow!  Qualms by deja vu gyp fox-kin thrice.
```

15-22 Yet another text-oriented bag problem. Two anagrams contain the exact same bag of letters (ignoring spaces and punctuation). Write a program that checks anagrams. As sample input, try:

```
multiple-word anagrams / plague raw mortal minds
Donald Ervin Knuth / hunt, drink, and love / halt unkind vendor!
Ronald Wilson Reagan / No, darlings—no ERA law.
```

15-23 A group of twenty faculty members at an unnamed university have spent a year engaged in mutual backscratching. They have agreed that whenever one faculty member acknowledges (in print) the inspiration or encouragement of another, the second must return the favor.

Near the end of the year, the faculty gather to total up their accounts. Prof. A notes that she has thanked Profs. E, G, M, and N; Prof. B records that she has acknowledged Profs. A, G, and R, and so on. Clearly some accounts are unbalanced—although B has thanked A, A has not thanked B for something else.

Write a program that helps the faculty determine who owes whom. Assume that input consists of twenty sets (for Professors A through T) of names (A through T again). Hint: use an **ARRAY** ["A".."T"] of appropriate sets.

15-24 Two engineers are comparing the merits of a number of building materials. They are having a difficult time of it because the tests are all equally important, and cannot be rank-ordered in any way. Worse yet, in any given test, each material either passed or failed.

Write a program that reads the results of ten tests on five materials. Create, for each material, a set of the tests it has passed. Write primitives that determine:

a) if any material has passed every test *all the others* have;

b) if any material has passed all the tests that *any single other* has, plus at least one more;

c) if any materials have passed none of the same tests;

d) if any materials have identical test results.

15-25 Here's a problem that's a little more computer science oriented. A *lexical analyzer* is employed in the first phase of preparing a program for compilation. It scans through the characters and words that form a program, and tries to determine what kind of *token* each one is.

Write a program that is able to recognize the tokens that appear in an assignment statement. Assume that:

a) *integers* are any sequence of digit characters;

b) *operators* are any of +, –, *, /;

c) *identifiers* are any sequence of letters;

d) *brackets* are (and),

e) *assignment* is :=.

Your program should echo, and print the type of, each token. For example, if input is '**`Number := ( 52 + 3 )`**' your output should be:

```
Number  identifier
:=  assignment
(  bracket
52  integer
+  operator
3  integer
)  bracket
```

15-26 *Wheel of Misfortune* is one of the many television shows that would be available if I were in charge. The rules are simple: an equation is shown, with blank lights taking the place of the factors of each term. The players take turns calling out digits; any correct numbers are lit up. The first player to guess the contents of the equation wins.

For example, the equation 27 + 23 = 2 * 25 would initially be shown as:

□□ + □□ = □ * □□

After the guess '2' it would be:

2□ + 2□ = 2 * 2□

A guess of '5' would yield:

2□ + 2□ = 2 * 25

Unfortunately, a numerically oriented Vanna White is not available. Write a program that takes her place.

15-27 Write a program that plays Tic-Tac-Toe. Number the boxes, then use one set to hold X choices, and another to hold the O's. Naturally, an array of sets should be used to hold, and check, winning positions.

Photo by Douglas Blau, TOPPS Chewing Gum, Inc.

. . . we've managed to describe a stack object without committing ourselves to a given underlying representation. We might consider several alternatives...' (p. 608)

Chapter 16 Modules with Structure and State

When you think about buying software, what comes to mind? *Leisure Suit Larry*? *Slime Demon Attack*? If so, you're thinking like a consumer, which is to say 'If it ain't plug and play, don't confuse me!'

But when programmers buy software, they usually buy tools. Instead of getting new systems, they buy pieces that help make existing systems better. And, for Modula-2 programmers, those pieces come in libraries.

Chapter 16 describes some of the strange and wonderful things that libraries can contain. 16-1 talks about libraries that make abstract *structures* available. 16-2 goes further afield, and looks at modules that maintain *data* on behalf of a program. The idea of going to the mall to buy a sorted list, a random number generator, or a source of the digits of π, may strike you as peculiar, but if you look at a magazine that specializes in computer programming (let me recommend *Computer Language* and *Dr. Dobb's Journal*) you'll find plenty of advertisements.

Most of the topics of Chapter 16 are beyond the usual limits of an introductory programming course. However, they're interesting and not at all hard to understand. Treat this chapter as a preview of the next course down the line, and enjoy.

Section 16-1

Abstract Structure Modules

LET'S TURN TO THE ABSTRACTION OF DATA structures. Defining a structure, as opposed to a type, means defining a relationship between values, rather than being primarily concerned with the values themselves. Abstracting the structure, in turn, invokes the same criteria we applied to abstracting types.

> A structure is only fully abstracted when we've separated and hidden the names and operations used to manipulate values from any underlying representation used to maintain the structure, or algorithms used to carry out the operations.

As a practical matter, we will cheat slightly, but regularly, by letting the abstract structure implementations rely on a little bit of knowledge—the type name is enough—about the stored values.*

Now, a data type (like Color.T or Cartesian.T) was defined abstractly in terms of its range of values, and the primitive operations allowed on those values. But a structure—a stack, for instance—is defined abstractly in terms of the operations allowed on the structure itself. For example, the abstract definition of a stack is:

stack rules

1. A new value can be added to the stack.
2. The newest of the remaining values on the stack can be retrieved and removed.

The queue is similar, but values are accessed in the reverse order:

queue rules

1. A new value can be added to the queue.
2. The oldest of the remaining values on the queue can be retrieved and removed.

Something very interesting is going on here. I haven't said anything about what kind of values are going to be held. Nor have I given any clue as to how they will be held. Nevertheless, differences in the operation of the stack and queue—one is a LIFO (last-in, first-out) structure, while the other is FIFO (first-in, first-out)—are as distinct as the difference between addition and subtraction.

Let's try a few more abstract definitions. A *ring buffer* holds and allows the inspection of values. It differs from stacks and queues in three ways. First, inspection is non-destructive; values aren't removed from the ring when it's inspected. Second, the ring's size is limited; it's intentionally set up to hold some fixed number of values. Third, the ring buffer does its own housekeeping. If it ever fills up, the oldest stored value is automatically removed to make room for a new value. I'll state the rules like this:

---

* This is why structure abstractions are sometimes called parameterized type abstractions. The type of the data it will hold is, in effect, supplied as an argument to the structure. Usually this magic is accomplished by building structures out of pointers to objects, rather than from the objects themselves.

ring buffer rules

1. The size of the ring can be defined.
2. A new value can be added to the ring.
3. The 'current' value can be inspected. This advances the current pointer to the next most recent value. By definition, the newest value is 'next to' the very oldest. In consequence, inspecting the whole ring returns us to our starting position.

The *set* and the *bag* are also commonly abstracted structures. They're a little different from the previous structures because supplemental, two-structure primitive operations (say, set intersection) are often provided as well. The set rules are:

set rules

1. A value can be added to the set if it isn't already included.
2. A value can be detected if it's in the set.
3. A value can be removed from the set.

A bag differs by letting more than one instance of a value be counted:

bag rules

1. An additional instance of a value can be added to the bag.
2. A value can be detected if it's in the bag.
3. One instance of a value can be removed from the bag.

> I've described the properties of five structures: stack, queue, ring buffer, set, and bag. In each case, I've illustrated the structure's properties without any reference whatsoever to the way it might be represented in Modula-2, or to the actual values it might hold. Instead, the structures are differentiated by their varying interpretations of concepts like *next* and *previous*, *first* and *last*, *oldest* and *newest*, *presence* and *absence*.

Surely abstracting structures cannot be so simple! In fact, abstracting them is really easy. The world of abstraction is a pleasant place, where counters are always correct and uninitialized pointers are ne'er to be found. But when it comes time to reify (make concrete) the abstract terms, more work must be done.

Common Structure Primitives

Each statement of rules so far has been pared down to the barest minimum required to distinguish one structure from the next. Were we to look at actual implementations, though, we'd find a few categories of primitive operations that are usually provided:

construction

- First, we will invariably find *construction* primitives for inserting or removing values from the structure. These change the content of the structure.

inspection

- Second, there may be *inspection* primitives that look at a 'current' value without removing it, or let us know if a particular value is present. These don't change the structure's contents.

iteration or repositioning

- Third, there may be an *iteration* or *repositioning* primitive that lets us change our current access point to the structure. We move around the structure but, again, we don't change it.

status

- Fourth, there are usually *status* primitives—Is it full? Is it empty? Was there an error?—that let us know something about the structure independent of its actual contents.

initialization

- Fifth, a structure may require an *initialization* primitive that prepares the structure to be used. This might have nothing to do with the structure as an abstraction, yet be required by its implementation.

Take a brief look at the structures (stack, ring buffer, and so on) described earlier. Not every primitive is required to support every structure, but the operations that might be carried out with any of them fit easily into the five categories.

The Stack Name Layer — the Definition Module

Given the abstract definition of a structure, we're ready for the second step in the design process—devising the name layer. Our goal is to make structure abstractions as readily available as type abstractions are. They should be *first-class citizens* of the definition world, which means that:

- we should be able to maintain multiple instances of structures, and
- we should be able to name them, and pass them as parameters.

Let's focus on one particular structure—a stack of CHAR values. Taking into account only the abstract definition of a stack, its name layer consists of:

stack name layer, first draft

the type name
a primitive for adding values
a primitive for retrieving values

convenience primitives

This is the bare minimum. However, additional operations are often provided as well for convenience or efficiency. After all, if a module is intended to be generally useful, we might as well include the features that will encourage it to be used. A second draft of a stack's name layer might include:

stack name layer, second draft

the type name
a primitive for adding values
a primitive for retrieving values
a primitive for inspecting the top value without removing it
a primitive that indicates if the stack is empty

In addition, the actual implementation may require primitives that have nothing to do with a structure's abstract definition. On the contrary, they are provided because of limits on abstraction imposed by the real-world considerations of writing Modula-2 code. The third draft of the name layer, below, includes proposed identifiers:

implementation-dependent primitives

stack name layer, third draft

the type name: T
an initialization primitive that prepares the stack for its first use: Initialize
a 'destruction' primitive that disposes of the stack: Remove
a construction primitive for adding values: Push
an inspection primitive for retrieving values: Pop
an inspection primitive for viewing the top value without removing it: Inspect
a status primitive that indicates if the stack is empty: Empty
a status primitive that indicates if the stack is full: Full

We formalize the name layer by laying out a definition module. For our current purposes, it's quite all right to delay the definition of type Stack.T. The particular representation we choose—record, array, pointer, or whatever—is just an implementation issue.

> Because we supply a sufficient set of primitive operations on Stack.T-type structures, we can reasonably expect the programmer to refrain from relying on Stack.T's underlying details.

definition of a stack object module

```
DEFINITION MODULE Stack;
  (*Makes a stack type and primitives available.*)
  TYPE T will be developed later;
  PROCEDURE Inspect (S: T; VAR Ch: CHAR);
    (*Returns, but doesn't pop, the stack's top value.*)
  PROCEDURE Push (VAR S: T; Ch: CHAR);
    (*Adds an item to the stack.*)
  PROCEDURE Pop (VAR S: T; VAR Ch: CHAR);
    (*Removes and returns the stack's top item.*)
  PROCEDURE Empty (S: T) : BOOLEAN;
    (*TRUE if the stack contains no items.*)
  PROCEDURE Full (S: T) : BOOLEAN;
    (*TRUE if an implementation-dependent limit on the stack has been reached.*)
  PROCEDURE Initialize (VAR S: T);
    (*Must be called before the first use of each stack variable.*)
  PROCEDURE Remove (VAR S: T);
    (*A convenient way to dispose of the stack—it might not be needed
     in this implementation (but could be if we used pointers).*)
END Stack.
```

Choosing a Representation

As you can see, we've managed to describe a stack object without committing ourselves to any given underlying representation. We might consider several alternatives for a stack of CHAR values, including:

alternative representations

- an array-based implementation, e.g. an **ARRAY OF** CHAR.

This idea is the simplest, but how will we know the location of the stack's 'top' value? Perhaps we should try:

- a record-based implementation, where each record holds an array (as above), along with a field that locates the current top.

This approach is fine within the usual limitation imposed by arrays—the maximum height of the stack will have to be fixed as part of the array's type definition. If that is unacceptable, we might consider:

- a pointer-based implementation; the stack is a special instance of a linked list.

I've taken the second approach below, and have repeated the definition module without comments.

settling the definition Stack.T

```
DEFINITION MODULE Stack;
   (*Makes a stack type and primitives available.*)
   CONST TOP = 100;  (*the maximum stack height*)
   TYPE DataTYPE = ARRAY [1..TOP] OF CHAR;
(*The next identifier, T, is the only one of these definitions that will be referred
 to by an importing program module. All the rest are expected to remain hidden.*)
      T = RECORD
            Info: DataTYPE;
            Position: INTEGER[0..TOP]
          END;
   PROCEDURE Inspect (S: T; VAR Ch: CHAR);
   PROCEDURE Push (VAR S: T; Ch: CHAR);
   PROCEDURE Pop (VAR S: T; VAR Ch: CHAR);
   PROCEDURE Empty (S: T) : BOOLEAN;
   PROCEDURE Full (S: T) : BOOLEAN;
   PROCEDURE Initialize (VAR S: T);
   PROCEDURE Remove (VAR S: T);
END Stack.
```

primitives define the structure

> Note that only one of the identifiers in the shaded section—T—is intended to be used by an importing program module. A careful supply of primitive procedures has made the use of TOP, DataTYPE, Info, and Position unnecessary.

The Stack Code Layer — the Implementation Module

Because the definition of TOP and T are automatically known within the implementation module of Stack, all the code layer must contain are the full procedure and function declarations.

implementation of a multiple-stacks library

```
IMPLEMENTATION MODULE Stack;
PROCEDURE Inspect (S: T; VAR Ch: CHAR);
   BEGIN
      Ch := S.Info[S.Position]
   END Inspect;
PROCEDURE Push (VAR S: T; Ch: CHAR);
   BEGIN
      IF NOT Full(S)
         THEN S.Position := S.Position + 1;
              S.Info[S.Position] := Ch
         ELSE (*take care of the error condition*)
      END
   END Push;
PROCEDURE Pop (VAR S: T; VAR Ch: CHAR);
   BEGIN
      IF NOT Empty(S)
         THEN Ch := S.Info[S.Position];
              S.Position := S.Position – 1;
         ELSE (*take care of the error condition*)
      END
   END Pop;
PROCEDURE Empty (S: T) : BOOLEAN;
   BEGIN
      RETURN S.Position = 0
   END Empty;
PROCEDURE Full (S: T) : BOOLEAN;
   BEGIN
      RETURN S.Position = TOP
   END Full;
PROCEDURE Initialize (VAR S: T);
   BEGIN
      S.Position := 0
   END Initialize;
PROCEDURE Remove (VAR S: T);
   BEGIN
      S.Position := 0
      (*Can you imagine the code that a linked-list representation would use?*)
   END Remove;
BEGIN
END Stack.
```

A program that imports Stack:

```
IMPORT Stack;
```

can go on to declare stacks as variables, and as value and variable parameters. For example, procedure ReverseCopy, below, creates a mirror image of its Original stack argument.

copying a reversed stack

```
(*IMPORT Stack;*)
PROCEDURE ReverseCopy (Original: Stack.T; VAR Reversed: Stack.T);
  (*Creates a reversed stack copy.*)
  VAR Item: CHAR; (*One of the stacked items.*)
  BEGIN
    Stack.Initialize (Reversed);
    LOOP
      (**)IF Stack.Empty(Original) THEN EXIT END;
      Stack.Pop (Original, Item);
      Stack.Push (Reversed, Item)
    END
  END ReverseCopy;
```

Were Stack.T based on a pointer implementation, by the way, this simple approach would not suffice. Can you figure out why?

Modification Problems

1. Write a test harness for the Stack module. A good way to approach the problem of testing a module like this is to write a 'menu' procedure that lets the user call primitives as she desires. The method is limited (because variables have to be hard-wired in place) but it's a good beginning.
2. Using Stack as a model, write an abstract structure module for any of the structures (queue, ring buffer, set, or bag) described at the start of this chapter.
3. Can you stack stacks? Why not? Modify Stack so that it holds values of type Stack.T.

A Structure that Holds an Abstract Type

The stack abstraction we just created was limited to the storage of CHAR values. There is no reason, though, to limit the definition of an abstract structure in this manner. After all, the stack is defined in abstract terms that have nothing to do with the type of value being stacked.

Suppose, for instance, that we wanted to stack the Cartesian coordinates defined in Chapter 12. All that's required is some quick editing of Stack. The definition module of CartesianStack would begin like this:

```
DEFINITION MODULE CartesianStack;
  (*Makes a stack of Cartesian coordinates and primitives.*)
  IMPORT Cartesian;
  CONST TOP = 100;  (*the maximum stack height*)
  TYPE DataTYPE = ARRAY [1..TOP] OF Cartesian.T;
(*The next identifier, T, is the only one of these definition that will be referred
 to by an importing program module. All the rest are expected to remain hidden.*)
       T = RECORD
             Info: DataTYPE;
             Position: INTEGER[0..TOP]
           END;
  PROCEDURE Inspect ( S: Cartesian.T; VAR Ch: CHAR);
  PROCEDURE Push (VAR S: Cartesian.T; Ch: CHAR);
  PROCEDURE Pop (VAR S: Cartesian.T; VAR Ch: CHAR);
       . . .  (*definition continues*)
END CartesianStack.
```

stacking Cartesian coordinates

> Drawing a distinction between abstraction of values and structures makes it easier to mix and match our tools. Complex structures like CartesianStack are clearly built from distinct elements. All the standard advantages of modularity apply, especially the fact that the stack type and coordinate type can be written and tested separately.

Section 16-2 Objects and State

THE MODULES WE'VE WRITTEN SO FAR have not really taken advantage of two module features. They haven't had or required initialization code, and, more importantly, they have not used local variables.

local variables and initialization code

```
IMPLEMENTATION MODULE Example;
VAR X, Y, Z: INTEGER;
PROCEDURE  headings and bodies follow
  . . .
BEGIN
  X := 1;
  (*This initialization code is carried out at the first IMPORT of the
   module. Y and Z can be initialized as well. All variables will persist
   for the lifetime of the client program, although their values may change.*)
END Example.
```

The implementation module sets up its own working environment. Local variables (like X, Y, and Z) are allocated when the module is first imported. The module's code runs automatically—it's generally employed to initialize the local variables. However, a module's local variables exist for the entire lifetime of the program that imports the module in the first place.

Let's look at a simpleminded example. It begins with the definition module of HideIt:

HideIt definition

```
DEFINITION MODULE HideIt;
   (*Demonstrates the use of a 'hidden' module variable.*)
   PROCEDURE Next ( ) : INTEGER;
   (*Returns the hidden variable's value.*)
END HideIt.
```

A call of Next() represents some INTEGER value. Exactly *what* value that is will be determined within the implementation module:

HideIt implementation

```
IMPLEMENTATION MODULE HideIt;
   (*Demonstrates the use of a 'hidden' module variable.*)
VAR Hidden: INTEGER;
PROCEDURE Next ( (*VAR SIDE-EFFECT: Hidden*) ) : INTEGER;
   (*Increments and returns the hidden variable's value.*)
   BEGIN
      Hidden := Hidden + 1;
      RETURN Hidden
   END Next;
BEGIN
   Hidden := 0   (*Initialize the hidden variable.*)
END HideIt.
```

comment the side-effect

Pay special attention to the manner in which procedure Next is commented. Because it includes an assignment to a variable that has not been passed as a parameter, Next contains a *side-effect*. Under normal circumstances, side-effects are avoided like plague. However, in this case, the side-effect is necessary if Hidden is going to remain hidden. Module HideTest, below, demonstrates the use of HideIt.

demonstrating HideIt

```
MODULE HideTest;
   (*Demonstrates the operation of HideIt.*)
IMPORT HideIt;  (*At run-time, Hidden is initialized here.*)
IMPORT IO;
BEGIN
   IO.IntOut ( HideIt.Next( ) , 5);
   IO.IntOut ( HideIt.Next( ) , 5);
   IO.IntOut ( HideIt.Next( ) , 5)
END HideTest.
```

```
    1    2    3
```

Objects and Internal States

As simple as it is, Hidelt has an important difference from the modules we've written in the past. Hidelt has a *state*, or internal value. It retains it by managing an *object*—the local variable Hidden.

Now, what is an object? In the everyday conversation of computer science, the word 'object' means *data object*.

object defined

> A data object is an entity (or 'thing') that represents data. An *object module* maintains data for the benefit of the program that imports it. The module provides utility routines that interact with an object that is kept hidden from the rest of the program.

As the Hidelt example showed, hidden objects are created and exist entirely within the module. As a consequence, the distinguishing feature of modules that hold hidden objects is that they have *internal states*—values that exist independently of a calling program.

hiding objects vs. exporting them

In contrast, the type and structure abstraction modules we've written previously didn't have internal states. Even though they supplied definitions that were the basis of main-program objects, the modules themselves had no independent life.

Before we look at some examples of object modules, let me draw your attention to a point that is subtle, but worth noting. Until now, we've always passed data—objects, in our new terminology—to and from procedures. Indirectly, the procedures have carried the objects into and out of modules.

passing objects vs. passing messages

When a module hides an object, though, the main data structures *don't* move. They remain fixed within the object. So instead of sending objects to a procedure, we send procedure calls to the object. Although the procedure calls may carry or return individual items of data, the calls act as *messages*. Object-oriented programming languages (but not the object-oriented design process) are based on this idea.

At first glance, object modules may seem to supply an 'anything goes' category for module design. But when we look carefully at the sorts of information it pays to hide, we'll find there are clear distinctions between one object and the next. Three basic candidates for concealment within modules are:

- *Structures* that hold data—stacks, queues, and the like.
- *Processes* that produce data—say, random number generators.
- *Devices* that transmit or represent data—terminals or filesystems.

> In general, the module acts as an object *manager*. This means that the module provides services that go beyond the underlying definition of any particular data abstraction the module hides.

structures

Structures should be quite familiar by now. It's not hard to imagine writing a module that maintains a single structure—just one stack, perhaps. In practice, though, modules may hold structures that are far more complicated. For example, a *database management system* has just one instance of its structure—there's just one database. However, that single data base will probably have an internal structure that is quite complex, and is built of many structures itself.

processes

The process is a new concept. We'll look at one kind of process—a *stream generator* that produces (i.e. generates) a sequence (or stream) of data values. The HideIt example developed above generated a stream of values, but a random number generator is a more typical process. Processes of this sort are not merely library calls to procedures or functions. They qualify as object modules because producing the stream requires that we remember a current state, e.g. the value that currently heads the stream.

devices

The third class of hidden objects we will discuss are collectively known as *devices*. The term 'device' is itself a device. It provides a convenient means for us to refer to computer system components that are described in abstract terms, yet are at some level, real, physical things. Some typical devices are:

- The screen of a CRT terminal, a printer, or some other output machine.
- A keyboard, mouse, pointer, or other hand-held input machine.
- The computer's file system, i.e. the hard disks that serve as permanent memory.
- *Window managers*: programs that split high resolution screens into essentially independent terminals.

Let's take a brief look at two modules that have internal state. First, we'll hide a structure—a stack. Then, we'll put a *stream generating process*—a random number generator—into a module too.

Hiding a Structure: OneStack

Consider the definition module shown below. It paves the way for a single stack of CHAR values:

supporting a single stack

```
DEFINITION MODULE OneStack;
  (*Supports the basic operations on a stack of CHAR values.*)
  PROCEDURE Push (Ch: CHAR);
  PROCEDURE Pop (VAR Ch: CHAR);
  PROCEDURE Empty () : BOOLEAN;
END OneStack.
```

> Since just one stack is made available, there is no need to make any of its implementation details public. The definition module only exports operations on a hidden data structure—we remain blissfully ignorant of its details.

The implementation module that underlies OneStack is shown below. It relies on a locally declared array to hold the stacked characters, and built-in initialization code to set the stack's initial height to zero. Notice that procedures Push and Pop both check to be sure that their actions won't cause a crash. Different means of dealing with the error, if it's detected, are discussed with exception handling in section 16-3.

implementing OneStack

```
IMPLEMENTATION MODULE OneStack;
  (*Supports a stack of up to 100 CHAR values.*)
CONST TOP = 100;                                   (*All of these details*)
TYPE DataTYPE = ARRAY [1..TOP] OF CHAR;            (*are hidden from the*)
VAR Info: DataTYPE;                                (*module's ultimate user.*)
     Height: INTEGER[0..TOP];
PROCEDURE Push (Ch: CHAR (*VAR SIDE-EFFECTS: Height, Info*) );
  BEGIN
    IF Height < TOP
      THEN Height := Height + 1;
           Info[Height] := Ch
      ELSE (*Do something to handle the error—what would you suggest?*)
    END (*IF*)
  END Push;
PROCEDURE Pop (VAR Ch: CHAR (*VAR SIDE-EFFECTS: Height, Info*) );
  BEGIN
    IF NOT Empty()
      THEN Ch := Info[Height];
           Height := Height - 1
      ELSE (*Do something to handle the error—what would you suggest?*)
    END (*IF*)
  END Pop;
PROCEDURE Empty( (*VALUE SIDE-EFFECT: Height*) ) : BOOLEAN;
  BEGIN
    RETURN Height = 0
  END Empty;
BEGIN
  Height := 0; (*Initialize the stack.*)
END OneStack.
```

As in the HideIt module example, assignments or other references are made directly to Info and Height. They are not passed as parameters because if they were, OneStack couldn't keep its objects hidden. After all, if Info were passed as a parameter to Push in the implementation module, it would ultimately have to be passed in a program that imported OneStack too.

commenting side-effects

> When side-effects are employed in implementation modules, clearly comment their use with 'pseudo-parameter' comments. This kind of documentation helps undo one of the risks of side-effects—that they may go unnoticed in a subsequent program modification.

OneStack was deliberately put together in a no-frills implementation. Indeed, I could have even done away with Empty by requiring any programmer who uses OneStack to count items as they go on and come off! However, I did, in effect, say tough luck to any user who might want an easy way to learn the stack's maximum legal height, or its current top value, or its present height, or its full contents. After she imports the module:

importing OneStack

```
(*in a program module*)
IMPORT OneStack;
```

only the calls Push, Pop, and Empty are available publicly.

Hiding a Process — Stream Generation

An application of modules that has attracted attention recently involves the packaging of a source of data. Module HideIt, despite being simpleminded, was a rather good example of this application.

> A *stream generator* produces a sequence of data values without relying on external inputs. As usual, we find ourselves using external modules for abstraction. We separate the availability of values from the underlying process that creates them.

some familiar stream generators

There is a veritable torrent of stream generators. Some familiar INTEGER-based streams include the cardinal numbers, the odd numbers, Fibonacci numbers, powers of two, and primes. Less familiar flows include the number of numbers between perfect squares, along with the numbers of composites between primes, and the Tribonacci (you figure it out) numbers.*

What about REAL numbers like π and e, or the square root of 2? Can they be streams too? Well, although we've had occasion to name limited portions of these numbers as constants, we can also treat them as infinitely long streams of single digits. A hidden process can be invoked, when and if necessary, to supply the next digit and provide whatever level of decimal accuracy is required.

In fact, there is a fascinating book that contains 2,372 different INTEGER sequences, including such arcane sequences as the number of pieces that may be obtained in n slices not only from a razor-thin pancake, but from a cake (which, having some thickness, can be sliced from the side) as well! It's N.J.A. Sloane's *A Handbook of Integer Sequences*, Academic Press, 1973.

Basic Generation Rules

Stream generation is of interest to us because of the amount of work that will be hidden within an implementation module. Some initialization code will always be needed to generate the stream's first value, and stream modules will have to keep variables hidden, but active, between calls.

A stream generator is generally defined in terms of the rules below. Their simplicity makes the stream a powerful abstraction: we can talk about doing things to or with a stream without regard for the stream's actual contents, much less the means by which they are generated.

stream generation rules

1. We can inspect the value that's currently at the head of the stream.
2. We can remove the value that's currently at the head of the stream. This implies that the next value will be prepared for a subsequent inspection.
3. We can initialize the stream, or jump to a different point along the stream.

In order to simplify the discussion I'm going to treat an additional rule as a concrete given for each module, rather than as a 'parameterizable' option the programmer can change with a procedure call:

4. We can install a rule for generating the next value.

In other words, I'll build this into the module, rather than letting the calling program specify the rule.

Consider the functions and procedures that might be supplied by a stream generating module:

outlining a stream's definition module

```
DEFINITION MODULE of a typical stream generator;
(*The contents of a stream generator's definition module.*)
  PROCEDURE Head () : INTEGER;
    (*Rule 1: Represents—but doesn't extract—the head of the stream.*)
  PROCEDURE Get () : INTEGER;
    (*Rule 2: Gets (and, implicitly, replaces) the current head of the stream.*)
  PROCEDURE Initialize (Value: INTEGER);
    (*Rule 3: Replaces the head's current value with Value, or the Valueth item.*)
END of the definition
```

The implementation module that matches this definition hides a great deal: the object that 'remembers' the stream's current value, additional procedures, and initialization code. Typical contents would be (the hidden parts are shaded):

outlining a stream's implementation module

```
IMPLEMENTATION MODULE of a typical stream generator;
(*The contents of a stream generator's implementation module.*)
VAR Current: INTEGER;
    (*Holds the current head-of-stream value.*)
PROCEDURE Reset (Value: INTEGER; (*VAR SIDE-EFFECT: Current*));
    (*Sets or resets the stream head as necessary.*)
PROCEDURE View ( (*VALUE SIDE-EFFECT: Current*) ) : INTEGER;
    (*Returns the value of the current stream head.*)
PROCEDURE Generate (*VAR SIDE-EFFECT: Current*) ;
    (*Implements the built-in stream generating rule.*)
PROCEDURE Head () : INTEGER;
    (*Represents—but doesn't extract—the head of the stream.*)
PROCEDURE Get () : INTEGER;
    (*Gets and represents the current head of the stream.*)
PROCEDURE Initialize (Value: INTEGER);
    (*Replaces the head's current value with Value, or
     with the Valueth item in the stream.*)
BEGIN
    Reset (an appropriate argument)          (*initialize the stream*)
END the implementation module
```

As you read the outlined implementation module, note the limited use of side-effects—the use, particularly in assignment, of a variable that has not been passed as a parameter. The private subprograms Reset, View, and Generate all access Current directly. Although side-effects are not generally desirable, they're needed here to help hide information. As you can see, each of the publicly available calls has a private call that contains a side-effect associated with it.

Let me also draw your attention to the identifiers Head, Get, and Initialize. When you program stream generators, try to make consistent use of simple, descriptive names. This serves the same purpose as the consistent usage of identifiers like T, In, and Out served for type abstractions. The module's name describes the stream; operation names describe themselves:

use consistent names

A module that contains:

```
IMPORT Random          IMPORT Prime          IMPORT Fibonacci
```

should refer to like operations consistently:

```
    Random.Head            Prime.Head            Fibonacci.Head
    Random.Get             Prime.Get             Fibonacci.Get
    Random.Init            Prime.Init            Fibonacci.Init
```

A Random Number Generator

A particularly useful stream generator provides randomly distributed numerical values. The mathematical basis of our stream generation rule will be the *linear congruential* method, introduced by D. Lehmer in 1949. In Lehmer's method, we take a starting number, called the *seed*, and subject it to these operations:

```
Seed := (MULTIPLIER*Seed) MOD MODULUS;
```

one random formula

Lehmer's method produces a stream of numbers that ranges from 0 through MODULUS less 1. If we then use REAL division to divide the value of Seed by MODULUS, we wind up with a number that falls between 0.0 and 0.999... . Technically, the stream is said to be *pseudorandom*. Although its numbers are essentially randomly and uniformly distributed, the sequence is predictable and eventually repeats.

Indeed, the stream of random numbers is circular. Initializing (or reinitializing) the stream to a particular seed value is, for all practical purposes, a matter of selecting a starting point for making the circuit. Any value returned by the generator is the seed for its successor.

Choosing specific values for MULTIPLIER and MODULUS is a task fraught with peril, since not every pair leads to a long sequence before numbers start to repeat. The method I've used in module Random, below, is based on a MULTIPLIER of 16807 and a MODULUS of 2147483647, or $2^{31} - 1$. The need to avoid arithmetic overflow complicates the algorithm a bit. I've used the standard type LONGINT for the same reason.*

The definition module of Random appears below. Notice how closely it parallels the stream generator outlined earlier:

definition of Random

```
DEFINITION MODULE Random;
   CONST MODULUS = 2147483647;
   PROCEDURE Head () : LONGINT;
     (*Represents the seed of the upcoming random stream.*)
   PROCEDURE Get () : LONGINT;
     (*Gets the next random number from the stream.*)
   PROCEDURE Initialize (Value: LONGINT);
     (*Provides a new seed for the upcoming random stream.*)
END Random.
```

MODULUS is defined publicly—it can be imported by a calling program module—so that the value returned by Random.Get can be normalized to an appropriate range. For example, since:

$$0.0 \leq (\text{Random.Get}() \div \text{MODULUS}) < 1.0$$

a value from 1 through 100 can be produced by multiplying the result above by 100, truncating the product, and adding one.

The remaining constants, as well as the starting seed, are private, hidden within Random's implementation module. Were it necessary to find the starting seed's value, a call of Random.Head() would suffice. Note that unless the module user goes to the trouble of calling Initialize to supply a new seed value—a new head for the pseudorandom stream—the random numbers generated won't change. This is a lifesaver during debugging, since we know what numbers to expect. At run-time some means of getting a new seed should be employed.

---

* See *Random Number Generators: Good Ones Are Hard to Find*, Park and Miller, Communications of the ACM, October 1988, p. 1192. The method I use here is the one they recommend as a minimal standard for random number generation.

implementation of Random

```
IMPLEMENTATION MODULE Random;
CONST DEFAULTseed = 1;
VAR Seed: LONGINT;
PROCEDURE Reset (Value: LONGINT (*VAR SIDE-EFFECT: Seed*));
   BEGIN
      Seed := Value
   END Reset;
PROCEDURE View ( (*VALUE SIDE-EFFECT: Seed*)) : LONGINT;
   BEGIN
      RETURN Seed
   END View;
PROCEDURE Generate (VAR Seed: LONGINT);
   (*Algorithm 2 from CACM, pg. 1195, vol. 31, no. 10, (October, 1988).*)
   CONST MULTIPLIER = 16807;    (*Don't forget that MODULUS = 2147483647*)
         QUOTIENT = 127773;     (*MODULUS DIV MULTIPLIER*)
         REMAINDER = 2836;      (*MODULUS MOD MULTIPLIER*)
   VAR Low, Hi, Test: LONGINT;
   BEGIN
      Hi := Seed DIV QUOTIENT;
      Low := Seed MOD QUOTIENT;
      Test := (MULTIPLIER * Low) – (REMAINDER * Hi);
      IF Test > 0
         THEN Seed := Test
         ELSE Seed := Test + MODULUS
      END;
   END Generate;
PROCEDURE Get () : LONGINT;
   (*Gets the next pseudo-random number from the stream.*)
   VAR Temp: LONGINT;
   BEGIN
      Temp := View ();  (*We can't return the current value ...*)
      Generate(Seed);   (*... until we have generated the next one.*)
      RETURN Temp
   END Get;
PROCEDURE Head () : LONGINT;
   (*Represents the seed of the upcoming random stream.*)
   BEGIN
      RETURN View()
   END Head;
PROCEDURE Initialize (Value: LONGINT);
   (*Provides a new seed for the upcoming random stream.*)
   BEGIN
      Reset (Value)
   END Initialize;
BEGIN
   Initialize (DEFAULTseed)          (*initialize the stream*)
END Random.
```

Modification Problems

1. In their article, Park and Miller suggest that a good test for the generator is to see if the 10,000th number it cranks out is 1,043,618,065. Try it (be careful about an off-by-one error). Note that the IO library I've provided allows you to print LONGINT values with IntOut.

2. Built-in random number generators generally return a value between 0.0 and 1.0. Add a GetReal procedure to Random that returns a LONGREAL value in that range.

3. On the other hand, you might enjoy modifying Random so that it only returns values in the INTEGER range (the easiest way is to reject values larger than MAX(INTEGER) from inside the call of Generate). Add a GetInBounds procedure to Random that takes a lower and upper bound, and returns an INTEGER that falls within those bounds.

Section 16-3 Program Engineering: Exception Handling

The most problematic area of data abstraction involves dealing with errors. With justification, we can view most kinds of abstraction as a means of protecting data from mistakes in our programs. However, we must find ways to protect our programs from errors in data or data handling as well.

Types and structures each have their own characteristic problems. Three error conditions that typically arise in the course of dealing with abstracted types are:

common type abstraction errors

- *Incoming* errors are caused by employing values that are outside the legal domain of the abstraction, just as CHAR values are outside the domain of INTEGER.
- *Outgoing* errors are the consequence of producing values that are outside of the legal range of results defined for the type, just as 5–7 falls outside the range of CARDINAL.
- *Result* errors occur when a value cannot be calculated correctly for some reason; for example, the operation isn't defined (e.g. division by zero), an operand is known to be undefined, or an intermediate value is outside the legal range (e.g. a FRACTION multiplication that reduces the operands *after* the operation might overflow needlessly).

A similar group of problems are associated with abstracted structures:

common structure abstraction errors

- *Status* errors are attempts to retrieve values that are unavailable; for instance, trying to get a value from an empty stack.
- *Implementation* errors are operations that unintentionally overpower a particular implementation; for example, adding a component to a stack that's full.
- *System-related* errors, such as opening too many files, or exhausting disk space, are similar to implementation errors. But whereas implementation

errors are 'soft,' and can be fixed by improving an implementation, system errors are 'hard,' and are more difficult to detect and work around.

Errors have different degrees of severity, ranging from no immediate effect to crash-and-burn catastrophe. For this reason, errors are usually known by the less value-laden name *exceptions*.

exceptions

> An exception is a condition that lies outside of normal program operation. It might lead to incorrect results or abnormal program termination, but doesn't necessarily have to.

In formal terms, the exception is detected by being *raised*, and is then *handled*. Automatic, detailed exception handling is important in real-time systems that must continue to run, in known states, no matter what happens. Although Modula-2 does not include special exception-handling features (yet), a few other languages (notably PL/I, CLU, and Ada) do. However, most Modula-2 systems do have libraries that are set up in a manner that allows you to customize exception handling.

The Severity of Exceptions

Whether an exception is fatal, is merely worth noticing, or can be ignored entirely will depend on the application. For example, consider an abstract data type used to represent strings. What should happen if a string contains a control character? In some applications this could be disastrous, while in others it is of no concern whatsoever. Similarly, an input string that is too long to be stored in a single variable might be skipped entirely, or read in part, or it might cause a program halt—it all depends.

Let's leave aside the question of 'proper' handling of exceptions, and just look at alternative means of dealing with the exception and/or informing the user. The first question is whether or not (and if so, to what extent) normal operation should be interrupted. We might:

kinds of exceptions

1. Halt the program if the exception can neither be fixed nor ignored. We'll call this a *fatal* exception.
2. Halt the current procedure or module if the exception cannot be recovered from, but can have its damaging effects limited. We'll call this a *non-propagating* (non-spreading) exception.
3. Repair or ignore the exception and continue as though nothing had happened. We'll call this a *recoverable* exception.

The category an exception will fall into isn't always a foregone conclusion. For example, consider the exception that would occur if we tried to read a CHAR value into an INTEGER variable. Alternative approaches to handling this exception include:

- Treating it as a fatal exception by halting program execution with an error message as soon as the problem is recognized. This is what a Pascal program would do.

alternatives in handling exceptions

- Treating it as a non-propagating exception by *a*) skipping additional CHAR values until a space or control character is encountered, then *b*) setting an error status variable, e.g. Done := FALSE; but otherwise ignoring the situation. This is similar to what the IO library does.
- Treating it as a recoverable exception by repeatedly attempting the input operation until it is successful. In effect, this is what an underlying system-level procedure does if it attempts to read before the user has had a chance to actually type something in.
- Using a combination of methods—say, assigning a default value like MAX(INTEGER) to the variable, and printing an error message that attests to this action, while continuing program operation.

The last approach is often relied on when useful information can still be obtained by continuing to execute the program. Especially good examples are provided by text formatting programs and compilers. Instead of blowing up when impossible conditions are encountered, the program assumes some default for the current stretch of garbled input. It continues processing on the assumption that it can recover (albeit imperfectly), since text and program features are relatively independent from section to section.

Preventing Exceptions

avoiding exceptions

The best method of dealing with exceptions, of course, is to handle them before they occur. Responsibility for this is split between the programmer as module user, and the programmer as module writer. As an importer of libraries, the programmer must take advantage of built-in tools. 'There are none so deaf,' goes the saying, 'as those who will not hear.' By the same token, there are none so certain to crash programs as those who forget to check Empty(Queue) before calling Remove(Item).

As the author of a library, though, the programmer must remember that nobody likes a person who goes through life saying 'I told you so!' The fact that an abstraction's implementation is hidden from the client means that run-time errors can be obscure, if not downright inexplicable. It's the implementor's job to detect and express exceptions in terms of the abstraction, rather than letting the Modula-2 system report on them in terms of the abstraction's implementation.

Thus, an error message like `Denominator too large for this implementation` means something. However, automatic system messages that refer to `INTEGER OVERFLOW` or `SUBRANGE EXCEEDED` or `RANGE ERROR LINE 487` are describing a representation that has intentionally been kept hidden from the clients—the messages are useless.

minimizing exposure

Through careful (and sometimes clever) programming, a library module's exposure to exceptions can be minimized. For example, consider the problem of arithmetic with fractions. Saying that:

$$2/4 \times 2/4 = 4/16 = 1/4$$

is certainly correct, and is easily represented. But while the final result of this operation should be arrived at just as easily:

$$5000/10000 \times 5000/10000 = 25000000/100000000 = 1/4$$

the intermediate product is likely to cause problems. Allowing a crash is unthinkable, of course, but returning an 'overflow' message is merely a weasely dodge. The simple step of reducing fractions when and where possible *before* operations means more work for the library programmer, but it leads to a reliable library.

Modula-2 Details

Remember . . .

- An abstract data structure is defined in terms of the relationship between stored values, rather than by specifying the values themselves. The relationships are given by the structure's *primitive operations.*
- Typical primitives for structures include operations for *construction, inspection, iteration* or *repositioning, status*, and *initialization.* The primitives should permit all defined operations on the structure, and make it unnecessary to rely on any underlying details of the structure's realization in Modula-2.
- A module that manages *data objects* is called an *object module.* As I've used the term, a data object is an entity that is entirely hidden from the rest of a program—it's created and maintained entirely within a library module. Thus, object modules are said to have *internal state.* Note that the objects they maintain are just 'second-class' citizens, since they can't be named, passed as parameters, duplicated, etc.
- A hidden *structure* holds data by applying well-defined relationships between stored values. The structure that's hidden may be very simple (say, a stack or a queue), or it can be complex and composed of more than one variable (like a database management system).
- Hidden *processes* produce data by providing a well-defined source of values. A stream is a sequence of values; a *stream generator* packages that sequence, and separates the availability of values from the underlying process that creates them. Common stream primitives include operations to generate values, to inspect the stream, and to initialize—or reinitialize—the stream.
- It's also often convenient to represent a real, physical device (like a terminal screen or computer filesystem) as a data object whose state and implementation are hidden within a module.
- Object modules almost invariably rely on side effects, since calls of primitives usually involve inspection or change of hidden variables that can't be passed as

parameters. It's generally a good idea to comment these side effects to make the implementation code easier to understand and modify.

- *Exceptions* are errors. However, since they are often due to limitations of an implementation, they may not be avoidable. When you write library modules, try to provide means of handling reasonably foreseen exceptions: preventing their propagation (spread), recovering if possible, and halting the program (with an explanatory error message) if necessary.

Zen Study Guide

Big Ideas

The Big Idea of Chapter 16 is:

1. *Sending values to processes, and sending processes to values, are two sides of the same coin.* The traditional approach to studying programming begins with procedures and their arguments. We see the action as the fixed object, and the value as a transient state that varies from call to call. However, we can just as easily view data as the mountain, and procedure calls as an assortment of messengers.

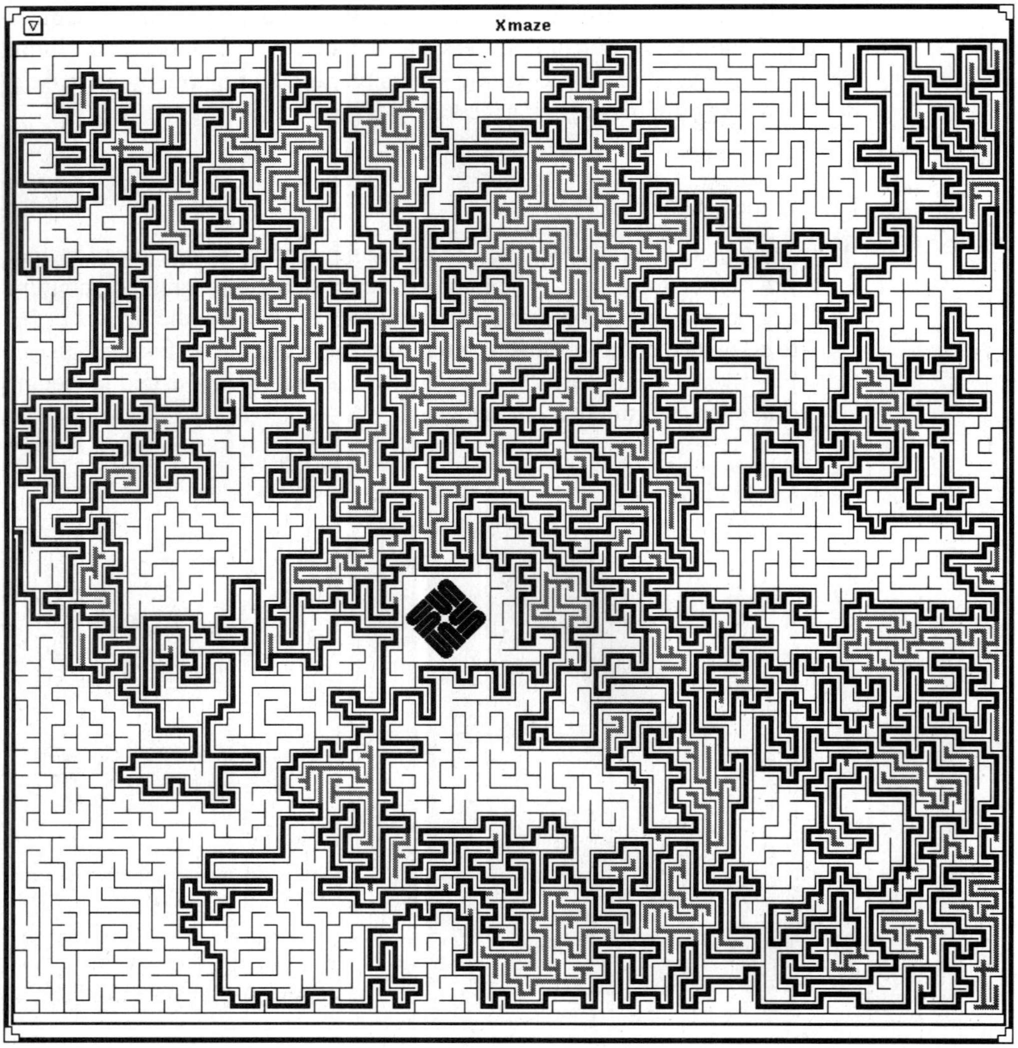

'Now, how does one get out of a maze? . Well, if you're Theseus, you trail a string behind you . . .' (p. 641) Gray paths indicate false trails.

Chapter 17 Recursive Programming Methods

'There is always more than one way to skin a cat,' as the expression goes, and nowhere is this more true than in programming. Although we've spent many pages discussing multiple and overlapping uses of different control statements, and on the tradeoff between algorithms and data structures, there is still one unmentioned tool left in our bag of tricks—*recursion*.

A recursive subprogram calls itself. Now, this property alone might seem unremarkable, and indeed, hundreds of programming languages, from FORTRAN to BASIC, have thrived without allowing such references. We will see, though, that recursion adds enormous power to Modula-2. It is as much a technique for structuring ideas and thinking as it is for structuring programs.

I've split the discussion into three parts. 17-1 introduces the idea of the stack, and deals strictly with recursion as a substitution for looping. It can probably be read with ease following Chapter 6. 17-2, in turn, adds the array type to the picture to allow the development of more elaborate algorithms. It can follow Chapter 8. Finally, 17-3 presents recursion in the context of pointers, and describes the definition of recursive data types; it can follow Chapter 13.

Recursion may seem awkward at first. But one reader in ten will leave this chapter indignant that recursion wasn't introduced in Chapter 1.

Section 17-1

Recursive Procedures and Functions

HOW CAN YOU PRINT A SENTENCE IN REVERSE without using arrays or pointers? Think about it for a minute, then consider this deceptively simple program:

recursive sentence-reverse program

```
MODULE Recur;
   (*Uses recursion to read a sentence and echo it in reverse.*)
IMPORT IO;
PROCEDURE StackTheCharacters;
   VAR TheCharacter: CHAR;
   BEGIN
      IO.CharIn (TheCharacter);
      IF (TheCharacter <> "." ) THEN
         StackTheCharacters          (*The recursive call.*)
      END;
      IO.CharOut (TheCharacter)
   END StackTheCharacters;
BEGIN
   IO.StringOut ("Enter a sentence that ends with a period.");
   IO.LineOut;
   StackTheCharacters;          (*The first call.*)
   IO.LineOut
END Recur.
```

```
Enter a sentence that ends with a period.
How I long for the Jurassic.
.cissaruJ eht rof gnol I woH
```

Program Recur takes an ordinary sentence of text as input, then prints it in reverse. It manages this magic with a procedure that calls itself.

recursion defined

> A procedure or function that calls itself is said to be *recursive*.

When procedure StackTheCharacters is first called, it reads a character value for its local variable TheCharacter. Then, unless the first character was a period, StackTheCharacters is called again. However, the final statement of the first call, which is IO.CharOut (TheCharacter), is still pending while the next instance of the procedure call tries to read a character again.

Now, leaving uncompleted statements behind is nothing new. We do it almost every time we call a procedure or function. Sooner or later we'll get back to this output statement, and print the value of the first character.

What happens in the second call of StackTheCharacters? Well, it's déjà vu all over again. Once again, a new local variable (TheCharacter) is created. Once more, a letter is read so that TheCharacter can have a value. And, once more, if recursive calls

the letter we read wasn't a period, StackTheCharacters is called. As before, the final call of IO.CharOut (TheCharacter) is still waiting to be executed.

Modification Problems

1. Modify Recur so that it reverses a number by reading it one digit character at a time. Spot the end of the number by spotting a carriage return.
2. What would happen to program Recur if the output statement preceded the shaded recursive call, rather than following it?

The Recursive Stack

Since it's hard to visualize what happens in a series of identical procedure calls, computer scientists use a metaphor to help explain recursion.

the stack

> The variables and pending statements of a partially executed program or subprogram are said to go on a *stack* within the computer.

When we jump out of a subprogram (by calling another procedure or function) all currently active variables, along with any pending statements, are added to the stack. The number of times this can occur—the height of the stack—is only limited by the memory resources of the computer.

StackTheCharacters is a procedure, and not a loop, but I can describe it in a way that's very much like a loop specification:

a recursive specification

Goal: Print a sentence in reverse.
Stacking Plan: Read a character.
Bound: We've read a period.
Unstacking Plan: Print a character.

Suppose that I 'unfold' program Reverse, and show every statement that actually does something. If we only look at the input statements—the stacking plan—executed during procedure StackTheCharacters, we'll see something like the series shown below. Although we keep creating local variables named TheCharacter, Modula-2's scope rule (that the most local variable takes precedence) gives the current input value to the most recently created instance.

building the stack

```
IO.StringOut ("Please enter a sentence.");
IO.CharIn (TheCharacter);          (*Reading in "T".*)
   IO.CharIn (TheCharacter);          (*Reading in "h".*)
      IO.CharIn (TheCharacter);          (*Reading in "i".*)
         . .     (*Intermediate calls...*)
            IO.CharIn (TheCharacter);          (*Reading in "m".*)
               IO.CharIn (TheCharacter);          (*Reading in "e".*)
                  IO.CharIn (TheCharacter);          (*Reading in ".", the last call.*)
```

The chain of calls to StackTheCharacters continues until we read the period that ends the sentence. At last we'll get past the **IF** statement! We can finally complete the last invocation of StackTheCharacters by printing the character we just read—the period.

Since this particular invocation of StackTheCharacters is finished, the program returns to where it was when the procedure call was made—to the calling procedure, and eventually to the main program. The computer executes the statements of the unstacking plan:

undoing the stack

```
                    IO.CharOut (TheCharacter);     (*Printing the ".".*)
                  IO.CharOut (TheCharacter);     (*Printing the "e".*)
                IO.CharOut (TheCharacter);     (*Printing the "m".*)
                  ..        (*Intermediate calls...*)
          IO.CharOut (TheCharacter);     (*Printing the "i".*)
        IO.CharOut (TheCharacter);     (*Printing the "h".*)
      IO.CharOut (TheCharacter);     (*Printing the "T".*)
  IO.LineOut          (*This is the last statement in the main program.*)
```

Now, a stack is actually created in any series of procedure or function calls—the sequence need not be recursive. The stack can, and usually does, grow and shrink during the course of a program. However, it will always be empty when a program ends normally.*

Procedure StackTheCharacters is a good introduction to recursion because it uses the stack in an obvious way. Each time a recursive call is made in StackTheCharacters, local variables and uncompleted statements are left on the stack.

However, recursive calls don't have to leave *anything* useful on the stack. For instance, program UnDigit, below, leaves almost no undone business. Its recursive procedure, ReverseDigits, prints a single digit and then calls itself:

> *Goal: Print a positive* INTEGER *in reverse.*
> *Stacking Plan: Print the 'ones' digit.*
> *Bound: There are no more digits to print.*
> *Unstacking Plan: —*

But there's no 'unstacking' plan. When the very last call is made, all that's left is the **END** of each of the previous calls.

end recursion

> This kind of recursion is called *end* or *tail* recursion. When the procedure makes its recursive call, no unfinished statements are left on the stack.

---

* This simplified explanation of stacks helps explain how the computer keeps track of scope within a program. *All* identifiers defined in a subprogram are put on the stack when the subprogram is invoked. Then, when an identifier is used, the computer looks down the stack for the identifier's most recent definition or declaration. Although the stack may contain several different usages of a single name, the most recent definition is the first one found.

recursive INTEGER-reverse program

```
MODULE UnDigit;
    (*Recursively reverses the digits of a positive INTEGER.*)
IMPORT IO;
VAR Number: INTEGER;
PROCEDURE ReverseDigits (Number: INTEGER);
    BEGIN
        IO.IntOut (Number MOD 10, 1);  (*Print the 'ones' digit.*)
        (*If there are more digits, divide away the 'ones' digit and pass the result.*)
        IF (Number DIV 10) <> 0 THEN
            ReverseDigits (Number DIV 10)
        END
    END ReverseDigits;
BEGIN
    IO.StringOut ("Please enter a positive integer.");
    IO.LineOut;
    IO.IntIn (Number);
    ReverseDigits (Number);
    IO.LineOut
END UnDigit.
```

```
Please enter a positive integer.
26407
70462
```

Procedures that use end recursion are usually easy to write iteratively (using a looping statement). For example:

iterative reversing procedure

```
PROCEDURE IterativeReverse (TheNumber: INTEGER);
    BEGIN
        LOOP
            IO.IntOut (TheNumber MOD 10, 1);
            TheNumber := TheNumber DIV 10;
            (**)IF TheNumber = 0 THEN EXIT END
        END
    END IterativeReverse;
```

Modification Problems

1. UnDigit would print the number **100** as 001. Fix the program so that leading zeros aren't printed.

2. Reversing a REAL value is harder than reversing an INTEGER. Try to modify UnDigit so that it reverses the whole portion and fractional portion of a REAL separately, so that **123.456** is printed as 321.654.

3. I usually check addition by re-adding the figures in reverse. Turn Recur or UnDigit into a program that reads and stacks a line's worth of INTEGER values. Add them before you put them on the stack, then add them as they come off. Are the sums the same?

Self-Check Questions

Q. Is the procedure in Print recursive? Should it be? What does it do?

```
MODULE Print;
IMPORT IO;
PROCEDURE Echo;
   VAR TheCharacter: CHAR;
   BEGIN
      IO.CharIn (TheCharacter);
      IO.CharOut (TheCharacter);
      IF TheCharacter <> "." THEN Echo END
   END Echo;
BEGIN
   IO.StringOut ("Enter a sentence that ends with a period.");
   IO.LineOut;
   Echo;
   IO.LineOut
END Print.
```

A. Procedure Echo offers another demonstration of end recursion. Although a stack is created, it doesn't contain any statements to be executed. The series of TheCharacter local variables is saved, but gets thrown away when the recursive calls end. The stack is created, but not used. A **LOOP** would do the job perfectly well, because the program just echoes a line of input.

```
Enter a sentence that ends with a period.
They had a hot time in the Cretaceous.
They had a hot time in the Cretaceous.
```

Recursive Functions

Like procedures, functions can be written recursively. Function Sum uses a series of recursive calls to add a series of numbers from 1 to its argument Limit.

recursive summing function

```
PROCEDURE Sum (Limit: INTEGER) : INTEGER;
   (*Recursively sums the positive numbers 1 through Limit.*)
   BEGIN
      IF Limit = 1
         THEN RETURN Limit
         ELSE RETURN Limit + Sum(Limit – 1)
      END
   END Sum;
```

The stack that Sum produces is a bit peculiar. Rather than variables or statements, it contains a series of partially evaluated expressions.

stacking recursive function calls

> The general outline of Sum is typical of recursive functions—the recursive call occurs while an expression is being calculated. Thus, the stack serves to delay the expression's evaluation.

Let's assume that Sum is called with 5 as its first Limit argument. Here's the stack of partial evaluations that's built during the sequence of calls:

```
RETURN 5 + Sum (5–1) (*first*)
            RETURN 4 + Sum (4–1) (*second*)
                        RETURN 3 + Sum (3–1) (*third*)
                                    RETURN 2 + Sum (2–1) (*fourth*)
                                                RETURN 1 (*fifth*)
```

None of the stacked expressions can be calculated until Sum gets a nonrecursive value—the fifth call, which doesn't depend on calling Sum again. Let's follow the sequence of **RETURN**s as they're able to be made:

```
                                            RETURN 1 (*fifth*)
                                RETURN 2 + 1 (*fourth*)
                      RETURN 3 + 3 (*third*)
           RETURN 4 + 6 (*second*)
RETURN 5 + 10 (*first*)
```

the bound call

> The last of a series of recursive calls is the *bound* call. The circumstances that give rise to the bound call form the exit condition of the recursion.

infinite recursion

By our definition of the function, Sum's bound call occurs when its argument is 1. An *infinite* recursion occurs if the exit condition can't be met, and the bound call is never made.

Recursive functions are usually specified by making a recursive definition of the function itself. This takes a little getting used to. Typically, the definition has two stages:

recursive definition

- First, define the function for its bound case.
- Then, define the function in terms that bring it one step closer to the bound.

I can express the same idea in terms of bounds and plans as:

recursive specification

Goal: Evaluate a function call for some argument.
Stacking plan: Call the function with a reduced argument.
Bound: The argument can't be reduced any further.
Unstacking plan: Evaluate the expression.

As was the case with loop specifications, the recursion ends because we reach the bound. The stacking plan *state the function with a reduced argument* means *take the smallest step you can toward the bound.* Usually, this means subtracting 1 from the recursive function call's argument.

For example, consider function Sum. Its bound is that the parameter Limit equals 1. In the bound case, Sum actually equals Limit, because the sum of the numbers from 1 through 1 is just 1.

But how can we define the function in a way that takes small steps toward the bound? By recognizing that if Limit is greater than 1, then Sum(Limit) can be expressed as Limit plus Sum(Limit – 1).

Sum(Limit) *equals* Limit + Sum(Limit – 1)

This restatement brings us one step closer to the ultimate recursive bound. If I put the two parts together, we have:

the recursive definition

If Limit equals 1, the sum of 1 to Limit is Limit.

If Limit exceeds 1, the sum of 1 to Limit equals Limit plus the sum of 1 to Limit less 1.

Or, stated as a recursive specification:

the recursive specification

Goal: Recursively evaluate the call Sum(Limit).
Stacking plan: Add Limit *to* Sum(Limit – 1).
Bound: Limit *equals one.*
Unstacking plan: Evaluate the stacked calls.

If you reconsider the **IF** statement from function Sum, you'll see that it says exactly the same thing in Modula-2.

```
IF Limit = 1
   THEN Sum := Limit
   ELSE Sum := Limit + Sum(Limit – 1)
END
```

A Few Examples

Once you get the hang of it, it's fun to make up recursive definitions of some of the basic functions. For example, how would you use recursion to raise a number X to the *n*th power, where *n* is an INTEGER?

problem: exponentiation

The English definition of a Power function is easy—'multiply X by itself *n* times.' But that sounds like an invitation to a **FOR** loop. Getting a recursive definition starts by considering the bound case. What is X raised to the first power? Why, it's X, of course. And if *n* exceeds one, we can define the function in a way that brings us one step closer to the bound:

the recursive definition

If n equals 1, then X to the n equals X.

If n is greater than 1, then X to the n equals X times X to the n–1.

Restating this as a recursive specification gives us:

the recursive specification

Goal: Recursively evaluate the call Power(X, n).
Stacking plan: Multiply X *by* Power(X, n – 1).
Bound: n *equals one.*
Unstacking plan: Evaluate the stacked calls.

As you read function Power, below, note that it requires two value parameters. At each step of the recursive algorithm we must know the values of both X and n. However, as you can see from the algorithm we described above, X remains constant while n steadily declines.

recursive power function

```
PROCEDURE Power (X: REAL; n: INTEGER) : REAL;
  (*Recursively calculates X^n. Assume n>0.*)
  BEGIN
    IF n = 1
      THEN RETURN X
      ELSE RETURN X * Power (X, n – 1)
    END
  END Power;
```

Another definition that's easy to state recursively in English is that of the nth Fibonacci number. The Fibonacci sequence begins 1, 1, then each subsequent value is the sum of the previous two: 1, 1, 2, 3, 5, 8 and so on. The recursive definition is:

problem: recursive Fibonacci definition

If n is 1 or 2, the nth Fibonacci number is 1.

If n is 3 or more, the nth Fibonacci number is the sum of the previous two.

A specification that's a bit more like Modula-2 is:

the recursive specification

Goal: Recursively evaluate the call Fibonacci(Which).
Stacking plan: Add Fibonacci(Which – 1) *to* Fibonacci(Which – 2).
Bound: Which *equals* 1 *or* 2.
Unstacking plan: Evaluate the stacked calls.

I've put the recursive function in the test harness TestFibs, on the next page.

Although I won't show it here, both Power and Fibonacci can easily be implemented as iterative functions. However, recursive solutions can be much more elegant than their iterative counterparts. I'd hate to say that shortness is a virtue in itself, but recursive subprograms can often be written more briefly and clearly than iterative ones. When you become comfortable with recursion, and start to recognize 'standard' recursive algorithms, you'll appreciate the ease with which such algorithms can be put into programs.

recursive Fibonacci program

```
MODULE TestFibs;
   (*Tests a function that recursively generates Fibonacci numbers.*)
IMPORT IO;
VAR Test1, Test2, Test3, Test4: INTEGER;
PROCEDURE Fibonacci (Which: INTEGER) : INTEGER;
   BEGIN
      IF (Which=1) OR (Which=2)
         THEN RETURN 1
         ELSE RETURN Fibonacci(Which - 1) + Fibonacci(Which - 2)
      END
   END Fibonacci;
BEGIN
   IO.StringOut ("Pick four Fibonacci numbers.");
   IO.LineOut;
   IO.IntIn (Test1);
   IO.IntIn (Test2);
   IO.IntIn (Test3);
   IO.IntIn (Test4);
   IO.StringOut ("The four Fibonacci numbers are:");
   IO.IntOut (Fibonacci(Test1), 6);
   IO.IntOut (Fibonacci(Test2), 6);
   IO.IntOut (Fibonacci(Test3), 6);
   IO.IntOut (Fibonacci(Test4), 6)
END TestFibs.
```

```
Pick four Fibonacci numbers.
1  7  15  23
The four Fibonacci numbers are:     1     13    610 28657
```

Modification Problems

1. Modify Power so that it calculates factorials, rather than powers. For instance, 5 factorial is 5×4×3×2×1. Then, write a test harness similar to TestFibs.
2. A Fibonacci-like sequence can actually begin with any two numbers. Modify Fibonacci so that it works properly for the sequence that starts 2, 2, 4, 6, 10 . . .

Self-Check Questions

Q. What is the intention and effect of function RawPower, below? What restriction of Power does it remove?

```
PROCEDURE RawPower (X: real; n: INTEGER) : REAL;
  (*Recursively calculates Xⁿ. Assume what?*)
  BEGIN
    IF n = 0
          THEN RETURN 1
       ELSIF n > 0
          THEN RETURN RawPower (X, n – 1)
       ELSE RETURN 1 / RawPower (X, –n)
    END
  END RawPower;
```

A. RawPower is considerably more robust than Power. It is still a recursive procedure whose algorithm is nearly identical to Power's. However, it can raise X to a power smaller than 1. In other words, n can be any INTEGER, positive, negative, or zero.

Q. What is the largest number of times the innermost **ELSE** will be entered, above?

A. At most, the call Power (X, –n) will occur once.

The Towers of Hanoi

Our first examples kept recursion simple by minimizing the numbers of statements and variables involved. The final introductory procedure will be the most complicated, even though it's just intended to help us play a game:

problem: towers of Hanoi

The *Towers of Hanoi* game is played with three pegs, and a pile of disks of different sizes. We start with all the disks stacked on one peg, like this:

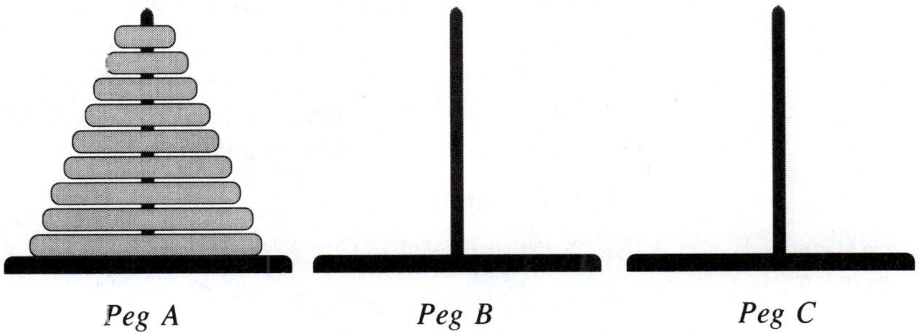

The object of the game is to move the entire stack from peg A to peg C, while obeying two rules:

the rules

1. Only one disk can be moved at a time.
2. A larger disk can never go on top of a smaller one.

Write a program that gives step-by-step instructions for moving a stack of height *n* from peg A to C.

The original version of this game is said to have been played on a set of three diamond needles and sixty-four golden disks. The end of the game was supposed to mark the end of the world.

the 1 or 2-disk problem

Let's try to get a handle on how the moves are made for stacks of various heights. Clearly a height of 1 is trivial—we move the disk directly from A to C. What about a height of 2? We put the top disk out of the way—to B. Then the bottom disk goes to C, and the smaller disk from B to C.

the 3-disk problem

With a stack of height 3 it gets interesting. Let's suppose, though, that we restate the problem (as I'm liable to do when I'm up to something). Instead of moving 3 disks from A to C, let's move 2 disks from A to B—we already know how to move two disks from one peg to another. Next, move the third disk directly to C. Finally, make another two-disk move, from B to C. We've switched all three disks.

the 4-disk problem

How about starting with 4 disks? Once more, let's begin by restating the problem. Can we move 3 disks from A to B? Sure—it's essentially the same as moving them from A to C. Then we switch the fourth disk directly from A to C, and, finally, transfer 3 disks from B to C.

developing a recursive statement

As you can probably gather, I've insisted on restating the problem in a particular way each time so that we can develop a special insight. We begin to solve the Towers of Hanoi problem for a stack of height n by trying to solve it for a stack of height $n - 1$. This solution must wait until we solve for $(n - 1) - 1$, and so on. Eventually we get to the trivial case of n equaling 1, and can begin to work our way back up.

induction

Almost without realizing it, we've used a high-priced method of thinking called *induction*. We start by solving a simple case of the Towers of Hanoi problem—a tower of height one or two. Then, we show that even if we start with a larger number, we can always work our way down to a problem that we know how to solve.

This is the heart of what will become our recursive solution to the problem. We can move n disks from peg A to peg C by first moving $n - 1$ of them to peg B. The recursive statement of the solution is:

the recursive Hanoi algorithm

1. Move $n - 1$ disks from A to B.
2. Move 1 disk from A to C.
3. Move $n - 1$ disks from B to C.

In steps 1 and 3, of course, we will use the remaining peg as an auxiliary 'holding' peg. Since the empty peg changes, it might be easier to refer to FromPeg, ToPeg, and UsingPeg rather than A, B, and C. A recursive specification is:

> *Goal: Solve the Tower of Hanoi problem for height* N.
> *Stacking plan: Move* N – 1 *disks from* FromPeg *to* UsingPeg.
> *Bound: There's just one disk left (it goes from* FromPeg *to* ToPeg*).*
> *Unstacking plan: Move* N – 1 *disks from* UsingPeg *to* ToPeg.

In understanding program Hanoi, below, it may help to imagine that the three pegs are arranged in a circle rather than in a line. The particular pegs that FromPeg, ToPeg, and UsingPeg represent will change (they'll actually seem to rotate). However, we'll always eventually find ourselves fulfilling the second step of our algorithm, and announcing a particular move.

Towers of Hanoi program

```
MODULE Hanoi;
  (*Recursively solves the Towers of Hanoi problem. Moves disks from A to C.*)
IMPORT IO;
VAR Height: INTEGER;
PROCEDURE Move (Height: INTEGER; FromPeg, ToPeg, UsingPeg: CHAR);
  (*Recursive procedure for determining moves. Keep this order—from,
   to, using—in mind when you read the recursive calls.*)
  BEGIN
    IF Height = 1
      THEN IO.StringOut ("Move a disk from  ");
           IO.CharOut (FromPeg);
           IO.StringOut ("  to  ");
           IO.CharOut (ToPeg);
           IO.LineOut
      ELSE Move (Height – 1, FromPeg, UsingPeg, ToPeg);
           IO.StringOut ("Move a disk from  ");
           IO.CharOut (FromPeg);
           IO.StringOut ("  to  ");
           IO.CharOut (ToPeg);
           IO.LineOut;
           Move (Height – 1, UsingPeg, ToPeg, FromPeg)
    END (*IF*)
  END Move;
BEGIN
  IO.StringOut ("How many disks are you going to start with?  ");
  IO.IntIn (Height);
  Move (Height, "A", "C", "B")
END Hanoi.
```

```
How many disks are you going to start with? 4
Move a disk from A to B
Move a disk from A to C
Move a disk from B to C
Move a disk from A to B
Move a disk from C to A
Move a disk from C to B
Move a disk from A to B
Move a disk from A to C
Move a disk from B to C
Move a disk from B to A
Move a disk from C to A
Move a disk from B to C
Move a disk from A to B
Move a disk from A to C
Move a disk from B to C
```

Modification Problems

1. Why will the world end when the sixty-four disk stack is moved? Modify Hanoi so that it simply counts steps, rather than printing them. For instance, moving four disks took fifteen steps. Have it list the number of steps required for stacks of height one, two, three, and so on, as high as you can go.

Section 17-2 Recursive Array Programming

MANY INTERESTING PROGRAMMING PROBLEMS INVOLVE recursive array manipulation. Arrays are useful because they store large amounts of data, while recursion is handy for describing what we want done with it.

The recursion we'll see is different from our previous examples. Before, we used recursion to solve problems that were stated 'one-dimensionally,' so to speak. We would recursively progress along one particular path until we found the end of the sentence (or evaluated an expression, or reached the beginning or end of a series). Then, we'd tumble all the way back to the start of the sequence of recursive calls, and the program would be finished.

Now, though, we'll use recursion to allow algorithms that require movement in several 'dimensions,' some of which may be false starts, and others of which might help lead in the right direction.

backtracking algorithms

Backtracking algorithms involve a series of trial and error solutions. We travel toward a solution until we know we're heading in the wrong direction, then backtrack to where we think we took the incorrect turn. Then, we try again.

The children's game of 'Hot and Cold' uses a classic backtracking algorithm. One player takes guesses as to what the second player is thinking of. The second player tells the first if a guess is hot (close) or cold (far from the truth). Since in-between degrees of warmth are allowed (warmer, cooler, 'now you're boiling to death!'), the first player is continually guessing and backtracking.

Backtracking algorithms are particularly suited to recursive programming methods. Each recursive call creates a context, or environment, associated with one step toward a solution. The sequence of steps that got us there is preserved in the series of recursive calls. Were we to look at the stack of recursive calls, we'd generally find it growing and shrinking repeatedly as we search further toward—or back away from—potential solutions.

Maze Searching and Recursive Backtracking

Searching through a maze is an ideal problem for recursive solution. I'll state the problem like this:

problem: maze searching

Imagine that you are trapped in the center of a maze. Find all paths to the outside world.

Let's assume that the maze is given to us as a pattern of characters that mark walls, and blanks that mark potential paths to the exit. We'll always start in position 6,6. For example, this is a 12 by 12 maze, with our starting position (at 6,6) marked with a '**s**':

```
+++ ++++++++
+++    +++++
+++ ++  ++++
+++   +  +++
+++++ ++ +++
+++++s++ +++
+++++ ++++++
+++++   ++++
++++  + ++++
+   ++  ++++
+ +    +++++
+ ++++++++++
```

The maze

```
+++ ++++++++
+++ssss+++++
+++s++ss++++
+++sss+ss+++
+++++s++s+++
+++++s++s+++
+++++ ++++++
+++++   ++++
++++  + ++++
+   ++  ++++
+ +    +++++
++ +++++++++
```

A false path

First, let's agree on how we'll store the maze. An easy approach is to hold it in a two-dimensional array of CHAR that goes from 1 to MAXCOL horizontally, and from 1 to MAXROW vertically. When we initially read in our data, we'll put a **+** in each 'wall' location, and leave potential paths blank. As we search the maze, we'll mark each step with an '**s**'. We'll know that we're at an exit if we find ourselves on row 1 or MAXROW, or at column 1 or MAXCOL. The Modula-2 type definition is:

data type definition

```
TYPE ArrayTYPE = ARRAY [1..MAXROW], [1..MAXCOL] OF CHAR;
VAR Maze: ArrayTYPE;
```

Now, how does one get out of a maze? Well, if you're Theseus, you trail a string behind you on the way in, then trace it back out again. Even if you don't have any string, a simple rule—stick to one wall, and always turn right—will eventually get you out of the maze.

Working with a computer, though, imposes certain limitations. Perhaps there's no Minotaur chasing us, but there's no string, either. Moreover, the path may be too narrow for a 'stick to a wall' rule. We may not be able to turn right, and if we reach a dead end, we'll have to back out (as shown in the right-hand illustration, above). We're more in the position of the monster than the hero!

Fortunately, the phrase 'back out' gives us a hint. Recall that one effect of recursive procedure calls is to create a stack of partially executed subprograms. Each one can have variables and pending statements associated with it. If each step down a path involves a recursive procedure call—building the stack—we can also imagine that exiting the sequence of procedure calls will back us out along our original path.

using the recursive stack

How can we state a maze searching method recursively? Let's try some inductive thinking as a warmup exercise. Suppose that we're one step away from an exit. Can we get to the door? Sure—just take one step. It may take several tries (once in each direction) but we'll certainly find the way out.

Suppose that we're two steps away. Can we get to the spot that's only one step away? Yes. If we take one step in any direction, we have a one in four chance of landing in the one-step-away spot. Obviously, we can cover all the bases by stepping left, up, right, and down in turn.

using induction

Now comes the clever step—our inductive leap. What if we're N steps away? Can we get to be one step away? Of course. No matter what N starts at, we can always get to $N - 1$. Since N keeps going down, it will eventually reach 2, and we'll be home free. No matter how many steps away we are, we can always get to a position we're sure we know how to solve.

Being clever was easy compared to stating the algorithm recursively:

first refinement

To search a maze (from the current spot) . . .
mark the current spot as part of the way out;
IF *we're at the exit*
  THEN *print the maze*
  ELSE *search a maze (from a new starting position)*

I've added a new step to the mental warmup we worked on. Each time we call the *search a maze* procedure we mark our present position. The maze itself is passed as a value parameter to each call of the procedure. Note that the path of positions we've marked grows one step longer on each call. You can think of it as a snapshot of our current state—where we are, and how we got there.

The key to the algorithm is finding the new starting position. If we can—if there's no wall there—we'll *search a maze* starting one step to the left. Then, if we can, we'll *search a maze* starting one step up. Then we'll search to the right, and finally down, each time making sure that we're not running into a wall. The shaded section below corresponds to the **ELSE** part shaded above:

second refinement

To search a maze (from the current spot) . . .
mark the current spot as part of the way out;
IF *we're at the exit*
  THEN *print the maze*
  ELSE **IF** *we can, search a maze (starting one step left);*
    IF *we can, search a maze (starting one step up);*
    IF *we can, search a maze (starting one step right);*
    IF *we can, search a maze (starting one step down);*

Once more, the maze, with our current position marked on the pathway out, is passed as a value parameter to the *search a maze* procedure. In effect, we are always searching a maze that has already been partially searched. The computer maintains a stack of copies of the maze—one for each of the partially completed procedure calls. Each copy shows the path that led to that particular location. If we're at the exit, the current copy of the maze shows the way out.

If we don't get to an exit, nothing happens at all. Suppose we take a left turn into a dead end. Since we can't go left, forward, or right, that particular invocation of the procedure ends, and its copy of the maze (with an incorrect path out) is removed. The completed program is shown below.

maze searching program

```
MODULE Threader;
  (*Recursively finds and prints all exit paths from a maze.*)
IMPORT IO;
CONST MAXROW = 12;
      MAXCOL = 12;
      OPEN = " ";
      THEWAYOUT = "s";
TYPE ArrayTYPE = ARRAY [1..MAXROW], [1..MAXCOL] OF CHAR;
VAR Maze: ArrayTYPE;
PROCEDURE StoreTheMaze (VAR Maze: ArrayTYPE);
  (*Reads in the maze.*)
  VAR i,j: INTEGER;
      EndOfLineChar: CHAR;
  BEGIN
    FOR i := 1 TO MAXROW DO
      FOR j := 1 TO MAXCOL DO
        IO.CharIn ( Maze[i, j] )
      END;
      IO.CharIn (EndOfLineChar)    (*which we'll ignore*)
    END
  END StoreTheMaze;
PROCEDURE PrintTheMaze (Maze: ArrayTYPE);
  (*Prints the maze contents, showing the exit path.*)
  VAR i, j: INTEGER;
  BEGIN
    FOR i := 1 TO MAXROW DO
      FOR j := 1 TO MAXCOL DO
        IO.CharOut ( Maze[i, j] )
      END;
      IO.LineOut
    END;
    IO.LineOut   (*Space between solutions.*)
  END PrintTheMaze;
PROCEDURE AtAnExit (Row, Col: INTEGER) : BOOLEAN;
  (*Tells whether or not we are on the border of the maze.*)
  BEGIN
    RETURN (Row = 1) OR (Row = MAXROW) OR (Col = 1) OR (Col = MAXCOL)
  END AtAnExit;
```

(*program continues*)

```
PROCEDURE Explore (Maze: ArrayTYPE; Row, Col: INTEGER);
   (*Recursive procedure for searching the maze.*)
   BEGIN
      Maze [Row, Col] := THEWAYOUT;
      IF AtAnExit (Row, Col)
         THEN PrintTheMaze (Maze)
         ELSE  (*inspect the possible pathways*)
(*up*)         IF Maze [Row – 1, Col] = OPEN THEN Explore (Maze, Row – 1, Col) END;
(*right*)      IF Maze [Row, Col+1] = OPEN THEN Explore (Maze, Row, Col+1) END;
(*down*)       IF Maze [Row+1, Col] = OPEN THEN Explore (Maze, Row+1, Col) END;
(*left*)       IF Maze [Row, Col – 1] = OPEN THEN Explore (Maze, Row, Col – 1) END
      END
   END Explore;
BEGIN
   IO.StringOut ("Reading the starting maze.");
   IO.LineOut;
   StoreTheMaze (Maze);
   IO.LineOut;
   IO.StringOut ("Solutions to the maze are:");
   IO.LineOut;
   Explore (Maze, 6, 6)     (*Start searching from the center.*)
END Threader.
```

```
Reading the starting maze.
+++ ++++++++
+++    +++++
+++ ++  ++++
+++   +  +++
+++++ ++ +++
+++++s++ +++
+++++ ++++++
+++++   ++++
++++  + ++++
+   ++  ++++
+ +    +++++
+ ++++++++++

Solutions to the maze are:
+++s++++++++
+++s   +++++
+++s++  ++++
+++sss+  +++
+++++s++ +++
+++++s++ +++
+++++ ++++++
+++++   ++++
++++  + ++++
+   ++  ++++
+ +    +++++
+ ++++++++++

+++ ++++++++
+++    +++++
+++ ++  ++++
+++   +  +++
+++++ ++ +++
+++++s++ +++
+++++s++++++
+++++sss++++
++++  +s++++
+sss++ss++++
+s+ssss+++++
+s++++++++++
```

Modification Problems

1. It's not likely that you'll want to type in an entire maze each time you run Threader. Modify the program so that it reads its maze from an existing file.
2. How many dead ends does Threader encounter? Modify the program so that it prints dead-end paths, rather than successful ones.
3. Why search for a path to an exit? Why not reverse the problem: start at an exit, and search for a path to the center? Modify Threader to do it.

Section 17-3 Recursion and Pointers

IN MOST PEOPLE'S MINDS, RECURSION FOLLOWS A three-step progression. It starts as curiosity, moves on to convenience, and finally winds up as a necessity.

The final transition often takes place when recursion encounters the pointer. The reason is that pointers can be used to build recursively defined data structures. The best example of a recursively defined data structure is a *tree*. For example:

a general tree

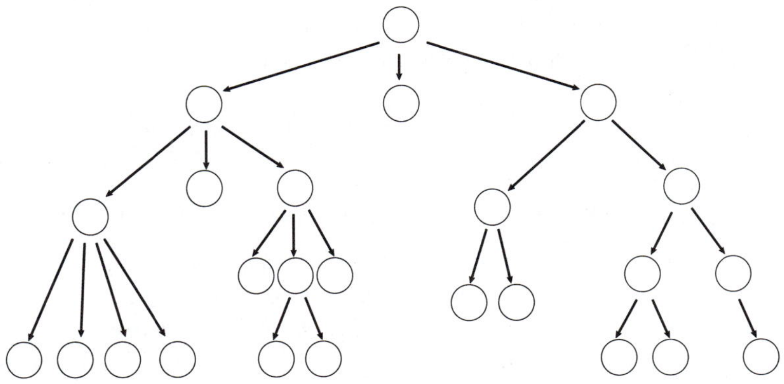

A recursive definition of a tree would describe it as a node that's linked to one or more trees, or to nothing. Now, the subtrees must be *distinct*; no two trees can share the same node. If you look carefully at the picture above, a whole forest is visible. Each labeled node points to one or more subtrees.

tree terminology

The *root* of a tree is its first (topmost) node. The nodes that each element points to are its *children*; it is the *parent*. Finally, a node that has no children is called a *leaf*.

binary trees

Limiting each node to a maximum of two children (i.e. to two potential subtrees) creates a *binary*, or two-part tree. This is a structure we'll explore in detail. The basic node definition is:

```
TYPE NodePOINT = POINTER TO NodeTYPE;
     NodeTYPE= RECORD
                    Data: DataTYPE;
                    LeftPtr, RightPtr: NodePOINT
               END;
VAR CurrentPtr: NodePOINT;
```

Recursive Tree Searching

Suppose that I start by describing the recursive steps of one tree-searching algorithm in English. Assume that CurrentPtr starts by addressing the root.

To Search a Tree . . .

recursive tree-searching algorithm

1. If CurrentPtr's left child isn't NIL, point CurrentPtr at the left child and search the tree.
2. If CurrentPtr's right child isn't NIL, point CurrentPtr at the right child and search the tree.
3. Print the value stored in the current node.

In practice, each time we return to action 1 (the equivalent of making a recursive procedure call) the values associated with the current node will be saved, along with any pending actions.

> Using recursion allows backtracking without backward pointers.

I can rephrase the same algorithm in terms of its bound and plans:

recursive specification

Goal: Print every node's stored value.
Stacking Plan: Visit the left subtree.
Visit the right subtree.
Bound: Reaching a leaf, or node that has no subtrees.
Unstacking Plan: Print the current node's stored value.

I've implemented the algorithm in the recursive procedure InspectTree, below. It assumes that CHAR values are stored in the tree, and that the tree isn't empty.

binary tree inspection procedure

```
PROCEDURE InspectTree (CurrentPtr: NodePOINT);
  (*Visits every node of a binary tree.*)
  BEGIN
    IF CurrentPtr↑.LeftPtr <> NIL THEN
      InspectTree (CurrentPtr↑.LeftPtr)
    END;
    IF CurrentPtr↑.RightPtr <> NIL THEN
      InspectTree (CurrentPtr↑.RightPtr)
    END;
    IO.CharOut (CurrentPtr↑.Data)
  END InspectTree;
```

The effect of InspectTree can be described as:

> Go down the tree as far as possible, trying to go left, but going right if necessary. Print this node's value. Back up one node, then go down the tree again, following the same strategy—left if possible, right if necessary—until you come to a dead end, or a node you've already visited. Inspect this node, then repeat the search process. When there are no more nodes to search—each child has been visited already—the root has been found, and the entire tree has been inspected.

If you have difficulty imagining the operation of a recursive procedure, stepping through an example on a small tree may help. Looking at the boundary cases of a large tree is also useful.

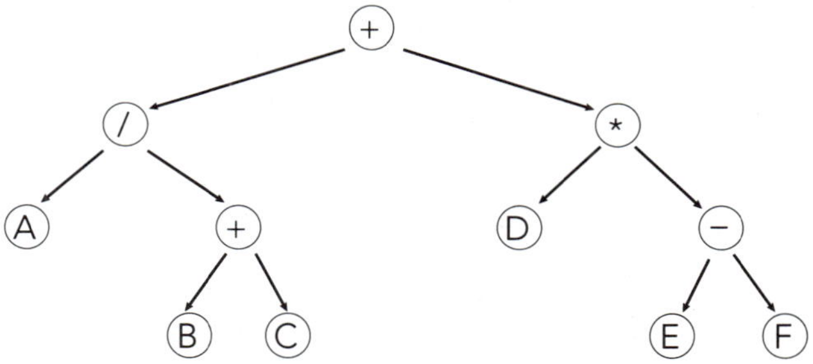

postorder (RPN) search

We can use InspectTree to search the binary tree above. Its nodes will be searched in the following order, called *postorder*. *Postfix* and *reverse Polish* (or *RPN*) are other names for this particular notation.

```
A B C + / D E F − * +
```

If you own a stack-type calculator you'll recognize this as an arithmetic expression, equivalent to `A/(B+C)+D*(E − F)`. In postorder search, a node's subtrees are inspected before the node itself is. As a result, the root is looked at last.

Self-Check Questions

Q. Suppose that I reorder the three statements of InspectTree like this:

```
(*First variation.*)
IO.CharOut (CurrentPtr↑.Data);
IF CurrentPtr↑.LeftPtr <> NIL THEN InspectTree (CurrentPtr↑.LeftPtr) END;
IF CurrentPtr↑.RightPtr <> NIL THEN InspectTree (CurrentPtr↑.RightPtr) END

(*Second variation.*)
IF CurrentPtr↑.LeftPtr <> NIL THEN InspectTree (CurrentPtr↑.LeftPtr) END;
IO.CharOut (CurrentPtr↑.Data);
IF CurrentPtr↑.RightPtr <> NIL THEN InspectTree (CurrentPtr↑.RightPtr) END
```

Are the variations legal? What do they do when applied to the example tree above?

A. Both variations work perfectly well. They change only the order in which values are printed. The first yields a *preorder* search—first a node is inspected, and then its subtrees. Applying the preorder algorithm to the tree above will give us: `+ / A + B C * D - E F`

The second modification produces an *inorder* search. The left subtree is searched, then the node, and finally the right subtree. The search path followed is: `A / B + C + D * E - F`

Programming Binary Trees

The applications of binary trees are unexpectedly diverse. Consider the tree below. Can you guess what it represents?

a mystery tree

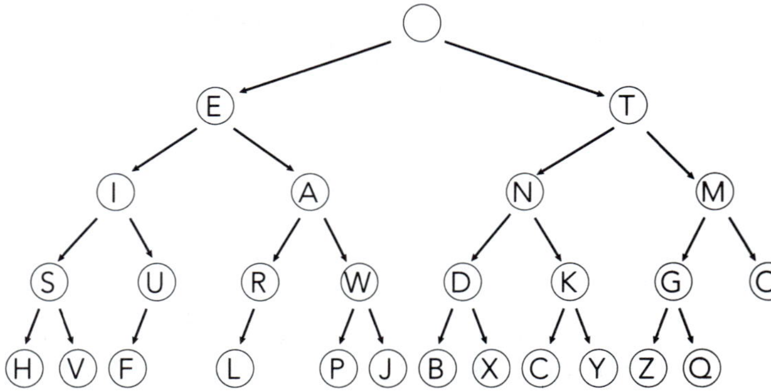

Perhaps the type definition of each node will help.

```
TYPE NextNodePOINT = POINTER TO CodeNodeTYPE;
     CodeNodeTYPE = RECORD
                       Letter: CHAR;
                       Dot, Dash: NextNodePOINT
                    END;
VAR RootPtr: NextNodePOINT;
```

As you may guess, the tree represents the Morse code. Procedure Decode, below, uses the code tree to translate a file of Morse into the letters it stands for. Its argument is RootPtr, which points to the root of the stored code-tree. Since we never backtrack (and don't need a stack) Decode wasn't written recursively. However, we do have to maintain a pointer to the root of the code tree, to start all over again for each new letter.

Morse code decoding procedure

```
PROCEDURE Decode (RootPtr: NextNodePOINT);
  (*Decodes Morse code input. Each full letter must be followed by a blank.*)
  VAR CurrentPtr: NextNodePOINT;
      InputCharacter: CHAR;
  BEGIN
    CurrentPtr := RootPtr;
    LOOP
      IO.CharIn (InputCharacter);
      (**)IF NOT IO.Done() THEN EXIT END;
      CASE InputCharacter OF
        "." : CurrentPtr := CurrentPtr↑.Dot
       | "-" : CurrentPtr := CurrentPtr↑.Dash
       | " " : IO.CharOut (CurrentPtr↑.Letter);
               CurrentPtr := RootPtr
      END (*CASE*)
    END; (*LOOP*)
    IO.CharOut (CurrentPtr↑.Letter)
  END Decode;
```

```
.--. .- -.-. -.-  -- -.--  -... --- -..-
.-- .. - ....  ..-. .. ...- .  -.. --- --.. . -.
.-.. .. --.- ..- --- .-.  .--- ..- --. ...
PACK MY BOX WITH FIVE DOZEN LIQUOR JUGS
```

binary tree applications

Morse can be stored in a binary tree because the dot/dash code is essentially a series of yes/no questions. Surprisingly, most data can be stored and retrieved using binary trees. Interactive computer guessing games are a good example. The computer plays by trying to guess something; say, the name of a singer the player is thinking of. Although singers may have many characteristics, considering only one at a time reduces the description to a string of binary (two-way) choices. For instance:

```
Think of a singer.
Is the singer female?
yes
Is her hair blonde?
no
Is her hair brown?
yes
Is she Tiffany?
no                    etc.
```

The program relies on two kinds of stored data—characteristics, and (ultimately) names. The most crucial set of facts—the relationship between characteristic and name—is contained in the binary tree that *holds* the information.

The program begins at the tree's root and asks the question stored as a string in that node. Whether the left or right node is visited next visited depends on the answer—sometimes a further question is required (and the process starts again), and sometimes we reach a leaf or final node (and with it, a name).

Incidentally, such programs can learn as they play. For instance, the following interaction might take place if a leaf has supplied an incorrect guess:

```
I guessed wrong.  What singer were you thinking of?
Ricky Lee Jones
Type in an additional question I should have asked.
Does she have any talent?
Is the correct answer yes or no?
yes                           etc.
```

Internally, a new node is added to the stored data structure.

Trees of Words

embedding information

A subtle aspect of understanding tree structures is recognizing the relation between the way data is stored and the way that it's retrieved again. This is especially true when a hand-drawn representation of a tree's stored data doesn't (at first glance) show its order or purpose. Consider this tree:

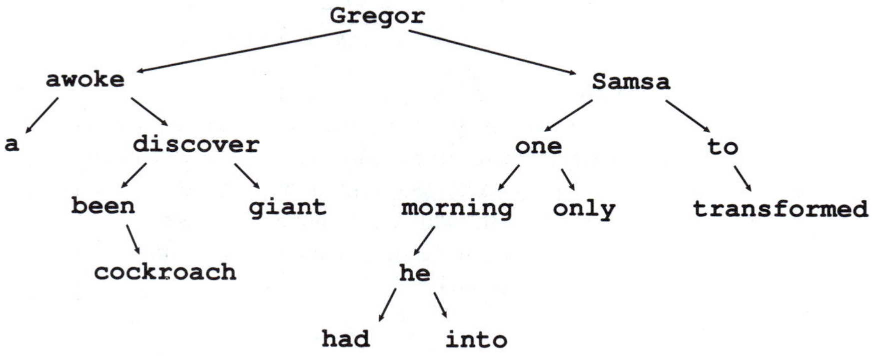

The tree stores the words of this sentence:

`Gregor Samsa awoke one morning only to discover he had been transformed into a giant cockroach`

in alphabetical order (disregarding capitalization). The first word, '`Gregor`', goes to the root of the tree. The second word follows the first alphabetically, so it's stored in the right child. The third word precedes the first alphabetically, so it goes into the left-hand node. The fourth word, '`one`', comes after the first, but before the second. It goes to the root's right child's left child. The final resting place of each word is determined by traveling down the tree, turning left or right, or making a new node as necessary.

As you might imagine, we can describe the ordering algorithm recursively.

To Build an Alphabetically Ordered Tree . . .

binary tree building

1. If the current node is NIL, store the new word there and stop.
2. If the new word precedes the word in the current node, point to the left child and build an alphabetically ordered tree.
3. If the new word follows the word in the current node, point to the right child and build an alphabetically ordered tree.
4. If the new word is the same as the word in the current node, stop.

In procedure AddAWord, below, the final **ELSE** (which represents step 4) isn't necessary, and could be omitted. I've just included it to show you how to touch all the bases.

AddAWord relies rely on two library procedures. Since we're allocating new nodes, we must import the Storage library. And, as I've done on rare occasion, I assume that a STRINGS library can be imported, and that it provides primitives for string comparison.

binary tree building procedure

```
(*Assume IMPORT Storage; IMPORT STRINGS;*)
TYPE StringTYPE = ARRAY [0..15] OF CHAR;
     WordPOINT = POINTER TO WordStoreTYPE;
     WordStoreTYPE = RECORD
                         Word: StringTYPE;
                         Before, After: WordPOINT
                     END;
 . . .  (*Other definitions and declarations.*)
PROCEDURE AddAWord (VAR Current: WordPOINT; NewWord: StringTYPE);
  (*Adds the string NewWord to an alphabetically ordered binary tree.*)
  BEGIN
    IF Current=NIL
        THEN Storage.Allocate (Current, SIZE (WordStoreTYPE));
             Current↑.Word := NewWord;
             Current↑.Before := NIL;
             Current↑.After := NIL
      ELSIF STRINGS.LessThan (NewWord, Current↑.Word)
        THEN AddAWord (Current↑.Before, NewWord)
      ELSIF STRINGS.LessThan (Current↑.Word, NewWord)
        THEN AddAWord (Current↑.After, NewWord)
      ELSE (*The word is a duplicate—NewWord=Current↑.Word.*)
    END
  END AddAWord;
```

AddAWord is probably the most complicated recursive procedure you will have to deal with. Note that AddAWord is an end recursion—the stack isn't used to store values or pending statements. It could easily be written iteratively.

It will come as a welcome surprise to find that a job that seems complicated, like printing the contents of a tree in alphabetical order, is really pretty easy. All it takes is an inorder traversal, which was one of the possible variations on procedure InspectTree, which we wrote a few pages back. The output of PrintlnOrder, below, assumes that CurrentWord currently addresses the root of the `Gregor Samsa awoke...` tree. (I broke the output into two lines myself.)

binary tree inspection procedure

```
PROCEDURE PrintInOrder (CurrentWord: WordPOINT);
  (*Prints the nodes of a non-NIL alphabetically ordered binary tree in order.*)
  BEGIN
    IF CurrentWord↑.Before <> NIL THEN
      PrintInOrder (CurrentWord↑.Before)
    END;
    IO.StringOut (CurrentWord↑.Word);
    IF CurrentWord↑.After <> NIL THEN
      PrintInOrder (CurrentWord↑.After)
    END;
    IO.LineOut
  END PrintInOrder;
```

```
a awoke been cockroach discover giant gregor had
he into morning one only samsa to transformed
```

Modification Problems

1. Write a test harness for AddAWord. Use it to build a tree of words, and then modify PrintlnOrder to print the tree's contents in reverse.
2. Modify AddAWord so that it counts appearances of words as well as organizing them.

Self-Check Questions

Q. Will all structures produced by AddAWord look tree-like? Draw the trees that would be produced by these input sentences:

`a big cat did everything`

`zesty young xylophones wed violins`

A. A quick perusal reveals that the sample sentences are in alphabetical and reverse-alphabetical order. They produce *degenerate* trees—trees that can't be distinguished from ordinary lists.

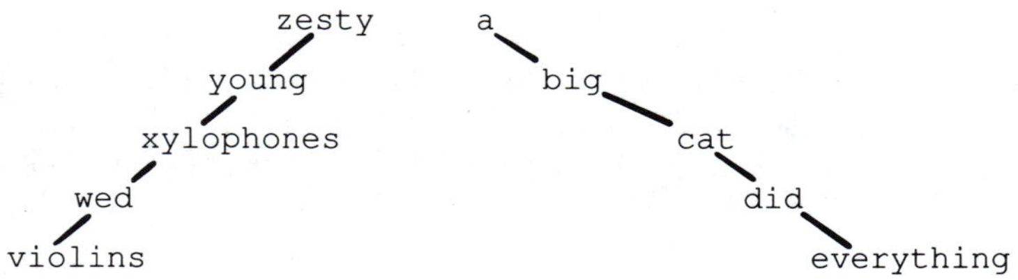

Modula-2 Details

Remember . . .

- A recursive procedure or function calls itself. However, there must always be a check that can prevent the recursive call from occurring, and allow the subprogram to end.
- Uncompleted recursive calls, and the variables that are associated with each call, are described as being *stacked.* A series of recursive calls that accidentally doesn't end eventually leads to stack overflow, and a program crash.
- *End* or *tail* recursions don't use the stack to hold data, and are often replaced by simpler iterative statements.
- Recursion is useful for *backtracking* algorithms because the stack can hold a snapshot of our data at each step of the search for a solution. If a trail fails to pan out, we can backtrack—end part of a sequence of recursive calls—until we reach the step that went astray.
- A *binary tree* is a linked structure that obeys two key restrictions. First, each node points to zero, one, or two additional nodes. Second, all of the nodes in any subtree are totally isolated from the rest of the tree—two trees may have a common root, but they can't share any nodes.
- Recursion is useful for tree-searching algorithms because it allows backtracking without backward pointers. The most common binary tree searches are *preorder*, *inorder*, and *postorder*. They differ in the order in which specific nodes are inspected.

Zen Study Guide

Big Ideas

The Big Ideas of Chapter 17 are:

1. *Many, many recursive subprograms follow a small handful of stereotypical models.* The best way to learn recursive programming is to focus on the similarity between problems, and to learn to state problems in a way that makes their similarities apparent.

2. *Recursion isn't necessary, but it's useful.* Recursive programs are a good example of trading design time for programming time, since we can take advantage of the computer's stack to support more sophisticated algorithms and implementations. Unfortunately, since we can't employ the primitives we're used to—there aren't any 'push' or 'pop' operators for the recursive stack—programs may be harder to design *at first.* In the long run, though, recursion will become one more tool in your bag of programming tricks.

Zen Study Guide

Questions

17-1 Pick your favorite arithmetic operation (addition, say), then ask the basic recursive function question: *Use recursion to implement this operation.*

17-2 An indirect approach to testing understanding of recursion is to ask how deep the recursive stack will become for a particular call. For instance, calling our recursive function Sum(5) led to a stack that was four calls deep. Take three examples from the text, and ask *How deep will the stack of recursive calls be?* for each.

17-3 What is the last call in a series of recursive calls called? What kind of recursion doesn't use the stack? Make up three *What is . . .* questions.

17-4 Pick three recursive procedures or functions, then ask *Write these* without *using recursion.*

17-5 We've seen that if we supply the recursive definition of a procedure or function, then writing the subprogram is pretty easy. It's also not too hard to start with a recursive calculation, then rephrase it as a recursive definition. However, what *is* pleasantly difficult is to write a definition or subprogram that has a misstated bound condition (an off-by-one error works nicely), then ask *Where's the bug?* Try three.

17-6 Consider program Threader. If you look at its recursive procedure Explore, you'll see that four one-word comments (*up, right*, and so on) are invaluable for helping visualize what the program does. You can make up a rather interesting variation on *What's the output?* in three steps:

a) get rid of the comments,

b) rearrange the **IF** statements,

c) ask *Which of these solutions was printed first?*

Try it. Can you pry a similar question out of program Hanoi?

17-7 What's a convenient way to see if somebody understands the basic definition of a binary tree? By drawing three trees, then asking *Which of these is a binary tree?*

Exercises Warmups and Test Programs

17-8 Crash alert—how many recursive calls can you make before you run into trouble? Does the size of the recursive procedure have any effect on the maximum number of calls? What about the size or number of its parameters or local variables?

17-9 Write a recursive procedure that, for no immediately obvious reason, reads a sequence of positive numbers, sums them in reverse, and returns the sum.

17-10 Write a recursive definition of INTEGER division.

17-11 Write a procedure that uses recursion to print the contents of a linked list in reverse.

17-12 Write a recursive procedure that reverses the contents of a linked list without modifying any of its pointers.

Exercises Programming Problems

Recursion that involves loop equivalents.

17-13 The basic stacking problems discussed in the text involved reading, and holding, a sequence of input characters. However, it's just as easy to recursively build a stack that holds the digits of a number. Write recursive procedures that:

a) count the number of digits in an INTEGER;

b) print an INTEGER with no leading blank spaces;

c) print an INTEGER with commas in appropriate places.

Hint: remember that (n **MOD** 10) yields the rightmost digit of n, and (n **DIV** 10) effectively 'strips' the digit off.

17-14 The standard example of a recursive function (always, by law, accompanied by a note attesting to its impracticality), is the computation of factorials:

If n equals 0, then $n! = 1$

If n exceeds 0, then $n! = n \times (n-1)!$

Implement a recursive Factorial function. Can you think of a way to prevent INTEGER overflow, or at least to spot it in advance?

17-15 The second ANSI standard example of recursion is Euclid's algorithm for computing two numbers' greatest common divisor, discussed in section 6-4. The recursive statement of his algorithm is:

if q equals 0 then GCD(p, q) = p

if q ≠ 0 then GCD(p, q) = GCD(q, p **MOD** q)

Write a recursive version of Euclid's algorithm.

17-16 Two ordered lists of numbers are to be merged. Unfortunately, one list is sorted from least to greatest, while the other is supplied from greatest to least. Write a program that uses recursion to merge the lists:

a) from the smallest value to the largest;

b) from the largest value to the smallest.

17-17 When the order in which expressions are evaluated is controlled by parentheses, recursion lets subexpressions be stacked until they are required. For example, the expression below must be evaluated 'in parts' if it is to be evaluated correctly:

```
(2 * ((3 * 4) + (6 – 5)))
(2 * (12 + 1))
(2 * 13)
26
```

An implicit requirement of a recursive program for evaluating such expressions is that *mutual recursion* be employed. A single procedure cannot handle the simultaneous tasks of matching parentheses and evaluating subexpressions. Instead, two procedures are required. One procedure does the work of skipping parentheses, reading operations, and performing calculations, while the second either reads a value *or* calls the first procedure again (if the upcoming character is another opening parenthesis instead of a value).

In words, I can describe the two procedures like this:

to evaluate an expression
skip a parenthesis;
read a value;
read the operator;
read a value;
skip a parenthesis;
perform the operation;

to read a value
IF *the next character is an opening parenthesis*
THEN *evaluate an expression*
ELSE *get the value*

Write a program that evaluates expressions in the form given above. Assume that all input is correct, contains no spaces, and is fully parenthesized.

Recursion that requires array types.

17-18 Write a function that recursively finds the product of the contents of a one-dimensional array of LIMIT components. The code itself is very easy—the hard part is thinking up a recursive statement of the problem. Hint: the product equals the current component's value times the solution for the rest of the array.

17-19 Suppose you choose an arbitrary point on a two-dimensional grid. How many different paths lead to this point from the lower-left corner of the grid? Write a recursive program that finds out. Restrict travel to steps toward the upper or right-hand side of the grid. Hint: treat this as a maze searching problem.

17-20 *Flood filling* is a technique used by graphics programmers to color an arbitrary shape—i.e. fill it with lit pixels. Suppose that we can rely on a function Inside(*x*,*y*) that lets us know if we are within the area we want to fill. The flood filling algorithm is:

```
to flood fill an area
    IF Inside(x, y) THEN
        light the current pixel;
        flood fill x, y+1;
        flood fill x, y-1;
        flood fill x+1, y;
        flood fill x-1, y;
```

The initial *x, y* point is chosen at random.

Define a two-dimensional array of on/off pixels, and an Inside function that determines if any particular pixel is inside a simple shape (a circle or rectangle) 'drawn' on the array. Then, write a recursive flood fill procedure to turn all enclosed elements on.

17-21 A messy programmer eating a pizza pie at her terminal manages to splatter various ingredients over a two-dimensional array. Each spot covers one or more contiguous array elements. As penance, she resolves to write a program that could count the number of splotches she had created.

Help her out by writing a recursive spot-counting program. Assume that your program's input will be a two-dimensional array of INTEGER, with spots marked by 1's, and empty areas filled by 0's. Hint: travel through the array with ordinary loops, but whenever you encounter a spot—an area of 1's—use a recursive procedure to search (and renumber) it.

Recursion that uses pointer types.

17-22 A *leaf* of a binary tree is a node that has no children. Are the leaves of a binary tree listed in the same relative order by preorder, postorder, and inorder searches? Write a program that searches a tree using each of the search paths, saves the 'output' of each search in a linked list, then compares the order of leaves in each list. (Hint: start by creating an ordered list of leaves only.)

17-23 The *internal path length* of a tree is defined to be the sum of the number of paths to each and every node. By definition, counting starts just above the root. The distance to the root is 1, the distance to each of the root's children is 2, each child's children have path lengths of 3, and so on.

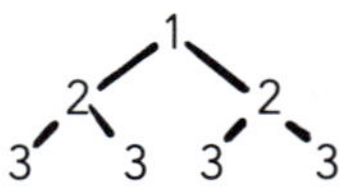

path length 17

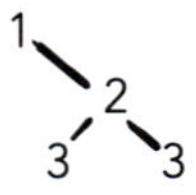

path length 9

Write a subprogram that is passed a pointer to the root of a tree, and returns its internal path length.

17-24 Many chemical substances have more than one molecular form, although each form shares the same atomic formula. These *isomers* can have surprisingly different physical properties, due to geometrical rearrangement of chemical bonds. A simple example of this is found in *optical* isomerism. Two substances that have the same formula and internal bonds, but are mirror images of each other, are optical isomers. They are identical except in the manner they reflect polarized light.

As a problem simpler than writing a isomer-analysis problem, write a procedure that creates a mirror-image of a binary tree. In the mirror image, left subtrees will become right subtrees, and so on. How difficult is it to show that two trees are identical except for being reflections?

'Even high-level languages like Modula-2 provide facilities . . . for low-level work. Fortunately, they are dressed up in formal attire.' (p. 659)

Chapter 18 Bits and Bytes

Bit-twiddling—the ability to deal with computers at the lowest possible level—used to be the exclusive province of those who were willing to forego such pleasures as sun, exercise, and sleep.

Now, though, we live in happier times. Even high-level languages like Modula-2 provide facilities of one sort or another for low-level work. Fortunately, they are dressed up in formal attire. We can work low-level deeds in high-level terms.

Chapter 18 presents two different ways of dealing with the basic unit of computer currency—the *word.* We'll start with *binary files.* These files contain information, but not necessarily in the sequence of characters we're accustomed to. Instead, data comes clumped in the (usually) 16 or 32 bits that make up one word. It's up to us to decide how those bits are going to be interpreted. Two procedures from the FILES module, FILES.WordIn and FILES.WordOut, will be our tools.

In 18-2, we'll use BITSETs to take single words apart. In a sense, it's coincidental that the set operators can also be used to perform operations of the bits of a single word. However, low-level bit manipulations exactly like these are the basis of computer arithmetic—it's all done by moving bits around.

As in Chapter 16, the bits and bytes we'll discuss here are beyond the demands of introductory programming courses. If you get a kick out of them, let me recommend the PC Magazine book *DOS Power Tools* as a good guide to the bit-twiddling you can do in the privacy of your own CPU.

Section 18-1
Binary Files

bits and bytes

SUPPOSE THAT I BEGIN BY EXPLAINING A FEW TERMS that I think you've probably heard of: *bit*, *byte*, and *word*.

A *bit* is a binary digit—a zero or a one. A *byte*, in turn, is a group of eight bits—an eight-digit binary number. Most computer systems use one byte to represent a single character. That's because *a*) eight bits are *long* enough to number 128 different characters and still have one bit left over for error checking, and *b*) eight bits are *short* enough to be transmitted over relatively low-speed wires in a reasonable amount of time. Bytes are standard; they're always eight bits long.

Now, even though most people associate bits and bytes with computers, the *word* is really the basic unit of computing. For example, each value of type INTEGER or CARDINAL is usually stored in a single word, while REAL values typically require two to four words. Structured values, like records or arrays, get as many words as they need.

computer words

I say that words are computing's basic unit because the hardware operations built into each computer system usually operate on full words, rather than smaller bits or bytes. However, the exact size of words varies from system to system. On most small computers a word contains two or four bytes—either sixteen or thirty-two bits.

All this preliminary material becomes relevant when we think about files. Thus far, all of our programs' input and output has consisted of CHAR values. Even when programs have read or printed numbers, they've worked one character—which is to say, one byte—at a time. The IO library has been hard at work translating bytes into digit characters (and then numbers) and vice versa.

However, there are really *two* kinds of files:

character files of bytes

- *Character files* that contain a whole number of bytes, since each character requires exactly one byte of storage.

binary files of words

- *Binary files* that contain a whole number of words, since non-character input and output is almost invariably handled one word's worth of bits at a time.

This means that files of characters are really a special case. All other files, regardless of the Modula-2 type of the values they hold, can be seen or thought of as containing computer words.

Files and SYSTEM.WORD

Here comes the interesting part. When we know that a file holds characters, we can inspect it or add to it in either of two ways: by using IO, or by using FILES.CharIn and FILES.CharOut. The file is a sequence of bytes.

Now, some values are easily translated to and from a CHAR representation. For instance, the INTEGER value 417 can be translated into the sequence of CHAR values "4", "1", and "7".

But other values aren't. When numbers are long, and there are a great many of them, all this translation is a waste of computer time and storage space.

When values are stored in structured types, matters are much worse. Each value in, say, a record, must be extracted before its character representation can be found. Life would be much simpler if we could store, in a file, the exact same sequence of words that the computer uses *without translating it into characters first*.

But how can we put a record into a file? How can we read an array from a file? All it takes it one insight:

> Values of any Modula-2 type, simple or structured, can be stored in a file as a sequence of computer words.

As you might imagine, there are tools that are capable of 'output'-ing variables into a file one computer word at a time, then reading them the same way later. In this text, we'll use two special-purpose procedures from the FILES library: FILES.WordIn, and FILES.WordOut.*

definition of the FILES library

```
DEFINITION MODULE FILES;
   (*This book's file I/O library.*)
   IMPORT SYSTEM;   (*to get type SYSTEM.WORD*)
   IMPORT FileSystem;   (*to get type FileSystem.FILE*)
   TYPE T = FileSystem.FILE;   (*to rename type FileSystem.FILE*)

      . . .        other definitions

   PROCEDURE WordIn (VAR F: T; VAR Val: SYSTEM.WORD);
     (*Read a single WORD-size value from F.*)
   PROCEDURE WordOut (VAR F: T; Val: SYSTEM.WORD);
     (*Append a single WORD-size value to F.*)
END FILES.
```

How do these procedures work? Well, SYSTEM is one of Wirth's standard libraries; it's found in every Modula-2 system. Among other things, SYSTEM makes available the type identifier WORD.

SYSTEM.WORD

> A parameter of type SYSTEM.WORD is assignment compatible with *any* argument that is one computer word long. An open array parameter **ARRAY OF** SYSTEM.WORD is assignment compatible with *any* argument that is any number of computer words long.

WORD as a parameter type

When SYSTEM.WORD is used as the type of a parameter, it matches any argument that is one computer word long, no matter what its type. WORD is only used for matching the type of an argument, though. A WORD-type variable can be assigned to, but it has no mathematical properties.

* Recall that even though every Modula-2 system will have a library like FILES, its name, and the procedure names it provides, aren't standardized.

Let's look at the new tools from FILES in action. The basic application is a *file copy* program. Procedure FileCopy, below, reads one word at a time from file Source, then immediately echoes each word to Destination. FileCopy *could* be used to copy an ordinary file of characters. However, it's more powerful than that. It can also duplicate data files, executable files, encrypted files . . . but don't let me give you any ideas.

copying an arbitrary file

```
(*IMPORT FILES;*)
(*IMPORT SYSTEM;*)
PROCEDURE FileCopy (Source: FILES.T; VAR Destination: FILES.T);
  (*Copies a file of an arbitrary type. Assume that Source and
   Destination have both been opened.*)
  VAR Current: SYSTEM.WORD;
  BEGIN
    LOOP
      FILES.WordIn (Source, Current);
      (**)IF FILES.eof(Source) THEN EXIT END;
      FILES.WordOut (Destination, Current)
    END
  END FileCopy;
```

Modification Problems

1. In FileCopy, variable Current has type SYSTEM.WORD. However, it can have any type that is allocated a single word, say, INTEGER. Modify FileCopy so that it reads INTEGER values entered by the user, but still uses FILES.WordOut to store them in Destination. How much longer does each output number make the Destination file?

2. Suppose you've done problem 1, above. Change FileCopy so that it reads words from Destination, then prints INTEGERs on the screen.

3. FileCopy is the simplest sort of process filter. Make it a bit more complex by having it double the value of every word it reads before it's echoed. As above, you should assume that INTEGER values are being read.

Self-Check Questions

Q. Suppose that you used file Destination to hold a few random INTEGER values (say, 35404, 44105, and 42529), as proposed in the modification problems. What would you find if you inspected Destination with an ordinary text editor?

A. The sequence of bits in Destination can be interpreted in two ways—in word-length chunks (the way we copied them), or in byte-length units (the way an editor will show them on screen). However, the stored values are only meaningful when they're read and interpreted the way we stored them. For instance:

```
as INTEGERs:      35404         |         44105         |         42529
as bytes:   01000101  01001100  01010110  01001001  01010011  0100001
as CHARs:     "E"     |   "L"    |   "V"    |   "I"    |   "S"    |   "!"
```

As you can see, even though byte-length chunks have *some* character connotation, it's entirely meaningless.

I/O of REALs and **RECORD**s

So far we've talked about storing the binary, word-oriented equivalents of variables that take exactly one word of storage. But how can we store a variable that requires more than one word of space? For example:

- A REAL usually requires two to four words.
- An array-type value requires enough room for each component, times the total number of components. An array of ten INTEGERs would probably need ten words.
- A record-type value requires as many words as the total of its fields require.

Open array parameters solve the problem. **ARRAY OF** SYSTEM.WORD, used as the type of a parameter, matches *any* argument, regardless of its type or structure.

Just as an INTEGER can also generally be viewed as a single SYSTEM.WORD value, 'longer' variables can also be seen as being arrays of words.

Consider procedures ObjectIn and ObjectOut, below. They take arguments of *any* type, because any argument can be stored in an **ARRAY OF** SYSTEM.WORD. ObjectIn reads a value from a file, one word at a time, into its parameter Object. Similarly, ObjectOut writes its argument, one word at a time, into a file.

definition of ObjectIn

```
(*Assume IMPORT FILES; IMPORT SYSTEM;*)
PROCEDURE ObjectIn (VAR F: FILES.T;
                    VAR Object: ARRAY OF SYSTEM.WORD);
  (*Reads the next HIGH(Object) + 1 words from F.*)
  VAR i: INTEGER;
  BEGIN
    FOR i := 0 TO HIGH(Object) DO
      FILES.WordIn (F, Object[i])
    END
  END ObjectIn;
```

The value HIGH(Object) + 1 equals the length, in words, of whatever variable is supplied as an argument to Object.

definition of ObjectOut

```
(*Assume IMPORT FILES; IMPORT SYSTEM;*)
PROCEDURE ObjectOut (VAR F: FILES.T;
                         Object: ARRAY OF SYSTEM.WORD);
  (*Writes the next HIGH(Object) + 1 words to F.*)
  VAR i: INTEGER;
  BEGIN
    FOR i := 0 TO HIGH(Object) DO
      FILES.WordOut (F, Object[i])
    END
  END ObjectOut;
```

Let's use ObjectIn and ObjectOut to modify the contents of a file of REAL values that have been stored as words. (We might store them that way to save space.) In pseudocode, we'd like to:

pseudocode

get a value as a series of words;
modify it as a REAL;
store it again as a series of words;

Suppose that Percentage is a REAL-valued variable. We can carry out the pseudocode with:

doubling the numbers

```
(*Assume that Data and Results have been opened.*)
FILES.ObjectIn (Data, Percentage);
Percentage := 2 * Percentage;
FILES.ObjectOut (Results, Percentage);
```

We can carry out the same sort of operations on a structured type, too. Suppose I make these definitions:

```
TYPE FractionTYPE = RECORD
                        Numerator, Denominator: INTEGER
                    END;
VAR Unit: FractionTYPE;
    FracData, FracResults: FILES.T;
```

We can use ObjectIn and ObjectOut to get a record from FracData, double the value of Unit.Denominator, and the return the record to FracResults.

modifying a record

```
ObjectIn (FracData, Unit);     (*Read the whole Unit.*)
Unit.Numerator := Unit.Numerator * 2;  (*Double the Numerator field.*)
ObjectOut (FracResults, Unit);     (*Write the whole Unit.*)
```

Once its value has been obtained from FracData, the Unit variable is just an ordinary record.

When you work with files of structured values, it's important to always, *always* read or write the entire structured variable. To us, it would make sense to assume that Numerator and Denominator are input and output in that order. However, only the compiler knows for sure.

Self-Check Questions

Q. Is this statement legal? Assume the situation described above.

ObjectIn (FracData, Unit.Numerator);

A. Yes, it's legal, but it could be meaningless. ObjectIn will read *any* number of words we tell it to, and store them in its second argument. However, it doesn't guarantee that the values we read should be interpreted as an INTEGER.

Section 18-2 BITSETs

THE SETS WE USED IN CHAPTER 15 WERE THE Modula-2 equivalents of mathematical sets. We defined sets with many different kinds of elements, but always treated them as collections of distinct, meaningful values of one type or another. However, Modula-2 takes advantage of the set type to allow an entirely different family of applications.

definition of BITSET

The BITSET is a standard type whose definition is equivalent to:

TYPE BITSET = **SET OF** [0 .. *WordLength* – 1];

where *WordLength* is the number of bits in one word of memory.

In effect, a bitset is a small set of CARDINAL values. Recall that even though the 8-bit byte is commonly thought of as being the basic computer unit, most internal system operations are actually defined in terms of words. BITSET and the set operators supply a convenient data type abstraction for working with single values of type SYSTEM.WORD.

BITSET is part of Modula-2 so that operations on the individual bits of a word can be carried out clearly, cleanly, and explicitly. For example:

set operations as word operations

Useful operations on words	*Available operations on sets*
turn every bit 'off'	assign { } to the set
turn every bit 'on'	assign it {0 .. *WordLength* – 1}
see if a flag (i.e. a bit) is 'on'	use the **IN** operator
change individual bits	use the set + and – operators
reverse all current bit states	subtract the set from {0 .. *WordLength* – 1}

Two standard procedures that are defined for *all* sets, but are particularly useful for BITSETs, replace the + and – operators:

INCL, EXCL

INCL (S, X); Include element X in the set S: INCL (SampleSet, 1);.

EXCL (S, X); Remove element X from the set S: EXCL (SampleSet, 3);.

Of course, assignments that involve more elements can be made as well.

default BITSETs

> By definition, an expression in braces that is not preceded by a type identifier is assumed to be of type BITSET. {1, 3, 5}, and {0..7} both have type BITSET. { } is an empty BITSET, while {0 .. *WordLength* – 1} is a 'full' BITSET.

In summary, a value of type BITSET takes exactly one word of storage space. This means that we have several different ways of treating a word of memory—different abstractions, each with different primitive operations.

three views of a word

As an INTEGER	*As a* BITSET	*As a* SYSTEM.WORD
has various arithmetic operations defined for it	*has operations for individual bits*	*can only be assigned to or copied*

But why do we need to operate on words in the first place? I can describe three general programming requirements that use words in an intimate manner, and are likely to require variables of type BITSET.

bitset applications

1. *Fast operation.* We can save time by carrying out operations in a wholesale manner. Since each of the components of an **ARRAY OF** BITSET is really a set, the array can be initialized, reversed, and so on, not just one bit at a time, but *WordLength* bits at once.
2. *Compact storage.* When a very large number of values are being held, it may be necessary to rely on a storage scheme that allows just one bit for each value. An **ARRAY OF** BITSET can, for all practical purposes, be crammed with more information than any other data type.*
3. *Memory access.* Finally, some low-level operations, mainly of interest to system programmers, may require inspecting or changing the values of specific bits, or flags, in specific memory locations. For example, a description of the properties of your terminal screen (which you'd need to know for a general-purpose graphics program) is usually saved in such a register.

We won't deal with consideration 3, access to specific locations. However, I will describe an application where fast operation is necessary (graphics), and write a program that relies on compact storage (checking the primality of some very large numbers).

* Incidentally, it's reasonable to wonder why an **ARRAY OF** BOOLEAN isn't just as compact as an **ARRAY OF** BITSET. **After all,** BOOLEAN values, like bits, are just binary (two-value) variables. The answer is that many compilers will set aside a full word of memory for each BOOLEAN variable, even though most of the word goes to waste. This tradeoff is usually justified because it simplifies access to stored variables—the compiler is always able to refer to a variable with an internal address or 'word number' without any bit offsets.

BITSETs Are Fast: Graphics

An everyday application that requires extremely fast operation is graphics. Suppose that the simplest kind of monitor, a black and white terminal screen, measures 5" by 8". Allowing a medium resolution of eighty dots per inch requires 480×640=307,000 pixels (*pic*ture *el*ements), or points on the screen. Since each pixel is either off or on just one bit, with value zero or one, is sufficient to represent the pixel.

Now, if each pixel requires a single bit, memory requirements for representing the screen aren't so bad—only 307,000 bits/8, or about 38,000 bytes. In fact, a special area of memory (or a separate memory altogether) called a *bitmap* is usually set aside for the express purpose of remembering the state of each screen pixel. There's a one-to-one correspondence, or mapping, between pixels on the screen, and stored components of the bitmap.

bitmaps

Unfortunately, treating each pixel as an individual variable can cause problems when a large number of pixels have to be changed. Computers are very fast, but hundreds of thousands of assignments will still require more than the flicker of an eye. The potential for slowdown obviously increases when high-resolution screens of a million or more bits are involved, or when pixels are expected to hold color or grayscale information as well.

Thus, any benefit in increased resolution we might from having more individually addressable bits would seem to exact an unavoidable toll in computational overhead. What kinds of operations are most costly?

time-consuming operations

- Clearing the screen, e.g. to all white.
- Reversing the screen, e.g. from black on white to vice versa.
- Block moves, e.g. moving a section from one screen area to another.
- Erasing, or filling in, large areas.

In each case, the operation itself is quite simple. However, having to inspect the state of, and perhaps change, so many bits individually can be objectionably time-consuming. Even though we seldom worry about efficiency, it's finally an issue.

As you might expect, there is a way out. Suppose that instead of imagining the bitmap as a large two-dimensional array of bits, we treat it as a relatively smaller array of bitsets.

bits vs. bitsets

rows and columns of bits *rows and columns of bitsets*

an array of bits *an array of bitsets*

There are still the same number of rows of bits. However, dividing each row into bitsets drastically reduces the number of easily assignable array components. The exact reduction depends on the number of bits in a word, but it will usually be at least 16.

Defining a Bitmap Display

Let's see exactly why the array of bitsets is so convenient. Suppose we define the data type that underlies a bitmap as shown below. Note that the constant definitions are all offset by one, so that later definitions can run from 0 through, say, *N* (rather than from 0 through *N–1*).

definitions for a bitmap

```
CONST WORDSIZE = 16 – 1;           (*Offset by one so we can count from zero.*)
      BITHEIGHT = 480 – 1;         (*Number of screen rows—height in bits.*)
      BITWIDTH = 640 – 1;          (*Number of screen columns—width in bits.*)
      FULL = {0..WORDSIZE};        (*We'll assume that a pixel set to 1 is 'on' . . . *)
      EMPTY = { };                 (* . . . while one that's 0 is 'off'.*)
TYPE BitmapTYPE = ARRAY [0..BITHEIGHT],
        [0..(BITWIDTH + 1) DIV (WORDSIZE + 1)] OF BITSET;
VAR Screen: BitmapTYPE;          (*The array that represents the whole screen.*)
    Background: BITSET;          (*Useful for quick bitset initialization.*)
    y: [0..BITWIDTH];            (*e.g. 0 . . 639; one particular column.*)
    x, Row: [0..BITHEIGHT];      (*e.g. 0 . . 479; one particular row.*)
    Block: [0..((BITWIDTH+1) DIV (WORDSIZE+1)) – 1];
       (*e.g. 0 . . 39; one of the forty bitsets that make up a row.*)
    Bit: [0..WORDSIZE];          (*e.g. 0 . . 15; one bit in a bitset.*)
```

The time-consuming screen operations listed earlier are much quicker when based on this data type. In the statements below, note that a Block indicates the number of one bitset on any row. Since the individual elements of each bitset are assigned to in parallel, WORDSIZE bits at a time, the full 307,200 bits of the bitmap can be changed in a parsimonious 19,200 assignments. For instance:

screen manipulations

```
Screen [Row, Block] := EMPTY;           (*Blank this block of pixels.*)
Screen [Row, Block] := FULL;            (*Light this block of pixels.*)
Background := {0,2,4,6,8,10,12,14};     (*Initialize a background pattern.*)
Screen [Row, Block] := Background;      (*Set this block to the background.*)
Screen [Row, Block] := FULL – Screen[Row, Block];
                                        (*Reverse this block of pixels.*)
```

Self-Check Questions

Q. Suppose we want to light the single pixel at screen coordinates x,y—at row x and column y. (Assume that the row and column counts begin with zero.) How might we do it?

A. We have to begin by determining which of the forty bitsets in each screen row holds column y. For example a y value of 15 would be found at bitset 0, bit 15, while y=20 is bitset 1, bit 5. y=31, in turn, heads the 'twoth' bitset—bitset 2, bit 0. I'll call the bitset's number Block:

```
(*Which of the forty bitsets is screen column y in?*)
Block := y DIV (WORDSIZE + 1);
```

Then, since we already know the row coordinate (it's row x) we can add the proper bit to the proper bitset:

```
(*Add one more bit to the proper bitset on row x.*)
INCL (Screen [x, Block], y MOD (WORDSIZE+1) );
```

BITSETs Are Compact: Generating Primes

Let's put aside considerations of time, and consider a form of computing that will be limited by space instead—generating prime numbers. Primes were the motivation behind one of the earliest known algorithms, a method of generating primes attributed to Eratosthenes (ca. 276–194 B.C.E.). It is:

problem: sieve of Eratosthenes

> Begin with a sequence of numbers, 1 through *n*. Starting with 2, travel through the sequence and remove every multiple (greater than one) of the starting number. All the numbers that remain will be prime.

Thus, we remove all multiples of 2, then of 3, and so on. Since every number with factors is removed, only primes remain.

A sieve based on an array of whole numbers isn't hard too imagine. We can initialize the sieve by storing each number from 1 to LIMIT in the corresponding component of the Sieve array:

```
FOR every Count from 1 to LIMIT
    Sieve[Count] := Count    (*Store 1..LIMIT in Sieve[1] .. Sieve[LIMIT].*)
END
```

Then, a **FOR** loop does the actual sifting:

first refinement

```
FOR every Count from 2 to LIMIT
    set every multiple of Sieve[Count] to zero
END;
(*Postcondition: every remaining component that isn't zero is prime.*)
```

There are various tricks we can employ to improve this algorithm.

accelerating the sieve

- If Sieve[Count] happens to equal zero—and this will happen more and more frequently as we sift out composite numbers—then all multiples of Sieve[Count] have already been set to zero, too. They all share at least one common factor.
- Must we remove multiples of *every* value from 2 through LIMIT? No. After we reach the square root of LIMIT, we'll find that all larger composite numbers have already been removed. For instance, suppose LIMIT equals 50. All multiples of 8, say, were removed when we sifted out multiples of 2, 3, and 5. All multiples of 11 vanish with multiples of 2 and 3, and so on.

A final pseudocode that incorporates the two speedups is:

final sifting pseudocode

```
FOR every Count from 1 to LIMIT
    Sieve[Count] := Count          (*initialize the sieve*)
END;
FOR every Count from 2 to the square root of LIMIT DO
    IF Sieve[Count] <> 0 it's a new prime, so
        FOR every multiple of Sieve[Count] that's less than LIMIT
            set that component to zero
END all the loops;
(*Postcondition: every non-zero component that remains is prime.*)
```

An Array of Bits

But despite the fine work we've done with the sifting algorithm, the implementation I suggested has a near-fatal flaw.

LIMIT is limited

> The biggest array that we can declare is likely to be only be $2^{WORDSIZE}$ components long. If WORDSIZE is 16, a sieve based on our plan can only find prime numbers smaller than 2^{16}, or 65,536.

This isn't our fault. It's due to the way that compilers usually deal with allocating memory for arrays.

Now, 2^{16} was a pretty big number in Eratosthenes' day. Some years later, though, Pietro Antonio Cataldi of the University of Bologna occupied himself by trying to determine whether $2^{17} - 1$, or 131,071, was prime or not. Let us see if, by clever use of bitsets and a fast computer, we can manage to determine what Cataldi knew in the year 1588.*

The insight that leads to bitsets is quite simple: *look at state*. Do we really need to store entire numbers in the array of potential primes? No. All we care about is each number's prime/not prime state. If we can figure out a way to keep track of them, a single bit for each number would be just what we need.

* The numbers I refer to are called *Mersenne numbers*, and have the form $2^p - 1$, where p is prime. Cataldi went on to determine that $2^{19} - 1$, or 524,387, was prime.

Suppose that instead of working with an array of whole numbers, we create an array of BITSET. Each bitset can hold the status of not just one number, but of a whole set of numbers. If the computer's word size is sixteen bits, then each BITSET can represent the prime/not prime state of sixteen numbers.

More to the point, a paltry ten thousand bitsets can represent the numbers 1 through 160,000—plenty of room to solve Cataldi's problem. Consider this definition:

an array of bitsets

```
CONST LIMIT = 10000;
TYPE BitSieveTYPE = ARRAY [0..LIMIT-1] OF BITSET;
VAR Sieve: BitSieveTYPE;
```

Sieve contains 10,000 components, but since each component is a bitset, we can actually inspect or change the state of $16 \times 10{,}000 = 160{,}000$ bits.

wow!

One serious problem remains. Although we have sidestepped the overall size limitation, we are still in a pickle when it comes to talking about the very large numbers we'll be working with. The solution lies in describing the numbers we're sieving with a more compact notation—as powers of 2 for input and output, and in terms of *word* and *bit offsets* internally. Suppose that all counting begins at zero, and that each word contains one bitset. Then:

offsets

Word and offset	*Overall bit number*	*As a power of* 2
word 0, *bit* 0	1	2^0
word 0, *bit* 15	16	2^4
word 1, *bit* 0	17	2^4+1
word 1, *bit* 15	32	2^5
word 65,535, *bit* 15	1,048,576	2^{20}

As a rule, for numbers greater than 2^4 we'll find that:

Word and offset	*As a power of* 2	*Why it's interesting*
word $2^{n-4}-1$, *bit* 15	2^n	integral powers of 2
word 2^{n-4}, *bit* 0	2^n+1	includes *Fermat* numbers
word $2^{n-4}-1$, *bit* 14	2^n-1	includes *Mersenne* numbers

The key is to work in terms of powers of 2. As it turns out, we'll find that many of the numbers you might be interested in checking the primality of are expressed this way too.

Implementing the Sieve

A final version of the sieve program appears on the next few pages. Use it as a reading example—in the real world, understanding programs is often as important as knowing how to put them together.

Sieve of Eratosthenes program

```
MODULE Sieve;
   (*Uses bitsets to implement a large Sieve of Eratosthenes.*)
IMPORT IO;
CONST WORDSIZE = 16;
        LIMIT = 10000;        (*Comfortably larger than 131,071.*)
        SqrtLIMIT = 100;        (*Multiples of larger primes will be gone already.*)
        POSSIBLE = {0..WORDSIZE – 1};
                (*an 'all-on' bitset for fast initialization *)
TYPE BitSieveTYPE = ARRAY [0..LIMIT] OF BITSET;
VAR Sieve: BitSieveTYPE;
    CurrentWord, TempWord: CARDINAL;
    CurrentBit, TempBit: [0..WORDSIZE – 1];
     (*The 'Current' variables represent*)
     (*the current number, as well as the jump size.*)
PROCEDURE Increment (VAR Wcurrent, Bcurrent: CARDINAL;
                                    Woffset, Boffset: CARDINAL);
   (*'Add' a word and bit offset to current word and bit positions.*)
   VAR Temp: CARDINAL;
   BEGIN
      Temp := Bcurrent + (Boffset + 1);  (*may be larger than WORDSIZE*)
      Bcurrent := Temp MOD WORDSIZE;
      Wcurrent := Wcurrent + Woffset + (Temp DIV WORDSIZE)
   END Increment;
BEGIN  (*main*)
   FOR CurrentWord := 0 TO LIMIT DO
      Sieve[CurrentWord] := POSSIBLE        (*Turn on each bit in every bitset.*)
   END;
   CurrentWord := 0;          (*Don't forget that the 'zeroth' number is 1.*)
   CurrentBit := 0;          (*Sifting will start with multiples of 2.*)
   LOOP  (*until we reach SqrtLIMIT*)
      Increment (CurrentWord, CurrentBit, 0, 0);  (*Advance by 1 bit.*)
      IF CurrentBit IN Sieve[CurrentWord] THEN
         (*If this 'number' is prime, remove its multiples.*)
         TempWord := CurrentWord;          (*Counters for the inner loop.*)
         TempBit := CurrentBit;
         LOOP  (*until we come very close to LIMIT*)
            (*Increment the current position 'number'.*)
            Increment (TempWord, TempBit, CurrentWord, CurrentBit);
            (**)IF (LIMIT – CurrentWord) < TempWord THEN EXIT (*inner*) END;
            (*Remove the multiple from the sieve.*)
            EXCL (Sieve[TempWord], TempBit)
         END  (*inner LOOP*)
      END; (*IF*)
      (**)IF CurrentWord = SqrtLIMIT THEN EXIT (*outer*) END;
   END; (*outer LOOP*)
```

(program continues)

```
(*The sieve of primes is complete. Now all that remains is to check certain bits.
  If they're still 'on', the numbers they represent are prime.
  Recall that 2^n - 1 is word 2^(n-4) - 1, bit 14.*)
IF (*bit*) 14 IN (*word*) Sieve[8192 - 1]  (*Check 2^17 - 1, where 2^13 = 8,192.*)
   THEN IO.StringOut ("131,071 is prime!")
   ELSE IO.StringOut ("131,071 isn't prime!")
END;
IO.LineOut
END Sieve.
```

```
131,071 is prime!
```

Modification Problems

1. Modify program Sieve to find out how many primes there are below 100,000.

2. Marin Mersenne (1588–1648), close friend of Descarte, Pascal, and Fermat, lent his name to describe numbers of the form $2^p - 1$, *where p* is prime. If the number (we'll refer to it as M_p) is also prime, the number is called a *Mersenne prime*. Which Mersenne numbers are prime? Which are composite?

'In many ways, what we call 'software engineering' is not really engineering at all—our methods are merely updated versions of 'Take the Chisholm Trail' and 'Git along, little dogie.'' (p. 684)

Chapter 19 Software Engineering

Writing programs is usually just that—writing. But when programs become large enough to be described (respectfully!) as *software*, we usually begin to think about programming in terms of construction.

Now, most people have an intuitive sense of the process of construction. How is a building built? Plans are prepared, a big hole is dug, a foundation laid, and so on. We can even imagine how an old building might be tested to see if it's still safe. We know this sort of sequence even if we're not civil engineers because even as kids, we see evidence of building construction around us all the time.

Software construction, in contrast, is invisible. Worse yet, it often runs against intuition. Software is a wonderful construction material—weightless, powerful, capable of prodigious feats—but it is a mistake to treat code too literally, as though it were actually bricks and mortar. A software system may be weightless, but it can still cause some damage if it collapses around your ears. Time after time, organizations that depend on software have been stunned to find just how incomplete a 'finished' software product can be.

Chapter 19 describes the different stages of the software *life cycle*: analysis, specification, design, implementation, testing, and maintenance. Then, we'll take a closer look at the problem of deciding whether or not a program is *correct*.

Section 19-1
The Software Life Cycle

SOFTWARE ENGINEERING IS THE SCIENCE of the development of *software systems*—programs, large and small, that will be used to solve real problems. Like more traditional engineering endeavors, the study of software engineering is largely motivated by a desire to avoid repeating mistakes—mistakes that were only belatedly recognized as involving engineering at all. The opening paragraphs of a seminal text on software engineering (written not two decades ago) describe the situation:

> 'No scene from prehistory is quite so vivid as that of the mortal struggles of great beasts in the tar pits. In the mind's eye one sees dinosaurs, mammoths, and sabertoothed tigers struggling against the grip of the tar. The fiercer the struggle, the more entangling the tar, and no beast is so strong or so skillful but that he ultimately sinks.
>
> 'Large-system programming has over the past decade been such a tar pit, and many great and powerful beasts have thrashed violently in it. Most have emerged with running systems—few have met goals, schedules, budgets. Large and small, massive or wiry, team after team has become entangled in the tar. No one thing seems to cause the difficulty—any particular paw can be pulled away. But the accumulation of simultaneous and interacting factors brings slower and slower motion. Everyone seems to have been surprised by the stickiness of the problem, and it is hard to discern the nature of it. But we must try to understand it if we are to solve it.'*

the tar pit

Dozens of software systems have fallen into the tar pit Brooks describes. Sometimes the development effort is entirely unsuccessful; a satisfactory system isn't constructed. Other systems may be accepted by the end user, but be late, over budget, poorly suited to the intended application, or hamstrung by inefficient use of system resources. In either case, the programmer and user are equally dissatisfied with the end result.

Why wasn't the software problem immediately obvious? Well, until the mid 1960's, the cost of hardware greatly outweighed the cost of software, and the capacity of hardware was easily overwhelmed by the capability of software. But in the late '60's, a series of studies pointed up two unexpected facts:

- *Software was hard and expensive to produce.* People were shocked to find that the average professional programmer only managed to produce, on average, *twelve* debugged lines of code *per day*!
- *The cost of producing software was only the tip of the iceberg.* Even more astonishing was the discovery that some eighty percent of the cost of software was incurred after it was placed in service.

The concept of a *software life cycle* developed in response to these findings. We can pick out distinct phases in the life of every program:

* *The Mythical Man-Month*, by Frederick P. Brooks, Jr. Addison-Wesley Publishing Co., 1975, pg. 4.

stages in the life cycle

1. *Analysis* of the problem.
2. *Specification* of the software's abilities.
3. *Design* of the software.
4. *Implementation* or coding.
5. *Testing* of the completed system.
6. *Maintenance* and evolution of the system.

We'll look at each of these stages in turn. But first, I'll digress briefly to caution against a too literal reading of their descriptions.

Two Models: Waterfall and Spiral

As presented above, the software life cycle is often described as the *waterfall* model of program development. Water only flows downhill; in the waterfall model, software only proceeds forward from step to step.

As you may be able to tell, notions about the predictability of the software life cycle were strongly influenced by everyday perceptions about more ordinary kinds of construction. If plans for buildings are completed before construction starts, why shouldn't plans for software be completed as well?

theory meets practice

> Unfortunately, while the waterfall model accurately describes many facts about software at various points in its lifetime, it's a misleading tool for planned software development. In practice, reassessment and revision, especially in the early stages, are almost invariably required. In other words: *Stop the life cycle, I want to get off.*

An alternative to the waterfall model, called the *spiral* model, is often used to guide project work. In a phrase, the spiral model is based on the idea that you get smarter as you go along. Planning is *never* sufficient, because you only gain the knowledge needed for effective planning through experience.

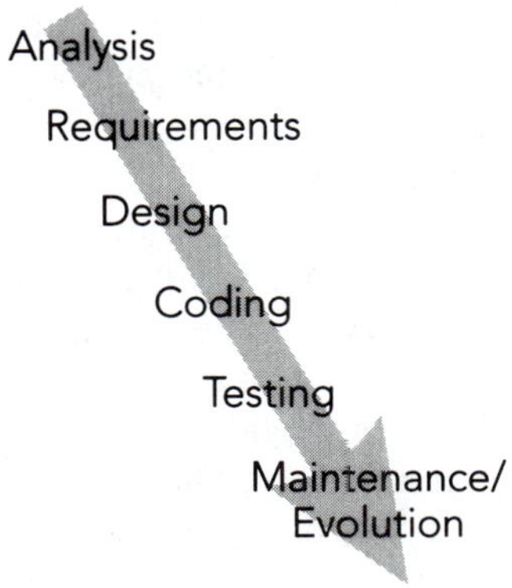

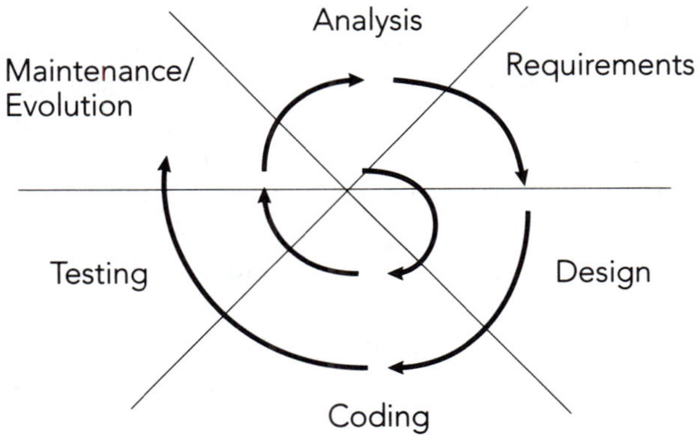

The 'waterfall' depicted on the left allows just one try. The spiral on the right implies that at least two trips through the entire process will be required.

Analysis

Analysis of the problem is probably the most difficult step in the software life cycle. The real-life problems faced by potential computer users are a far remove from the neatly prepared exercises presented in computer science textbooks. This is not because real-life problems are any more difficult; indeed, the reverse is quite often the case. Instead, it is because customers and programmers often speak very different languages. As a result, simple problems can be misunderstood, or difficult ones understated.

Analysis tries to answer the question 'What should the software do?' in a manner that will be meaningful to the software designer. Simply saying that the software should 'keep our books,' or 'monitor our test equipment,' or 'give sample Modula-2 examinations' doesn't give much direction. Instead, the analyst must work with the end user to answer many questions about the system. The system is pictured both as it is now, and as it may come to be:

questions for analysis

- What will the system's input be? What output should the system produce?
- Will these requirements change? How seriously? How often?
- What sort of people will use the system? Can they be specially trained?
- Will there be errors in input? In stored data? How should errors be fixed?
- What kind of equipment is available? What can be obtained?
- How fast should the system work? How reliable does it have to be?
- Will the system grow? In what directions?

software prototypes

One current area of computer science research that's intended to help with the problems of analysis is *rapid software prototyping*. The idea of building prototypes, or test models, is common to many engineering disciplines. A prototype is a scaled-down model that can be examined and tested before any commitment is made to a final design. Prototyping is especially appropriate for software systems, since it helps the user and designer communicate their understanding of what problems are, and how they can be solved.

Software prototyping relies on the idea that many software components—routines for input, output, data sorting—are more or less independent of the systems they're found in. These building blocks can be joined, with a minimum of new code, to rough out a prototype of the end-user's system. This working model can be used to help give the user an idea of a computer's capabilities, as well as to give the software analyst a feel for the users' needs.

Modula-2 is particularly well suited for prototyping. The module system was designed to make it easy to alter details of a procedure's implementation, or to change its 'look and feel' entirely. You might want to explore your own Modula-2 system's libraries to get a sense of just how many built-in tools it already has.

Requirements Specification

The specification of a software system's requirements is a formal statement of its capabilities, capacities, and constraints. Whereas analysis is meant to determine in general terms what the proposed system was supposed to do, the requirements specification states in detail what the finished system *will* do.

A requirements specification will be referred to throughout the entire software development process. It can act as the contract between programmer and end user, and is often the only point of contact between the two groups. If a feature or requirement isn't specified in the requirement, it's not liable to show up in the final product. The requirements specification will also generally be the standard against which the final system is tested.

Specification of software requirements is a more difficult task than it might appear to be at first glance. Consider the variety of areas that a specification has to define:

details of the specification

- It must state the specific abilities of the system—the commands that will be available to the user. These are the system's *functional requirements.*
- It has to specify the assumptions that will be made about the system's input, users, response time, data—its operating environment.
- It must define the system's limitations—how many users will there be, how much data need be handled, etc.
- It has to describe any special hardware requirements, or any restrictions imposed by hardware limitations.
- It must specify possible modifications to the system that have to be allowed for in the system's design.
- It should describe the nature and extent of documentation that are supposed to accompany the system. A preliminary users' manual may be required as well.

The functional specification is the most visible part of a requirements document. In small systems, stating functional requirements is a pretty straightforward task. A list of allowed commands may be all that is needed. In more complex systems, though, a list of commands may not be enough—we'll need a better mechanism for describing the big picture.

A variety of schemes have been developed to help in stating specifications. Most of them are elaborate charting systems that try to fix the action of a program on paper. In general, specification systems are object-oriented. They describe a program in terms of its data—how it is stored and transferred, and what can be done with it in the process.

Once we begin to talk about data, programs can't be far away. Before we begin to code, though, it's a good idea to spend some time on design.

Design

Analysis and specification help determine *what* should be done. Design, in contrast, specifies *how* it should be done. The end result of the design phase is a *software blueprint* that can be implemented with a minimum of difficulty.

software blueprint

Although it is inconceivable that one would embark on the construction of, say, a building without a detailed plan of action in hand, software projects used to be undertaken routinely without the barest hint of advance planning. Widespread use of the **goto** statement, which allowed unstructured jumps to arbitrary locations in a program, made matters even worse. Not only wasn't software understood before it was written, but it couldn't be followed *after* it was written either!

But how do we go about designing software, and how is a software design shown? *Pseudocoding* is quite familiar to us by now, and it seems like a self-evident idea. But in the early 1970's, pseudocode was just an experimental alternative to flowcharting (drawing a series of boxes, connected by arrows, that showed a program's flow of control). A number of formalized pseudocoding systems were defined as *program design languages*, or PDL's. Pascal, Modula-2's predecessor, was one of the first languages to incorporate many of the techniques that PDL's suggested.

pseudocode

Another idea that we take for granted is *structured design*. Indeed, it has become a generic term for programming with an emphasis on modularity, supported by minimal use of the **goto** (and, of course, by languages that encourage modular programming and don't *need* the **goto**). It might seem obvious to us that writing procedures and using parameters is the right thing to do. But that wasn't always so. Many of the ideas of structured program design were first formalized in a series of design texts and seminars put together by Larry Constantine and Ed Yourdon.*

structured design

I've already pointed out that structured design is best suited for goal-oriented programming. In recent years, *object-oriented* design has been used to complement structured design. Object-oriented design, which we've used in our own work with definition and implementation modules, focuses on managing data rather than on accomplishing specific goals. It's a marvelous tool for splitting up the tasks of large programming projects.

Although it's an overused and imprecise term, *abstraction* is the key idea behind all of the design schemes. We want to separate design from implementation—*what?* from *how?* Thus, regardless of the system used to design and specify code, the design phase has two goals. First, any ambiguities located in the requirements specification should be found and clarified. Second, there should be a detailed guide prepared for the next step—coding.

* Such as *How to Manage Structured Programming*, by Ed Yourdon, YOURDON Inc., 1976 (and dedicated 'to my first, third, and fourth wives...').

Coding

For most of the history of computing, coding the main program has been thought of as the programmer's main activity. Surprisingly, surveys consistently show that program coding occupies only about 20% of the time and effort involved in producing software systems. Nevertheless, carefully made specifications and designs are all for naught if they are not well-implemented.

The software engineer looks at the coding phase of programming in several different ways. First, there's the code itself. What language, or languages, should be used? What should modules contain? How long should subprograms be? What rules should be followed for defining identifiers? How should the code be laid out? How efficient must it be? How detailed should comments be? Are any kinds of programming tricks forbidden?

Next, there's the programming staff. How can work be divided? What are the responsibilities of individual coders? How closely should programmers be supervised? How should proposals for coding be reviewed? How much communication should there be between programmers, and how can it be arranged? How can we estimate the difficulty of specific program segments?

Finally, there are the methods used to produce the code. What electronic tools are there for coding support? How are different versions of programs maintained? How should debugging or testing code be built into the software? Will different hardware—terminals, printers, interactive equipment—have any effect on programmer productivity?

Of all the phases of software development, coding is probably the least formalized. One reason is that the management of coding efforts has turned out to differ from other kinds of management in unexpected ways. A particularly instructive example comes from attempts to apply general notions of manpower to coding. Now, in most sorts of organized activity—drafting, or claims processing, or chopping wood—the volume of work accomplished grows in rough proportion to the amount of effort expended. As a consequence, doubling effort doubles results—or halves completion time, more or less.

The unanticipated results of applying this rule to programming gave Brooks the title of his book: *The Mythical Man-Month.* He found that most project managers treated software production just like other sorts of production. When a project fell behind schedule, they would add additional programmers. To their astonishment, they often found that adding help made matters worse! Extra programmers only made the project fall further behind schedule. Brooks characterized this experience rather cynically in Brooks' Law: *Adding manpower to a late software project makes it later.*

the mythical man-month

On close inspection we can recognize the two characteristics of software production that give the law its grain of truth. First, new staff must be trained. Even if they are expert programmers, the current project must be explained; they have to be brought up to speed in the project's goals, rules, coding strategies, etc. Second, they must communicate as they work, since a program's modules can never be made entirely independent of each other. A group of n

people can meet in $n!/2(n-2)!$ different pairs; which means that doubling a group from three programmers to six increases the number of two-person meetings they can hold by a factor of *five*.

A variety of solutions have helped improve the situation. Insistence on modular software design reduces the amount of communication needed between groups, and helps reduce the impact of delays in software production. Layered design lets programming groups work independently, but with the assurance that they share a common goal.

programming teams

A very effective approach toward programmer management, suggested by Brooks and others, is the creation of *programming teams*, in which outstanding 'superprogrammers' are given sufficient staff support to avoid unnecessary distractions. Such teams include a 'copilot' who helps the superprogrammer with design, a 'language lawyer' who specializes in knowing the ins and outs of a particular language or programming system, an 'editor' in charge of documentation, a 'program clerk' who can handle secretarial details that require technical expertise, etc.

Testing

The test phase of the software life cycle can involve as much time and effort as the coding phase. No matter how carefully a program is planned and coded, it will still contain bugs and imperfections. In addition, large software systems may not even be fully assembled until the test phase begins. A ship's maiden voyage is traditionally a shakedown cruise; a program's first run marks the start of the test period.

module testing

The test phase has several goals. Most obviously, we want to find bugs introduced during the coding process. The attempt to find this sort of bug relies on methodical testing of individual program modules. This is usually known as *module testing*. Modula-2 is an especially effective language in this regard, because libraries can be compiled and tested separately.

integration testing

Next, modules are put together to form systems or subsystems. This is sometimes called *integration testing*, since we are integrating the activity of different modules. The bugs found during integration testing are generally due to design errors. These will typically involve the interface between separate modules. Code may have been implemented correctly, but programmers may have labored under false impressions of the input their particular modules could expect, or the output they were supposed to produce. If problems found in integration testing involve basic data storage methods they can be very serious, and involve a considerable amount of code rewriting.

acceptance testing

In the final testing step, the completed system is presented to its end users for *acceptance testing*. Well-managed software systems will usually work their way into this phase slowly by *alpha* and *beta* site testing. The system is distributed to a limited number of sites for feedback and refinement before it is presented for final validation and acceptance. Unfortunately, problems that come up in acceptance testing sometimes date back to the original analysis and specifications phases. The program does what is called for in the design, but the design itself may be incomplete or incorrect from the end user's point of view.

Is it possible to write software so well that it doesn't require testing? Although this was a fond hope for a number of years, the idea that programs could be proved to work as they were being written is fading fast. We'll look further at the issue of formal program *correctness* in section 19-2.

Maintenance and Evolution

Software may outlive its usefulness, but it never wears out. Once an individual or organization has put effort into learning (or adapting to) a particular software system, there is a great tendency to prefer modification of existing software over the acquisition of new software. Even armed with this understanding, it will probably come as a great surprise to find that maintenance and modification of a program can cost two to four times as much as its original coding—or up to 80% of the costs contained in the entire software life cycle.

correction

There are a number of motivations for modifying software once it is presumed to be complete. A first category involves *correction.* Although the rate at which bugs are found declines drastically, they usually appear throughout a software system's lifetime. Occasionally they will fall into the 'bug or feature?' column, especially when a user employs a poorly documented or unintended command.

perfection

A second motivation for modification is the desire to improve the system's usefulness—*perfective* changes. Surprisingly, a successful system may require the most modifications, since it may be widely adopted in environments other than the one it was originally intended for. The addition of new features, or fine-tuning and improvement (perhaps by using new algorithms) of existing features are typical of this kind of modification and evolution.

adaptation

A third category of modifications are sometimes called *adaptive* changes. These are mandated by changes in the system's operating environment. They may be caused by improvements in the hardware system, or by changes in external software the system relies on, for instance, a change in the computer's operating system.

The understanding that maintenance accounts for such a large portion of software costs has been one of the prime motivating factors in the development of software engineering. Well-understood systems—say, automobiles—are designed with maintenance in mind. As a result, their construction is fairly modular, and points that need to be checked regularly are easily accessible.* Software is only recently developing this kind of self-awareness. Embedded debugging tools are a perfect example of built-in aids for program maintenance.

* There are notable exceptions. Consider, for example, the Chevy Monza, which required that the engine be partly removed in order to change the spark plugs!

A Retrospective View

Imagine, if you will, an enterprising young group of cowboys in the old West. As they prepare to drive their herd to market, they mentally review the tools at their disposal. There are different trails to take, and scouting reports on the likelihood of encountering rustlers or maurauders along alternative routes.

They are also able to make estimates about changing conditions: how much the herd will shrink during the trip because of disease or accident; how much weight the average steer will lose, and how much can be replaced by a last-minute stop at the waterhole; how many cowpokes will mutter disgruntled comments and have to be shot on the spot.

If and when problems do arise, they also know how to deal with them. They know where to find box canyons for containing runaway herds, and how far away the Cavalry is in case a rescue is needed. They know the cries all proper cowboys know—'Come a ti-yi-yipee,' and 'Yipee-i-ay-ky-o.'

Indeed, if our young cowboys had thought of it, they might have fancied themselves more, much more, than simple cowpokes. 'Beef Transport Engineers' would have seemed to be a more appropriate term, certain to garner the professional respect every cowboy thought was his due.

Fifty years from now, computer scientists may mock the software engineers of the 1990's for having presented themselves with the same seriousness as our hypothetical cowhands. After all, a large software project is very much like a herd of cattle. No individual steer can match the cowboy's strength or guile—any dogie that acts up can be reined in for special attention.

> But when a project stampedes, the notion that it can be controlled is illusory. Instead, we gain a sense of control by going with the flow. We give in gracefully to forces that are beyond us, and try to guide runaway projects into the nearest box canyon.

In many ways, what we call 'software engineering' is not really engineering at all—our methods and estimates are merely updated versions of 'Take the Chisholm Trail, pardner,' and 'Git along, little dogie.' Rather than engineering (that is, a well-organized and well-understood set of underlying principles) we have *methodologies*—a collection of programming techniques that usually work. To paraphrase Chamfort's comment on philosophy (and eighteenth-century medicine), we can offer a great many drugs and a few very good remedies, but almost no cures.

If we look at our methods, it's easy to see just how deep our ignorance runs. We do not really know a 'best' size for modules; we only know that big modules are confusing. We don't have any idea how long software will last; we only know that it lasts longer than it's supposed to. We don't understand the best way to make changes or improvements in software; we only know that it's an awfully good idea to keep records that are detailed enough to let improvements be undone.

As Fred Brooks himself has pointed out (recall that I cited his *Mythical Man-Month* at the beginning of this chapter), the very fact that his twenty-year-old book continues to be relevant is rather depressing. Yes, problems are solved, but they tend to be the surface problems. We get better at avoiding superficial programming errors, but progress in dealing with the inherent complexity of software systems is slow.

Perhaps the best thing to do is to take the advice I've given in the last few pages to heart. In a sense, *all* programming is itself a vast software project. The languages that we program in, and the schemes we use to design and test our programs, are, in this view, prototypes. Some may be more or less successful, but it is unrealistic to imagine that we're putting together systems that wouldn't be better off for being tossed out and rewritten from scratch at least once every decade.

How to Schedule Successfully

A manager asked a programmer how long it would take him to finish the program on which he was working.

'I will be finished tomorrow,' the programmer promptly replied.

'I think that you are being unrealistic,' said the manager. 'Truthfully, how long will it take?'

The programmer thought for a moment. 'I have some features that I wish to add. This will take at least two weeks,' he finally said.

'Even that is too much to expect,' insisted the manager. 'I will be satisfied if you simply tell me when the program is complete.'

The programmer agreed to this.

Several years later, the manager retired. On the way to his retirement lunch, he discovered the programmer asleep at his terminal. He had been programming all night.

From *The Tao of Programming*, by Geoffrey James, © 1987, InfoBooks, Santa Monica, Ca. 90406

Section 19-2 Program Correctness

THE STUDY OF PROGRAMMING HAS ALWAYS OCCUPIED an uneasy boundary between science and art. Somehow, it seems obvious that programming is a scientific pursuit. But although we can easily make sage observations about the characteristics of programs, we can seldom enunciate any underlying principles that will help us generate new ones. Where programming is concerned, computer science has had about as much relation to science as computer art has borne to art!

Nevertheless, it is only natural that, as scientists, we continue the search for a more formal basis of programming. A particularly interesting area of research is involved with attempting to define a language and method for reasoning about program *correctness*. This section introduces the notion of a program *proof*, and shows how mathematical modes of thinking have some use, but also some shortcomings, when they're applied to the writing of programs.

What Is Correctness?

How do we show that a program works? A few weeks ago we might have given the obvious reply—*run it.* By now, though, we've probably learned to be a bit more cautious. We understand that, at best, running a program (and checking its results) shows that it works for a particular set of data. Making up a broader range of test data gives us a stronger feeling that the program will always work, but even testing is usually limited to ferreting out bugs whose symptoms we can imagine in advance.

Let's look at some of the methods used to gain assurance about the correctness, or reliability, of programs. Now, it's tempting to think that we can simply prove that any program will always work, just as we might prove that a mathematical theorem is correct. However, we'll find that confidence in the correctness of programs is much like confidence in the empirical correctness of engineering methods, rather than the more abstract notions of eternal correctness in mathematics.

What do I mean when I talk about developing confidence? Well, confidence lets us ride in airplanes, or cross bridges. We can't prove that a bridge won't ever fall down once it's in place, but we can feel certain enough about it to trust the bridge with our lives. A combination of tests join to give us this confidence. The bridge may follow the same design as other structures. We can build models for aerodynamic testing in a wind tunnel, and employ mathematical formulas for the design of structural members. We may even go so far as to stress randomly selected beams, cables, and the like to the point of destruction, in order to establish minimum strength levels.

what is confidence?

Program tests are merged in the same manner to give us confidence in code. Some features are trusted because they've worked in similar programs, while others are allowed because the programmer and her peers believe that they will work. Parts of a program may have to undergo exhaustive testing by being run on carefully gathered real data, while for others, artificially manufactured data is good enough. Finally, some portions of the code may be so crucial that we have to try to prove, on paper, that they will always work.

Let's consider some of the less formal methods before we see what a program proof looks like. The first sort of testing most programs undergo can be called *bench testing*. The programmer explains her work to another programmer, or small group of programmers, in a structured walkthrough, or walking tour, of the code. This kind of examination is useful for two reasons. First, programmers less intimately involved with actual coding may spot conceptual errors that have escaped previous notice. Second, the discussion can lead to useful suggestions for tests that can be made at later stages of production.

bench testing

Static analysis of the program is a step that's usually reserved for very large systems. Static analyzers are programs that examine the source code (e.g. the Modula-2 version) of a program without actually running it. They're able to spot certain kinds of errors that aren't always found by the compiler. One kind of error is the use of uninitialized variables in assignments, or as arguments to value

static analysis

parameters. A more interesting error that can be found through static analysis is the existence of unreachable code segments that won't ever be run, no matter what program input is.

trace tools

Trace tools give us a window into program execution as the program runs. A simple kind of trace will print a message every time a subprogram is entered; arranging for a count and display of totals is barely more difficult. A trace tool might also keep track of changes in the value of a particular variable, or group of variables. In the hands of an expert, a trace is an invaluable tool for spotting potential errors in program design. Most Modula-2 systems have interactive debugging facilities that have trace tools built in.

Finally, *data testing* is a method we should be quite familiar with. Builders of large systems often create *data generators*; programs that automatically produce data that has a set of characteristics specified by the programmer. The output of a data generator might be sophisticated data intended to make a program follow every possible execution path, or it might simply be a long, randomly produced sequence of five-letter words. With huge amounts of test input available, it's no surprise to find programmers also creating automated tools for checking raw program output as well.

mutation testing

The methods we've mentioned here give only a rough overview of the kinds of tests that can be performed. To give you an idea of what a fertile field program testing is, consider a totally unexpected variation called *mutation testing*. In this approach, as exhaustive a set of test data as possible, with known results, is prepared. Next, programs are systematically mutated by having small errors introduced: a plus sign might be changed to a minus sign, or a constant might be increased by 1. The mutated program is then run on the original test data. If it works (i.e. it has the same results as the original program), we can conclude either that there is something very wrong with our original program, or that our test data is too weak to be useful. What an idea!

limits on confidence

It's easy to imagine that a large program might require *all* of the different testing approaches described here to give us confidence that the program will really work. Even then, though, our faith in the program depends largely on its prospective application. We have greater faith in less important programs because we don't pay a high price for their failure. An interrupted video game may be annoying, but the manufacturer's desire to bring it to market will probably outweigh concern about some minor residual bugs. It's easy enough to refund the user's quarter if she's unlucky enough to find the bug the hard way.

A program that controls a weapons system, on the other hand, is a different matter entirely. The recognition that no one test method is sufficient to guarantee that a program is correct and error-free is a cause for alarm, particularly when there is no way to undo a mistake. Many computer scientists believe that there isn't *any* way to test such systems adequately.

Program Proofs

In recognition of the limits of program testing, computer scientists have tried to develop other methods of gaining confidence in programs.

> A *program proof* is a 'paper' analysis of a program that attempts to formally verify that the program will always produce a correct result.

In one sense, a program proof *is* like a mathematical proof. A mathematical proof tries to justify the correctness of a mathematical statement—a theorem. A program proof tries to give us the same sort of assurance about a sequence of code statements—a program.

However, there is also an important difference between the two kinds of proofs. A mathematical proof tries to show that following a certain sequence of steps will result in an irrefutable conclusion. A program proof, in contrast, tries to show that the conclusion reached by following a series of steps will always be correct. The mathematical sort of proof works well when we want to show that, in principle, an algorithm will work. However, proving an actual implementation—a completed program—requires a different sort of tack.

assertions

> The proof of a program is based on a series of *assertions* about the values of program variables and data.

An assertion is a statement that we expect to be true. Typically, we'll use BOOLEAN-valued expressions (like a<>b) to make assertions.

In general, we'll find that assertions come in pairs—there's an assertion right before a program action, then one immediately following it. We can think of the opening assertions as giving *preconditions*, while the closing assertions state *postconditions*. If you note that one statement's postcondition can be the next statement's precondition, you can begin to picture how program proofs are established. First, we make assertions about the effect of each statement:

preconditions, postconditions

```
(*assertion1*)     Statement1;     (*assertion2*)
(*assertion2*)     Statement2;     (*assertion3*)
(*assertion3*)     Statement3;     (*assertion4*)
```

Then, by applying simple rules of logic, we can remove intermediate assertions:

```
(*precondition*)     Statement1;
                     Statement2;
                     Statement3;     (*postcondition*)
```

In practice, we usually work in the opposite manner, by starting with the outlying assertions, and attempting to develop assertions for parts of the program, then parts of those parts, etc.

verifying sequences

Now, it's pretty easy to see how to make some kinds of assertions. Suppose that we want to divide A by B and save the result in C. The closing assertion is a check on the operation—our assertion that the answer is correct is A=B*C. However, if we want to be assured that we will survive the division we need an opening assertion as well—that B isn't 0. The sequence of assertions (in comment brackets) is:

```
(* B<>0 *)
C:= A/B;
(* A=B*C *)
```

verifying loops

Looping statements are more interesting because assertions before and after the loop aren't sufficient to make a proof. Why not? Well, if the assertions are to actually prove anything, they must also establish that we arrive at the loop's end—that the loop isn't infinite!

variant and invariant assertions

> A loop's *invariant* assertion is a statement about the loop that is true both before and after each iteration of the loop. Its companion is a *variant* assertion whose truth will change between the loop's initial and final iterations. This is sometimes called the loop's *bound function.*

These two assertions serve complementary purposes in a loop proof. The invariant assertion makes a statement about the correctness of the loop's action, which is why it must always be true. The truth of the variant assertion, in contrast, is changed by the loop's action. It helps assure us that the loop will eventually be terminated, which is why it's also known as a bound function. The invariant assertion helps make sure that the loop doesn't do the *wrong* thing, while the variant assertion ensures that it does do the *right* thing.

We've relied on these ideas many times already, but in a reversed form. Formal loop proofs focus on the loop invariant, whereas we have keyed on the loop's bound function. I've done things this way because it's more intuitive—we can do some constructive reasoning about loop design without having to go the whole nine yards. However, both approaches ultimately make the same point about the loop's contents.

problem: proving division by subtraction

For instance, suppose that we want to do *integer* division by repeated subtraction. The segment below implements an algorithm most people learn in second or third grade. I've used the **WHILE** statement (described in section 6-5) to code it because that's how formal proofs are usually presented:

```
Remainder := Dividend;
Quotient := 0;
WHILE Remainder >= Divisor DO
    Remainder := Remainder – Divisor;
    Quotient := Quotient + 1
END
```

developing the invariant

If this loop is correct, we should arrive at proper values for Quotient and Remainder. First, what's our invariant assertion? Well, both before and after each loop iteration there should be a special relationship between the dividend, divisor, quotient, and remainder:

```
(* Dividend = (Divisor*Quotient)+Remainder *)
```

Since Dividend isn't changed within the loop, we don't have to worry about monkey business that would require us to save Dividend's original value, and make our invariant more complicated. While the invariant relation is true, we can be confident that our loop, at the very least, isn't doing the wrong thing.

need for a bound

However, the invariant relation isn't enough. Suppose that the loop's action made no assignments to either Quotient or Remainder. Although the invariant assertion would still stay true, it wouldn't assure us that the loop would ever end. We need to state some sort of bound that is approached by the loop's action, but which can act as a threshold beyond which the loop won't venture.

The variant assertion:

```
(*Remainder >= Divisor, and Remainder declines.*)
```

works for us. Its truth is potentially changed on each iteration of the loop—each time we change the value of Remainder. Eventually it becomes false; we pass the bound or threshold, and the loop is terminated. It is no accident that it forms the entry condition of the loop. Complete with assertions, the loop is:

```
Remainder := Dividend;
Quotient := 0;
(* Dividend = (Divisor*Quotient)+Remainder *)
WHILE Remainder >= Divisor DO
    Remainder := Remainder – Divisor;
    (* Remainder >= Divisor, and Remainder declines.*)
    Quotient := Quotient + 1
    (* Dividend = (Divisor*Quotient)+Remainder *)
END
```

As usual, I've made everything look easy by coming up with the correct answer on our first try. Let's consider a false proof, though. Suppose that we had chosen as our variant assertion the relation Quotient<=Dividend. We would still make progress toward loop termination, since we increment Quotient on each pass through the loop. Since it provides an upper bound on the loop, it's reasonable to think that the assertion is a good bound function.

a false proof

Unfortunately, it's the *wrong* bound function. Before we reach the limit it sets, we'll have allowed the invariant assertion to become incorrect. Our attempted proof would fail, even though we set a threshold and approached it.

Difficulties in Proving Programs

The idea that we can prove that a program is correct is intensely appealing, since it would greatly increase our confidence in programs. True, verification seems complicated at first; but then again, so do mathematical proofs. Unfortunately, the promise of program proving has not been realized in as full a manner as was originally hoped. Let's investigate the reasons.

Two conditions have to be satisfied if we want a program proof to work. First, the action of a program statement can't undermine our assertions about what the statement will do. Although this notion seems obvious (we clearly wouldn't have a statement that directly contradicts an assertion), there are subtle difficulties that are easy to overlook. For instance, mathematical proofs don't have to worry about whether or not the axioms of mathematics will apply, but program proofs do. The machine code that takes care of computer arithmetic may never have been formally proven—and it may not always obey the rules!

correctness

Second, the assertions we make have to define the entire 'universe' of the program. Any necessary assertions that are left out cause gaping holes in the program proof that may not be detected until the program fails. For instance, suppose that a routine should sort three variables into increasing order. It's not enough to prove that the variables are in order when the routine ends. We can't ignore the possibility that the routine might have accidentally given all the variables the *same* value.

completeness

> The need to satisfy these two conditions—correctness and completeness—make program proofs very difficult to develop.

In mathematical proofs, small errors will not necessarily have a negative impact on the proof as a whole. It may be that an individual step is incorrect or misstated. However, this sort of error won't always invalidate the overall goal of the proof—a fact can be true even if our explanation of it is faulty. Mathematical proofs usually fail because of larger conceptual errors.

For programs, though, the smallest step is vitally important to the conclusion of the program proof. The tiniest untested assumption about a data value in a program can suffice to undo an elaborate program proof.

limits of proofs

> More importantly, though, the knowledge that such errors can occur in proofs without being detected by expert computer scientists tends to undermine our confidence in the absolute reliability of program proofs at all.

Like a good trial attorney, we can cause a proof to collapse simply by demonstrating a reasonable doubt about its correctness. As a result, programs are actually proved only in a limited set of cases; and then only for relatively small program segments that are written with eventual proof in mind.

automated verification

Incidentally, you may be tempted to suggest that, since keeping track of small details is so important, program proving would be a perfect job for a computer. Why not write a program that could automatically check the correctness of a proof?

It's a good idea until you imagine what such a program's first job would be. Obviously we'd want to run it on itself. But what will we make of the answer? Suppose that the program announces that its own code is correct. Can we trust it? Worse yet (in a much more likely outcome) suppose that the automated verifier announces that its own code is *wrong*. Oh no! Back to the drawing board...

Even though few programs are formally verified, the techniques used to prove programs are widely applied as an aid to program development. In one aspect, the assertions that we might use to prove a program will turn up as detailed comments about the state of variables before and after a procedure or function call, or on the successful completion of a loop. Proof methods are also employed during the development of especially confusing program segments. In particular, ideas of loop invariants and bounding conditions are used to help assure ourselves that algorithms are correctly stated in the first place.

benefits of program proving

Another application of proof techniques is manifested by features built into programming languages. The availability of subrange types is a good example. A variable declared to have a subrange type is essentially a variable accompanied by an assertion—that is constantly checked—about the range of the variable's values. *Exception handling*—the specification of special notification procedures to be followed when errors are encountered—is another feature of programming languages like Modula-2.

Finally, programs can be written in a 'rough proof' form. You see, for many kinds of programs it almost seems easier to write the proof first, then tailor the code to fit the verification. In practice, when programs must be proved the development of proof and code usually go hand in hand. Even when the code is not intended to be exhaustively proved, writing code as though it will be verified can lead to less error-prone programs.

Where is Reliability Found?

There was once a warship equipped with a sophisticated computer. One day, the ship shot down an airliner. A representative of the navy said: 'Our system is too advanced to make mistakes. Even though the system had never been tested, it was not the system that erred—it was one of the junior officers.'

A programmer heard this and could not believe his ears. 'Just so might a murderer say, 'I am blameless, for it was not I who slew, but my hand.' A computer system consists of three parts: the hardware, the software, and the human. If the human errs, the system has failed.

From *Computer Parables*, by Geoffrey James, © 1989, InfoBooks, Santa Monica, Ca. 90406

'. . . the tools that will let us get away with carrying out low-level machine operations also have the effect of subverting the protection of Modula-2's type checking.' (p. 702)

Chapter 20 Details High and Low

In the interests of giving you a complete picture of Modula-2, this chapter provides a brief introduction to a number of topics that are not merely beyond the scope of this book. They are *way* beyond the scope of this book.

However, one of Wirth's philosophies of programming language design is that languages should be small enough for ordinary humans to comprehend in a reasonable amount of time—even at the price of leaving out ever more powerful commands. I agree wholeheartedly,

Now, if I could, I'd love to have three or four more chapters that went into these 'details' in depth. Alas, the world isn't ready for the 1,000-page introductory Modula-2 textbook. However, I think we can squeeze in a brief description of the rest of Modula-2's features.

If you want to learn more about Modula-2, let me recommend Kim King's reference book, *Modula-2: A Complete Guide*. It's a particularly good book because Kim explains why as well as how. Niklaus Wirth's introduction and reference, *Programming in Modula-2*, is harder to understand, but it should also be on your shelf if you're planning to continue in computer science.

Section 20-1

Modules, Etc.

LET'S BEGIN WITH MODULES. FIRST, MODULES can declared be within the body of a program. Although the modules we've written have been used as external libraries, *local* modules can be declared as well. The purpose of local modules is to establish an entirely new scope that's totally separate from the enclosing program.

local modules

Programs can import specific procedures and variables from their own local modules, while encapsulating and hiding other processes or values. Frankly, this is an extremely specialized application, and we won't explore it further.

module priority

Second, modules can be given *priorities*. This is also a very specialized application that is mainly of interest to operating systems programmers. I mention it only to point out that Modula-2 contains all the tools necessary to write a complete computer operating system—not just the application programs that run on top of the operating system.

Third, we can import procedures, types, and so on *by name* from a library. This is a feature you'll see often:

```
FROM IO IMPORT StringIn, StringOut;
```

FROM imports

The imported identifiers don't have to be *qualified*, or preceded by their library names:

```
StringOut ("This is a legal call of IO.StringOut.");
```

However, identifiers from IO that were not imported are *not* known.

There are a few reasons for selectively importing identifiers. First, the **FROM** . . . **IMPORT** documents the names from a given library that are going to be used. Second, unqualified names are obviously shorter and more convenient to use if a library name is long. Third, the **FROM** form makes it easier to substitute one library (say, DoublePrecisionLib) for another (like MathLib∅) that has the same functions, but has been implemented differently. Finally, selective import restricts the use of a library to a specific list of names.

why not use **FROM**?

But there are also reasons for not making selective imports. Personally, I prefer not to use selective import because of its effect on library design. **FROM** distinguishes between identifiers at the procedure level, rather than at the module level. In contrast, importing complete modules:

```
IMPORT Color;
IMPORT Fraction;
IMPORT Pixel;
```

lets us use simple, consistent names like T, In, Out, and so on within our libraries. Procedures that do the same things are encouraged to have the same short, easy-to-remember names. We don't have to worry about name clashes, because we'll always use the module name as a qualifier.

In contrast, a general practice of using selective imports forces us to devise unique names for the contents of each library. Instead of trying for consistency, we're forced to seek different ways to describe the same actions. This leads to the sort of name pollution I try to avoid.

In any case, the method I've used has been meant to advance the goals of this text—introducing programming in a clear and consistent manner. But it's important to remember that goals in other working environments may be different, and may reasonably call for different tools or standards.

Other Standard Types

Recall from our discussion of type SYSTEM.WORD that most Modula-2 systems set aside one word of storage for values of type INTEGER, and two to four words for each REAL. These standards limit the overall size of INTEGER values, and the decimal accuracy of REAL numbers.

Types LONGINT and LONGREAL increase the legal size of these variables. Typically, twice as many words are set aside for each variable. This has a dramatic effect on the size of legal values, incidentally. Suppose that an ordinary INTEGER is one 16-bit word long, while a LONGINT gets 32 bits. The LONGINT isn't twice as large as the INTEGER—it's 2^{16} times as big!

Although the long types are a part of Modula-2, their implementation and relation to the standard lengths has always been somewhat vague. I assume, for instance, that any reasonable Modula-2 system will allow an INTEGER to be assigned to, or appear in an expression with, a LONGINT. However, you never know. Sigh.

Procedure Types

Procedures can be 'stored' in variables and parameters. That makes it possible to write procedure calls like:

```
PlotPoints (Function (x), LowerBound, UpperBound);
```

where Function, complete with parameters, can be replaced by different calls:

```
PlotPoints (Projection (1), LowerBound, UpperBound);
PlotPoints (Mathematical (LIMIT), LowerBound, UpperBound);
```

The goal of a procedure's type definition is to clearly specify *a*) the number and type of its arguments, and *b*) its own type, if it's a function. For example, the type definition of a function of type REAL that had one INTEGER argument would be:

procedure type definitions

```
TYPE OperatorPROC = PROCEDURE ( INTEGER ) : REAL;
  (*The underlying procedure type definition.*)
VAR Action: OperatorPROC;
  (*An example of a procedure-typed variable.*)
PROCEDURE PlotPoints (Calculate: OperatorPROC; Low, Hi: INTEGER);
  (*An example of a procedure with a procedure-typed parameter.*)
```

After this definition and declaration, Action can be assigned any outer-level procedure of type REAL that takes one INTEGER argument. Similarly, PlotPoints can be passed any function of type REAL that has one INTEGER argument.

> The standard type PROC matches any procedure that has no parameters.

Procedure types are very useful for an application that lies just beyond the boundary of Chapter 16: creating *generic*, general-purpose data structures. For example, let's think about an ordered list structure whose primitives must include a StoreInOrder procedure. If such a procedure is really going to be general purpose, its operation can't depend on the type of the objects that are being stored.

This has two implications. First, the list has to hold pointers to objects, rather than the objects themselves. That's the only way the list's contents can be type-independent. Second, the StoreInOrder procedure needs a way to find the relative order of values it doesn't know anything about! This little miracle is accomplished by passing a procedure to the sorted-list library, which StoreInOrder can use to compare two unknown values. In other words, StoreInOrder uses a comparison procedure that's not 'installed' until final program run-time.

The **RETURN** Statement

The **RETURN** statement is familiar; we've used it whenever we've written a function:

```
PROCEDURE Even (Number: INTEGER) : BOOLEAN;
  BEGIN
    RETURN NOT Odd(Number)
  END Even;
```

Here, **RETURN** ends the function call, and specifies the value it's going to represent.

However, **RETURN** can actually appear in *any* procedure *or* module as well. Its effect is to end the procedure or module immediately. Thus, it's generally used in extraordinary cases, say, if faulty data is encountered. Unlike HALT, though, **RETURN** doesn't stop all processing.

Section 20-2 Record Variants

THE RECORDS WE DEFINED IN CHAPTER 10 EACH HAD a fixed contingent of fields. However, Wirth enhanced Modula-2 records by allowing the definition of *record variants*. They let the number and type of fields in a single record change during the course of a program. This means that two variables can be of an identical record type, yet have different numbers or types of fields.

A typical application that would call for record variants might involve holding measurements of different plane figures:

why do we need variants?

Shape	*Required Dimensions*
Square	Side
Rectangle	Length, Width
Rhomboid	Side, AcuteAngle
Trapezoid	Top, Bottom, Height
Parallelogram	Top, Side, ObtuseAngle

shortcomings of records

We could easily define five different records—one for each shape. Or, it might be more convenient to define a single record with enough fields to record the dimensions of *any* of the shapes. However, the first solution makes it difficult to define general-purpose subprograms, since each record type will require its own procedures and functions. The second solution is grossly inefficient, and a program that stored hundreds or thousands of such records might run into space trouble.

Record variants come to the rescue. If you examine the list of shapes, it's easy to imagine a WhatShape field that could tell us which fields are actually required in the rest of the record, and which are superfluous.

the tag field

> The idea that the value of one field could or should determine the rest of the structure is the basis of record variants. One field is designated to be a *tag* or marker field—a field whose value tags or marks the proper group of *variant* fields.

DimensionTYPE is defined below as a record with variants. WhatShape is the tag field, and the record contains five shaded groups of variant fields. Notice how the vertical bar is used to separate each group of field definitions.

a record with variants

```
TYPE ShapeTYPE = (Square, Rectangle, Rhomboid, Trapezoid, Parallelogram);
     DimensionTYPE = RECORD
                       CASE WhatShape: ShapeTYPE OF      (*tag field*)
                         Square: Side1: REAL;
                       | Rectangle: Length, Width: REAL;
                       | Rhomboid: Side2: REAL; AcuteAngle: [0..360];
                       | Trapezoid: Top1, Bottom, Height: REAL;
                       | Parallelogram: Top2, Side3: REAL;
                           ObtuseAngle: [0..360]
                       END (*the WhatShape variant*)
                     END; (*DimensionTYPE*)
VAR FourSidedObject: DimensionTYPE;
```

Each variant's fields must be unique. No field identifier can appear in more than one group. The tag field, in contrast, is shared by each of the variant groups.

Until the tag field has a value, the remainder of the record variant's structure is undefined. At this point, we can only make an assignment to the tag field, WhatShape.

```
FourSidedObject.WhatShape := Rectangle;
```

activating variants

Once WhatShape has been given a value, the fields associated with that value are created. The assignment above activates a certain group of fields—in this case, Length and Width. As long as the value of WhatShape is Rectangle, these are the only fields that FourSidedObject will contain. We can make the assignments:

```
FourSidedObject.Length := 4.3;
FourSidedObject.Width := 7.5;
```

but an attempted assignment to a field in one of the other variant groups (say, Top1 or ObtuseAngle) is an error—it does not exist.

What if the value of the tag field changes? If we now say that:

```
FourSidedObject.WhatShape := Parallelogram;
```

we find ourselves able to access three new, but as yet undefined, fields—Top2, Side3 and ObtuseAngle. The former variant fields Length and Width simply don't exist any more—they've been deactivated and replaced.

Let me caution you to take what I've just said with a grain of salt. In theory, the Modula-2 compiler will provide run-time checks validity of assignments by inspecting the tag field. In practice, making those checks is up to us. For example, consider function Area, below. A single variable of type Dimensions is passed as a parameter, then dissected within the routine. If we were using five different records, we'd have to write five different subprograms. But since Dimensions is defined as a record variant, just one declaration suffices.

using the active variant

```
PROCEDURE Area (Object: DimensionTYPE) : REAL;
  (*Computes an area that depends on an active variant.*)
  VAR Temp: REAL;
  BEGIN
    WITH Object DO
      CASE WhatShape OF  (*Recall that WhatShape is the tag field.*)
          Square: Temp := Side1 * Side1;
        | Rectangle: Temp := Length*Width;
        | Rhomboid: Temp := (Side2*Side2)*MathLib0.sin(AcuteAngle);
        | Trapezoid: Temp := (Top1+Bottom)/(2*Height);
        | Parallelogram: Temp := (Top2*Side3)*MathLib0.sin(ObtuseAngle)
      END (*CASE*)
    END; (*WITH*)
    RETURN Temp
  END; (*Area*)
```

An **ELSE** part can be provided for default variants. For example:

default fields

```
TYPE ELSEexample = RECORD
                     CASE Tag: TagTYPE OF
                        Constant1: Field1: TYPE1
                     | Constant2: Field2: TYPE2
                     | ...
                     | ConstantN: FieldN: TYPEn
                       ELSE AllOtherFields: TheirTYPE
                     END (*the variant part*)
                   END (*the record definition*)
```

If the value of Tag isn't one of the listed constant values, then AllOtherFields are declared. Note that Tag should still be defined.

> A tag field need not be specified for a record variant. However, the colon and tag-type identifier still have to be given:
>
> **CASE** :TYPEidentifier **OF**
>
> The rest of the variant part's definition proceeds normally.

This 'feature' is error-prone and rarely used. Since the tag field is optional, it can't be checked by the compiler. Thus, *you* should always declare a tag field, and check its value before trying to access fields of a variant part.

fixed and variant parts

The variant fields of the DimensionTYPE record were *disjoint*, which means that they only shared the tag field. However, Modula-2 permits records that have several shared fields, as well as more than one variant part. In the example below, the Year and ExpirationDate fields form RegistrationTYPE's fixed part. They, along with the tag fields VehicleTYPE and Fees, are shared by every variant.

```
TYPE ModelTYPE = (Motorcycle, Car, Truck);
     FeeClassTYPE = (Commercial, Personal, City, State);
     RegistrationTYPE = RECORD
                          Year: INTEGER[1915..1993];
                          ExpirationDate: INTEGER[1991..1994];
                          CASE VehicleTYPE: ModelTYPE OF
                             Motorcycle: EngineSize: INTEGER[50..1200];
                          | Car: Cylinders: 2..8; SmogRequired: BOOLEAN;
                          | Truck: Axles: 2..10; Weight, Tare: INTEGER)
                          END; (*the VehicleTYPE case*)
                          CASE Fees: FeeClassTYPE OF
                            Commercial: WeightClass: INTEGER;
                                        UserFee: REAL;
                          | Personal: Amount: REAL;
                          ELSE ExemptionAppliedFor: BOOLEAN
                          END; (*the Fees case*)
                        END; (*RegistrationTYPE*)
```

The current value of the tag fields VehicleTYPE and Fees determine which group of variant fields will be accessible.

Many applications require storage of specialized and general information in a record. Examples of records with both fixed and variant fields include employment records, library records, medical records, and the like.

Section 20-3 The SYSTEM Module

HIGH-LEVEL PROGRAMMING LANGUAGES MAY BE restrictive, but unlike machine languages, they're usually not impenetrable. Thanks to the translation done by compilers we can program in a language that, if not English, is at least English-like. By worrying about machine-level details for us, compilers make life pretty easy.

why use low-level operations?

On occasion, though, we will want to manage the machine-level details ourselves. For example, on many microcomputer systems, new hardware attachments are controlled by the values stored in particular *words* or locations in computer memory. Each computer's documentation lists the word locations that correspond to physical plugs or sockets, and explains the exact effect of turning individual bits on and off. As a result, we'll want to inspect or make changes directly to particular locations in computer memory—the very epitome of bit-twiddling.

A second reason for delving into machine details arises when we want to escape Modula-2's type rules. For example, we might wish to create a data structure, like a stack or list, that holds values whose types are not specified. In other words, we might want to create a stack that holds a special kind of pointer that can address *anything*, regardless of type. This, too, requires low-level work.

Methods of carrying out operations like these aren't usually introduced in a first programming course, and for good cause. The applications described above require access to the computer in a way that might make programs non-portable. Moreover, the tools that will let us get away with carrying out low-level machine operations also have the effect of subverting the protection of Modula-2's type checking.

However, if we must get down to the real nitty-gritty, we can. Now, Wirth faced a serious dilemma when he tried to supply solutions for the problems posed above. How is it possible to describe non-standard features, which may vary from machine to machine, in standard terms?

In the big scheme of things, it's relatively easy to invent a new programming language, and to write a compiler for it. However, it is quite difficult indeed to define the new programming language in a manner that is *a*) simple enough to understand while still being *b*) precise enough to ensure that *other* people can write compilers too.

the problem

Wirth's problem was that he didn't want to *require* features to be built into Modula-2 that might cause programs to be non-portable (because some compilers had the features, while others didn't). At the same time, he didn't want to *forbid* features that might turn out to be essential; for example, a way to deal with a hand-held mouse. The inability to deal with such situations had caused serious problems for earlier languages, in particular, for Wirth's earlier language Pascal.

the solution

A large part of Wirth's solution was to permit a special kind of module called a *pseudo-module*. In effect, a pseudo-module has a definition part, but no implementation part. It's really part of the compiler—you can't write it, but the compiler-writer can.

Modula-2 has just one 'standard' pseudo-module, called SYSTEM. Its definition module looks like this:

the pseudo-module SYSTEM

```
DEFINITION MODULE SYSTEM;
  TYPE WORD;  (*Matches any type that is one word (usually two bytes) long.*)
       ADDRESS;  (*Matches any pointer type or memory location address.*)
       PROCESS;  (*Formerly used as a procedure 'type,' but since superseded.*)
  PROCEDURE ADR (AnyVariable: AnyTYPE): ADDRESS;
    (*The address (memory location) of AnyVariable during program execution.*)
  PROCEDURE TSIZE (AnyTYPE): CARDINAL;
    (*The number of bytes or words required to store a variable of type AnyTYPE.*)
  PROCEDURE NEWPROCESS (ProcedureName: PROC;
                          (*the name of a procedure*)
                        Location: ADDRESS;
                          (*where memory set aside for the procedure starts*)
                        ApproximateSize: CARDINAL;
                          (*how big the space should be, in bytes or words*)
                        VAR NewName: ADDRESS;
                (*the name we'll use to call procedure P in TRANSFER, below.*)
    (*Prepares an ordinary procedure to be invoked as a coroutine.*)
  PROCEDURE TRANSFER (VAR FromProcedure, ToProcedure: ADDRESS);
    (*Suspends FromProcedure and resumes election of ToProcedure.*)
  ...  (*There may be other declarations as well.*)
END SYSTEM.
```

> This description includes the minimum contents of SYSTEM. Any particular implementation of Modula-2 may contain additional types or procedures as well.

SYSTEM is a contrived solution, but it does the job: it provides a standard way to do non-standard things. The appearance of this line in a program:

```
IMPORT SYSTEM;
```

has the effect, as intended, of alerting program maintainers to the possibility of non-standard code.

Using SYSTEM

Let's take the briefest of looks at what SYSTEM does. First, there are two things to ignore: type PROCESS and function TSIZE. They date from the second edition of Wirth's Modula-2 reference; the tasks they accomplish are handled differently now.*

Type WORD, and the meaning of computer words, were described in detail in Chapter 18. To recap, recall that a word is usually the smallest practical storage unit used in assembly language programming on a particular computer. Values of the basic types (in particular, of INTEGER and CARDINAL) are generally stored as single words.

Structured types and more complex values (like REAL) usually consist of more than one word. As parameters to Modula-2 procedures, they can always be matched by the open array parameter **ARRAY OF** WORD.

SYSTEM.WORD

> WORD solves the problem of allowing the input and output of arbitrary kinds of values. Any value can be read or written as a sequence of WORDs. Only assignment is allowed on WORD-type variables or parameters, though—they may not appear in arithmetic expressions.

The next type, ADDRESS, matches any pointer type. Now, I'm sure that you're scratching your head, and asking the following question: If WORD can match any type, why do we need an extra type for pointers?

The answer is found in the restriction that limits type WORD to assignment. Unlike WORD, ADDRESS *can* appear in arithmetic expressions.

SYSTEM.ADDRESS

> Variables of type ADDRESS are compatible with variables of any pointer type for purposes of assignment. However, they may also appear in expressions, where they are treated as values of type CARDINAL.

This is not the kind of application that you're liable to encounter. However, it is necessary in programs that play an active role in managing their own memory requirements, rather than letting the compiler do it for them.

Type ADDRESS also comes into play in one of the applications mentioned above: creating a data structure that can hold values whose types aren't specified. Such structures are called *generic*. For example, a generic stack might hold numbers, or characters, or records—after all, the kind of value being held doesn't affect the definition of the stack.

Why is ADDRESS useful? Well, rather than organizing the stack as a sequence of values, it can be arranged as a sequence of pointers to values. Instead of pushing and popping specific values, with specific types, the stack primitives can be written to push and pop pointers of type ADDRESS:

* Although people tend to view programming language standards with a respect that is usually applied to the tablets of Moses, languages are seldom fully formed when they are first proposed. They are certainly not carved in stone; at this writing, Wirth's standard is in its fourth revised edition.

```
PROCEDURE Push (Thing: ADDRESS);
```
(**Push the* Thing *pointer—and, implicitly, whatever it points to, onto a stack.**)

Coroutines

Finally, we arrive at procedures NEWPROCESS and TRANSFER. They are intended to give the programmer an extraordinary kind of control over her programs—the ability to temporarily *suspend* a procedure's execution, and then *resume* it later.

SYSTEM.NEWPROCESS, SYSTEM.TRANSFER

> NEWPROCESS prepares an ordinary procedure to be treated as a *coroutine* that can be suspended and resumed. TRANSFER causes the switch: it names the coroutines to be suspended and resumed.

How is a procedure invoked? In the past, we have called procedures by name. Using the name of a procedure as a statement:

```
SolveAllMyProblems;
```

causes the code of the procedure to execute. Its local variables are allocated and the procedure's statements take place. When the procedure is done, the local variables go away, and we return back to the main program (or the calling procedure).

However, there is another way to call a procedure—by treating it as a *process*. When it's handled as a process, we can temporarily suspend the procedure before it's done, without losing either local variables or our place in the procedure. We can come back later and pick up where we left off.

suspending execution

> *Coroutines* are procedures that call each other in a way that lets them take turns executing. Coroutines are suspended rather than being exited.

Coroutines are unlike recursive procedures because there's only one instance of each coroutine. Instead, two or more procedures can take turns; each suspends its own operation while restarting another's.

Let's take a quick look at 'how?' before we ask 'why?' Making procedures behave like coroutines takes calls to two standard procedures: NEWPROCESS and TRANSFER.

1. First, prepare each procedure to execute by creating a new process:

NEWPROCESS (*the first procedure's name and various other arguments*);
NEWPROCESS (*the second procedure's name and various other arguments*);
(**Similar calls for any other coroutine procedures.**)

2. Second, suspend the main program, and call the first coroutine:

   TRANSFER (main *(the current process), the new process*);

3. Third, call TRANSFER again, as needed, to go from one coroutine to another. These calls are found *within* the coroutines themselves.

   TRANSFER (*the current process, the new process*);

Thus, nobody ever makes just one TRANSFER. Instead, two or more processes take turns. In effect, one procedure executes while the other goes into suspended animation. That's why they're called *co*routines—they cooperate in sharing the computer.

procedure call vs. process transfer

Let's compare this approach to an ordinary procedure call. There, the procedure is ready to execute simply by being declared. The current main program or procedure is suspended during the procedure call. Finally, we return to the main program or calling procedure when the called procedure is complete.

In contrast, a coroutine process transfer requires a call of NEWPROCESS in addition to the ordinary procedure declaration. Then, the current main program or procedure is suspended by the call of TRANSFER. If we call TRANSFER again, to return to the main program or calling procedure, the new procedure is itself suspended. We can return to it later if we want.

Thus, when a procedure is called, we always begin at the beginning of the procedure. In contrast, when a *process* is called, we can pick up from where we left off—it's being resumed, rather than invoked. TRANSFER halts the current procedure without getting rid of local variables or parameters, then restarts the new procedure from wherever it had been left off previously. In effect, two or more coroutines act like customers of a timeshared computer system. Each coroutine gets a slice of time, then passes the ball to its fellow.

Section 20-4 Other Low-Level Details

Although SYSTEM is a convenient repository for most of Modula's system-dependent facilities, the language includes other ways to get down to the machine level.

- The SIZE function describes the amount of memory required to store a value.
- Type transfer functions cause a value of one type to be treated as though it had a different type.
- Absolute addressing allows variables to be allocated at a specific area of memory.

Let's look at each in turn.

The Standard Function SIZE

The SIZE function is one we've already encountered. In Chapter 13, we used SIZE to help allocate pointer types. Formally:

SIZE(V) *or* SIZE(T): represents, as an INTEGER, the amount of memory occupied by a particular variable whose name is V or by a value whose type is T.

Thus, SIZE is given an argument that is either a variable identifier *or* a type name. It returns, as an INTEGER, the size of its argument.

SIZE is necessary for working with pointers. Recall that after importing the semi-standard library:

```
IMPORT Storage;
```

we used SIZE in conjunction with pointer allocation and deallocation:

```
Storage.Allocate (CurrentPtr, SIZE(PointedToTYPE));
    (*CurrentPtr addresses an anonymous variable of the proper size.*)
        . . .
Storage.Deallocate (CurrentPtr, SIZE(PointedToTYPE));
    (*Return the anonymous variable to the system for reallocation.*)
```

The sixty-four dollar question, of course, is what the value that SIZE returns means. Is it a size in bits? Bytes? Words?

> The basic unit of computer measure is usually the word. However, some systems may return the number of bytes, or even bits, as the result of SIZE. Check your system documentation to be sure.

Type Transfer Functions

Recall that one of the features of which Modula-2 is most proud is its strong type checking. Every value has a type, every variable has a type, every operator must have operands of a certain type, and so on. Except for certain explicitly defined cases, types cannot be intermingled: a variable can't be assigned a value of a different type, and values with disparate types can't appear in a single expression.

Nor does the programmer know much about how the computer stores values—what is the *internal representation*—of a given type. The compiler takes care of translating values from their stored internal form to whatever form a program requires for input, output, or arithmetic operations.

The intention of strong type checking is laudable. It helps prevent errors caused by inadvertently confusing different kinds of values. However, on occasion the programmer will have a goal that is equally honorable, but which is prevented by strong type checking. She may want to:

- see, bit by bit, just how a value is stored (and maybe even change some bits), or

why they´re useful

- make a statement legal by subverting the type rules, so that a stored value of one type is temporarily treated as though its type were different.

For example, she might wish to treat an INTEGER value as though it were a BITSET. After all, BITSETs can be manipulated with the built-in bit operators (see Chapter 18) to carry out some kinds of mathematical operations with extraordinary speed.

Similarly, she might want to treat a value with a pointer type as though it were a CARDINAL. As a CARDINAL she could carry out arithmetic operations that were forbidden for pointers, but which might be needed to build some kinds of complex data structures.

> A *type transfer function* overrides the type of its argument. In other words, it allows expressions and assignments that would ordinarily be illegal under Modula-2 type rules. In Modula-2, any type name can be used as a transfer function. However, values of the type, and its argument's type, must take the same amount of storage space.

The last point is so important that I'll repeat it. The type of the transfer function and the type of its argument are going to be different. However, values of these types must occupy the same amount storage space in computer memory. The expression SomeTYPE (SomeValue) is only legal if:

SIZE (SomeTYPE) *equals* SIZE (SomeValue)

INTEGER and CARDINAL are sure to qualify, but CHAR and REAL almost certainly don't.

In practice, transfers usually take place between types INTEGER, CARDINAL, POINTER, and BITSET. Why these? Because they usually take the same amount of memory for storage:

SIZE(INTEGER) *equals* SIZE(CARDINAL) *equals* SIZE(BITSET) *equals* SIZE(PointerTYPE)

Now, it's important to recognize that type *transfer* is not the same thing as type *conversion*. In a type conversion, one value is translated into another. We can convert the INTEGER value 3 into the REAL value 3.0, despite the fact that INTEGER and REAL store values in different amounts of space. Typical type conversion functions include FLOAT, CHR, TRUNC, ORD, and VAL.

A type transfer function, in contrast, does not convert its argument into an 'equivalent' value with a new type. It merely treats the memory that its argument occupies as though it had been filled according to the new type's rules. Any type name can be used as a type transfer function.

Thus, some kinds of type transfers are sensible, while others aren't. For instance, using BITSET as a transfer function doesn't pose any risk because we're just getting a bit pattern—we don't assume that it has any particular value.

In contrast, a transfer between INTEGER and CARDINAL may be very dangerous indeed. Within reasonable bounds (such as the range of positive INTEGER values) the transfer yields the same value for both types on most computer systems. But suppose that some CARDINAL variable represents the largest possible CARDINAL value. Indeed, on most systems, a call of INTEGER(MAX(CARDINAL)) represents –1!

Let me close by suggesting a few avenues of exploration. First, you might want to probe type boundaries, and see what happens when values that cross the bounds are transferred:

```
IO.IntOut ( INTEGER (MAX(INTEGER)+1) , 1);
IO.IntOut ( INTEGER (MAX(CARDINAL) , 1);
```

Second, you might want to see if INTEGER and CARDINAL values within the same range (i.e. of low positive numbers) can be transferred without any surprises.

Finally, you might want to assign values of a few different types to BITSET variables, to get an idea of exactly how values are stored. In particular, explore the representation of very small and very large INTEGER and CARDINAL values.

Absolute Addressing

One of the lowest level details permitted by Modula-2 lets the programmer specify the exact location in memory she wants a variable to occupy. This feature is sometimes called *absolute addressing*, and it is not required to be available on all Modula-2 systems.

Why override the compiler, and insist that a variable occupy a particular location in memory? It's usually done to allow the programmer and computer to share information. We're not giving just any old location, after all. The goal of absolute addressing is to let us peek into, and possibly change, a segment of memory the computer uses for some specific purpose.

For example, when you type letters at a keyboard, they are stored in a segment of memory called the *input buffer*. By declaring variables that coincide with the input buffer, you can inspect input before reading them as part of the input stream. In fact, by changing the input buffer's contents, it's possible to stuff input back into the input stream *after* it has been read.

Another typical application of absolute addressing is to permit very fast output. For example, a certain kind of terminal screen, called a *bitmapped display*, turns individual pixels on or off in response to changes in value in a particular area of memory (the *bitmap*). We can declare an array variable that coincides with the bitmap, then control the screen by making assignments to the array.

Finally, certain computer systems will reserve particular memory locations to indicate the presence of various kinds of internal hardware. A program that knows where to look (via absolute addressing) can tell a good deal about what kind of equipment your computer has attached to it.

Appendix A: Syntax Charts

These syntax charts are taken from Modula-2: A Complete Guide*, by Kim King (© 1988, D.C. Heath). He, in turn, worked from the originals in* Programming in Modula-2*, by Niklaus Wirth (© 1985, 1988, Springer-Verlag). The charts reflect the terminology used in Wirth's formal definition of Modula-2 syntax.*

1. Compilation Units

1.1 *CompilationUnit*
1.2 *DefinitionModule*
1.3 *Definition*
1.4 *ProgramModule*

2. Declarations

2.1 *Declaration*
2.2 *ConstantDeclaration*
2.3 *TypeDeclaration*
2.4 *VariableDeclaration*
2.5 *ProcedureDeclaration*
2.6 *ProcedureHeading*
2.7 *FormalParameters*
2.8 *FPSection*
2.9 *Block*
2.10 *ModuleDeclaration*
2.11 *Priority*
2.12 *Import*
2.13 *Export*

3. Types

3.1 *Type*
3.2 *SimpleType*
3.3 *Enumeration*
3.4 *SubrangeType*
3.5 *ArrayType*
3.6 *RecordType*
3.7 *FieldListSequence*
3.8 *FieldList*
3.9 *Variant*
3.10 *SetType*
3.11 *PointerType*
3.12 *ProcedureType*
3.13 *FormalTypeList*
3.14 *FormalType*

4. Statements

4.1 *StatementSequence*
4.2 *Statement*
4.3 *Assignment*
4.4 *ProcedureCall*
4.5 *ActualParameters*
4.6 *IfStatement*
4.7 *CaseStatement*
4.8 *Case*
4.9 *CaseLabelList*
4.10 *CaseLabels*
4.11 *WhileStatement*
4.12 *RepeatStatement*
4.13 *LoopStatement*
4.14 *ForStatement*
4.15 *WithStatement*

5. Expressions

5.1 *ConstExpression*
5.2 *ExpList*
5.3 *Expression*
5.4 *Relation*
5.5 *SimpleExpression*
5.6 *AddOperator*
5.7 *Term*
5.8 *MulOperator*
5.9 *Factor*
5.10 *Set*
5.11 *Element*

6. Names

6.1 *IdentList*
6.2 *Ident*
6.3 *QualIdent*
6.4 *Designator*

7. Literals

7.1 *Number*
7.2 *Integer*
7.3 *Digit*
7.4 *OctalDigit*
7.5 *HexDigit*
7.6 *Real*
7.7 *ScaleFactor*
7.8 *String*

1. Compilation Units

1.1 ***CompilationUnit***

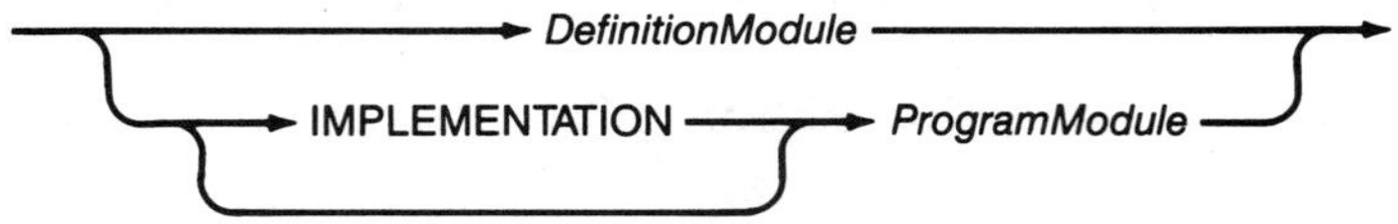

1.2 ***DefinitionModule***

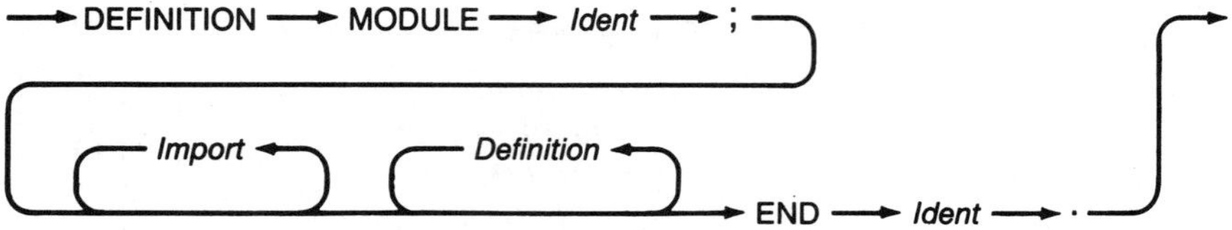

1.3 ***Definition***

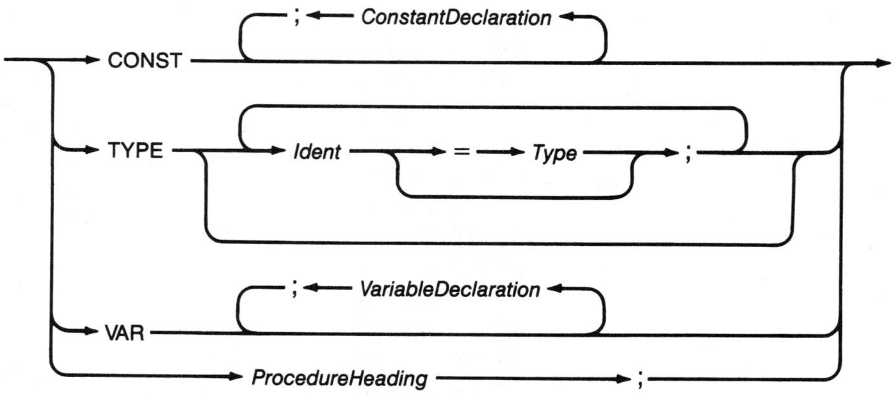

1.4 ***ProgramModule***

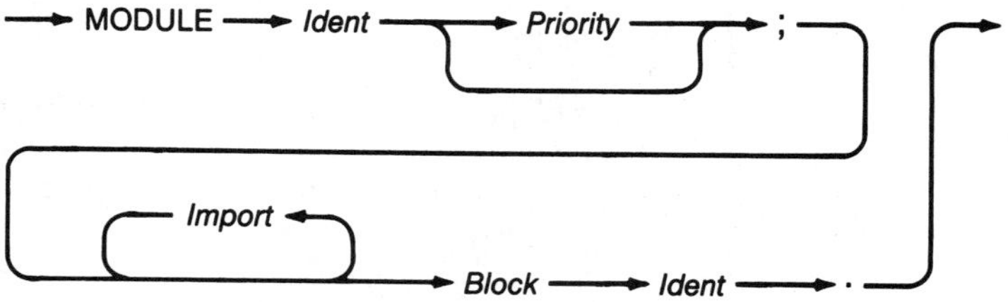

2. Declarations

2.1 **Declaration**

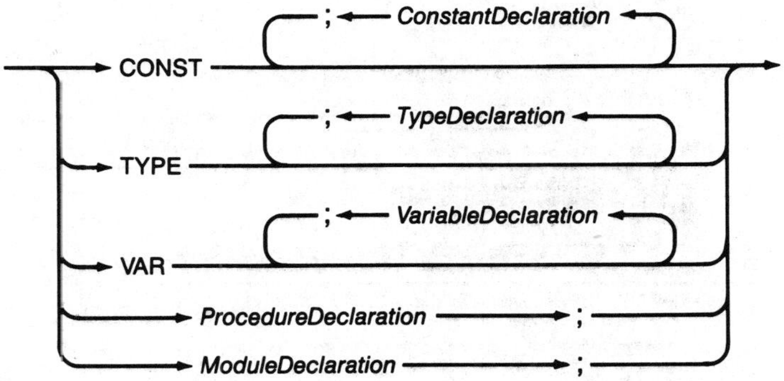

2.2 **ConstantDeclaration**

2.3 **TypeDeclaration**

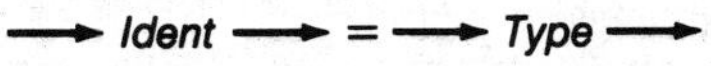

2.4 **VariableDeclaration**

2.5 **ProcedureDeclaration**

2.6 **ProcedureHeading**

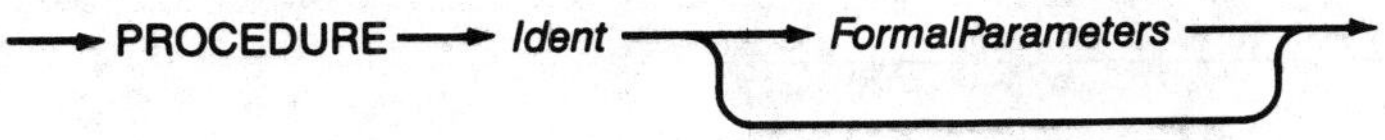

2.7 **FormalParameters**

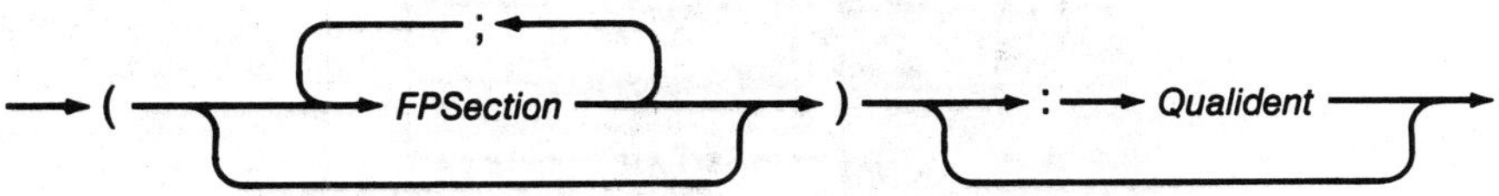

2.8 ***FPSection***

2.9 ***Block***

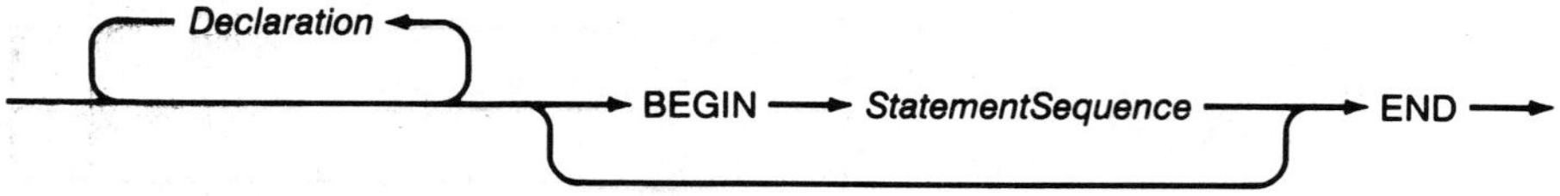

2.10 ***ModuleDeclaration***

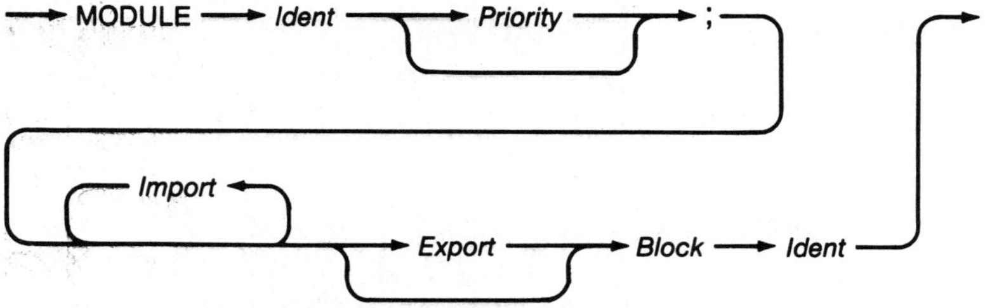

2.11 ***Priority***

2.12 ***Import***

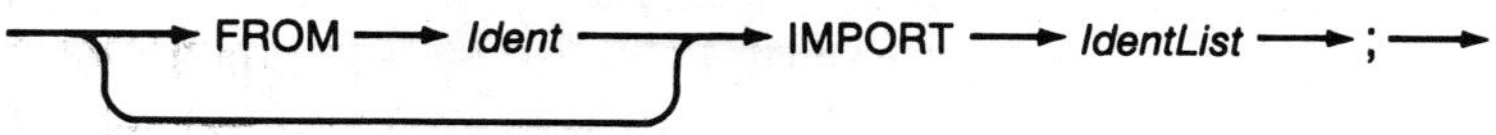

2.13 ***Export***

3. Types

3.1 ***Type***

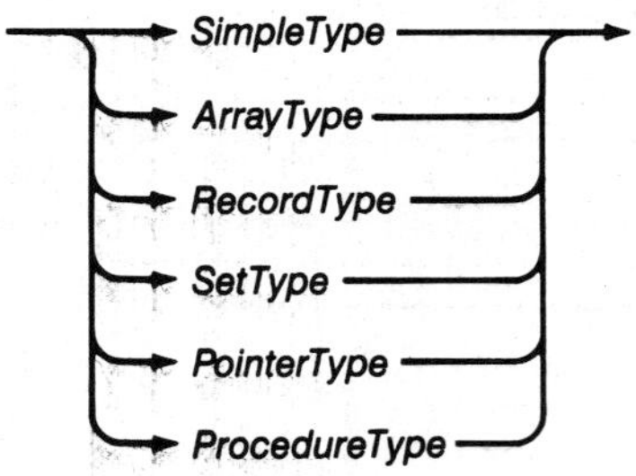

3.2 ***SimpleType***

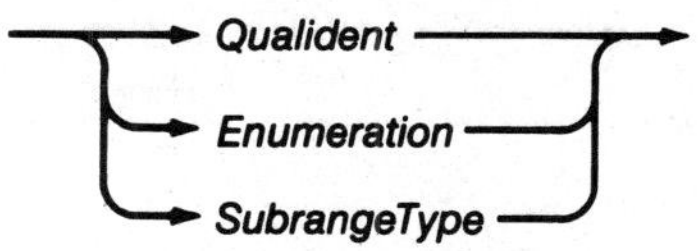

3.3 ***Enumeration***

3.4 ***SubrangeType***

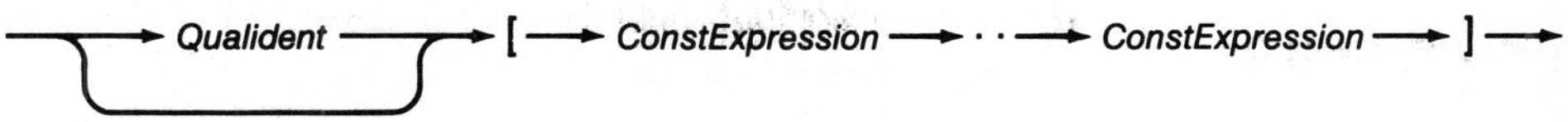

3.5 ***ArrayType***

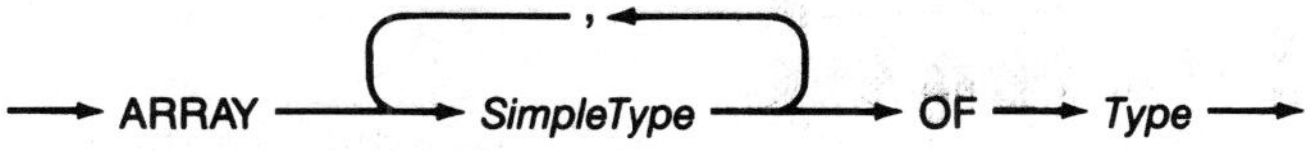

3.6 ***RecordType***

3.7 ***FieldListSequence***

3.8 ***FieldList***

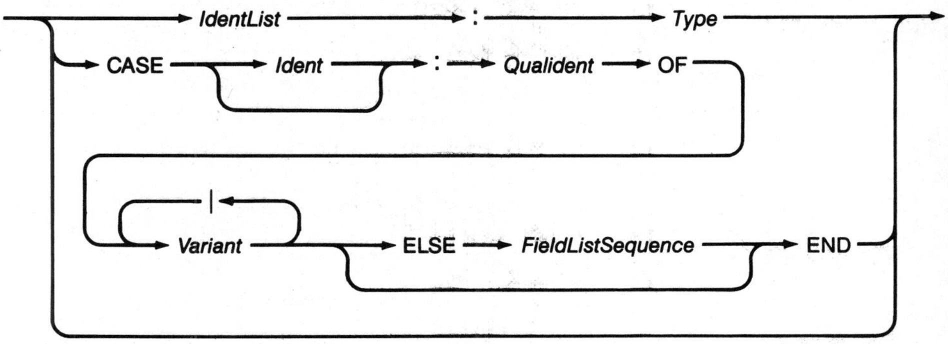

3.9 ***Variant***

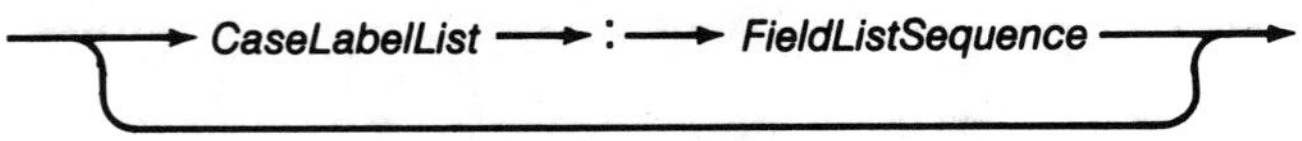

3.10 ***SetType***

3.11 ***PointerType***

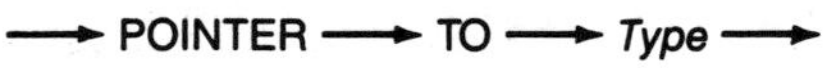

3.12 ***ProcedureType***

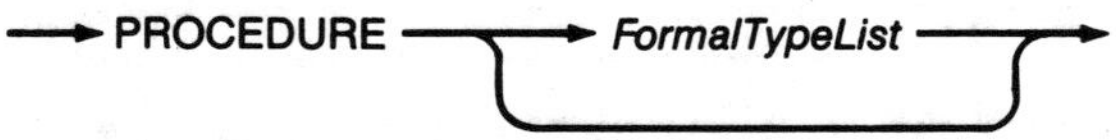

3.13 ***FormalTypeList***

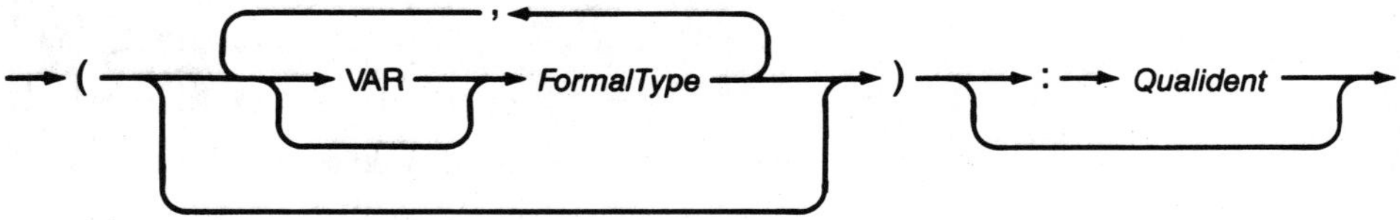

3.14 ***FormalType***

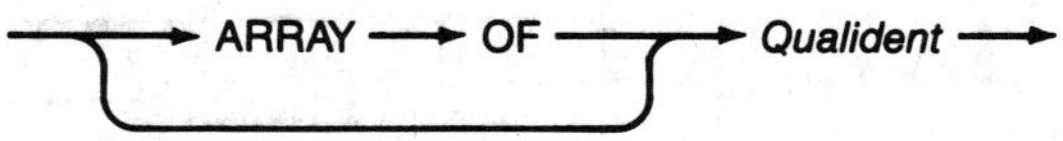

4. Statements

4.1 ***StatementSequence***

4.2 *Statement*

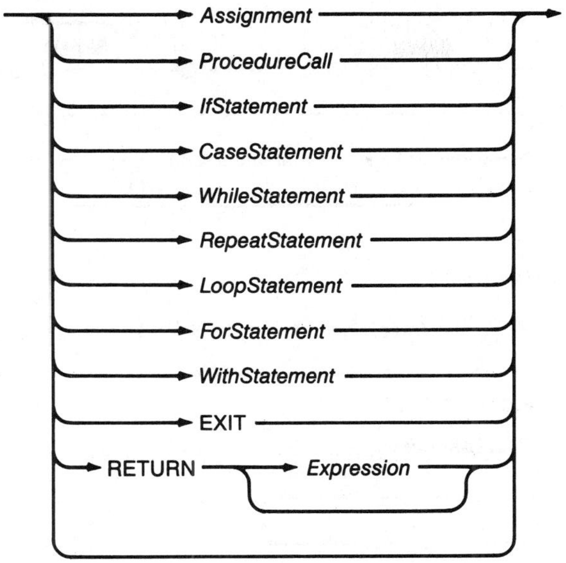

4.3 *Assignment*

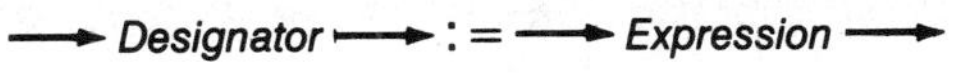

4.4 *ProcedureCall*

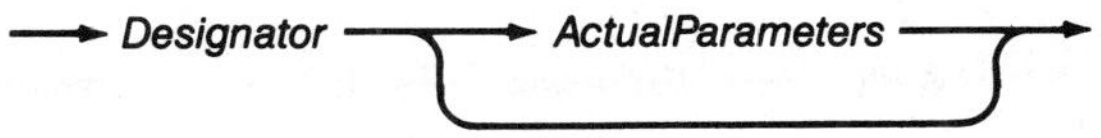

4.5 *ActualParameters*

4.6 *IfStatement*

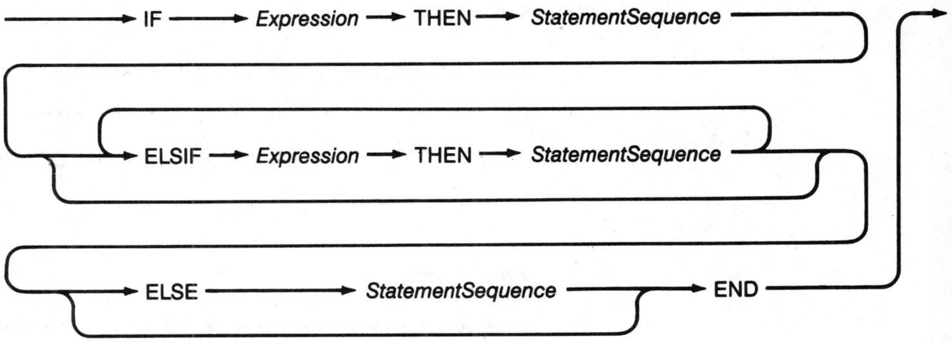

4.7 ***CaseStatement***

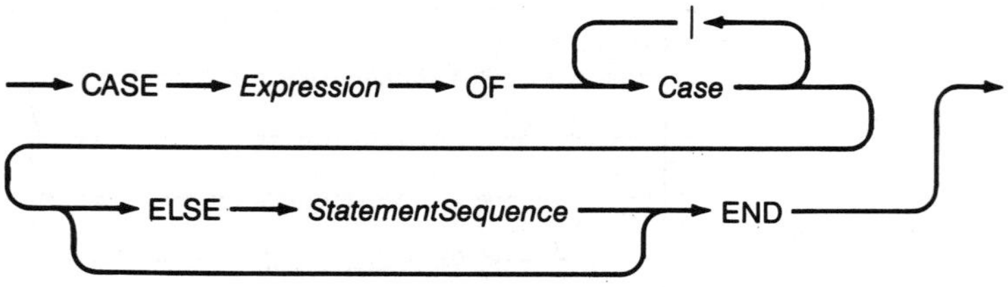

4.8 ***Case***

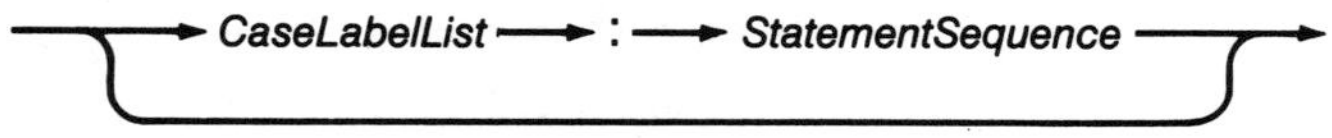

4.9 ***CaseLabelList***

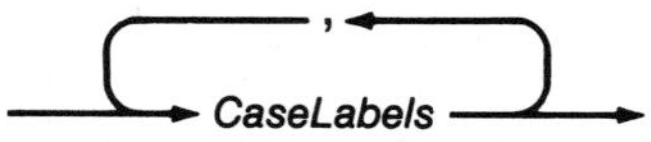

4.10 ***CaseLabels***

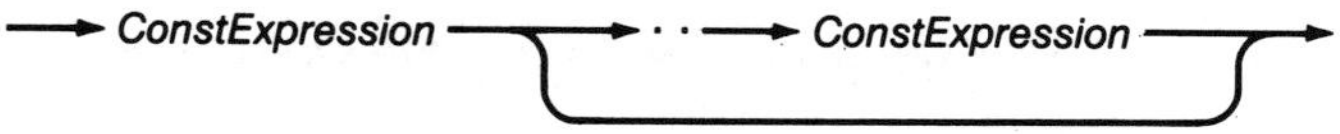

4.11 ***WhileStatement***

4.12 ***RepeatStatement***

4.13 ***LoopStatement***

4.14 ***ForStatement***

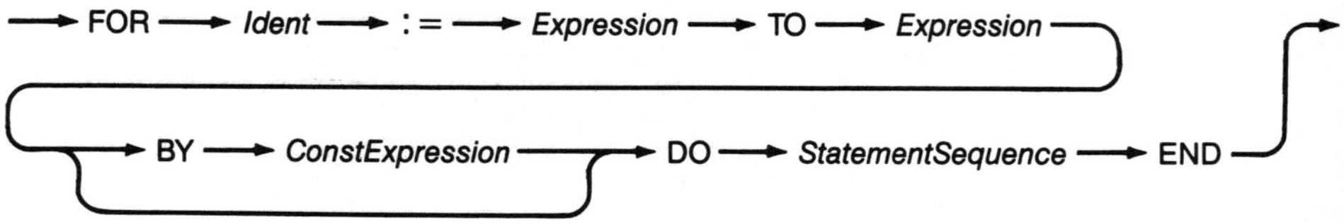

4.15 ***WithStatement***

5. Expressions

5.1 ***ConstExpression***

5.2 ***ExpList***

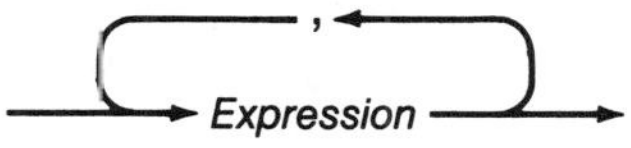

5.3 ***Expression***

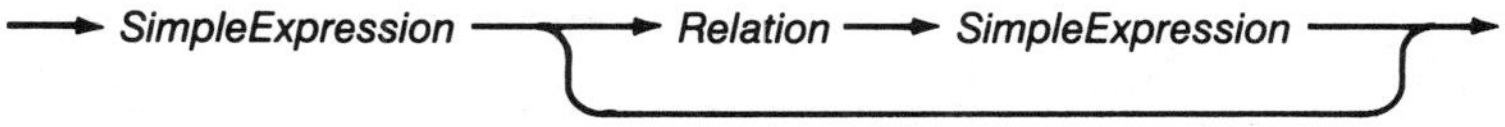

5.4 ***Relation***

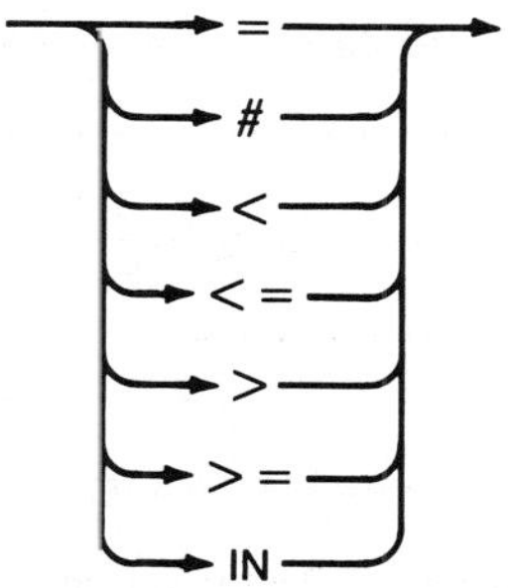

5.5 ***SimpleExpression***

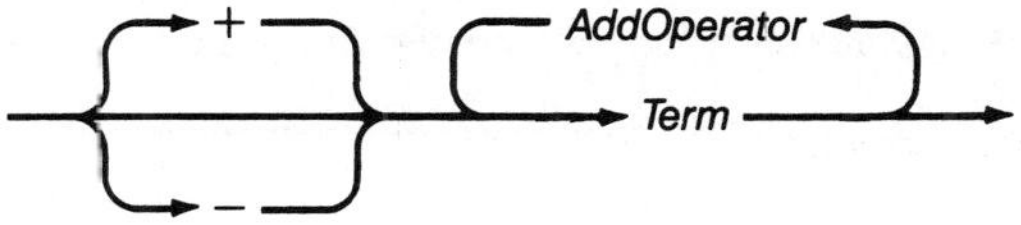

5.6 ***AddOperator***

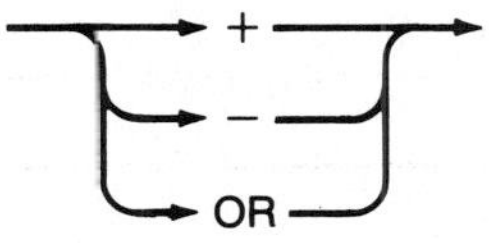

5.7 ***Term***

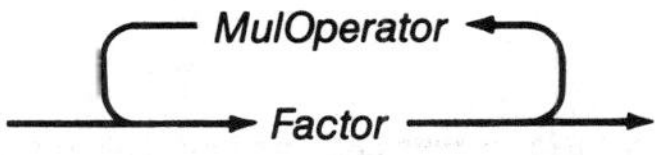

5.8 ***MulOperator***

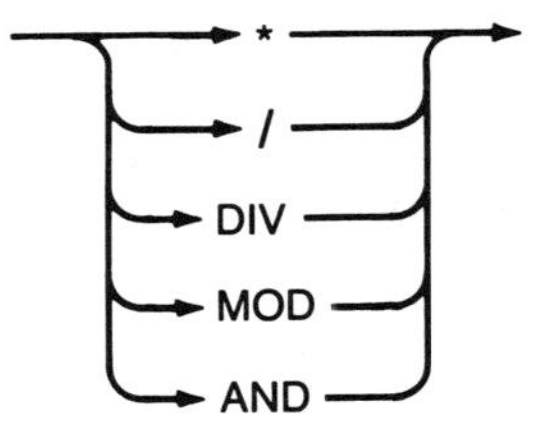

5.9 ***Factor***

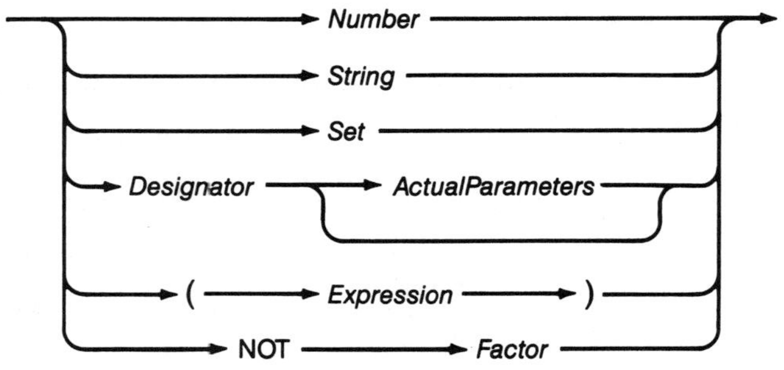

5.10 ***Set***

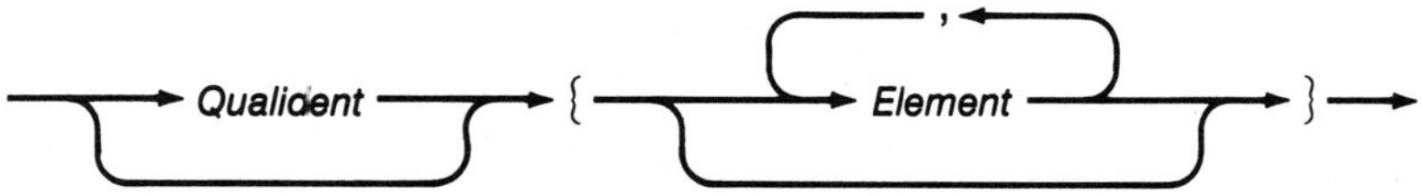

5.11 ***Element***

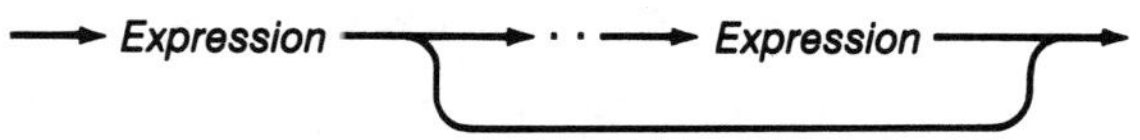

6. Names

6.1 ***IdentList***

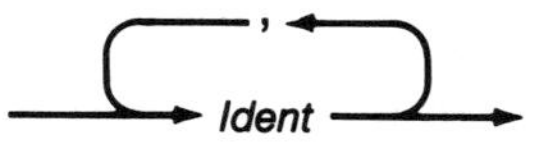

6.2 ***Ident***

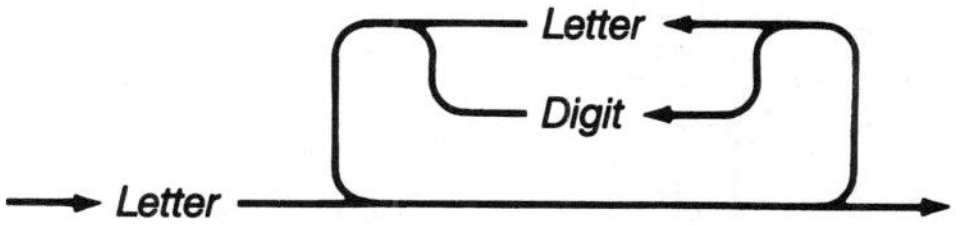

6.3 ***Qualident***

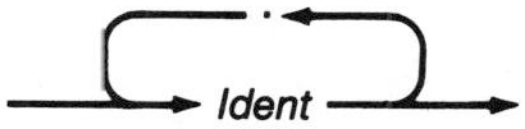

6.4 ***Designator***

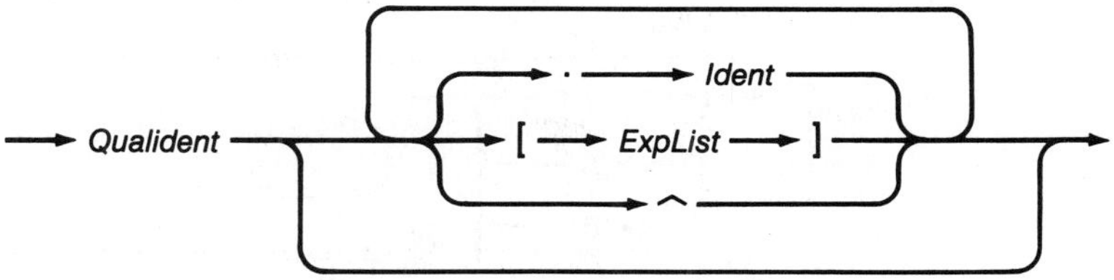

7. Literals

7.1 ***Number***

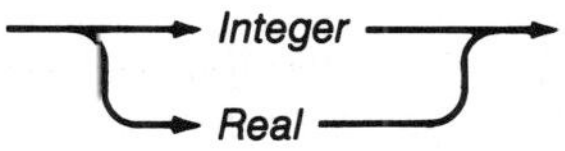

7.2 ***Integer***

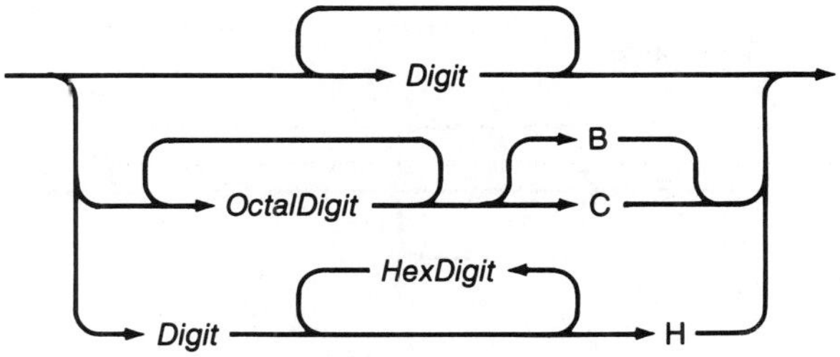

7.3 ***Digit***

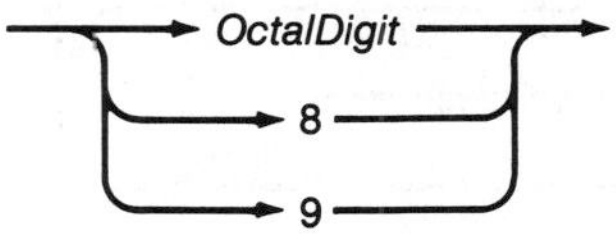

7.4 ***OctalDigit***

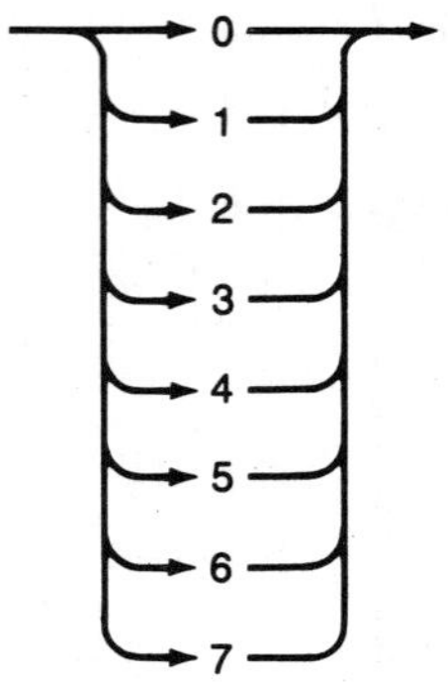

7.5 ***HexDigit***

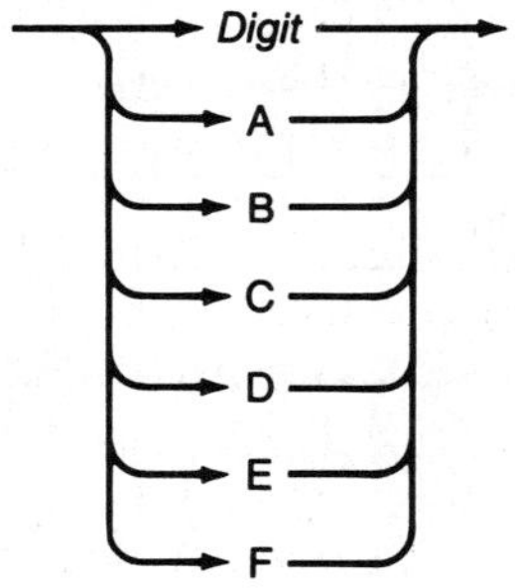

7.6 ***Real***

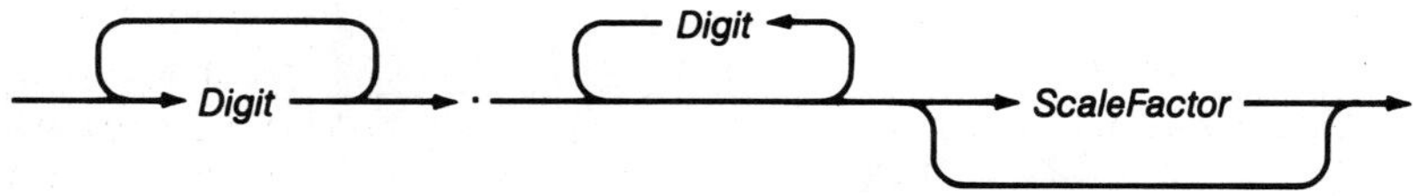

7.7 ***ScaleFactor***

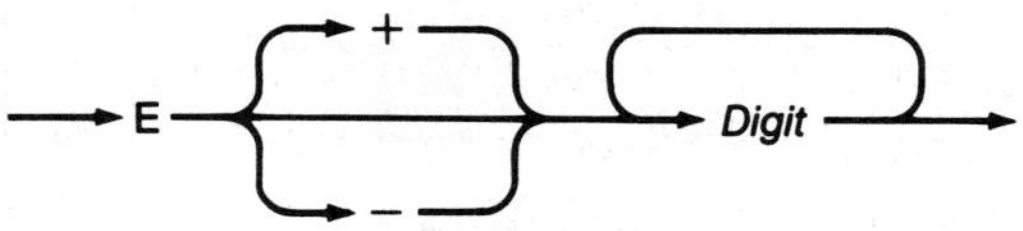

7.8 ***String***

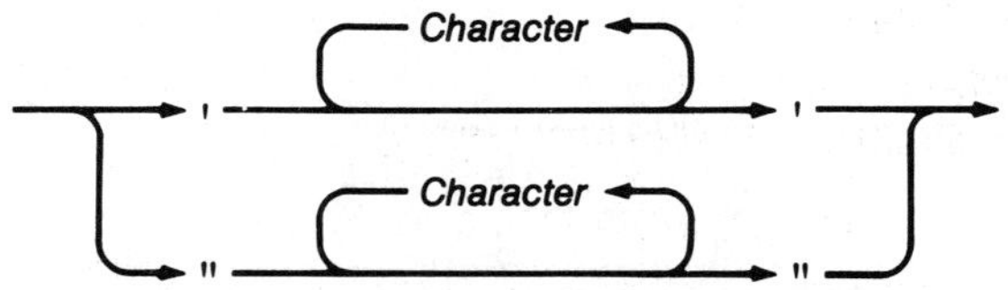

FOR *Oh My! Modula-2!* STUDENTS ONLY—

Special Savings on Upgrades to Stony Brook's Commercial Modula-2 Products

Save $55 on Stony Brook QuickMod, complete commercial product

Save $100 on Stony Brook Professional Modula-2

Use this reply card to upgrade to the complete commercial product, Stony Brook QuickMod, for just $40 plus shipping, or to Stony Brook Professional Modula-2 for just $195 plus shipping.

With your upgrade, you become a fully licensed Stony Brook user. You'll receive:

- the complete Stony Brook software package, including sample modules, and the complete documentation set: User's Guide, Language Reference, and Runtime Library
  (Note that purchasers of the student version of QuickMod will have received the same software as the commercial QuickMod product, minus the sample modules.)
- ongoing information about upgrades
- full technical support

This offer is available to *Oh My! Modula-2!* users only. You must be a registered student, and you must use this card to order. Only one order per card will be accepted. No telephone orders accepted. *NO PURCHASE ORDERS ACCEPTED*. This offer is valid in the U.S. only.

__ **Yes, I want to upgrade to Stony Brook QuickMod.** Enclosed is my check for $45 ($40 plus $5 shipping). Or charge my VISA or MasterCard as indicated below.

__ **Yes, I want to upgrade to Stony Brook Professional Modula-2.** Enclosed is my check for $200 ($195 plus $5 shipping). Or charge my VISA or MasterCard as indicated below.

__ Disk size 5 $\frac{1}{4}$" __ Disk size 3 $\frac{1}{2}$"

(California residents, please add sales tax.)

Name: ______________________________

Address: ______________________________

____

Phone number: ______________________________

School or Educational Institution: ______________________________

Charge this purchase: __ MasterCard __ Visa

Account #: ______________________________

Expiration Date: ______________________________

Signature: ______________________________

Mail to:

Stony Brook Software
187 E. Wilbur Road, Suite 9
Thousand Oaks, California 91360

Appendix B: Definition Modules

These definition modules describe all identifiers used within the text. Implementations for specific Modula-2 systems may include additional features.

FILES Definition

```
DEFINITION MODULE FILES;
    (*The basic FILES definition module.*)

 IMPORT SYSTEM;
    (*We'll use type System.WORD to match non-CHAR values.*)
 IMPORT FileSystem;
    (*The module that underlies the local file system.
    This import will certainly vary between computers.*)

 CONST EOL = FileSystem.EOL;
    (*EOL will usually appear at the end of lines of characters.*)

 TYPE T = FileSystem.File;
    (*Rename FileSystem.File as FILES.T. Within this module, we refer to T.*)

 PROCEDURE Done () : BOOLEAN;
    (*A general-purpose 'was the last operation successful?' function.*)

 PROCEDURE eof (F: T) : BOOLEAN;
    (*TRUE if the last attempted read from F was unsuccessful.*)

 PROCEDURE OpenRead (VAR F: T; Name: ARRAY OF CHAR);
    (*Prepare to read F, starting with its first component.*)

 PROCEDURE OpenWrite (VAR F: T; Name: ARRAY OF CHAR);
    (*Prepare to add to F. Remove any current contents.*)

 PROCEDURE Close (VAR F: T);
    (*The file system may require closing processing on open files.*)

 PROCEDURE CharIn (VAR F: T; VAR Ch: CHAR);
    (*Read one character, if one is available, from F.*)

 PROCEDURE CharOut (VAR F: T; Ch: CHAR);
    (*Append the value of Ch to file F.*)

 PROCEDURE WordIn (VAR F: T; VAR Val: SYSTEM.WORD);
    (*Read a single WORD-size value from F.*)

 PROCEDURE WordOut (VAR F: T; Val: SYSTEM.WORD);
    (*Append a single WORD-size value to F.*)

END FILES.
```

IO Definition

```
DEFINITION MODULE IO;
(*Basic IO definition module.*)

CONST EOL = 36C;
    (*The end-of-line, if 'read', equals EOL.*)

(*****FUNCTIONS that are declared as variables in Wirth's InOut*****)
 PROCEDURE termCH () : CHAR;
    (* termCH represents the character (usually a space or control
    character) that followed, and thus terminated, the last input string.
    It is stuffed back into the input stream for the next input call.*)

PROCEDURE Done () : BOOLEAN;
    (* Reports on success of last IO-oriented operation. *)

(*****I/O REDIRECTION PROCEDURES*****)
 PROCEDURE OpenInput (filename : ARRAY OF CHAR);
    (* Open input file 'filename'. Done () = FALSE if unsuccessful. *)

PROCEDURE OpenOutput (filename : ARRAY OF CHAR);
    (* Open output file 'filename'. Done () = FALSE if unsuccessful. *)

PROCEDURE CloseInput;
    (* Close the current input file and revert to standard input. *)

PROCEDURE CloseOutput;
    (* Close the current output file and revert to standard output. *)

(*****INPUT PROCEDURES*****)
PROCEDURE CharIn (VAR ch : CHAR);
    (* Read a character from the current input stream. *)

PROCEDURE StringIn (VAR s : ARRAY OF CHAR);
    (* Read a string terminated by a space or any control character.
    Leading blanks are echoed but ignored. *)

PROCEDURE IntIn (VAR num : INTEGER);
    (* Read a string, then convert it to an integer. *)

PROCEDURE CardIn (VAR num : CARDINAL);
    (* Read a string, then convert it to a cardinal. *)

PROCEDURE RealIn (VAR num : REAL);
    (* Read a string, then convert it to a real number. *)
```

```
(*****OUTPUT PROCEDURES*****)
PROCEDURE LineOut;
    (* Write a carriage return to the current output stream. *)

PROCEDURE CharOut (ch : CHAR);
    (* Write a character to the current output stream. *)

PROCEDURE StringOut (s : ARRAY OF CHAR);
    (* Write a string to the current output stream. *)

PROCEDURE IntOut (num : INTEGER;  size : CARDINAL);
PROCEDURE CardOut (num, size : CARDINAL);
    (* Write 'num' as a string of digits to the current output stream.
    If 'size' exceeds the number of digits in 'num', then the output
    is preceded by ('size' – number of digits) spaces. *)

PROCEDURE FloatOut (num : REAL; size, digits : CARDINAL);
    (* Write 'num' in scientific notation, right-aligned in a field
    of 'size' spaces, with 'digits' to the right of the decimal.*)

PROCEDURE FixOut (num : REAL; size, digits : CARDINAL);
    (* Write 'num' in fixed-point notation, with 'digits' to the right of the decimal.
    If 'size' exceeds (leading minus sign + leading digits + decimal point + 'digits'),
    then blanks precede the number. *)

(*****************************************************
The following unadvertised extras are available in most implementations of IO:

UNADVERTISED 'FEATURES':
    Procedure IntOut may be given a LONGINT argument.
    Procedures FixOut and FloatOut may be given LONGREAL arguments.
UNADVERTISED PROCEDURES:
    ReadLongReal (longreal ) reads a long real.
    ReadLn (string) returns a full line (EOL terminated).
    EchoFileInput (boolean) echos input that's been redirected from
    a file (normally only the keyboard echoes). Useful for debugging.
*****************************************************)

END IO.
```

Appendix C: Terms From the Text

Programming language design proceeds so quickly that descriptive words sometimes lag behind. This brief glossary lists words and phrases that are relatively new, and are employed with a specific meaning and intent in the text.

bound condition The specification of a loop's exit point. Bound conditions can be divided into four rough categories:

Sentinel bounds tie termination to the appearance of a specific marker value.

Count bounds explicitly restrict the number of loop iterations.

Limit bounds are characteristic of mathematical methods; e.g. the difference between successive approximations approaches zero.

Data bounds are common in algorithms (e.g. binary search) that methodically reduce the 'still relevant' portion of a large data space.

Bounds that are tied to achieving a loop's goal are described as *intentional* bounds, while additional bound conditions that are needed to guarantee loop termination are called *necessary* bounds. In practice, the bound categories are primarily useful for *a*) making the underlying similarity between seemingly disparate loops clear, and *b*) providing a checklist for the necessary bounds that guarantee termination.

cluster A group of related definitions and declarations within a program; e.g. a collection of procedure and variable declarations that, for readability, accompany separate type definition parts.

code layer See *layered design*.

confirmation comment A postcondition or post-assertion that's used to document the current program state. A confirmation comment follows a code segment to document what we believe we have accomplished. Ordinary comments, in contrast, usually precede the segment and announce what we intend to do.

data abstraction A type or structure that is defined in terms of its primitives and properties, rather than its implementation. The text makes the following distinction:

Type abstractions are defined by their values, and the primitives that operate on them: fractions, musical notes, colors.

Structure abstractions are defined by relationships *between* values, and the primitives that maintain that relationship: stacks, queues, lists.

data module A library module that supplies methods of storing data, and of acting upon stored data. The text distinguishes between libraries that provide *type* abstractions and *structure* abstractions (see *data abstraction*). Modules that actually allocate and manage encapsulated data objects for the benefit of a client program are said to have *internal state*; see *object library*.

filter A program that reads an input stream of data, and, by inspecting and/or modifying it, produces an output stream of information. The text refers to two broad classes of filters:

Process filters do something *to* values in the data stream: they capitalize letters, translate decimal to binary, etc.

State filters learn something *about* the data stream (i.e. they spot state transitions): they count words, or identify ascending sequences.

The distinction is intended to be useful in suggesting approaches to problem solving; e.g. modifying existing code. See also *stream*.

function A procedure that represents a result value; what Wirth refers to as a *function procedure*, as opposed to a *proper procedure*.

goal-oriented design A general term for traditional approaches to programming—functional decomposition, procedural abstraction, top-down design, stepwise refinement, structured programming—that focus on implementing action-oriented algorithms in imperative terms. Used in contrast to data first, *object-oriented* design techniques.

interface module A name slick. Interface modules typically customize calls of existing library software; e.g. they simplify parameter lists or translate units of measure.

internal state A library module that maintains data objects on behalf of a client program is said to have internal state. See *object library*.

layered design An approach to designing data abstractions that encourages a split between specification and implementation.

Name layer: a *requirements* specification for implementating the abstraction, and a *performance* specification for using it. The name layer usually consists of a commented set of *calling forms* (mock procedure headings) of the abstraction's primitives.

Code layer: the Modula-2 type definitions and subprogram declarations that implement the abstraction.

In the text, layered design is ultimately superseded by the introduction of definition and implementation modules; in effect, Modula-2 provides syntactic support for layered design.

layered implementation An approach to the design of library modules. Low-level libraries are written in as detailed and general-purpose a manner as possible, in the assumption that a final layer of *interface modules* will be written to make them more easily used.

library module An external module that can be imported. To help clarify the distinct advantages of encapsulation and information hiding, the text presents three general goals of library usage:

Name abstraction: these modules make identifiers available, but don't contain type definitions or implementations of algorithms; e.g. ASCII or ANSI. See *interface module.*

Algorithm abstraction: these modules implement algorithms, but don't hide type information; e.g. MathLib0. See *package module.*

Data abstraction: these modules hide both definitions and implementations. See *data abstraction module.*

loop specification A three-part description of a loop: *goal, bound*, and *plan.*

Goal: states the loop's intent: What problem does it solve?

Bound: states the loop's exit condition: Why will it terminate?

Plan: lists the specific actions the loop will take: What does it do?

A specification is useful because:

a) it clearly distinguishes what the loop is meant to accomplish from its reason for termination;

b) by isolating the bound condition, the specificaiton makes it easier to establish that the loop will, in fact, terminate;

c) it eases the task of positioning the precise exit point.

See also *bound condition.*

name layer See *layered design.*

object library A library that is intended to manage an object (i.e. maintain an internal state) for the benefit of an importing program. The text describes two common applications: hiding *structures*, like stacks, and hiding *processes*, like stream generators. See *stream generator.*

object-oriented design The 'data-first' approach to program design (in contrast to 'action-first' goal-oriented design). Object-oriented design focuses on the data abstractions a task will require; it presumes that a rich selection of data-handling primitives will ease the task of developing any goal-oriented procedural abstractions.

package module A library of procedures that is *not* intended to hide type information; e.g. a statistics package.

program An executable Modula-2 main program module, as opposed to a library module.

program engineering Software engineering in the small; the application of programming techniques that improve the quality of relatively brief programs.

programming riddle A programming exercise that never grew up.

standard practices Methods of programming in the small that are characteristic of a particular statement or type, regardless of the programming language it appears in.

stream The 'flow' of data through a program. The *standard input stream* typically comes from the keyboard, while the *standard output stream* goes to the monitor. However, both input and output streams can usually be *redirected* to files or other devices. Streams may be processed; see also *filter*.

stream generator A process that generates a sequence of data values; e.g. a random number generator. See also *object library*.

structure abstraction See *data abstraction*.

type abstraction See *data abstraction*.

Zen study A technique in which one learns to take examinations by learning to write them.

Appendix D: Terms For the Terminal Room

Terms in this glossary are taken from the famous MIT/Stanford Artificial Intelligence Lab (SAIL) Jargon file. It contains only items that are not likely to be found in the dictionary, but which should be understood by all would-be True Programming Gods. Compiled by Guy L. Steele Jr., Raphael Finkel, Donald Woods, and Mark Crispin, with assistance from the MIT and Stanford AI communities and Worcester Polytechnic Institute. Some contributions were submitted via ARPAnet from miscellaneous sites.

ARG *n.* Abbreviation for 'argument' (to a function), used so often as to have become a new word.

AUTOMAGICALLY adv. Automatically, but in a way that for some reason (typically because it is too complicated, or too ugly, or perhaps even too trivial), I don't feel like explaining to you. See MAGIC.

BAGBITER 1. *n.* Equipment or program that fails, usually intermittently. 2. BAGBITING: *adj.* Failing hardware or software. 'This bagbiting system won't let me get out of SpaceWar.' Synonyms: LOSER, LOSING, CRETINOUS, BLETCHEROUS, BARFUCIOUS, CHOMPER, CHOMPING.

BANG *n.* Common alternate name for exclamation point.

BAR *n.* 1. The second metasyntactic variable, after FOO. 'Suppose we have two functions FOO and BAR. FOO calls BAR . . .' 2. Often appended to FOO to produce FOOBAR.

BIGNUMS [from Macsyma] *n.* 1. Multiple-precision (sometimes infinitely extendable) integers and, through analogy, any very large numbers. 2. EL CAMINO BIGNUM: El Camino Real, a street through the San Francisco peninsula that originally extended (and still appears in places) all the way to Mexico City. It was termed 'El Camino Double Precision' when someone noted it was a very long street, and then 'El Camino Bignum' when it was pointed out that it was hundreds of miles long.

BINARY *n.* The object code for a program.

BLETCH 1. *interj.* Term of disgust. 2. BLETCHEROUS: *adj.* Disgusting in design or function. 'This keyboard is bletcherous!' Usage: slightly comic.

BOGUS *adj.* 1. Non-functional. 'Your patches are bogus.' 2. Useless. 'OPCON is a bogus program.' 3. False. 'Your arguments are bogus.' 4. Incorrect. 'That algorithm is bogus.' 5. Silly. 'Stop writing those bogus sagas.'

BRAIN-DAMAGED *adj.* Obviously wrong; cretinous; demented. There is an implication that the person responsible must have suffered brain damage, because he should have known better. Calling something brain-damaged is really bad; it also implies it is unusable.

BREAK *v.* 1. To cause to be broken (in any sense). 'Your latest patch to the system broke the TELNET server.' 2. (of a program) To stop temporarily, so that it may be examined for debugging purposes. The place where it stops is a BREAKPOINT.

BROKEN *adj.* 1. Not working properly (of programs). 2. Behaving strangely; especially (of people) exhibiting extreme depression.

BUG [from telephone terminology, 'bugs in a telephone cable,' blamed for noisy lines] *n.* An unwanted and unintended property of a program. See FEATURE.

CANONICAL *adj.* The usual or standard state or manner of something.

CATATONIA *n.* A condition of suspended animation in which the system is in a wedged (CATATONIC) state.

CHOMP *v.* To lose; to chew on something of which more was bitten off than one can digest. Probably related to gnashing of teeth. See BAGBITER.

CLOSE *n.* Abbreviation for 'close (or right) parenthesis,' used when necessary to eliminate oral ambiguity. See OPEN.

COKEBOTTLE *n.* Any very unusual character. MIT people complain about the 'control-meta-cokebottle' commands at SAIL, and SAIL people complain about the 'altmode-altmode-cokebottle' commands at MIT.

CONNECTOR CONSPIRACY *n.* The tendency of manufacturers (or, by extension, programmers or purveyors of anything) to come up with new products that don't fit together with the old stuff, thereby making you buy either all new stuff or expensive interface devices.

CRASH 1. *n.* A sudden, usually drastic failure. Most often said of the operating system, sometimes of magnetic disk drives. 'Three lusers lost their files in last night's disk crash.' A disk crash, which entails the read/write heads dropping onto the surface of the disks and scraping off the oxide, may also be referred to as a 'head crash.' 2. *v.* To fail suddenly. 'Has the system just crashed?' Also used transitively to indicate the cause of the crash (usually a person or a program, or both). 'Those idiots playing SpaceWar crashed the system.' Sometimes said of people.

CRETIN 1. *n.* Congenital loser. 2. CRETINOUS: *adj.* See BLETCHEROUS and BAGBITING. Usage: somewhat ad hominem.

CRLF (cur'lif or crul'lif) *n.* A carriage return (CR) followed by a line feed (LF).

CROCK *n.* An awkward feature or programming technique that ought to be made cleaner. Example: Using small integers to represent error codes without the program interpreting them for the user is a crock. Also, a technique that works acceptably but that is quite prone to failure if disturbed in the least; for example depending on the machine opcodes having particular bit patterns so that you can use instructions as data words too, which results in a tightly woven, but almost completely unmodifiable, structure.

CRUFTY *adj.* 1. Poorly built, possibly overly complex. 'This is standard old crufty IBM software.' Hence CRUFT, *n.* shoddy construction. 2. Unpleasant, especially to the touch, often with encrusted junk, like spilled coffee smeared with peanut butter and catsup. Hence CRUFT, *n.* disgusting mess. 3. Generally unpleasant.

CRUNCH *v.* 1. To process, usually in a time-consuming or complicated way. Connotes an essentially trivial operation that is nonetheless painful to perform. The pain may result from putting the triviality in a loop from 1 to 1000000000. 'FORTRAN programs do mostly number crunching.' 2. To reduce the size of a file by a complicated scheme that produces bit configurations completely unrelated to the original data, such as by a Huffman code. This meaning is usually used in the construction 'file crunch(ing)' to distinguish it from 'number crunch(ing).')

CUSPY [from the DEC acronym CUSP, for Commonly Used System Program, i.e., a utility program used by many people] *adj.* 1. Well-written. 2. Functionally excellent. A program that performs well and interfaces well to users is cuspy. See RUDE.

DAEMON (day'mun, dee'mun) *n.* A program that is not invoked explicitly, but lies dormant waiting for some condition(s) to occur. The idea is that the perpetrator of the condition need not be aware that a daemon is lurking (though often a program will commit an action only because it knows that it will implicitly invoke a daemon). For example, writing a file on the printer spooler's directory will invoke the spooling daemon, which prints the file.

DAY MODE See PHASE (of people).

DEADLOCK *n.* A situation wherein two or more processes are unable to proceed because each is waiting for another to do something. A common example occurs when a program that is communicating with an external device (say, a network) waits for a signal to begin, without realizing that the network itself is also waiting for a 'start' signal. This particular flavor of deadlock is called 'starvation.'

DEMENTED *adj.* Yet another term of disgust used to describe a program. The connotation in this case is that the program works as designed, but the design is bad. For example, a program that generates large numbers of meaningless error messages implying it is on the point of imminent collapse.

DIABLO (dee-ah'blow) [from the Diablo printer] 1. *n.* Any letter-quality printing device. 2. *v.* To produce letter-quality output from such a device.

DIDDLE *v.* To work with in a not particularly serious manner. 'I diddled with a copy of ADVENT so it didn't double-space all the time.' 'Let's diddle this piece of code and see if the problem goes away.' See TWEAK and TWIDDLE.

DOWN 1. *adj.* Not working. 'The up escalator is down.' 2. TAKE DOWN, BRING DOWN: *v.* To deactivate, usually for repair work. See UP.

ENGLISH *n.* The source code for a program, which may be in any language, as opposed to BINARY. Usage: slightly obsolete, used mostly by old-time hackers, though recognizable in context.

EPSILON [from standard mathematical notation for a small quantity] 1. *n.* A small quantity of anything. 'The cost is epsilon.' 2. *adj.* Very small, negligible; less than marginal. 'We can get this feature for epsilon cost.' 3. WITHIN EPSILON OF: Close enough to be indistinguishable for all practical purposes.

FAULTY *adj.* Same denotation as 'bagbiting,' 'bletcherous,' 'losing,' but the connotation is much milder.

FEATURE *n.* 1. A surprising property of a program. Occasionally documented. To call a property a feature sometimes means the author of the program did not consider the particular case, and the program makes an unexpected, although not strictly speaking incorrect, response. See BUG. 'That's not a bug, that's a feature!' A bug can be changed to a feature by documenting it. 2. A well-known and beloved property; a facility. Sometimes features are planned, but are called crocks by others.

FENCEPOST ERROR *n.* The discrete equivalent of a boundary condition. Often exhibited in programs by iterative loops. From the following problem: 'If you build a fence 100 feet long with posts ten feet apart, how many posts do you need?' (Either 9 or 11 is a better answer than the obvious 10.)

FLAG DAY [from a bit of Multics history involving a change in the ASCII character set originally scheduled for June 14, 1966] *n.* A software change which is neither forward nor backward compatible, and which is costly to make and costly to revert. 'Can we install that without causing a flag day for all users?'

FLAKEY *adj.* Subject to frequent lossages. See LOSSAGE.

FLAME *v.* To speak incessantly and/or rabidly on some relatively uninteresting subject or with a patently ridiculous attitude.

FLAME ON *v.* To continue to flame. See RAVE.

FLAVOR *n.* 1. Variety, type, kind. 'DDT commands come in two flavors.' 2. The attribute of causing something to be FLAVORFUL. 'This convention yields additional flavor by allowing one to...' See VANILLA.

FLUSH *v.* 1. To delete something, usually superfluous. 'All that nonsense has been flushed.' Standard ITS terminology for aborting an output operation. 2. To leave at the end of a day's work (as opposed to leaving for a meal). 'I'm going to flush now.' 'Time to flush.' 3. To exclude someone from an activity.

FOO 1. [from Yiddish 'feh' or the Anglo-Saxon 'fooey!'] *interj.* Term of disgust. 2. *n.* [from FUBAR (F----- Up Beyond All Recognition), from WWII, often seen as FOOBAR] Name used for temporary programs, or samples of three-letter names. Other similar words are BAR, BAZ (Stanford corruption of BAR), and rarely RAG. These have been used in Pogo as well. 3. Used very generally as a sample name for absolutely anything. The old 'Smokey Stover' comic strips often included the word FOO, in particular on license plates of cars.

FRIED *adj.* 1. Non-working due to hardware failure; burnt out. 2. Of people, exhausted. Said particularly of those who continue to work in such a state. Often used as an explanation or excuse. 'Yeah, I know that fix destroyed the file system, but I was fried when I put it in.'

FROB 1. *n.* (MIT) The official Tech Model Railroad Club definition is 'FROB = protruding arm or trunnion,' and by metaphoric extension any somewhat small thing. See FROBNITZ. 2. *v.* Abbreviated form of FROBNICATE.

FROBNICATE *v.* To manipulate or adjust, to tweak. Derived from FROBNITZ. Usually abbreviated to FROB. Thus one has the saying 'to frob a frob.' See TWEAK and TWIDDLE. Usage: FROB, TWIDDLE, and TWEAK sometimes connote points along a continuum. FROB connotes aimless manipulation; TWIDDLE connotes gross manipulation, often a coarse search for a proper setting; TWEAK connotes fine-tuning. If someone is adjusting a knob on an oscilloscope carefully then he is probably tweaking it; if he is turning it but looking at the screen he is probably twiddling it; if he's just doing it because turning a knob is fun, he's frobbing it.

FRY *v.* 1. To fail. Said especially of smoke-producing hardware failures. 2. More generally, to become non-working. Usage: never said of software, only of hardware and humans. See FRIED.

FUDGE 1. *v.* To perform in an incomplete but marginally acceptable way, particularly with respect to the writing of a program. 'I didn't feel like going through that pain and suffering, so I fudged it.' 2. *n.* The resulting code.

FUDGE FACTOR *n.* A value or parameter that is varied in an ad hoc way to produce the desired result. The terms 'tolerance' and 'slop' are also used, though these usually indicate a one-sided leeway, such as a buffer which is made larger than necessary because one isn't sure exactly how large it needs to be, and it is better to waste a little space than to lose completely for not having enough. A fudge factor, on the other hand, can often be tweaked in more than one direction. An example might involve varying the coefficients of equation in an attempt to make it fit certain criteria.

GLASS TTY *n.* A terminal which has a display screen but which, because of hardware or software limitations, behaves like a teletype or other printing terminal. An example is the old ADM-3 (without cursor control). A glass tty can't do neat display hacks, and you can't save the output either.

GLITCH [from the Yiddish 'glitshen,' to slide] 1. *n.* A sudden interruption in electric service, sanity, or program function. Sometimes recoverable.

GOBBLE *v.* To consume or to obtain. GOBBLE UP tends to imply 'consume,' while GOBBLE DOWN tends to imply 'obtain.' 'The output spy gobbles characters out of a TTY output buffer.' 'I guess I'll gobble down a copy of the documentation tomorrow.' See SNARF.

GRIND *v.* 1. To format code, especially LISP code, by indenting lines so that it looks pretty. Hence PRETTY PRINT, the generic term for such operations. 2. To run seemingly interminably, performing some tedious and inherently useless task. Similar to CRUNCH.

HACK *n.* 1. Originally a quick job that produces what is needed, but not well. 2. The result of that job. 3. NEAT HACK: A clever technique. Also, a brilliant practical joke, where neatness is correlated with cleverness, harmlessness, and surprise value. 4. REAL

HACK: A crock (occasionally affectionate). *v.* 5. With 'together,' to throw something together so it will work. 6. To bear emotionally or physically. 'I can't hack this heat!' 7. To work on something (typically a program). In specific sense: 'What are you doing?' 'I'm hacking UNIX.' In general sense: 'What do you do around here?' 'I hack UNIX.' (The former is time-immediate, the latter time-extended.) More generally, 'I hack x' is roughly equivalent to 'x is my bag.' 'I hack solid-state physics.' 8. To pull a prank on. See definition 3 and HACKER (def. 6). 9. *v.* To waste time (as opposed to TOOL). 'Watcha up to?' 'Oh, just hacking.' 10. HACK UP (ON): To hack, but generally implies that the result is meanings 1 or 2. 11. *n.* HACK VALUE: Term used as the reason or motivation for expending effort toward a seemingly useless goal, the point being that the accomplished goal is a hack. For example, MacLISP has code to read and print roman numerals, which was installed purely for hack value. HAPPY HACKING: A farewell. HOW'S HACKING?: A friendly greeting among hackers. HACK HACK: A somewhat pointless but friendly comment, often used as a temporary farewell.

HACKER [originally, someone who makes furniture with an axe] *n.* 1. A person who enjoys learning the details of programming systems and how to stretch their capabilities, as opposed to most users who prefer to learn only the minimum necessary. 2. One who programs enthusiastically, or who enjoys programming rather than just theorizing about programming. 3. A person capable of appreciating hack value. 4. A person who is good at programming quickly. Not everything a hacker produces is a hack. 5. An expert at a particular program, or one who frequently does work using it or on it; example: 'A SAIL hacker.' (Definitions 1 to 5 are correlated, and people who fit them congregate.) 6. A malicious or inquisitive meddler who tries to discover information by poking around. Hence 'password hacker,' 'network hacker.'

HAIRY *adj.* 1. Overly complicated. 'DWIM is incredibly hairy.' 2. Incomprehensible. 'DWIM is incredibly hairy.' 3. Of people, high-powered, authoritative, rare, expert, and/or incomprehensible. Hard to explain except in context: 'He knows this hairy lawyer who says there's nothing to worry about.'

HANDWAVE 1. *v.* To gloss over a complex point; to distract a listener; to support a (possibly actually valid) point with blatantly faulty logic. 2. *n.* The act of handwaving. 'Boy, what a handwave!' The use of this word is often accompanied by gestures: both hands up, palms forward, swinging the hands in a vertical plane pivoting at the elbows and/or shoulders (depending on the magnitude of the handwave); alternatively, holding the forearms still while rotating the hands at the wrist to make them flutter. In context, the gestures alone can suffice as a remark.

HOOK *n.* An extraneous piece of software or hardware included in order to simplify later additions or debugging options.

INFINITE *adj.* Consisting of a large number of objects; extreme. Used very loosely as in: 'This program produces infinite garbage.'

JOCK *n.* Programmer who is characterized by large and somewhat brute force programs. The term is particularly well-suited to systems programmers.

KLUGE (kloodj) alt. KLUDGE [from the German 'kluge,' clever] *n.* 1. A Rube Goldberg device in hardware or software. 2. A clever programming trick intended to solve a particular nasty case in an efficient, if not clear, manner. Often used to repair bugs. Often verges on being a crock. 3. Something that works for the wrong reason. 4. *v.* To insert a kluge into a program. 'I've kluged this routine to get around that weird bug, but there's probably a better way.' Also KLUGE UP. 5. KLUGE AROUND: To avoid by inserting a kluge. 6. *n.* A feature that is implemented in a RUDE manner.

LOGICAL *adj.* Understood to have a meaning not necessarily corresponding to reality. For example, if a person who has long held a certain post (e.g., Les Earnest at SAIL) left and was replaced, the replacement would for a while be known as the 'logical Les Earnest.' The word VIRTUAL is also used.

LOSE *v.* 1. To fail. A program loses when it encounters an exceptional condition. 2. *adj.* To be exceptionally unaesthetic. 3. Of people, to be obnoxious or unusually stupid (as opposed to ignorant). 4. DESERVE TO LOSE: *v.* Said of someone who willfully does the wrong thing; humorously, if one uses a feature known to be marginal. What is meant is that one deserves the consequences of one's losing actions. 'Boy, anyone who tries to use DOS deserves to lose!'

LOSER *n.* An unexpectedly bad situation, program, programmer, or person. Especially 'real loser.'

LOSSAGE *n.* The result of a bug or malfunction.

LPT (lip'-it) *n.* Line printer, of course.

LUSER See USER.

MAGIC *adj.* 1. As yet unexplained, or too complicated to explain. (Arthur C. Clarke once said that magic was as-yet-not-understood science.) 'TTY echoing is controlled by a large number of magic bits.' 'This routine magically computes the parity of an eight-bit byte in three instructions.' 2. (Stanford) A feature not generally publicized that allows something otherwise impossible, or a feature formerly in that category but now unveiled. Example: The keyboard commands that override the screen-hiding features.

MARGINAL *adj.* 1. Extremely small. 'A marginal increase in core can decrease GC time drastically.' 2. Of extremely small merit. 'This proposed new feature seems rather marginal to me.' 3. Of extremely small probability of winning. 'The power supply was rather marginal anyway; no wonder it blew out.' 4. MARGINALLY: adv. Slightly. 'The ravs here are only marginally better than at Small Eating Place.'

MISFEATURE *n.* A feature that eventually causes problems, possibly because it is not adequate for a new situation that has evolved. It is not the same as a bug because fixing it involves a gross philosophical change to the structure of the system involved. Often a former feature becomes a misfeature because a tradeoff was made whose parameters subsequently changed (possibly only in the judgment of the implementors). 'Well, yeah, it's kind of a misfeature that file names are limited to six characters, but we're stuck with it for now.'

MODE *n.* A general state, usually used with an adjective describing the state. 'No time to hack; I'm in thesis mode.' Usage: in its jargon sense, MODE is most often said of people, though it is sometimes applied to programs and inanimate objects.

MODULO prep. Except for. From mathematical terminology: one can consider saying that 4=22 'except for the 9's' (4=22 mod 9). 'Well, LISP seems to work okay now, modulo that GC bug.'

MUMBLE *interj.* 1. Said when the correct response is either too complicated to enunciate or the speaker has not thought it out. Often prefaces a longer answer, or indicates a general reluctance to get into a big long discussion. 'Well, mumble.' 2. Sometimes used as an expression of disagreement. 'I think we should buy it.' 'Mumble!' Common variant: MUMBLE FROTZ.

MUNG (variant: MUNGE) [recursive acronym for Mung Until No Good] *v.* 1. To make changes to a file, often large-scale, usually irrevocable, occasionally accidental. 2. To destroy, usually accidentally, occasionally maliciously. The system only mungs things maliciously.

N *adj.* 1. Some large and indeterminate number of objects: 'There were N bugs in that crock!'; also used in its original sense of a variable name. 2. An arbitrarily large (and perhaps infinite) number. 3. A variable whose value is specified by the current context. 'We'd like to order N wonton soups and a family dinner for N-1.' 4. NTH: *adj.* The ordinal counterpart of N. 'Now for the Nth and last time...' In the specific context 'Nth-year grad student,' N is generally assumed to be at least 4, and is usually 5 or more.

NIGHT MODE See PHASE (of people).

OBSCURE *adj.* Used in an exaggeration of its normal meaning, to imply a total lack of comprehensibility. 'The reason for that last crash is obscure.' 'FIND's command syntax is obscure.' MODERATELY OBSCURE implies that it could be figured out but probably isn't worth the trouble.

OPEN *n.* Abbreviation for 'open (or left) parenthesis,' used when necessary to eliminate oral ambiguity. To read aloud the LISP form (DEFUN FOO (X) (PLUS X 1)) one might say: 'Open def-fun foo, open eks close, open, plus eks one, close close.' See CLOSE.

PARSE [from linguistic terminology] *v.* 1. To determine the syntactic structure of a sentence or other utterance (close to the standard English meaning). Example: 'That was the one I saw you.' 'I can't parse that.' 2. More generally, to understand or comprehend: 'It's very simple; you just kretch the glims and then aos the zotz.' 'I can't parse that.'

PATCH 1. *n.* A temporary addition to a piece of code, usually as a quick and dirty remedy to an existing bug or misfeature. A patch may or may not work, and may or may not eventually be incorporated permanently into the program. 2. *v.* To insert a patch into a piece of code.

PHASE (of people) 1. *n.* The phase of one's waking-sleeping schedule with respect to the standard 24-hour cycle. This is a useful concept among people who often work at night according to no fixed schedule. It is not uncommon to change one's phase by as much as six hours/day on a regular basis. 'What's your phase?' 'I've been getting in about 8 PM lately, but I'm going to work around to the day schedule by Friday.' A person who is roughly 12 hours out of phase is sometimes said to be in 'night mode.' (The term 'day mode' is also used, but less frequently.) 2. *v.* CHANGE PHASE THE HARD WAY: To stay awake for a very long time in order to get into a different phase. 3. CHANGE PHASE THE EASY WAY: To stay asleep, etc.

PHASE OF THE MOON *n.* Used humorously as a random parameter on which something is said to depend. Sometimes implies unreliability of whatever is dependent, or that reliability seems to be dependent on conditions nobody has been able to determine. 'This feature depends on having the channel open in mumble mode, having the foo switch set, and on the phase of the moon.'

PUNT [from the punch line of an old joke: 'Drop back 15 yards and punt'] *v.* To give up, typically without any intention of retrying.

RANDOM *adj.* 1. Unpredictable (closest to mathematical definition); weird. 'The system's been behaving pretty randomly.' 2. Assorted; undistinguished. 'Who was at the conference?' 'Just a bunch of random business types.' 3. Frivolous; unproductive; undirected (pejorative). 'He's just a random loser.' 4. Incoherent or inelegant; not well organized. 'The program has a random set of misfeatures.' 'That's a random name for that function.' 'Well, all the names were chosen pretty randomly.' 5. Gratuitously wrong, i.e., poorly done and for no good apparent reason. For example, a program that handles filename defaulting in a particularly useless way, or a routine that could easily have been coded using only three ac's, but randomly uses seven for assorted non-overlapping purposes, so that no one else can invoke it without first saving four extra ac's. 6. In no particular order, though deterministic. 'The I/O channels are in a pool, and when a file is opened one is chosen randomly.' *n.* 7. A random hacker; used particularly of high school students who soak up computer time and generally get in the way. 8. (occasional MIT usage) One who lives at Random Hall. J. RANDOM is often prefixed to a noun to make a 'name' out of it (by comparison to common names such as 'J. Fred Muggs'). The most common uses are 'J. Random Loser' and 'J. Random Nurd' ('Should J. Random Loser be allowed to gun down other people?'), but can be used as an elaborate version of RANDOM in any sense.

RANDOMNESS *n.* An unexplainable misfeature; gratuitous inelegance. Also a hack or crock that depends on a complex combination of coincidences (or rather, the combination upon which the crock depends). 'This hack can output characters 40-57 by putting the character in the accumulator field of an XCT and then extracting 6 bits — the low two bits of the XCT opcode are the right thing.' 'What randomness!'

RAVE *v.* 1. To persist in discussing a specific subject. 2. To speak authoritatively on a subject about which one knows very little. 3. To complain to a person who is not in a position to correct the difficulty. 4. To purposely annoy another person verbally. 5. To evangelize. See FLAME.

REAL USER *n.* 1. A commercial user. One who is paying 'real' money for computer usage. 2. A non-hacker. Someone using the system for an explicit purpose (research project, course, etc.). See USER.

REAL WORLD, THE *n.* 1. In programming, those institutions at which programming may be used in the same sentence as FORTRAN, COBOL, RPG, IBM, etc. 2. To programmers, the location of non-programmers and activities not related to programming. 3. A universe in which the standard dress is shirt and tie and in which a person's working hours are defined as 9 to 5. 4. The location of the status quo. 5. Anywhere outside a university. 'Poor fellow, he's left MIT and gone into the real world.' Used pejoratively by those not in residence there. In conversation, talking of someone who has entered the real world is not unlike talking about a deceased person.

RIGHT THING, THE *n.* That which is 'obviously' the correct or appropriate thing to use, do, say, etc. Use of this term often implies that in fact reasonable people may disagree. 'Never let your conscience keep you from doing the right thing!' 'What's the right thing for LISP to do when it reads '(.)'?'

RUDE *adj.* 1. (of a program) Badly written. 2. Functionally poor, e.g. a program that is very difficult to use because of gratuitously poor design decisions. See CRUFTY.

SACRED *adj.* Reserved for the exclusive use of something (a metaphorical extension of the standard meaning). 'Accumulator 7 is sacred to the UUO handler.' Often means that anyone may look at the sacred object, but clobbering it will break whatever it is sacred to.

SEMI *n.* 1. Abbreviation for 'semicolon' when speaking. 'Commands to GRIND are prefixed by semi-semi-star' means that the prefix is ';;*,' not 1/4 of a star. 2. Prefix with words such as 'immediately,' as a qualifier. 'When is the system coming up?' 'Semi-immediately.'

SLURP *v.* To read a large data file entirely into core before working on it. 'This program slurps in a 1K-by-1K matrix and does a fast Fourier transform.'

SNARF *v.* To grab, esp. a large document or file for the purpose of using it either with or without the author's permission. Variant: SNARF (IT) DOWN.

SOFTWARE ROT *n.* Hypothetical disease the existence of which has been deduced from the observation that unused programs or features will stop working after sufficient time has passed, even if 'nothing has changed.' Also known as 'bit decay.'

SOFTWARILY adv. In a way pertaining to software. 'The system is softwarily unreliable.' The adjective 'softwary' is NOT used.

SUPERPROGRAMMER *n.* See 'wizard,' 'hacker.' Usage: rare. (Becoming more common among IBM and Yourdon types.)

SWAPPED *adj.* From the use of secondary storage devices to implement virtual memory in computer systems. Something that is SWAPPED IN is available for immediate use in main memory, and otherwise is SWAPPED OUT. Often used metaphorically to refer to people's memories ('I read the AWK documentation every few months to keep the information swapped in.') or to their own availability ('I'll swap you in as soon as I finish looking at this other problem.').

SYSTEM *n.* 1. The supervisor program on the computer. 2. Any large-scale program. 3. Any method or algorithm. 4. The way things are usually done. Usage: a fairly ambiguous word. 'You can't beat the system.' SYSTEM HACKER: one who hacks the system (in sense 1 only; for sense 2 one mentions the particular program: e.g., LISP HACKER).

THEORY *n.* Used in the general sense of idea, plan, story, or set of rules. 'What's the theory on fixing this TECO loss?' 'What's the theory on dinner tonight?' ('Chinatown, I guess.') 'What's the current theory on letting losers on during the day?' 'The theory behind this change is to fix the following well-known hack...'

THRASH *v.* To move wildly or violently, without accomplishing anything useful. Swapping systems which are overloaded waste most of their time moving pages into and out of core (rather than performing useful computation), and are therefore said to thrash.

TOOL *v.* To work; to study. See HACK (def. 9).

TTY *n.* Terminal of the teletype variety, characterized by a noisy mechanical printer, a very limited character set, and poor print quality. Usage: antiquated (like the TTY's themselves). Sometimes used to refer to any terminal at all; sometimes used to refer to the particular terminal controlling a job.

TWEAK *v.* To change slightly, usually in reference to a value. Also used synonymously with TWIDDLE. See FROBNICATE and FUDGE FACTOR.

TWIDDLE *n.* 1. tilde (ASCII 176, '~'). Also called 'squiggle,' 'sqiggle' (sic--pronounced 'skig'gul'), and 'twaddle,' but twiddle is by far the most common term. 2. A small and insignificant change to a program. Usually fixes one bug and generates several new ones. 3. *v.* To change something in a small way. Bits, for example, are often twiddled. Twiddling a switch or knob implies much less sense of purpose than toggling or tweaking it; see FROBNICATE.

UP *adj.* 1. Working, in order. 'The down escalator is up.' 2. BRING UP: *v.* To create a working version and start it. 'They brought up a down system.'

USER *n.* A programmer who will believe anything you tell him. One who asks questions. Identified at MIT with 'loser' by the spelling 'luser.' See REAL USER.

VANILLA *adj.* Ordinary flavor, standard. See FLAVOR. When used of food, very often does not mean that the food is flavored with vanilla extract! For example, 'vanilla-flavored wonton soup' (or simply 'vanilla wonton soup') means ordinary wonton soup, as opposed to hot and sour wonton soup.

VAXEN [from 'oxen,' perhaps influenced by 'vixen'] *n.* The plural of VAX (a DEC machine).

VIRTUAL *adj.* 1. Common alternative to LOGICAL, but never used with compass directions. 2. Performing the functions of. Virtual memory acts like real memory but isn't.

WEDGED *adj.* To be in a locked state, incapable of proceeding without help. Often refers to humans suffering misconceptions. 'The swapper is wedged.' This term is sometimes used as a synonym for DEADLOCKED.

WIN 1. *v.* To succeed. A program wins if no unexpected conditions arise. 2. BIG WIN: *n.* Serendipity. Emphatic forms: MOBY WIN, SUPER WIN, HYPER-WIN (often used interjectively as a reply). See LOSE.

WINNER *n.* 1. An unexpectedly good situation, program, programmer, or person. 2. REAL WINNER: Often sarcastic, but also used as high praise.

WIZARD *n.* 1. A person who knows how a complex piece of software or hardware works; someone who can find and fix his bugs in an emergency. 2. A person who is permitted to do things forbidden to ordinary people, e.g., a 'net wizard' on a TENEX may run programs that speak low-level host-imp protocol; an ADVENT wizard at SAIL may play Adventure during the day.

ZERO *v.* 1. To set to zero. Usually said of small pieces of data, such as bits or words. 2. To erase; to discard all data from. Said of disks and directories, where 'zeroing' need not involve actually writing zeros throughout the area being zeroed.

Appendix E: Index to Programs

Chapter 5 The IF Statement

Chapter 6 The LOOP Statement

Chapter 7 New Types and the CASE Statement

Chapter 8 The ARRAY Type

Chapter 9 Sorting and Searching: An Excursion

Chapter 10 The **RECORD** Type

Chapter 11 Introduction to Library Modules

Chapter 12 Interfaces, Packages, Data Abstractions

Chapter 13 The Pointer Types

Chapter 14 Files

Chapter 15 The SET Type

Chapter 16 Modules with Structure and State

Chapter 17 Recursive Programming Methods

Chapter 18 Bits and Bytes

Index

ASCII Control Characters

Name	**nul**	soh	**stx**	etx	**eot**	enq	**ack**	bel	**bs**	ht	**lf**	vt	**ff**	cr	**so**	si	**dle**
Keyboard Character	**^@**	^A	**^B**	^C	**^D**	^E	**^F**	^G	**^H**	^I	**^J**	^K	**^L**	^M	**^N**	^O	**^P**
Position (Octal)	**000**	001	**002**	003	**004**	005	**006**	007	**010**	011	**012**	013	**014**	015	**016**	017	**020**
Position (Decimal)	**0**	1	**2**	3	**4**	5	**6**	7	**8**	9	**10**	11	**12**	13	**14**	15	**16**

Name	**dc1**	dc2	**dc3**	dc4	**nak**	syn	**etb**	can	**em**	sub	**esc**	fs	**gs**	rs	**us**	del
Keyboard Character	**^Q**	^R	**^S**	^T	**^U**	^V	**^W**	^X	**^Y**	^Z	**ESC**	^\	**^]**	^^	**^_**	DEL
Position (Octal)	**021**	022	**023**	024	**025**	026	**027**	030	**031**	032	**033**	034	**035**	036	**037**	177
Position (Decimal)	**17**	18	**19**	20	**21**	22	**23**	24	**25**	26	**27**	28	**29**	30	**31**	127

ANSI / VT100 Escape Sequences
(Not all systems will support all escape sequences.)

Escape sequence	*Effect on the screen*
esc [*X* **;** *Y* **H**	*move the cursor to screen coordinates X, Y*
esc [*N* **A**	*move the cursor up by N lines*
esc [*N* **B**	*move the cursor down by N lines*
esc [*N* **C**	*move the cursor right by N spaces*
esc [*N* **D**	*move the cursor left by N spaces*
esc [2 J	*erase the screen, then go to the upper left corner*
esc [K	*erase from the current position to the end of the line*
esc [s	*save the current cursor position in a hidden variable* (**esc7** *on VT100*)
esc [u	*go to the position saved by the call above* (**esc8** *on VT100*)

esc [*val* **m** *or* **esc [** *val* **;** *val* **. . . ;** *val* **m**

give the display one or more attributes. Legal values for 'val' are:

	General characteristics	*Foreground color*		*Background color*
0	*reset to terminal defaults*	**30**	*Black*	**40**
1	*high intensity characters*	**31**	*Red*	**41**
2	*faint intensity*	**32**	*Green*	**42**
3	*italic characters*	**33**	*Yellow*	**43**
4	*underline characters*	**34**	*Blue*	**44**
5	*blink on*	**35**	*Magenta*	**45**
6	*blink rapidly*	**36**	*Cyan*	**46**
7	*reverse video*	**37**	*White*	**47**
8	*don't show letters printing*			

For example, **esc [0;1;5;31;43m** *sets bright, flashing, red letters on a yellow background.*